BASIC ALGEBRA FOR COLLEGE STUDENTS

BASIC ALGEBRA FOR COLLEGE STUDENTS

Corrinne Pellillo Brase
Arapahoe Community College
Littleton, Colorado
Charles Henry Brase
Regis College
Denver, Colorado

HOUGHTON MIFFLIN COMPANY BOSTON
Atlanta Dallas Geneva, Illinois Hopewell, New Jersey
Palo Alto London

To
Zelma
M.J.
Nell
Paul

Printed in the U.S.A.

Library of Congress Catalog Card Number: 75-26093

ISBN: 0-395-20656-1

CONTENTS

PREFACE

In the authors' teaching experiences with developmental algebra they have found students with very different backgrounds. There are students who never understood high school algebra or who never took high school algebra. There are many students who have some basic skills in algebra, but they have not worked with algebra for several years. Such students need work in selected topics and can go at a faster pace through material they already partially know. Finally there are students who are mathematically inclined. They may know very little algebra but are highly motivated and can progress rapidly.

The authors believe that this book can reach the needs of all these students. This is a partially programmed textbook. Each section of the book contains both a traditional lecture part and a programmed part. The student and the instructor may use as much or as little of a section as is required for mastery.

The lecture part is a concise statement of the material of the section. It includes an introduction, motivation, and examples. This is the material an instructor would normally present to a class during a lecture.

The programmed part is used to build skills and to reinforce lecture material. The programs use stimulus-response-type frames. Hints and prompts are plentiful in the beginning of each program. They are gradually reduced. Near the end of each program there are release frames, that is, frames in which the student is asked to demonstrate his skill in the presence of a minimal number of prompts.

The student can use the lecture part of each section for later review. In this way each program need not be reworked to determine the content of a section.

At the end of each section there are additional practice problems and a Selftest. The practice problems are available to reinforce skills. More complex problems are sometimes included at the end of a set. The Selftest measures basic skill mastery.

As a partially programmed text, this book can meet the needs of each student enrolled in developmental algebra. It can also be used as a source for independent study or as a reference book.

Students who have little or no background in algebra will not be overwhelmed. They will find the programmed material suitable to their level of understanding. The programs carefully develop all skills necessary for later work.

The student who already has some basic skill need not be bored by unnecessary review. These students may review the lecture part of a section and then go directly to the practice problems and Selftest. If they find the Selftest too difficult, they may go to the programmed part.

A student doing individualized work may start at the beginning of the book and do the chapter Pretest and section Selftests. If the tests are successfully completed, entire sections or chapters may be skipped. Otherwise, both the lecture part and the programmed part are available for self-instruction.

Finally, the text will benefit the instructor. For example, the programmed sections can be used to replace numerous detailed examples in the lecture. Of course a good lecture always includes some examples. But in a lecture presentation of many detailed examples, the instructor may often feel that the student is just sitting back and not really participating in the solutions. The student merely watches. The proper use of the stimulus-response-type program requires continuous student participation. The instructor, freed from presenting large numbers of examples, may use the lecture time to motivate and unify various topics.

To the Student:
This book is designed to make it possible for you to be your own instructor. If you are doing independent work, be sure also to read *To the Instructor.*

There are tests available to help you decide if you need to study a particular chapter or section. Each chapter of this book contains a Pretest at the beginning. When you start a new chapter, take the Pretest without studying for it. Answers are provided at the end of the test. A table on the Pretest designates the section of the chapter in which each problem is discussed.

If you miss a problem, you should do the designated section. A correct response indicates that you might be able to skip the related section. But first do the Selftest at the end of the designated section. If you can do most of the problems on the Selftest, it is safe for you to omit that section.

Most students will start at the beginning of each chapter and proceed through all of the topics. This will ensure a solid mastery and serve as preparation for later work. Students who are better prepared are encouraged to move at a quicker pace and skip sections in which they are already competent.

Within a chapter, each section contains several more parts. First there is a lecture part. This is a brief, to-the-point account of the material in that section. It is complete in itself. If you understand the lecture and can do all the practice problems at the end of the section, you may skip the programmed part. This is in keeping with the philosophy that students who learn quickly should go at a faster pace.

However, even the best students will find that they want to go through many, perhaps most, of the programs. The programs are designed to give solid mastery of the skills talked about in lecture. We think it is best to work through a program immediately after the lecture.

As you work through a program you are expected not only to read carefully, as you would any text, but also to think out the answers to the questions. Keep thinking as you go. You must be an active participant. *Write down* the answers.

Each page of a program has a column that contains a series of questions called frames. Another column contains the answers to the frames. The answers should be covered by one of the masks provided so that they are not visible. Each question will be relatively easy to answer in its context. But the correct answer relies upon your mastery of previously learned skills.

Always start at the beginning of a program. Read the questions carefully. Then circle your answer or write it in the space provided, whichever is appropriate; or write your answers on a separate answer sheet. Take time to think. Give the answer *you* think is correct. Then immediately check your answer with the answer given in the answer column. If your answer is correct, go on to the next question. If the answer is not correct, try to find a reason for the error. Do not keep going if you get a string of wrong answers. Go back over the previous frames until your difficulties are straightened out. A time lapse might help. If you have an instructor, seek help.

At the end of each section there is a Selftest. It serves as a quick review for the mastery of that section. At the end of each chapter there is a Posttest to serve as a check on the mastery of all skills included in the chapter. Answers for all Selftests and all chapter tests are provided in the text. The answers to even-numbered practice problems are given in the back of the book.

To the Instructor:

Each chapter contains these features:

1. a list of basic skills to be mastered in the chapter
2. a chapter Pretest (with answers immediately following)
3. each section contains
   a. a lecture portion
   b. a programmed portion
   c. a set of practice problems (with answers to the even-numbered problems at the end of the book)
   d. a Selftest (with answers immediately following)
4. a chapter Posttest (with answers immediately following)

The basic skills listed at the beginning of each chapter will help the student realize what is expected of him. The chapter Pretest and Posttest and the Selftests can be used for diagnostic purposes (see *To the Student* for the mechanics). The lecture portion of the section will give the student general direction, and can later be used for review or reference. The programmed part guides the student through the subject, but demands active participation. Practice problems are available for drill. Selftests measure the student's progress through a section. They serve as check points for both the instructor and student. If the student performs well, he or she can proceed. If not, more work is needed. Chapter Posttests include problems from the entire chapter and give a good overview of progress. They can also be used for review.

This text can be used in instructional settings ranging from the traditional lecture to the individualized style. In a lecture setting, the instructor can simply present the topics provided and have the students work through the program outside of class. This technique works particularly well if there is a large amount of material to be covered so that class time is limited for any one topic. If recitations are used, the programs and practice problems will provide good material for discussion.

With more class time, part of the hour could be used for lecture and the students might begin work on the programs in class. This format provides time for individual work with students. Entire class periods might be devoted to the programmed material. The class could participate orally, work in small groups, or work individually.

The material is also quite suitable for individualized instruction. An *Instructor's Manual* provides alternate forms of the chapter tests and Selftests. There is also a *Complete Solutions Manual to Odd-Numbered Problems* available.

Alternate Routes:
We have starred topics in the table of contents that may be omitted. No later sections depend on the material in the starred sections. The table of chapters with prerequisite chapters or sections makes it possible to pick alternate paths through the book.

Chapter	*Prerequisite Chapters or Sections*
1 (whole numbers)	
2 (integers)	Chapter 1
3 (polynomials)	Chapters 1, 2
4 (factoring polynomials)	Chapters 1, 2, 3
5 (fractions)	Chapters 1, 2, 3; Section 4.1
6 (linear equations)	Chapters 1, 2, 3; Section 4.1; Chapter 5
7 (exponents)	Chapters 1, 2, 3
8 (radicals)	Chapters 1, 2, 3; Section 4.1; Chapter 5
9 (functions, graphing)	Chapters 1, 2, 3; Section 4.1; Chapters 5, 6; Section 8.1
10 (systems of equations)	Chapters 1, 2, 3; Section 4.1; Chapters 5, 6; Section 8.1; Chapter 9
11 (quadratic equations)	Chapters 1, 2, 3, 4, 5, 6, 8

The order of topics can easily be changed if you cover the prerequisite sections listed in the table. Some alternate routes are shown in the figure.

Alternate Routes

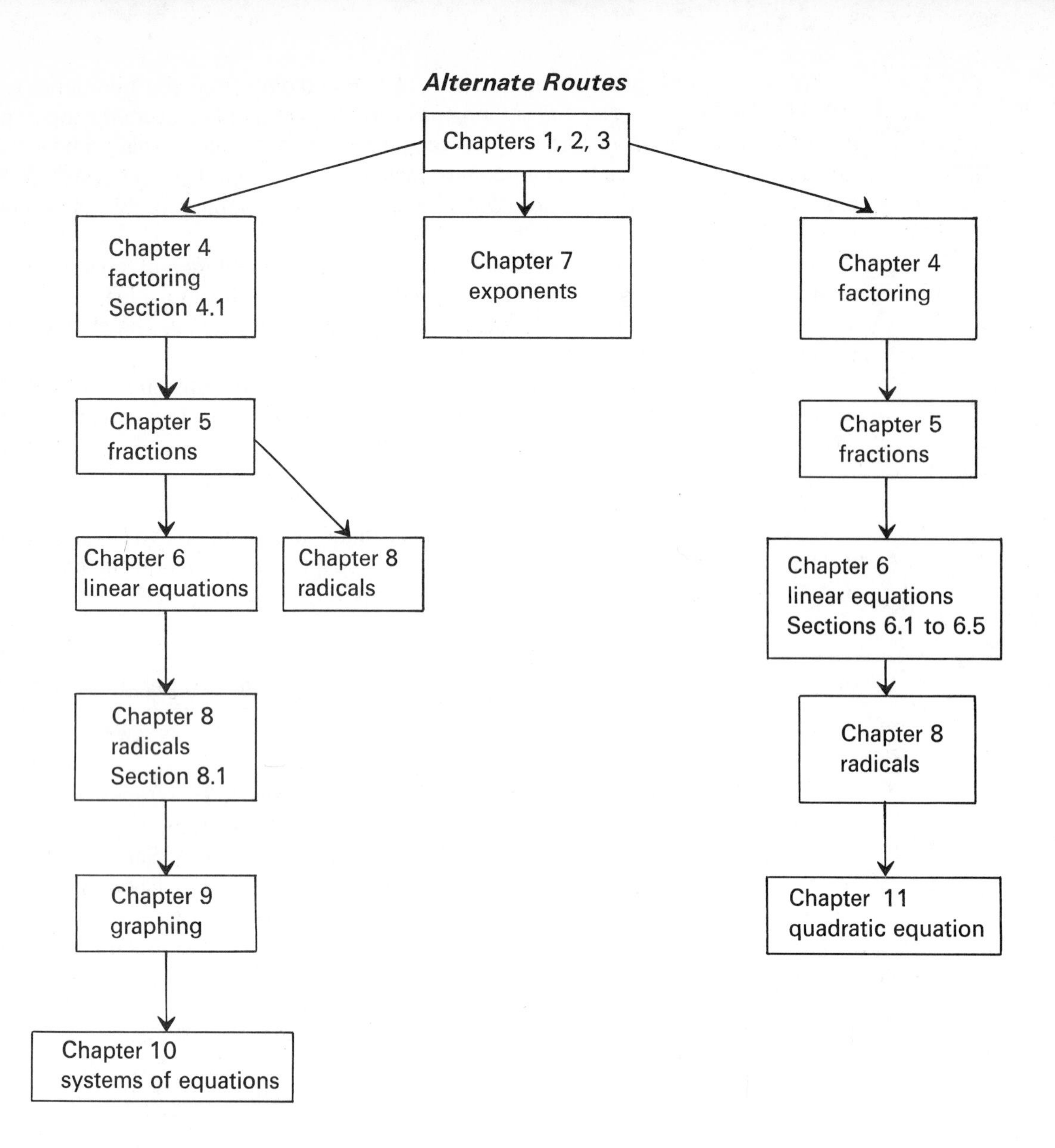

Acknowledgments

A special thanks to the students, faculty, and administration of Honolulu Community College. The cooperation and encouragement they gave us was most helpful. We also appreciate the many valuable comments given by Jerald Ball of Chabot College, Robert T. Carlson of Bunker Hill Community College, Philip J. Hippenstell of Harrisburgh Area Community College, Anita Tracy of the University of Connecticut, and James Wallace.

C.P.B.

C.H.B.

1

PROPERTIES OF WHOLE NUMBERS

BASIC SKILLS

Upon completion of the indicated section the student should be able to:

1.1 *a.* Use set notation to list elements in a set or to describe elements in a set.
b. Distinguish sets of whole numbers, counting numbers, even whole numbers, and odd whole numbers.
c. Do arithmetic with 0.
d. Identify the multiplicative and additive identity in the set of whole numbers.

1.2 *a.* Use and read the symbols $=$, $>$, $<$, $\geq$, $\leq$, $\neq$, $\not>$, $\not<$.
b. Identify examples of the trichotomy law, the transitive laws, and the reflexive and symmetric laws.

1.3 *a.* Recognize prime numbers.
b. Factor a number completely.
c. Find the greatest common divisor and the least common multiple of two numbers.

1.4 *a.* Recognize different notations for multiplication.
b. Carry out additions, subtractions, multiplications, and divisions in the proper order.
c. Identify or complete examples of the associative, commutative, and distributive laws.

PRETEST 1

1. What are the numbers of the set

$$\{x \mid x \text{ is a whole number between 10 and 25 that is divisible by 4}\}$$

2. Answer the following yes or no.

a. Is every counting number a whole number?
b. Is every whole number a counting number?
c. Is the set of counting numbers the same as the set of whole numbers?

3. Simplify each of the following, and note the one that is not a well-defined operation.

a. $\frac{0}{8}$

b. $0 \cdot 5$

c. $\frac{9}{0}$

d. $4 \cdot 1$

e. $5 + 0$

f. $12 - 0$

4. Match the law to the example that illustrates it.

a. reflexive	1. if $y < x$ and $x < 3$, then $y < 3$
b. symmetric	2. $5 = 5$
c. transitive	3. if $x \not> 3$ and $x \not< 3$, then $x = 3$
d. trichotomy	4. if $y = 7$, then $7 = y$

5. If $5 > x$ and $x > y$, is $5 > y$?

6. Factor 120 completely.

7. Find the greatest common divisor of 60 and 70.

8. Find the least common multiple of 9, 24, and 80.

9. Simplify $(50 \cdot 3 \div 5) \cdot 2$.

10. Complete the examples so that they illustrate the given law.

a. The associative law says that $3(2a) =$ _________.

b. By the commutative law, $4 + x =$ _________.

c. By the distributive law, $3(x + 4) =$ _________.

Each problem above refers to a section in this chapter as shown in the table.

Problems	*Section*
1 and 2	1.1
3	1.1 and 1.2
4	1.2
5–8	1.3
9 and 10	1.4

PRETEST 1 ANSWERS

1. 12, 16, 20, 24 **2.** *a.* yes *b.* no (for example, 0) *c.* no

3. *a.* 0 *b.* 0 *c.* not defined *d.* 4 *e.* 5 *f.* 12 **4.** *a.* 2 *b.* 4 *c.* 1 *d.* 3

5. yes **6.** $2 \cdot 2 \cdot 2 \cdot 3 \cdot 5$ **7.** 10 **8.** 720 **9.** 60

10. *a.* $(3 \cdot 2)a$ or $6a$ *b.* $x + 4$ *c.* $3x + 3 \cdot 4$ or $3x + 12$

1.1 SETS, COUNTING NUMBERS, AND WHOLE NUMBERS

In mathematics we use the word *set* to mean a collection of objects. The objects in the set are called *elements* or members of the set. Hawaii is a member of the set of all states in the U.S.A. The number 4 is an element in the set of numbers $\{4, 5, 6\}$.

The description of a set must allow us to decide if a given element is or is not a member of the set. The collection of all students enrolled in a college is a set since we can check enrollment lists to decide if any particular student is enrolled.

The collection of the five best paintings in the world is not a *well-defined set*, for there would be endless arguments as to whether to include a particular painting or not.

A straightforward way to describe a set is to list the elements of the set in the special brackets, $\{\quad\}$, called *braces*. The notation $B = \{a, b, c, d\}$ means that B is the set consisting of the elements a, b, c, and d.

We often make statements about some or all the elements in a set. We could specifically name the elements, but generally we use a letter that is not a member of the set to represent a typical element of the set. A letter so used is called a *variable*.

Consider the set $S = \{5, 10, 15, 20\}$. The statement "let t be an element of S" means that t represents any of the numbers 5, 10, 15, or 20. But t cannot be 25 since 25 is not an element of S. Since each element of S is divisible by 5, it is true that t is divisible by 5. We can put conditions on t so that it actually represents only some of the elements of S. For instance, if we say that t is an element of S that is divisible by 2, the condition "divisible by 2" means that t can be either 10 or 20. The restrictive condition does not allow t to be 5 or 15 even though these elements are in S. If t is an element in S such that $t + 8 = 18$, then the only element t represents is 10.

Another way to describe a set is to specify the conditions that a typical element must satisfy. We read the new notation as follows:

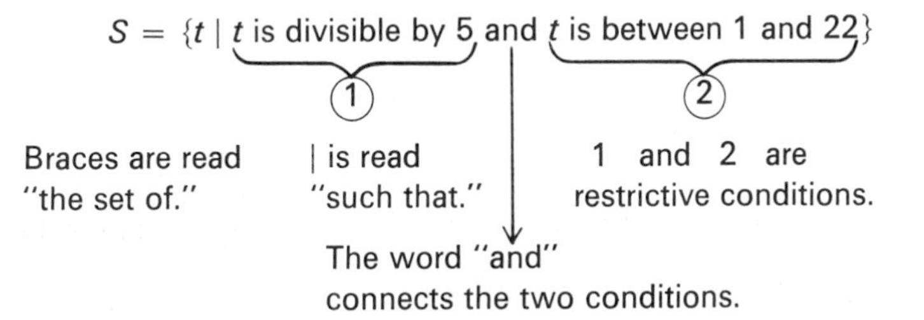
S = {t | t is divisible by 5 and t is between 1 and 22}
1
2
Braces are read "the set of."
| is read "such that."
1 and 2 are restrictive conditions.
The word "and" connects the two conditions.

The only elements that satisfy these two conditions are 5, 10, 15, and 20, for they are the only numbers between 1 and 22 that are divisible by 5. Thus the elements in S are 5, 10, 15, 20. The letter t is not actually a member of S. It is a variable that represents members of S.

Example 1 List the elements of the set Q.

$$Q = \{r \mid r \text{ is a number divisible by 2 and } r \text{ is between 3 and 11}\}$$

The elements of Q are 4, 6, 8, 10. The letter r represents any of these numbers.

Example 2 Let M be the set of numbers greater than 4 and less than 10. Write this set in two ways: *a.* Use the listing technique. *b.* Use the "such that" style.

a. $M = \{5, 6, 7, 8, 9\}$

b. $M = \{n \mid n \text{ is a number greater than 4 and less than 10}\}$

One special set is the *empty set*. It is a set that contains no elements at all. We denote this set by the symbol $\varnothing$ from the Scandinavian alphabet.

$\varnothing$ = the set of all unicorns since there are no unicorns

$\varnothing = \{x \mid x \text{ is a number greater than 4 and less than 2}\}$

An important set of numbers is the set of *counting numbers*. This set consists of the numbers we use to count whole objects. The number 1 is a counting number; the number 2 is a counting number; the number 3 is a counting number. The next counting numbers are 4, 5, 6, etc.

We use three dots, . . . , to stand for the words *et cetera* (etc.). The three dots at the end of a list mean that the list continues in the same pattern established by the given elements. So the set of counting numbers is $\{1, 2, 3, \ldots\}$. A counting number is *even* if it is divisible by 2. Thus the set of even counting numbers is $\{2, 4, 6, \ldots\}$. The counting numbers that are not even are *odd*. The set of odd counting numbers is $\{1, 3, 5, \ldots\}$.

One particular number that is not included in a set of counting numbers is zero. The set consisting of all the counting numbers and zero is the set of *whole numbers*. So the set $\{0, 1, 2, 3, \ldots\}$ is the set of whole numbers.

Special results occur when zero is combined with other numbers. If we add zero to any number a, the result is again a. Likewise, if zero is subtracted from any number a, the result is again a. Symbolically we write $a + 0 = a$, $0 + a = a$, and $a - 0 = a$ for all numbers a.

If a number a is multiplied by zero, the result is zero. Thus $a \cdot 0 = 0$ and $0 \cdot a = 0$ for any number a. (Recall that the centered dot stands for multiply. Thus $a \cdot 0$ means "a times 0.")

We use several notations to indicate division. All the following mean "a divided by b."

$a \div b$ — In each case, b is the *divisor* and a is the *dividend*.

$\frac{a}{b}$

a/b

$b \overline{)a}$

The result of a division is called the *quotient*.

$$a \div b = c$$

dividend (a) — divisor (b) — quotient (c)

Recall that $a \div b = c$ if and only if $b \cdot c = a$. The divisor times the quotient must yield the dividend. For example:

$12 \div 3 = 4$ because $3 \cdot 4 = 12$

$\frac{30}{5} = 6$ because $5 \cdot 6 = 30$

$16/8 = 2$ because $8 \cdot 2 = 16$

We must be a little cautious when division involves zero, for *0 can be a dividend but never a divisor*. Look at what happens:

$\frac{0}{6} = 0$ because $6 \cdot 0 = 0$

However, let's try to do the following division.

$\frac{12}{0} = \boxed{?}$ because $0 \cdot \boxed{?} = 12$

Clearly, there is no number that will work in $0 \cdot \boxed{?} = 12$.

Since we get into trouble whenever we try to divide by zero, we say that *division by zero is undefined*.

Summary of Facts about Zero

$a + 0 = a$

$0 + a = a$

$0 \cdot a = 0$

$a \cdot 0 = 0$

$\frac{0}{a} = 0$ if $a \neq 0$

$\frac{a}{0}$ is undefined

The number zero is called the *additive identity* because $0 + a = a$ and $a + 0 = a$ for any number a. Likewise, the number one is called the *multiplicative identity* since $a \cdot 1 = a$ and $1 \cdot a = a$.

PROGRAM 1.1

0. The programmed portions of this text are designed to help you master the skills of each section. But *you* must actively participate. The numbered comments, discussions, and questions on the outside of the page are called *frames*. Most frames will ask you a question. Answers to the questions are in the inside column. We suggest that you cover all the answers before you begin a page. Then try to answer the questions in the frame. Write your answer. Then, check each answer immediately with the answer given in the answer column.

You should be able to answer most of the questions correctly. But if you begin to miss two or three in a row, stop. Review the previous frames or ask for help before you continue.

Let's try some sample questions. The numbered comments and questions on the outside are called (frames\answers\exercises). What is the number of this frame? ____ To derive the most benefit from the program you should (study the answers first\cover the answers, give your own answers to each frame, and then check your answers with those in the inside column\copy the answers from the answer column into the blanks in the frames). If you miss several frames in a row, what should you do? ________________

frames

0

cover the answers, etc.

Review previous frames, or ask for help.

1. Mathematicians use the word *set* to mean a collection of things. The collection of states in the U.S. form a ____.

set

2. The things in a set are called *members* or *elements* of the set. Hawaii is an element of the set of all states. Dallas (is\is not) an element of the set of all states. Colorado (is\is not) an element of the set of all states.

is not

is

3. One way to describe a set is to list all the elements of the set within the braces, { }. The set consisting of the first four letters in the English alphabet is denoted by $\{a, \_\_, \_\_, \_\_\}$. The set consisting of the elements q and r is denoted by ________.

b; c; d

$\{q, r\}$

4. The notation $\{x, y, z\}$ means the set consisting of the elements ________.

x, y, z

5. The order in which the elements of a set are listed makes no difference. The sets $\{8, 10, 12\}$ and $\{10, 8, 12\}$ (have\do not have) the same elements. The sets (are\are not) the same.

have
are

6. Which sets are the same as $\{3, 5, 7\}$?

a. $\{5, 7, 3\}$ *b.* $\{3, 5, 7, 9\}$ *c.* $\{3, 5, 6\}$ *d.* $\{3, 7, 5\}$

a. and *d.*

7. When we wish to refer to a general element in a set, we use a *variable*. Consider the set of all male college students. Let us call a typical member of this set Joe College. Then Joe College is a variable. Joe College could be any male college student. In the statement "let x be an element of the set $\{3, 4, 5\}$," the variable is ____. The variable can be any of the numbers ____, ____, or ____. Can the variable represent 8? (yes\no)

x

3; 4; 5; no

8. In the expression "let n be an element of the set $\{2, 4, 5\}$," n is used as a ________. Which elements can n represent? ________

variable; 2, 4, or 5

9. We often make further restrictions on the variable so that it actually represents only some particular elements of the set instead of any element of the set. If x is an element of $\{10, 11, 12\}$ such that $x + 5 = 15$, then x stands for ____. Can x also represent 11? (yes\no)

10; no

10. Let y be an element of the set $\{2, 3, 4, 5, 10\}$ which is divisible by 2. The elements that y can represent are ________.

2, 4, or 10

11. It is sometimes inconvenient to list all the elements in a set. Another way to describe the set is to give the conditions that a typical element must satisfy in order to be in the set. The notation

$$\{x \mid x \text{ is a U.S. state}\}$$

means the set of all elements x such that x is a U.S. state. The vertical bar | means "such that."

$$\{t \mid t \text{ is a former U.S. President}\}$$

means the set of all elements t ________________________.
Is George Washington a member of this set? (yes\no) Is Winston Churchill a member of this set? (yes\no)

such that t is a former U.S. President
yes
no

12. The elements of

$$\{x \mid x \text{ is divisible by 2 and } x \text{ is between 1 and 11}\}$$

are 2, __, __, __, and __.

4; 6; 8; 10

13. The elements of

$$\{t \mid t \text{ is divisible by 3 and } t \text{ is between 1 and 11}\}$$

are ________. Is 15 a member of the set? (yes\no)

3, 6, and 9; no

14. One special set is the empty set. This set has no elements at all. Therefore, if the passenger pigeon is extinct, the set of all living passenger pigeons is the ________.

empty set

is

15. The set

$$\{x \mid x \text{ is a number greater than 10 but less than 5}\}$$

(is\is not) the empty set.

4

16. We use three dots, . . . , to mean et cetera. So three dots at the end of a list of numbers means we are to continue listing the numbers in the same pattern established by the given numbers. The next number in the list 1, 2, 3, . . . is ____.

is not; is

17. The number 5 (is\is not) an element of {1, 2, 3}. The number 5 (is\is not) an element of {1, 2, 3, . . .}.

yes; no

18. The set of *counting numbers* is the set {1, 2, 3, . . .}. Is 6 a counting number? (yes\no) Is 0 a counting number? (yes\no)

is; yes
no

19. A counting number is *even* if it is divisible by 2. So the set {2, 4, 6, . . .} (is\is not) the set of even numbers. Is 8 an even number? (yes\no) Is 5 an even number? (yes\no)

odd

20. The odd counting numbers are those counting numbers which are not even. The set {1, 3, 5, . . .} is the set of (odd\even) counting numbers.

2; 4; 6; 8; 10

21. The first five even counting numbers are ____, ____, ____, ____, ____.

1; 3; 5; 7; 9

22. The first five odd counting numbers are ____, ____, ____, ____, ____.

odd

23. The number 75 is an (even\odd) number.

even

24. The number 68 is an (even\odd) number.

is not

25. The number 0 is not a counting number. The set {0, 1, 2, 3, . . .} (is\is not) the set of counting numbers.

is

26. The set {1, 2, 3, . . .} (is\is not) the set of counting numbers.

0

27. The set {0, 1, 2, 3, . . .} is the set of *whole numbers*. The only element that is a whole number but not a counting number is ____.

is

28. Every counting number (is\is not) a whole number.

29. Special results occur when zero is used in the basic operations of arithmetic. Let m be a whole number. Then

$$m + 0 = m \qquad 0 + m = m \qquad \text{and} \qquad m - 0 = m$$

3; 2; 5

So $3 + 0 =$ ____, $0 + 2 =$ ____, and $5 - 0 =$ ____.

r

30. If we add or subtract 0 from any whole number r, the result is ____.

0
0

31. If we multiply any whole number n by 0, the product is 0. So $8 \cdot 0 =$ ____ and $0 \cdot 46 =$ ____.

0

32. If m is any whole number, then $m \cdot 0 =$ ____.

33. Compute the following:

a. $31 + 0 =$ ____ *b.* $31 \cdot 0 =$ ____ — *a.* 31; *b.* 0

c. $15 - 0 =$ ____ *d.* $0 \cdot 15 =$ ____ — *c.* 15; *d.* 0

34. We can divide 0 by any whole number m as long as m is not 0. The result is 0. So $0 \div 13 = 0$ (recall $0 \div 13$ means 0 divided by 13). What is $0 \div 6$?

____ — 0

35. $\frac{0}{45} =$ ____ $\left(\frac{a}{b} \text{ means } a \div b\right)$. — 0

36. $\frac{0}{2} =$ ____. — 0

37. We can never divide any number by 0. So $5 \div 0$ is not allowed, because 0 times the quotient does not equal the ________. — dividend

38. Which of the following operations are not allowed?

a. $17 \div 0$ *b.* $0 \div 25$ *c.* $0 \div 46$ *d.* $8 \div 0$ — *a.* and *d.*

39. The division $73 \div 0$ is not allowed because we cannot divide by ____. — 0

However, $0 \div 73$ is defined and it equals ____. — 0

40. Which of the following equal 0?

a. $\frac{0}{15}$ *b.* $\frac{1}{0}$ *c.* $\frac{0}{0}$ *d.* $\frac{0}{7}$

a. and *d.* (1/0 or 0/0 are not allowed because we can never divide by 0.)

41. The expression $n + 0$ (is\is not) identical to n. So we call 0 the *additive identity*. Likewise, $n \cdot 1$ (is\is not) identical to n. And we call 1 the *multiplicative identity*. Is 3 also an additive identity? (yes\no) Is 0 also a multiplicative identity? (yes\no)

is
is
no
no

42. The multiplicative identity is the number ____. Zero is the ________ identity. — 1; additive

PRACTICE PROBLEMS

1. Which of the following collections are well-defined sets?
   a. The letters of the English alphabet
   b. The ten best photographs
   c. Counting numbers greater than 12 but less than 3

2. Let $A = \{2, 5, 8, 9\}$.
   a. If n is an element of A, then what values can n represent?
   b. If t is an element of A and t is even, what values in A can t represent?
   c. If w is an element of A and w is odd, what values in A can w represent?
   d. If x is an element of A such that $x + 1 = 4$, then what values in A can x represent? Is there an element in $\{3, 4, 5\}$ that x might represent?

3. List the elements in each set.

   a. $\{x \mid x$ is a whole number between 4 and 10 that is divisible by $3\}$
   b. $\{x \mid x$ is a counting number divisible by $5\}$

4. Use set notation to describe the following sets.

   a. counting numbers
   b. whole numbers
   c. even whole numbers
   d. odd whole numbers

5. Which of the sets in Problem 4 have a multiplicative identity? Which have an additive identity?

6. Do the arithmetic if possible. Let x be a counting number.

   a. $17 \cdot 0$ *b.* $35 - 0$ *c.* $14 \div 0$ *d.* $0 \div 6$
   e. $0 \cdot 0$ *f.* $0 \div 0$ *g.* $5 \div 5$ *h.* $8 + 0$
   i. $0 + 3$ *j.* $1 \cdot 9$ *k.* $x - 0$ *l.* $0 \cdot x$
   m. $x \div 0$ *n.* $0 + x$

SELFTEST

1. *a.* How many elements does the set $\{a, b, 7, 19\}$ have?
   b. List each of the elements.

2. Let t be an element of the set $\{0, 1, 5, 9\}$ such that $t + 1 = 10$. Find t.

3. List the elements in the following set.

$$\{x \mid x \text{ is greater than 2 but less than } 10\}$$

4. Label each of the following as a counting number, a whole number, both, or neither.

   a. 3 *b.* $^1/_2$ *c.* 0 *d.* 1.34 *e.* 6 *f.* $^4/_5$

5. *a.* List five even numbers larger than 100.
   b. List five odd numbers larger than 1,000.
   c. How many even numbers are there?
   d. How many odd numbers are there?

6. Answer each of the following as true or false. Let a, b, and c be any whole numbers.

   a. $a + 0 = a$ *b.* $a \cdot 0 = 0$
   c. $a \div b = c$ if and only if $a = b \cdot c$ and $b \neq 0$
   d. $5 \div 0 = 0$ *e.* $0 \div 0 = 0$
   f. $0 \div 5 = 0$ *g.* $1 \cdot a = a$

7. *a.* How many elements does the empty set have?
   b. List two examples of the empty set.
   c. Is $\{0\}$ the empty set?

SELFTEST ANSWERS

1. *a.* 4 elements *b.* They are $a, b, 7, 19$ **2.** t is 9 **3.** 3, 4, 5, 6, 7, 8, 9
4. *a.* both *b.* none of these *c.* whole *d.* none of these *e.* both *f.* none of these
5. *a.* Your numbers should be greater than 100 and have as last digit 0, 2, 4, 6, or 8; the first 5 such numbers are 102, 104, 106, 108, 110. *b.* These five numbers should all be more than 1,000 and

end in one of the digits 1, 3, 5, 7, or 9; the first five such numbers are 1,001, 1,003, 1,005, 1,007, 1,009. *c.* and *d.* There are infinitely many even and odd numbers.

6. *a.* T *b.* T *c.* T *d.* F *e.* F *f.* T *g.* T

7. *a.* No elements. *b.* Empty set = set of all numbers which are both smaller than 1 and bigger than 1; empty set = set of all cows with wings. *c.* $\{0\}$ is not the empty set since 0 is a member of this set.

1.2 LAWS OF EQUALITIES AND INEQUALITIES

Given any two numbers, we can compare their relative "size." The number 8 is equal to 8. The number 25 is greater than 10. But 30 is less than 56. We use special symbols to indicate the relations between numbers.

Symbol	*Meaning*	*Example*
$=$	is equal to	$3 = 3$
$>$	is greater than	$12 > 5$
$<$	is less than	$4 < 9$
$\geq$	is greater than or equal to	$3 \geq 3$; $6 \geq 2$
$\leq$	is less than or equal to	$4 \leq 4$; $8 \leq 9$

A bar through a symbol negates the symbol or means "is not."

$\neq$	is not equal to	$4 \neq 5$
$\not>$	is not greater than	$2 \not> 8$
$\not<$	is not less than	$3 \not< 0$

Notice that the larger value is always next to the larger end of the $>$ or $<$ symbol. So we write $108 > 3$ since 108 is the greater value. In $4 < 9$, 9 is the greater value. We can express the fact that 5 is greater than 1 in two ways. The direct way is to write $5 > 1$. But we can also say "1 is less than 5" or $1 < 5$. So $a < b$ and $b > a$ express the same relation between a and b.

Example 1 The elements in $\{x \mid x \text{ is a whole number and } x > 5 \text{ and } x < 10\}$ are 6, 7, 8, 9.

Example 2 The elements in $\{x \mid x \text{ is a whole number and } x \geq 5 \text{ and } x < 10\}$ are 5, 6, 7, 8, 9.

Example 3 The elements in $\{x \mid x \text{ is a counting number and } x \geq 0 \text{ and } x \leq 3\}$ are 1, 2, 3. We do not include 0 since 0 is not a counting number. We do include 3 since $x \leq 3$ allows x to equal 3.

Example 4 The elements in $\{x \mid x \text{ is a whole number, } x > 12, x \leq 15, \text{ and } x \neq 13\}$ are 14 and 15. The condition $x \neq 13$ excludes 13.

Since the "greater than" and "less than" relations imply that the values are not equal, we often refer to these relations as *inequalities*. There are some general statements that are true about inequalities. These are called *laws*. We will list them here.

LAWS OF INEQUALITIES

Trichotomy Law: **If a and b are any two numbers, then one and only one of the following conditions is true**

$$a > b \quad \text{or} \quad a < b \quad \text{or} \quad a = b$$

The trichotomy law says that we can always compare the relative size of two numbers.

Example 5 Given 3 and 15, is $3 > 15$, $3 < 15$, or $3 = 15$ true? The only true statement is that $3 < 15$.

Example 6 Given x and 9, and $x \neq 9$, and $x \not< 9$, then by the trichotomy law, $x > 9$ must be true.

Example 7 Two people, Joan and Sam, weigh themselves. Let us say that Joan has weight a pounds and Sam has weight b pounds. The trichotomy law says that exactly one of the following conditions holds.

$a > b$ or Joan is heavier than Sam
$a < b$ or Sam is heavier than Joan
$a = b$ or Joan and Sam have the same weight

Transitive Law of >: **If $a > b$ and $b > c$, then $a > c$.**

Transitive Law of <: **If $a < b$ and $b < c$, then $a < c$.**

Example 8 Three people A, B, and C measure their heights. Say A has height a feet, B has height b feet, and C has height c feet. If A is taller than B (that is, $a > b$) and B is taller than C (that is, $b > c$), the transitive law says that A is taller than C (that is, $a > c$).

Example 9 If $x > y$ and $y > 2$, then by the transitive law, $x > 2$.

Example 10 If $y < 8$ and $8 < z$, then what is the relation between y and z? By the transitive law, $y < z$.

LAWS OF EQUALITY

Reflexive Law of =: **$a = a$ for any number a.**

Symmetric Law of =: **If $a = b$, then $b = a$.**

Transitive Law of =: **If $a = b$ and $b = c$, then $a = c$.**

The reflexive law says that any number is always equal to itself. Thus $4 = 4$, $7 = 7$, and $0 = 0$.

The symmetric law says that it does not matter which quantity we write on the left side of the equals symbol. So it does not matter if we write $3 = 2 + 1$ or $2 + 1 = 3$. Both expressions say that 3 and $2 + 1$ are equal. If $5 = x$, the symmetric law says that $x = 5$ as well.

The transitive law is especially convenient. For if we have a string of equalities, we can conclude that the first expression equals the last one.

Example 11 From the string of equalities

$$x + 7 = 16 = y$$

we conclude that $x + 7 = y$ by the transitive law.

Example 12 The string of equalities

$$x = 3 + 2 - 1 = 5 - 1 = 4$$

means that $x = 4$ by the transitive law.

Example 13 Suppose that Joe makes a salary of J dollars per year, Beth makes a salary of B dollars per year, and Tom makes a salary of T dollars per year. If Joe and Beth make the same salary (that is, $J = B$) and Beth and Tom also make the same salary (that is, $B = T$), then the transitive law assures us that Joe and Tom make the same salary (that is, $J = T$).

PROGRAM 1.2

1. The basic symbols we use to show relations between numbers are

Symbol	*Meaning*	*Example*
$=$	is equal to	$6 = 6$
$>$	is greater than	$9 > 5$
$<$	is less than	$3 < 7$

Fill in the appropriate $>$, $<$, or $=$ symbol. 8 ____ 2; 9 ____ 9; 4 ____ 7

$>$; $=$; $<$

2. We read $25 > 7$ as "25 is greater than 7." We read $14 < 19$ as "____ ________________."

14 is less than 19

3. When we use the "greater than" symbol as in $15 > 10$, the larger number is next to the wider part of $>$. In $5 < 7$, the larger number (is\is not) next to the wider part of the $<$ symbol.

is

4. So the larger number always goes next to the larger end of the $>$ or $<$ symbol. In $x > 7$, ____ is the larger value. In $y < 10$, ____ is the larger value.

x; 10

5. Both $x < 13$ and $13 > x$ say that ____ is the larger value. We can say that 10 is larger than y in two ways: $y < 10$ and ________.

13

$10 > y$

6. We can say that 4 is less than 10 in two ways: ________ and ________.

$4 < 10$; $10 > 4$

7. The elements in $\{x \mid x \text{ is a whole number, } x > 5, \text{ and } x < 10\}$ are 6, 7, ___, ___.

8, 9

8. A bar through a symbol negates the symbol.

Symbol	*Meaning*	*Example*
$\neq$	is not equal to	$5 \neq 4$
$\ngtr$	____________	$2 \ngtr 5$
___	is not less than	$7 \nless 3$

is not greater than

$\nless$

4 and 6

9. The elements in $\{x \mid x \text{ is a whole number}, x > 3, x < 7, \text{ and } x \neq 5\}$ are ________.

10. Two more symbols combine "equals" with "greater than" and "equals" with "less than."

is less than or equal to

Symbol	*Meaning*	*Example*
$\geq$	is greater than or equal to	$3 \geq 3$; $3 \geq 2$
$\leq$	________________	$5 \leq 5$; $5 \leq 6$

11. $x \geq 8$ is true if one of two situations occurs.

$x \geq 8$ is true if x equals 8
$x \geq 8$ is true if x is greater than 8

true; true; false

So $8 \geq 8$ is (true\false); $9 \geq 8$ is (true\false); $7 \geq 8$ is (true\false).

12. Similarly, $x \leq 12$ is true if one of two situations occurs.

$x \leq 12$ is true if x equals 12
$x \leq 12$ is true if x is less than 12

true; false; true

So $10 \leq 12$ is (true\false); $15 \leq 12$ is (true\false); $12 \leq 12$ is (true\false).

7

8; 9

13. The elements in $\{x \mid x \geq 7, x \leq 9, \text{ and } x \text{ is a whole number}\}$ are ____, ____, and ____.

8 and 9

14. The elements in $\{x \mid x \text{ is a whole number}, x \geq 8, \text{ and } x < 10\}$ are ________.

15. The *trichotomy law* says that if a and b are any two numbers, then one and only one of the following relations is true.

$$a > b \quad \text{or} \quad a < b \quad \text{or} \quad a = b$$

one

$y < 10$

By the trichotomy law, (none\one\two\all) of the following conditions hold(s) for any number y: $y > 10$, $y = 10$, or ________.

must be

16. Suppose x and y are numbers and $x \neq y$ and $x \not> y$. Then by the trichotomy law, $x < y$ (must be\need not be) true.

$y = 3$

17. Let y be a number such that $y \not> 3$ and $y \not< 3$. Then by the trichotomy law, ________.

one

18. If club A charges yearly dues a, and club B charges yearly dues b, then by the trichotomy law, (none\one\two\all) of the following conditions hold(s).

$a > b$; club A's dues are more than club B's

$a < b$

$a = b$; the dues for club A and club B are equal

____; club B's dues are more than club A's

____; ________________

$a > c$

19. The *transitive law of* $>$ says that if $a > b$ and $b > c$, then $a > c$. So if $a > 8$ and $8 > c$, then by the transitive law, ($a = c$ \ $a > c$ \ $a < c$) is true.

20. If $x > 100$ and $100 > y$, then by the transitive law, ________ is the relation between x and y.

$x > y$

21. Let J be Jean's weight; P be Pete's weight, and R be Ron's weight. If Jean weighs more than Pete (that is, $J > P$), and Pete weighs more than Ron (that is, $P > R$), then Jean weighs (more than\less than\the same as) Ron (that is, $J >$ ____).

more than

R

22. The *transitive law of* $<$ says that if $a < b$ and $b < c$, then $a < c$. So the transitive law of $<$ (is\is not) the same as the transitive law of $>$, except it uses $<$ throughout instead of $>$.

is

23. If $x < 7$ and $7 < y$, then ($x = y$ \ $x > y$ \ $x < y$) is true by the ________ law of $<$.

$x < y$; transitive

24. If $3 < x$ and $x < y$, then what is the relation between 3 and y? ________

$3 < y$

25. The *reflexive law of* $=$ says that any number is equal to itself. So the reflexive law says that $a =$ ____ for any number a.

a

26. It is true that $8 = 8$ by the ________ law of $=$.

reflexive

27. The *symmetric law of* $=$ says that if two expressions are equal, it does not matter which is written on the left side of the equals symbol. So if x and 10 are equal, we can write $x = 10$ or ________.

$10 = x$

28. The relation that y and 8 are equal can be expressed in two ways: ________ and ________.

$y = 8$;
$8 = y$

29. The *transitive law of* $=$ is similar to the transitive laws of $<$ and $>$. But it says that if $a = b$ and $b = c$, then $a = c$. So $x = 5$ and $5 = y$ imply that ________.

$x = y$

30. From

$$x + 5 = 16 = y$$

we can conclude that $x + 5 =$ ____ by the transitive law.

y

31. Match the examples to the laws they illustrate.

a. reflexive law of $=$ | *b.* symmetric law of $=$
c. transitive law of $>$ | *d.* transitive law of $<$
e. transitive law of $=$ | *f.* trichotomy law

Examples

If $x = 3$, then $3 = x$. Law ____. — *b.*

$65 = 65$. Law ____. — *a.*

If $x = 2 = y$, then $x = y$. Law ____. — *e.*

If $3 < y$ and $y < x$, then $3 < x$. Law ____. — *d.*

If $x \neq 2$ and $x \not< 2$, then $x > 2$. Law ____. — *f.*

If $x > 4$ and $4 > z$, then $x > z$. Law ____. — *c.*

PRACTICE PROBLEMS

1. Answer each of the following as true or false.

a. $9 > 5$ *b.* $4 \geq 4$ *c.* $3 < 2$ *d.* $6 \not> 12$

e. $7 \not> 9$ *f.* $5 \not< 2$ *g.* $12 \neq 18$ *h.* $1 \leq 1$

i. $x \leq x + 1$ *j.* $0 \leq 2$ *k.* $x = x + 1$ *l.* $0 > 0$

2. If x is a number such that $x \not> 3$ and $x \not< 3$, then what is the value of x?

3. If $x \not< 5$, then which one of the following conditions is true?

a. $x = 5$ *b.* $x > 5$ *c.* $x < 5$ *d.* $x \geq 5$

4. The trichotomy law says that if we put two weights in a balance (one on each side of the balance), then exactly one of three things will happen. What are the three possibilities?

5. If x and y are counting numbers such that $5 \geq x$ and $x \geq y$, then what is the largest possible value for y? What is the value of x when y has its largest possible value?

SELFTEST

1. Answer each of the following as true or false.

a. $5 > 4$ *b.* $6 \neq 7$ *c.* $14 < 12$

d. $5 \geq 5$ *e.* $8 \leq 9$ *f.* $25 \neq 25$

g. $3 \not> 4$ *h.* $0 \not< 3$ *i.* $0 \geq 3$

2. What does the trichotomy law say about two numbers a and b under each of the following circumstances?

a. if $a \neq b$ and $a \not> b$ *b.* if $a \neq b$ *c.* $a \neq b$ and $a \not< b$

3. Determine if each statement is true or false.

a. If $3 = x$ and $x = y$, then $3 = y$

b. If $3 = x$ and $3 = y$, then $x = y$

c. If $7 > x$ and $x > y$, then $7 > y$

d. $5 = 5$

e. If $x = 8$, then $8 = x$

f. $4 > 4$

g. If $4 > a$, then $a > 4$

h. If $10 > x$, then $x < 10$

i. If $x \not> y$ and $x \neq y$, then $x < y$

SELFTEST ANSWERS

1. *a.* T *b.* T *c.* F *d.* T *e.* T *f.* F *g.* T *h.* F *i.* F

2. *a.* $a < b$ *b.* $a > b$ or $a < b$ *c.* $a > b$

3. *a.* T *b.* T *c.* T *d.* T *e.* T *f.* F *g.* F *h.* T *i.* T

1.3 PRIME FACTORIZATION

We say that the number a *divides* the number b if a divides b without a remainder. Thus a divides b if there is a whole number c such that $ac = b$. So 6 divides 12 since $6 \cdot 2 = 12$. But 6 does not divide 13 since no whole number times 6 equals 13.

The number a is a *factor* of b if a divides b. So 6 is a factor of 12 since 6 divides 12. But 6 is not a factor of 13.

We know that for any number n, 1 divides n since $1 \cdot n = n$, so 1 is a factor of any number. In particular, 1 is a factor of 39, 2, and 0. Also if $n \neq 0$, then n divides n since $n \cdot 1 = n$. Thus n $(n \neq 0)$ is always a factor of itself. So 14 is a factor of 14; and 35 is a factor of 35.

Example 1 The factors of 14 are 1, 2, 7, and 14 since these are the only numbers which divide 14.

Example 2 The factors of 24 are 1, 2, 3, 4, 6, 8, 12, and 24.

Example 3 The factors of 29 are 1 and 29.

Counting numbers which have exactly two factors are called *prime numbers*.

Definition: **The counting number p is *prime* if $p \neq 1$ and the only factors of p are 1 and p.**

Example 4 In the set $\{1, 2, 3, 4, 5, 6, 7, 8, 9, 10\}$, the only prime numbers are 2, 3, 5, and 7. The number 1 is not prime since it violates the first condition. The numbers 4, 6, 8, and 10 are not prime since they each have 2 as a factor. And 9 is not prime since 3 is a factor of 9.

Example 5 What are the factors of 33? Is 33 prime? The factors of 33 are 1, 3, 11, 33, so 33 is not prime since it has factors other than 1 and 33.

Prime numbers are important because they serve as the basic building blocks for counting numbers. The Fundamental Theorem of Arithmetic tells us that we can start with prime numbers and use multiplication of prime numbers to construct all the other counting numbers greater than one. Furthermore there is essentially only one way to construct a given number from the prime numbers.

Fundamental Theorem of Arithmetic: **Every counting number can be written as a product of prime numbers in a unique way (except for the order of the factors).**

Example 6 Write 8 as a product of prime numbers.

$8 = 2 \cdot 4$ But 4 is not prime because $4 = 2 \cdot 2$.
$8 = 2 \cdot 2 \cdot 2$ Here each factor is prime.

Example 7 Write 15 as a product of prime numbers.

$15 = 3 \cdot 5$ Each factor is prime.

Example 8 Write 17 as a product of prime numbers.

17 is already prime.

Example 9 Write 490 as a product of primes.

$490 = 2 \cdot 245$ But 245 is not prime because $245 = 5 \cdot 49$.
$490 = 2 \cdot 5 \cdot 49$ But 49 is not prime because $49 = 7 \cdot 7$.
$490 = 2 \cdot 5 \cdot 7 \cdot 7$ Now all the factors are prime.

To *factor a number completely* means to write the number as a product of its prime factors. The same factor may be used several times as in Example 6. There we used 2 as a factor in the product three times. The order of factors is not important, but we usually write the factors from left to right in increasing order. The important thing about complete factorization is that each factor must be prime. For example, 12 can be factored as $1 \cdot 12$, $2 \cdot 6$, $3 \cdot 4$, and $2 \cdot 2 \cdot 3$; but $2 \cdot 2 \cdot 3$ is the only factorization which uses prime numbers exclusively. So the complete factorization of 12 is $2 \cdot 2 \cdot 3$.

The *greatest common divisor* (GCD) of two numbers a and b is the largest counting number that is a factor of a and also a factor of b. One way to find the greatest common divisor is to use the prime factorization of each of the given numbers.

Rule to Find the Greatest Common Divisor: **To find the greatest common divisor of a and b: *First* factor a and b into primes. *Then* compute the GCD as the product formed by using the common prime factors that appear in a and also in b. If the same prime appears several times as a factor of a and as a factor of b, we use that prime the *minimal* number of times it appears as a factor of a or b.**

Example 10 Find the GCD of 40 and 60.

$40 = 2 \cdot 2 \cdot 2 \cdot 5$	Factor 40 into primes.
$60 = 2 \cdot 2 \cdot \quad 5 \cdot 3$	Factor 60 into primes and line up common factors.
$\text{GCD} = 2 \cdot 2 \cdot \quad 5$	Bring down factors common to 40 and 60.
$= 20$	Form the product of those factors.

Example 11 Find the GCD of 30, 60, and 90.

$30 = 2 \cdot 3 \cdot 5$	Factor 30, 60, and 90 into primes and line up factors common to all three of the numbers.
$60 = 2 \cdot 3 \cdot 5 \cdot 2$	
$90 = 2 \cdot 3 \cdot 5 \cdot \quad 3$	
$\text{GCD} = 2 \cdot 3 \cdot 5 = 30$	Form the product of the common factors.

The number m is a *multiple* of n if n is a factor of m. So 12 is a multiple of 6 since 6 is a factor of 12. The multiples of 2 are elements in $\{0, 2, 4, 6, \ldots\}$ or the set of even whole numbers. We count by 8 to find the multiples of 8. They are elements in the set

$$\{0, 8, 16, 24, 32, \ldots\}$$

Definition: **The *least common multiple* (LCM) of a and b is the smallest counting number that is a multiple of a and also a multiple of b.**

Example 12 The multiples of 15 are 0, 15, 30, 45, 60, ... The multiples of 20 are 0, 20, 40, 60, ... The LCM of 15 and 20 is 60 since 60 is the smallest counting number that is in both sets of multiples.

Another way to find the least common multiple of a and b is to factor a and b into primes.

Rule to Find the Least Common Multiple: **To find the LCM of a and b: *First* factor a and b into primes. *Then* compute the LCM as the product formed by using all the primes that appear in a or in b. If the same prime appears as a factor of a and as a factor of b, we use that prime the *maximal* number of times it appears as a factor of a or b.**

Example 13 Find the LCM of 10 and 45.

$10 = 2 \cdot 5$	Factor 10 into primes.
$45 = \quad 5 \cdot 3 \cdot 3$	Factor 45 into primes and line up common factors.
$\text{LCM} = 2 \cdot 5 \cdot 3 \cdot 3$	Bring down all the factors of line 1 and bring down the factors of line 2 that are not in line 1.
$\text{LCM} = 90$	Form the product of these factors.

Example 14 Find the LCM of 12, 40, and 9.

$$\begin{array}{rl} 12 = & 2 \cdot 2 \cdot 3 \\ 9 = & 3 \cdot 3 \\ 40 = & 2 \cdot 2 \cdot 2 \cdot 5 \\ \hline \text{LCM} = & 2 \cdot 2 \cdot 3 \cdot 3 \cdot 2 \cdot 5 \\ \text{LCM} = & 360 \end{array}$$

Factor 12, 9, and 40 into primes and line up common factors.

Bring down *one* factor from *each column*. Form the product of these factors.

The next example will show an important relationship between the LCM and GCD.

Example 15 *a.* Compute the GCD and LCM of 30 and 50.

$$\begin{array}{rl} 30 = & 2 \cdot 3 \cdot 5 \\ 50 = & 2 \cdot 5 \cdot 5 \\ \hline \text{GCD} = & 2 \cdot 5 = 10 \\ \text{LCM} = & 2 \cdot 3 \cdot 5 \cdot 5 = 150 \end{array}$$

b. Compute $\dfrac{30 \cdot 50}{\text{GCD}}$. Does this value equal the LCM?

$$\frac{30 \cdot 50}{\text{GCD}} = \frac{30 \cdot 50}{10} = \frac{1{,}500}{10} = 150$$

c. In general, for numbers a and b

$$\text{LCM of } a \text{ and } b = \frac{a \cdot b}{\text{GCD of } a \text{ and } b}$$

PROGRAM 1.3

1. The number x divides the number y if there is a whole number w such that $xw = y$. So 7 divides 14 since $7 \cdot 2 = 14$. Also 3 is a divisor of 12 since $3 \cdot$ ____ $= 12$. — 4

2. But 5 does not divide 12 since no whole number times ____ equals 12. Does 5 divide 15? (yes\no) — 5; yes (since $5 \cdot 3 = 15$)

3. If x divides y, then x is a *factor* of y. The factors of 6 are 1, 2, 3, and 6 since these are the only numbers which divide 6. The factors of 8 are 1, ____, ____, and ____. — 2; 4; 8

4. The number 1 divides every number n since $1 \cdot n = n$ for every number n. So 1 is a ______ of every number. — factor or divisor

5. Every number n except 0 divides itself since $n \cdot 1 = n$. So if $n \neq 0$, then n is always a factor of ____. The factors of 3 are ____ and ____. — n; 1; 3

6. The factors of 15 are ____, ____, ____, ____. — 1; 3; 5; 15

7. The factors of 9 are ______. — 1, 3, 9

8. If a counting number has exactly two factors, it is *prime*. The number 5 has the factors ____ and ____, so 5 (is\is not) prime. — 1; 5; is

9. The counting number p is *prime* if

$$p \neq 1 \qquad \text{and} \qquad \text{the only factors of } p \text{ are 1 and } p$$

Is 1 a prime number? (yes\no) Is 7 a prime number? (yes\no)

no; yes

10. Let's find the primes in $\{1, 2, 3, 4, 5, 6, 7, 8, 9, 10\}$. Is 1 prime? (yes\no) What are the factors of 2? ________ Is 2 prime? (yes\no) Since 2 is a factor of 4, 6, 8, and 10, are any of these numbers prime? (yes\no) Are 3, 5, and 7 prime? (yes\no) The factors of 9 are ________, so 9 (is\is not) prime. Thus the prime numbers in the first ten counting numbers are ____, ____, ____, and ____.

no; 1 and 2; yes; no; yes; 1, 3, 9; is not; 2; 3; 5; 7

11. Which numbers are prime in $\{11, 12, 13, 14, 15, 16, 17, 18, 19, 20\}$? ____, ____, ____, ____.

11; 13; 17; 19

12. Does 55 have any factors other than 1 and 55? (yes\no) 5 and ____ are the other ________. Is 55 prime? (yes\no)

yes; 11; factors; no

13. *To factor a number* means to write it as a product of some of its factors. The factored forms of 6 are $1 \cdot 6$ and $2 \cdot 3$. The two factored forms of 21 are ________ and ________ (the order of factors does not matter).

$21 \cdot 1$; $3 \cdot 7$

14. The factored forms of 8 are $8 \cdot 1$, $2 \cdot$ ____, and $2 \cdot 2 \cdot$ ____.

4; 2

15. The factored forms of 12 are $1 \cdot 12$, $2 \cdot 6$, $3 \cdot 4$, and $2 \cdot 2 \cdot 3$. Which of these forms uses only prime factors? ________

$2 \cdot 2 \cdot 3$

16. The *Fundamental Theorem of Arithmetic* says that every counting number greater than one can be written as a product of prime numbers in a unique way (except for the order of the factors). So 15 (can\cannot) be written as a product of prime numbers. The factored forms of 15 are ________ and ________. The only factorization of 15 into prime factors is ________.

can; $1 \cdot 15$; $3 \cdot 5$; $3 \cdot 5$

17. The prime factorization of 24 is $2 \cdot 2 \cdot 2 \cdot 3$. Except for the order of factors, is there another factorization of 24 which uses only prime factors? (yes\no)

no

18. Write 6 as a product of prime numbers. ________

$2 \cdot 3$ or $3 \cdot 2$

19. *To factor completely* means to write a number as a product of prime numbers. The complete factorization of 25 is ____ $\cdot$ ____. Can either factor be broken up into smaller factors? (yes\no)

5; 5; no

20. Factor 20 completely: $20 = 2 \cdot 10$. But 10 (is\is not) prime, because $10 = 2 \cdot 5$. So use $2 \cdot 5$ in place of 10. Then $20 = 2 \cdot 2 \cdot 5$, and all the factors (are\are not) prime.

is not; are

21. To factor a number completely, we first factor it into a product of two numbers. If these numbers are prime, we are finished. If these factors are not prime, we factor each of them into a product of two numbers. We continue in this fashion until all the factors are ________.

prime

22. Factor 40 completely: 40 = ____ · 20. Is 20 prime? (Yes\No), because 20 = 2 · ____. So then 40 = 2 · 2 · 10. Is 10 prime? (yes\no) If not, factor 10 into prime factors. ________ So now 40 = 2 · 2 · ____ · ____. Are all these factors prime? (yes\no)

2; No
10; no
10 = 2 · 5; 2; 5
yes

23. Factor 50 completely: 50 = 5 · ____. Are both factors prime? (yes\no) So 50 = ____ · ____ · ____.

10; no
2; 5; 5

24. Factor 27 completely: ________________

27 = 3 · 9 = 3 · 3 · 3

25. The *greatest common divisor* (GCD) of two numbers a and b is the largest counting number that is a factor of a and also a factor of b. The factors of 10 are 1, 2, 5, and 10. The factors of 35 are ____, ____, ____, and ____. The largest number that is in both lists of factors is ____. Thus ____ is the greatest common divisor of 10 and 35.

1; 5; 7; 35
5; 5

26. We can use the following scheme to compute the GCD of any two numbers. Find the GCD of 20 and 36.

20 = 2 · 2 · 5 ← Factor 20 and 36 into primes.
36 = 2 · 2 · 3 · 3 ← Line up the factors that 20 and 36 have in common.

GCD = 2 · 2 ← Bring down the prime factors common to 36 and 20.
GCD = ____ ← Multiply the common factors.

So ____ is the largest factor that is common to 20 and 36.

4
4

27. If there should be no common prime factors, then the GCD is 1 and we are done. Do the numbers 10 and 7 have any prime factors in common? (yes\no) So the GCD of 10 and 7 is (7\1\0).

no
1

28. Find the GCD of 30 and 45.

30 = 2 · 3 · 5 ← Factor 30 and 45 completely and line up the
45 = 3 · 5 · 3 ← factors common to 30 and 45.

GCD = ____ · ____ ← Bring down only the factors common to 30 and 45.
GCD = ____ ← Multiply.

3; 5
15

29. Find the GCD of 40, 56, and 24.

40 = ________ ← Factor 40, 56, and 24 completely and line up common factors.
56 = ________ ←
24 = ________ ←

GCD = ________ ← Bring down only the factors common to 40, 56, and 24.
GCD = ____ ← Multiply.

2 · 2 · 2 · 5
2 · 2 · 2 · 7
2 · 2 · 2 · 3
2 · 2 · 2
8

30. Find the GCD of 28 and 42.

28 = 2 · 2 · 7
42 = 2 · 7 · 3
GCD = 2 · 7
GCD = 14

Answers	Frames
5; 10; 15	**31.** The number m is a *multiple* of n if n is a factor of m. The multiples of 6 are 0, 6, 12, . . . or the numbers we get by counting by 6. The multiples of 5 are 0, ___, ___, ___, ...
multiple	**32.** 56 is a (factor\multiple) of 7.
greater than or equal to	**33.** Every nonzero multiple of n is (greater than or equal to\less than or equal to) n.
multiple; factor	**34.** 81 is a (factor\multiple) of 9, and 9 is a (factor\multiple) of 81.
15 15	**35.** The *least common multiple* (LCM) of two numbers a and b is the smallest counting number that is a multiple of a and also a multiple of b. Multiples of 15 are 0, 15, 30, ... Multiples of 5 are 0, 5, 10, 15, ... The smallest counting number which appears in both sets of multiples is ___. Therefore, ___ is the least common multiple of 15 and 5.
3; 5; 2; 2; 2 120 120	**36.** We can use the following scheme to find the LCM. Find the LCM of 15 and 40. 15 = 3 · 5 ← Factor 15 and 40 completely and 40 = 5 · 2 · 2 · 2 ← line up common factors. LCM = ___ · ___ · ___ · ___ · ___ ← Bring down one factor from *each column*. LCM = ___ ← Multiply these factors. The least common multiple of 15 and 40 is ___.
2 · 2 · 3 2 · 3 · 5 2 · 2 · 5 2 · 2 · 3 · 5 60	**37.** Find the least common multiple of 12, 30, and 20. 12 = ___ ← Factor 12, 30, and 20 into primes and line up common factors. 30 = ___ 20 = ___ LCM = ___ ← Bring down one factor from *each column*. LCM = ___ ← Multiply.
60 = 2 · 2 · 3 · 5 42 = 2 · 3 · 7 LCM = 2 · 2 · 3 · 5 · 7 LCM = 420	**38.** Find the least common multiple of 60 and 42.
48 = 2 · 2 · 2 · 2 · 3 36 = 2 · 2 · 3 · 3 LCM = 2 · 2 · 2 · 2 · 3 · 3 LCM = 144 GCD = 2 · 2 · 3 GCD = 12	**39.** Find the least common multiple and the greatest common divisor of 48 and 36.

PRACTICE PROBLEMS

1. Answer each of the following as true or false. If your answer is true, then give the factors involved. (For example, 9 divides 27 is true since 9 · 3 = 27.)

a. 6 divides 9 *b.* 4 divides 20 *c.* 7 divides 126
d. 11 divides 64 *e.* 12 divides 88 *f.* 15 divides 25

2. List the prime numbers in the set $\{1, 2, 3, 4, 5, 11, 29, 31, 121, 200, 2{,}000\}$.

3. List all factors of each of the following numbers. (For example, the factors of 70 are 1, 2, 5, 7, 10, 14, 35, 70.)

   a. 12 *b.* 50 *c.* 75 *d.* 165

4. Write each of the following as a product of prime numbers.

   a. 14 *b.* 105 *c.* 286 *d.* 300

5. Find the greatest common divisor of each of the following pairs of numbers.

   a. 15, 6 *b.* 22, 4 *c.* 25, 75 *d.* 6, 14 *e.* 12, 18 *f.* 72, 98

6. Find the least common multiple of each of the pairs of numbers in Problem 5.

7. For each of the pairs in Problem 5, do the following:

   a. Take the product of the two given numbers.
   b. Take the product of the LCM and GCD of the pairs.
   c. Compare the above results.

8. Let A and B be given numbers. Let the GCD of A and B be G. Let the LCM of A and B be L. The rule for finding the GCD says we must take all prime factors which appear in A and also in B. The rule for finding the LCM says we must take all prime factors of B that are not already in A. From these last two statements it follows that

$$A \cdot B = G \cdot L$$

   Do the results of Problem 7 agree with the formula

$$A \cdot B = (\text{GCD of } A, B)(\text{LCM of } A, B)$$

SELFTEST

1. Which of the members of the set $\{1, 5, 7, 10, 12, 25\}$ divide 50?
2. List all the factors of 30.
3. What are the next ten prime numbers after 2?
4. Factor 84 completely.
5. Find the greatest common divisor of 18 and 30.
6. Find the least common multiple of 27 and 45.

SELFTEST ANSWERS

1. 1, 5, 10, 25 **2.** 1, 2, 3, 5, 6, 10, 15, 30 **3.** 3, 5, 7, 11, 13, 17, 19, 23, 29, 31
4. $84 = 2 \cdot 2 \cdot 3 \cdot 7$ **5.** 6 **6.** 135

1.4 PARENTHESES AND LAWS OF ADDITION AND MULTIPLICATION

Parentheses serve two main functions in mathematics. One function is to indicate multiplication. When a number or a variable is written next to a parenthesis symbol, then the variable and the expression in the parentheses are to be multiplied together. So $8(3)$ means $8 \cdot 3$ and $x(2 + 9)$ means $x \cdot 11$. If two expressions in parentheses are written next to one another, we multiply one expression by the other. Thus $(5)(4)$ means $5 \cdot 4$ and $(3 + 2)(8 + 2)$ means $5 \cdot 10$.

We also use parentheses to group expressions together and to indicate the order of the operations addition, subtraction, multiplication, or division. We do operations in the following order.

1. Do all operations in parentheses first.
2. Do all multiplications or divisions from left to right as they occur.
3. Do all additions or subtractions from left to right as they occur.

Example 1 $$2 + 3 \cdot 5 = 2 + 15 = 17$$

Since there are no parentheses, we do the multiplication first and then the addition.

Example 2 $$5(2 + 3) = 5 \cdot 5 = 25$$

The parentheses tell us to do the addition first and then to multiply the sum by 5.

Example 3 $$10 + 24 \div (2 + 4) = 10 + 24 \div 6 = 10 + 4 = 14$$

The parentheses tell us to do the addition $2 + 4$ first, then do the division $24 \div 6$. The last operation is to add 10 to the quotient of $24 \div 6$.

From our experience with addition and multiplication of whole numbers we observe several laws of addition and multiplication.

When we compute $3 + 8$ and $8 + 3$, we discover that both sums equal 11. Also when we compute $4 \cdot 2$ and $2 \cdot 4$, we see that both products equal 8. For both multiplication and addition of any two given numbers, it does not matter which number is written first. The *commutative law* is a formal statement of this result.

Commutative Law: **If q and r are any two numbers, then**

$$q + r = r + q \qquad \text{and} \qquad q \cdot r = r \cdot q$$

Example 4 $$7 + x = x + 7 \qquad \text{for any number } x$$

Example 5 $$5 \cdot x = x \cdot 5 \qquad \text{for any number } x$$

When we add or multiply several numbers together, we combine only two numbers at a time. For example, to compute $3 + 2 + 4$ we begin at the left and add 3 to 2 to obtain 5. Then we compute $5 + 4$ to obtain the final result 9. We use parentheses to indicate which two numbers should be combined first. Are the expressions $(3 + 2) + 4$ and $3 + (2 + 4)$ equal? The answer is yes since both expressions equal 9. Are the expressions $(4 \cdot 5) \cdot 9$ and $4 \cdot (5 \cdot 9)$ equal? Again the answer is yes since both equal 180. The *associative law* says that whenever we have a sum or product of three numbers, we can put parentheses around the first two or we can put parentheses around the last two. In either case, we get the same result.

Associative Law: **If q, r, and s are any numbers, then**

$$(q + r) + s = q + (r + s) \qquad \text{and} \qquad (q \cdot r) \cdot s = q \cdot (r \cdot s)$$

The shift of parentheses seems obvious. But let's see what happens numerically when we carry out the shift.

Example 6 $$(3 + 4) + 5 = 3 + (4 + 5)$$

by the associative law, $$7 + 5 = 3 + 9$$

Unless we actually perform the addition, it is not clear that $7 + 5$ and $3 + 9$ are equal. But the associative law guarantees us that they are.

Example 7 Likewise

$$(3 \cdot 4) \cdot 5 = 3 \cdot (4 \cdot 5)$$

by the associative law, $$12 \cdot 5 = 3 \cdot 20$$

By the associative law, we know that $12 \cdot 5$ equals $3 \cdot 20$ without bothering to check that they both equal 60.

The commutative law allows us to rearrange the order of numbers. The associative law lets us regroup the numbers. A combination of the two laws allows us to regroup and rearrange a sum or product in any way we please. So we might as well do the operations from left to right and omit the unnecessary grouping parentheses.

Example 8

$$\begin{aligned}(4 + 2) + (7 + x) + 8 &= 4 + 2 + 7 + x + 8 \\ &= 4 + 2 + 7 + 8 + x \quad \text{commutative law} \\ &= 21 + x\end{aligned}$$

Example 9

$$\begin{aligned}(6 \cdot x)(3 \cdot y) &= 6 \cdot x \cdot 3 \cdot y \\ &= 6 \cdot 3 \cdot x \cdot y \quad \text{commutative law} \\ &= 18 \cdot x \cdot y\end{aligned}$$

The *distributive law* involves both the addition operation and the multiplication operation. It allows us to write certain products as sums. The product $4(3 + 7)$ and the sum $4 \cdot 3 + 4 \cdot 7$ are equal. Thus

$$4(3 + 7) = 4(10) = 40 \quad \text{and} \quad 4 \cdot 3 + 4 \cdot 7 = 12 + 28 = 40$$

Therefore

$$4(3 + 7) = 4 \cdot 3 + 4 \cdot 7$$

Distributive Law: **If q, r, and s are any numbers, then**

$$q(r + s) = qr + qs$$

Example 10

$$3(4 + x) = 3 \cdot 4 + 3 \cdot x = 12 + 3x$$

Example 11

$$x(y + 7) = x \cdot y + x \cdot 7$$

Example 12

$$4x + 5x = (4 + 5)x = 9 \cdot x$$

Notice that x is a factor of $4x$ and of $5x$. The distributive law says we can "factor out" the x or write $4x + 5x$ as the product $(4 + 5)x$.

Example 13

$$4z + 12x = 4(z + 3x)$$

since 4 is a factor of $4z$ and of $12x$.

Example 14 The distributive law generalizes so that

$$3(x + y + z) = 3 \cdot x + 3 \cdot y + 3 \cdot z$$

When we wish to indicate that two variables are to be multiplied together, we often omit the multiplication sign. We also omit the multiplication sign when we multiply a variable with a numeral. Thus qr means $q \cdot r$. The expression $3r$ means $3 \cdot r$.

PROGRAM 1.4

1. We can indicate multiplication without using the $\cdot$ symbol or the $\times$ symbol. One way is to enclose one or both of the factors in parentheses. So $3(x)$ means $3 \cdot x$ and $(6)(4)$ means $6 \cdot 4$. Also $7(2)$ means ________ and $(4)(x)$ means ________.

   $7 \cdot 2$

   $4 \cdot x$

7 + 3 or 10

2. 4(5 + 3) means 4 times the sum 5 + 3 or 4 · 8. Also, 2(7 + 3) means 2 times ________.

x

y times z; 5 · x · y · z

3. We often omit the multiplication symbol when we multiply variables together or numbers with variables: $3x$ means 3 times ____; yz means ________; $5xyz$ means ________.

x times (4 + 5) or x times 9

4. $x(4 + 5)$ means ________________.

5. How do we read expressions that involve a combination of additions, subtractions, multiplications, or divisions?

a. First we do all operations in parentheses.
b. Then we do all multiplications or divisions from left to right as they occur.
c. Finally we do all additions or subtractions from left to right as they occur.

16

19

In 3 + 2 · 8, we compute 2 · 8 first and then add 3 to ____ to obtain the final result ____.

2 + 7

45

6. In 5(2 + 7), we compute ________ first. Then we multiply that result by 5 to obtain the final answer ____.

12 − 3 is in parentheses.

7. In (12 − 3) ÷ 3, why do we compute 12 − 3 first? ________________ ________________.

9; 3

8. (12 − 3) ÷ 3 = ____ ÷ 3 = ____.

3 ÷ 3

9. In 12 − 3 ÷ 3, what do we compute first? ________.

1; 11

10. So 12 − 3 ÷ 3 = 12 − ____ = ____.

11. Simplify:

10

10; 3

7

2 · 5 − (4 + 2) ÷ 2 = ____ − 6 ÷ 2
= ____ − ____
= ____

12

12. Simplify: 5 · 3 − 12 ÷ 4 = ____

35

13. Simplify: 5(3 − 2)7 = ____

25

14. Simplify: 20 ÷ (4 + 1) + 3(7) = ____

15. In the case of nested parentheses (parentheses within brackets or braces or both), we do the operation in the innermost set first and then work outward.

5 + 40

In 2[(5 + 40) ÷ 9], we compute ________ first.

45

5

10

16. So 2[(5 + 40) ÷ 9] = 2(____ ÷ 9)
= 2(____)
= ____

30 ÷ (3 · 5); 30 ÷ 15; 2

17. Simplify 30 ÷ [3(3 + 2)]: First ________. Then ________ = ____.

18. Simplify: $(30 \div 3)(3 + 2) =$ ________ $=$ ____

10 · 5; 50

19. Are $3 \cdot 7$ and $7 \cdot 3$ equal? (yes\no) Are $4 + 5$ and $5 + 4$ equal? (yes\no) The commutative laws say that for any numbers a and b

yes; yes

$$a \cdot b = b \cdot a \quad \text{and} \quad a + b = b + a$$

By the ________ law, $4 + a = a + 4$. By the commutative law, $x \cdot 3 =$ ________.

commutative

$3 \cdot x$ or $3x$

20. By the commutative laws, $25 + 7 =$ ________ and $10 \cdot 4 =$ ________.

$7 + 25$; $4 \cdot 10$

21. Does $25 \div 5$ equal $5 \div 25$? (yes\no) Is division commutative? (yes\no)

no; no

22. When we add or multiply several numbers together, we actually only combine two numbers at a time. We begin with the first two numbers on the left.

$$6 + 4 + 9 = \_\_\_\_ + 9 = \_\_\_\_$$

10; 19

$$7 \cdot 2 \cdot 3 = \_\_\_\_ \cdot 3 = \_\_\_\_$$

14; 42

23. Simplify:

$$(6 + 4) + 9 = \_\_\_\_ + 9 = \_\_\_\_$$

10; 19

$$6 + (4 + 9) = 6 + \_\_\_\_ = \_\_\_\_$$

13; 19

So $(6 + 4) + 9$ (equals\does not equal) $6 + (4 + 9)$.

equals

24. Are the expressions $(2 \cdot 3)4$ and $2(3 \cdot 4)$ equal? (Yes\No), because both equal ____.

yes

24

25. The *associative laws* say that for any numbers a, b, and c,

$$(a + b) + c = a + (b + c) \quad \text{and} \quad (a \cdot b) \cdot c = a \cdot (b \cdot c)$$

By the ________ law, $(4 + y) + x = 4 + (y + x)$. By the associative law, $7(5 \cdot x) =$ ________.

associative

$(7 \cdot 5) \cdot x$ or $(7 \cdot 5)x$

26. The associative law allows us to regroup the numbers. By the associative laws,

$$x + (3 + y) = \_\_\_\_\_\_\_\_$$

$(x + 3) + y$

$$8(x \cdot 4) = \_\_\_\_\_\_\_\_$$

$(8x)4$

In the above examples, did we change the order in which we wrote the numbers or variables? (yes\no)

no

27. By the ________ law, $7 + x = x + 7$. So the commutative law (does\does not) allow us to change the order of the numbers or variables.

commutative

does

28. The (associative law\commutative law) allows us to regroup the numbers or variables. The (associative law\commutative law) allows us to reorder the numbers.

associative law

commutative law

In the next frames we will simplify an expression by using the commutative and associative laws. The changes will be on the left and the reason for the change on the right.

29. *Change* — *Reason*

$10 + (x + 2) = 10 + (2 + x)$ — commutative law

associative — $= (10 + 2) + x$ — ________ law

12 — $= \_\_\_\_ + x$

30. *Change* — *Reason*

associative law — $(5 \cdot x) \cdot 7 = 5(x \cdot 7)$ — ________

commutative law — $= 5(7 \cdot x)$ — ________

$(5 \cdot 7)x$ — $= \_\_\_\_\_\_$ — associative law

$= 35x$

31. Simplify as much as possible by using the commutative and associative laws and arithmetic.

$12y$ — $4(y \cdot 3) = \_\_\_\_$

32. Simplify as much as possible by using the commutative and associative laws and arithmetic.

$11 + x$ — $(6 + x) + 5 = \_\_\_\_\_\_$

33. Since the associative law says that it does not matter if we group from the left or from the right, we often omit grouping parentheses.

x — $(4 + y) + x = 4 + y + \_\_\_\_$

$3;\ y$ — $7(3y) = 7 \cdot \_\_\_\_ \cdot \_\_\_\_$

34. Use the associative and commutative laws and arithmetic to simplify

$11 + x$ — $4 + x + 7$

and

$45y$ — $5 \cdot y \cdot 9$

35. We wish to determine if $4(3 + 2)$ and $4 \cdot 3 + 4 \cdot 2$ are equal.

$5;\ 20$ — $4(3 + 2) = 4 \cdot \_\_\_\_ = \_\_\_\_$

$12;\ 8;\ 20$ — $4 \cdot 3 + 4 \cdot 2 = \_\_\_\_ + \_\_\_\_ = \_\_\_\_$

yes — Are the expressions equal? (yes\no)

36. The *distributive law* says that for any numbers, a, b, and c,

$$a(b + c) = a \cdot b + a \cdot c$$

$4 \cdot y$ or $4y$ — By the distributive law, $4(x + y) = 4 \cdot x + \_\_\_\_$.

37. The distributive law allows us to write a product as a sum and vice versa. By the distributive law,

$6;\ 3x$ or $3 \cdot 2;\ 3x$ — $3(2 + x) = \_\_\_\_ + \_\_\_\_$

and

$5;\ y$ — $x(\_\_\_\_ + \_\_\_\_) = x \cdot 5 + x \cdot y$

$7r;\ 7 \cdot 6$ or $7r;\ 42$ — **38.** By the distributive law, $7(r + 6) = \_\_\_\_ + \_\_\_\_$.

$50 + 10x$ or $10 \cdot 5 + 10x$ — **39.** By the distributive law, $10(5 + x) = \_\_\_\_\_\_$.

40. The distributive law generalizes so that

$$n(r + s + t) = n \cdot r + n \cdot s + n \cdot t$$

and

$$5(4 + y + 3) = \_\_\_\_ + \_\_\_\_ + \_\_\_\_$$

20; $5y$; 15

41. $x(5 + z + 7 + t) = \_\_\_\_\_\_\_\_\_\_$.

$x \cdot 5 + xz + x \cdot 7 + xt$
or $5x + xz + 7x + xt$

42. Now let's use the distributive law to write a sum as a product. The product is in *factored* form. The factored form of $x \cdot 2 + x \cdot 3$ is ____$(2 + 3)$. (Use the distributive law.)

x

43. Write $15 + 10x$ in factored form. What value is the largest common factor of 15 and $10x$? ____ We write this value outside the parentheses and the sum of the remaining factors inside the parentheses.

5

$$15 + 10x = \_\_\_\_ (3 + 2x)$$

5

$$4x + 8y = 4(\_\_\_\_ + 2y)$$

x

44. $14x + 21y = 7(\_\_\_\_ + \_\_\_\_)$.

$2x$; $3y$

45. $9x + 18 = 9(\_\_\_\_\_\_\_\_)$.

$x + 2$

46. $y \cdot 3 + y \cdot 5 = \_\_\_\_(3 + 5) = \_\_\_\_(8)$.

y; y

47. $6x + 21 = \_\_\_\_(2x + 7)$.

3

48. Factor: $8 + 10x = 2(\_\_\_\_ + \_\_\_\_)$.

4; $5x$

49. Factor: $9x + 15 = \_\_\_\_(3x + 5)$.

3

50. Factor: $6x + 4 = \_\_\_\_\_\_\_\_$.

$2(3x + 2)$

51. Factor: $x \cdot 7 + x \cdot 4 = \_\_\_\_ (\_\_\_\_) = \_\_\_\_ \cdot x$.

x; $7 + 4$; 11

52. By the distributive law,

$$a(b + c) = \_\_\_\_\_\_\_\_ \quad \text{and} \quad qr + qs = q(\_\_\_\_\_\_\_\_)$$

$ab + ac$; $r + s$

PRACTICE PROBLEMS

1. Simplify:

a. $2 + 16 \div 2$

b. $4 \cdot 2 + 3 - 7$

c. $3(2 + 7) - 36 \div 6$

d. $20 + (4 \cdot 9 \div 3) - 2$

e. $26 \div (15 - 2) + 4 \cdot 3$

f. $(15 + 17) \div (8 \cdot 2)$

g. $(50 \cdot 2 \div 5)5$

h. $50 \cdot 2 \div (5 \cdot 5)$

i. $(36 - 4 \cdot 9) \div 6$

j. $36 - (4 \cdot 9 \div 6)$

2. Identify the law used.

a. $4 + a = a + 4$

b. $(4 + 5) + b = 4 + (5 + b)$

c. $16x = x \cdot 16$

d. $10x + 10y = 10(x + y)$

e. $(xy)z = x(yz)$

f. $5(x + 3) = 5x + 15$

g. $$\begin{aligned} 12 + (x + 18) &= 12 + (18 + x) \\ &= (12 + 18) + x \\ &= 30 + x \end{aligned}$$

3. The distributive law says that for numbers x, y, z,

$$x(y + z) = xy + xz$$

Is it also true that $(y + z)x = yx + zx$? *Hint*: Is it true that $(5 + 3)6 = 5 \cdot 6 + 3 \cdot 6$? Use the commutative and distributive laws to prove that $(y + z)x = yx + zx$.

4. Use the distributive law to write each sum as a product or each product as a sum.

a. $5(x + 3)$

b. $x(4 + y)$

c. $4x + 4y$

d. $9x + 20x$

e. $(x + 4)2$

f. $15x + 15 \cdot 2$

g. $2(x + 3 + y)$

h. $6(a + 2 + b)$

i. $3x + 3 \cdot 4 + 3y$

j. $4x + 2x + 3x$

k. $23(10 + x)$

l. $(5 + x)7$

5. Is division commutative? Use two examples to justify your answer.

6. Are the following equations true?

$$(18 \div 6) \div 3 = 18 \div (6 \div 3) \qquad (16 - 4) - 3 = 16 - (4 - 3)$$

7. Is division associative? Is subtraction associative?

SELFTEST

1. Simplify:

a. $6 + 3 \cdot 2$

b. $(6 + 3)2$

c. $[4 + 2(5)] \div 7$

d. $(2 \cdot 8 + 4)5$

e. $5(7 - 3) - 10$

f. $6(5 - 5) + 5$

2. Which law guarantees that for any number x, $4 + x = x + 4$?

3. By the associative law, $(a + 3) + 2 =$ ________________.

4. Use the distributive law to write these products as sums.

a. $2(3 + a)$ *b.* $x(4 + y)$ *c.* $(5 + x)y$ *d.* $q(r + s)$

5. Use the distributive law to write these sums as products.

a. $3 + 3a$ *b.* $24b + 7b$ *c.* $4x + 10$ *d.* $3x + 4x$

6. For numbers m and n, state the commutative law for addition and for multiplication.

SELFTEST ANSWERS

1. *a.* 12 *b.* 18 *c.* 2 *d.* 100 *e.* 10 *f.* 5 **2.** commutative law of addition

3. $a + (3 + 2)$ or $a + 5$ **4.** *a.* $6 + 2a$ *b.* $4x + xy$ *c.* $5y + xy$ *d.* $qr + qs$

5. *a.* $3(1 + a)$ *b.* $(24 + 7)b$ or $31b$ *c.* $2(2x + 5)$ *d.* $(3 + 4)x$ or $7x$

6. commutative law of addition, $m + n = n + m$; commutative law of multiplication, $mn = nm$

POSTTEST 1

1. Match the sets to their names.

1. $\{2, 4, 6, 8, \ldots\}$ — *a.* odd counting numbers
2. $\{1, 2, 3, 4, \ldots\}$ — *b.* even counting numbers
3. $\{0, 1, 2, 3, \ldots\}$ — *c.* counting numbers
4. $\{1, 3, 5, 7, \ldots\}$ — *d.* whole numbers

2. Let x be an element of the set $\{1, 5, 17, 20\}$ such that $x + 5 = 22$. What is x?

3. Answer each of the following true or false.

a. $8 > 3$

b. $^0/_2 = 0$

c. $1 \cdot 5 = 5$

d. $^1/_0 = 1$

e. $2 \leq 2$

f. $4 \nless 3$

g. $0 \cdot 4 = 0$

h. $x \cdot 0 = 0$

i. $1 \cdot x = x$

4. Match the law to the example that illustrates it.

a. reflexive — 1. if $x = 4$, then $4 = x$

b. symmetric — 2. $8 = 8$

c. transitive — 3. if $x > y$ and $y > 4$, then $x > 4$

d. trichotomy — 4. if $x \neq 2$ and $x \ngtr 2$, then $x < 2$

5. Factor 75 completely.

6. Find the greatest common divisor of 50 and 75.

7. Find the least common multiple of 18 and 4.

8. Simplify $(12 \cdot 3 - 6) \div 2$.

9. Simplify $6(x - 1) + 6$.

10. Complete the examples so that they illustrate the given law.

a. By the associative law, $4 + (a + 2) =$ ________.

b. By the commutative law, $9x =$ ________.

c. By the distributive law, $5(2 + x) =$ ________.

Each problem above refers to a section in this chapter as shown in the table.

Problems	*Section*
1 and 2	1.1
3	1.1 and 1.2
4	1.2
5–8	1.3
9 and 10	1.4

POSTTEST 1 ANSWERS

1. 1. *b.* 2. *c.* 3. *d.* 4. *a.*

2. $x = 17$

3. *a.* T *b.* T *c.* T *d.* F *e.* T *f.* T *g.* T *h.* T *i.* T

4. *a.* 2. *b.* 1. *c.* 3. *d.* 4.

5. $3 \cdot 5 \cdot 5$ **6.** 25 **7.** 36 **8.** 15 **9.** $6x$

10. *a.* $(4 + a) + 2$ *b.* $x \cdot 9$ *c.* $5 \cdot 2 + 5x$ or $10 + 5x$

INTEGERS

BASIC SKILLS

Upon completion of the appropriate section the student should be able to:

2.1 *a.* Recognize integers.
b. Graph a given integer on the number line.
c. Give the absolute value of an integer.
d. Add, subtract, multiply, or divide absolute values.
e. Give the additive inverse of an integer.

2.2 *a.* Add any two integers.
b. Add a series of integers.

2.3 *a.* Subtract one integer from another.
b. Simplify expressions that involve both addition and subtraction of integers.

2.4 *a.* Multiply two integers.
b. Multiply more than two integers.
c. Simplify expressions that involve addition, subtraction, and multiplication of integers.

2.5 *a.* Divide two integers.
b. Simplify expressions that involve addition, subtraction, multiplication, or division of integers.

PRETEST 2

1. Answer yes or no.
   a. Is every integer a whole number?
   b. Is every counting number an integer?

2. Which of the numbers below are integers or whole numbers or both or neither?
   a. 0
   b. -1
   c. $|-3|$
   d. -15
   e. 6
   f. $^1/_2$

3. Locate the integers 5, -1, 0, 3, -6 on the number line. Of these integers, which is the largest and which is the smallest?

4. What is the additive inverse of each of the following?
   a. 0 *b.* 2 *c.* -1
   d. -3 *e.* x *f.* -12

5. Simplify:
   a. $|25| - |-20|$
   b. $|-48| \div |8|$
   c. $-8 + 2$
   d. $15 - 25$
   e. $-450 + 50$
   f. $-11 + 15 - 12 + 4$

6. Simplify:
   a. $8 - (-5)$
   b. $-6 - 4$
   c. $-3 + 10 - 18$
   d. $-7 - (-3) + 9$
   e. $-8 - 0$
   f. $56 - 6 - (-6)$

7. Simplify:
   a. $4(-1)$
   b. $(-5)6$
   c. $(-8)(-2)$
   d. $(-9)(-3)(-2)$
   e. $(15)(0)$
   f. $(-43)(1)$

8. Simplify:

a. $75 \div (-3)$

b. $(10 - 25) \div 5$

c. $-8 \div (-4)$

d. $-(3 - 5)$

e. $0 \div (-7)$

f. $\dfrac{(6)(-2)}{-4}$

Each problem above refers to a section in this chapter as shown in the table.

Problems	*Section*
1–4	2.1
5	2.1 and 2.2
6	2.3
7	2.4
8	2.4 and 2.5

PRETEST 2 ANSWERS

1. *a.* no *b.* yes

2. *a.* both *b.* integer *c.* both *d.* integer *e.* both *f.* neither

3. Smallest −6; −1, 0, 3; Largest 5 (number line with points at −6, −1, 0, 1, 3, 5)

4. *a.* 0 *b.* −2 *c.* 1 *d.* 3 *e.* $-x$ *f.* 12

5. *a.* 5 *b.* 6 *c.* −6 *d.* −10 *e.* −400 *f.* −4

6. *a.* 13 *b.* −10 *c.* −11 *d.* 5 *e.* −8 *f.* 56

7. *a.* −4 *b.* −30 *c.* 16 *d.* −54 *e.* 0 *f.* −43

8. *a.* −25 *b.* −3 *c.* 2 *d.* 2 *e.* 0 *f.* 3

2.1 INTEGERS AND THE NUMBER LINE

Examples abound in which situations occur below a fixed reference point. Temperatures drop below zero and register as negative temperatures. A few miles off the Hawaiian Islands the ocean floor is 15,000 feet below sea level. Euclid compiled his famous *Elements* about the year 300 B.C.

In our number system there are also numbers that occur "below" a fixed reference point. The reference point is the number zero. The numbers "below" zero are the *negative numbers*. Certain of these negative numbers together with the whole numbers form the set of integers.

The *number line* serves as a useful model for the integers. We draw a line and label one point on the line as zero or the origin. We pick a unit of length, and beginning from the origin, we mark off consecutive unit lengths to the right of the origin. Again starting from the origin, we mark off consecutive unit lengths to the left of the origin. We label the units to the right of zero by the counting numbers which correspond to the number of unit lengths from zero. We label the first unit to the left of zero -1, the second unit to the left of zero -2, and so forth. For example, if we use the unit length |—| we obtain

−5 −4 −3 −2 −1 0 1 2 3 4 5

The labels we put on the unit lengths to the left of zero are the negative integers. The set of negative integers is $\{-1, -2, -3, \ldots\}$.

The negative integers together with the whole numbers form the set of integers. Thus the set of integers is

$$\{\ldots, -3, -2, -1, 0, 1, 2, 3, 4, \ldots\}$$

We cannot draw a line long enough to include all the integers since the line would have to be infinitely long. So we indicate that the line continues indefinitely by using an arrow. However, we can place an integer as large as we please on the line. All we do is make the line long enough. A more practical solution is to let the unit length be shorter.

Example 1 Place 3 on the number line.

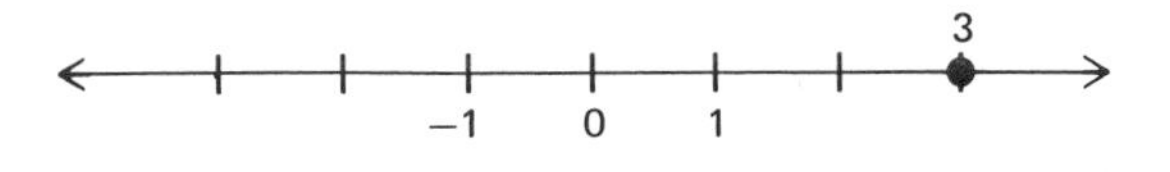

Example 2 Place -30 on the number line.

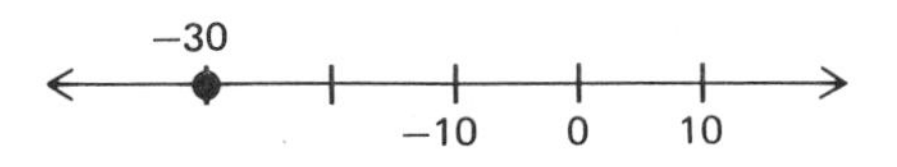

Now we can compare any two integers by finding their relative positions on the number line. As we proceed to the right of any number on the line, we come to larger numbers. As we proceed to the left of any number on the line, we encounter smaller numbers.

Example 3 Which is greater, -5 or 2?

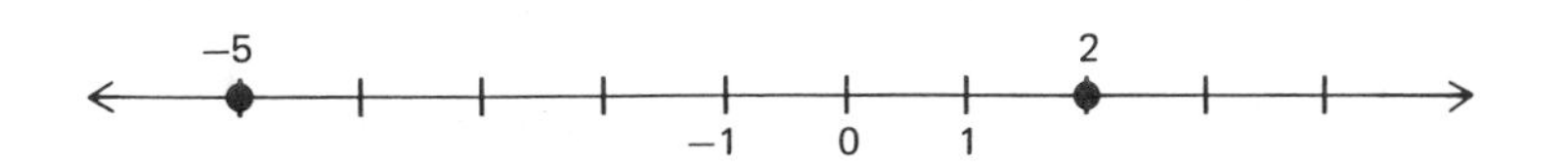

The number farthest to the right is 2, so

$$2 > -5$$

Example 4 Which is greater, -7 or -2?

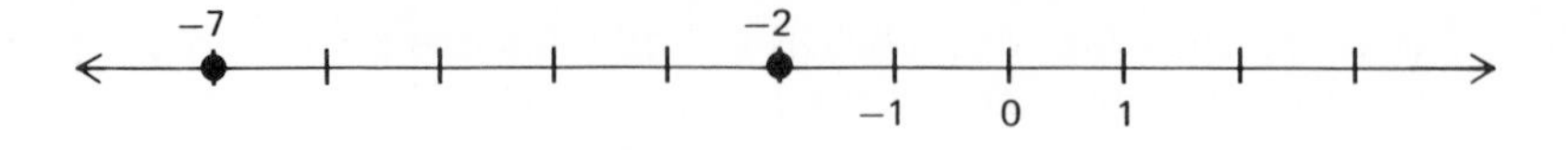

The number -2 lies to the right of -7, so

$$-2 > -7$$

If the number x is to the right of the number y, then y is to the left of x. So $x > y$ and $y < x$ express the same relative positions of x and y on the number line.

We can also visualize the concept of absolute value by using the number line. But first we need to know about distance. The *distance* between two points a and b on the number line is the number of unit lengths that lie between a and b.

Definition: **The *absolute value* of a number n is the distance between the origin and the number n on the number line. The notation $|n|$ means "absolute value of n."**

Example 5 What is the absolute value of 6?

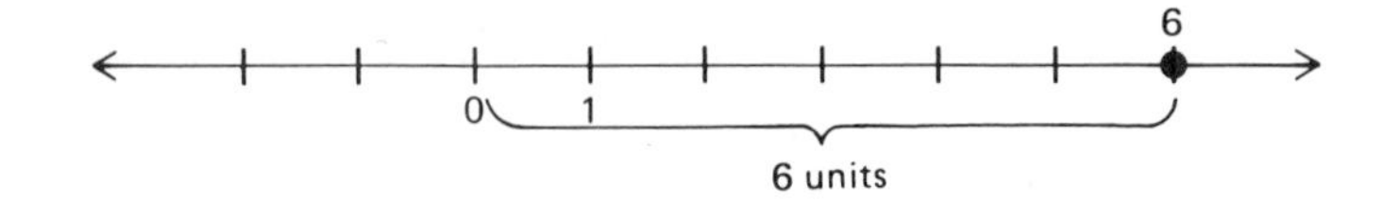

So $|6| = 6$.

Example 6 Find the absolute value of -3 and 3.

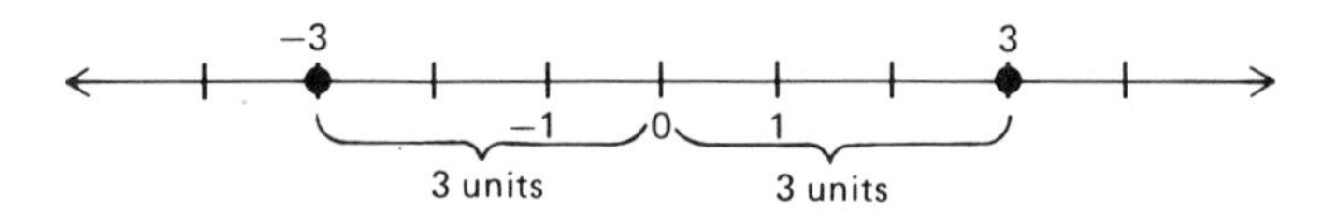

So $|3|$ and $|-3|$ both equal 3.

To do arithmetic with absolute values, we compute the absolute values first. So the arithmetic of absolute values is the arithmetic of whole numbers. We will use the arithmetic of absolute values in the rules for the arithmetic of integers.

Example 7 $|-7| + |3| = 7 + 3 = 10$

Example 8 $|20| - |-12| = 20 - 12 = 8$

Example 9 $|-3| \cdot |-2| = 3 \cdot 2 = 6$

Example 10 $|8| \div |-4| = 8 \div 4 = 2$

Another useful description of negative integers involves the notion of additive inverse. We will use properties of additive inverses to motivate the arithmetic techniques for integers.

Definition: **The *additive inverse* of a number r is a number written $-r$. The additive inverse satisfies the conditions**

$$r + (-r) = 0 \quad \text{and} \quad -r + r = 0$$

Furthermore, every number has *exactly one additive inverse*.

Thus each negative integer is the additive inverse of some counting number. The number -2 is the additive inverse of 2. Consequently

$$2 + (-2) = 0 \qquad \text{and} \qquad -2 + 2 = 0$$

The additive inverse of 35 is -35, and so

$$35 + (-35) = 0 \qquad \text{and} \qquad -35 + 35 = 0$$

Notice too that the counting numbers are the additive inverses of the negative numbers. For suppose n is a counting number. Then $-n$ is a negative number. The additive inverse of $-n$ is denoted by $-(-n)$. By definition, then,

$$-n + [-(-n)] = 0$$

But we know that $-n + n = 0$ as well. Thus n and $-(-n)$ must be equal since the additive inverse of n is unique, so we see that

$$-(-n) = n \qquad \text{for any counting number } n$$

Example 11 The additive inverse of -5 is 5 since $-(-5) = 5$.

Example 12 The additive inverse of 7 is -7.

Example 13 The additive inverse of -7 is 7 since $-(-7) = 7$.

It is sometimes convenient to consider pairs of numbers which are additive inverses of each other. Two numbers are additive inverses of each other if and only if their sum is 0. Thus 4 and -4 are additive inverses of each other since $4 + (-4) = 0$. Likewise $-x$ and x are additive inverses of each other. For $-x + x = 0$.

PROGRAM 2.1

1. The number line serves as a useful model for numbers since numbers correspond to points on the number line. To make a number line, we draw a line and mark an arbitrary point. We label this point zero. It is called the *origin*. Which point is the origin on the number line below?

The point above 0.

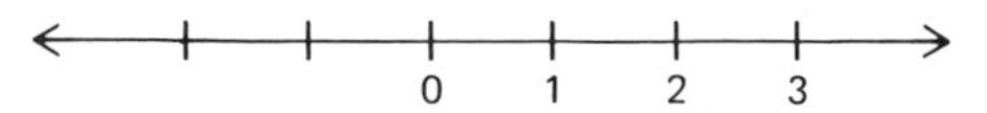

2. Next we pick a convenient unit of length and mark off consecutive unit lengths to the right of the origin. For example, let ├─┤ be the unit length. There are (5\8\11) unit lengths marked off on the line below.

8

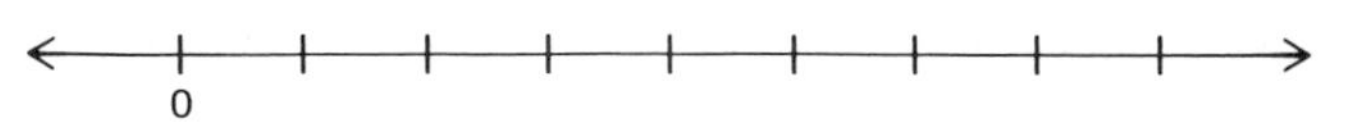

3. We label the first unit-length mark after the origin 1, the second mark 2, etc. The last mark should be labeled (6\8\10) on the line below.

6

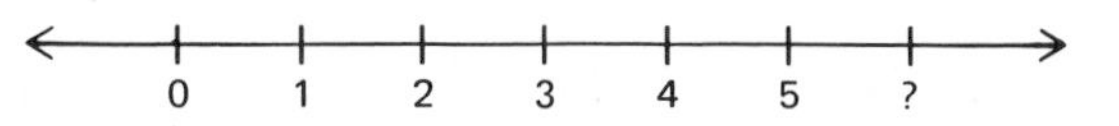

4. The interval between the origin and the point labeled 1 is the unit length for a given number line. Shade in the first and third unit lengths to the right of 0 on the line.

0 1 2 3

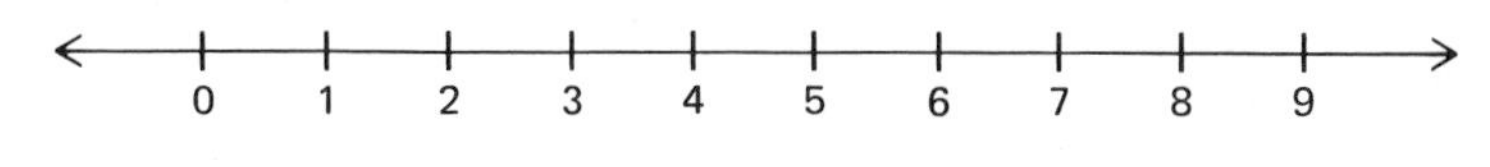

5. Compare these two number lines. They have different ________ ________.

unit lengths

6. The negative integers correspond to points at unit lengths to the left of the origin. We mark off consecutive unit lengths to the left of zero. We label the first unit length to the left of zero −1, the second one to the left −2, etc. Fill in the missing labels.

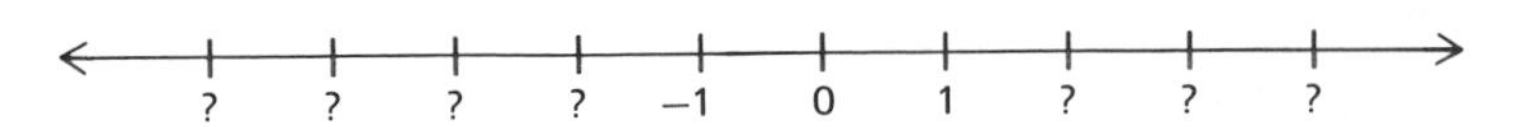

−5; −4; −3; −2; 2; 3; 4

7. The number −12 corresponds to a mark on the number line 12 unit lengths to the (left\right) of zero.

left

8. The number −12 is a (positive\negative) integer.

negative

9. The number −53 corresponds to a mark _____ unit lengths to the (left\right) of zero.

53; left

10. A mark 72 unit lengths to the left of zero corresponds to the number _____.

−72

11. Make a heavy dot to indicate the position of −3, −2, and 6 on the number line.

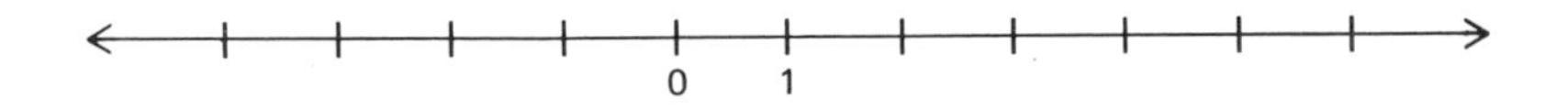

12. In order to place higher numbers on the number line, we can make the unit length smaller. Place 30 and −20 on the number line,

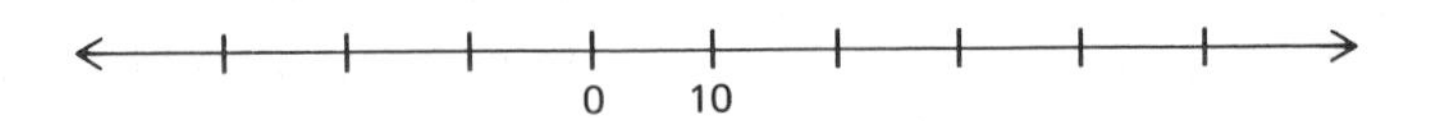

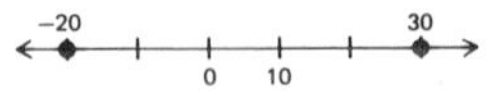

13. Infinitely many unit lengths extend to the left of the origin. So there are (finitely\infinitely) many negative numbers.

infinitely

14. We use the "dot dot dot" notation to "list" the negative integers. The tenth number in the list −1, −2, −3, . . . is _____.

−10

15. The set {−1, −2, −3, . . .} is the set of ________ ________.

negative integers

16. The set of *integers* is the set consisting of all the negative integers and all the whole numbers. The set

{. . ., −3, −2, −1, 0, 1, 2, 3, . . .}

is the set of (counting numbers\integers\whole numbers\negative numbers).

integers

17. The nonnegative integers are the (whole numbers\counting numbers).

whole numbers

18. The number 25 is a counting number, a whole number, and also an ________.

integer

19. The number −32 is a negative ________. It (is\is not) a whole number. It (is\is not) a counting number.

integer; is not
is not

yes

20. Is 0 an integer? (yes\no)

−3

21. We can compare the size of any two integers by looking at their relative positions on the number line. The number to the right on the number line is always the greater number. Which is greater, −5 or −3? ____ *Hint:* place −5 and −3 on a number line.

right

22. The number 16 is greater than 5 since 16 lies to the (right\left) of 5 on the number line.

greater than
right

23. The number −10 is (greater than\less than) −50 since −10 lies to the (right\left) of −50 on the number line.

right

24. $-9 > -14$ since −9 lies to the (right\left) of −14.

left

25. As we proceed to the left on the number line, we encounter smaller numbers. The number 5 is less than 10 because 5 is to the (right\left) of 10 on the number line.

less than
left

26. The number −15 is (greater than\less than) −10 because −15 lies to the (right\left) of −10 on the number line.

left

27. $-4 < -2$ since −4 lies to the (right\left) of −2 on the number line.

$<$

28. If n lies to the right of m on the number line, then m lies to the left of n. So if $n > m$, then m ($>$\$<$\$=$) n.

$<$; $>$; $<$

29. Fill in the appropriate $>$ or $<$ symbol.

−5 ____ −2, −4 ____ −8, and −6 ____ 3

2

30. The distance between any two points on the number line is the number of unit lengths between the points. The distance between 1 and 3 is ____.

−5 −4 −3 −2 −1 0 1 2 3 4 5 6

6

31. The distance between −4 and 2 is 6 since there are ____ unit lengths between −4 and 2.

−4 0 1 2

8

32. What is the distance between −8 and 0? ____

7

33. What is the distance between 0 and 7? ____

4
4

34. The *absolute value* of a number n is the distance between n and 0 on the number line. What is the distance between 4 and 0? ____ What is the absolute value of 4? ____

The distance is 9.
9

35. What is the distance between −9 and 0 on the number line? ________ ________ What is the absolute value of −9? ____

36. Use the number line below to find two numbers which have absolute value 5. They are ____ and ____. — 5; -5

37. The notation $|n|$ means absolute value of n. The absolute value of 3 is written $|3|$. The absolute value of 5 is written ____. — $|5|$

38. The absolute value of -12 is written ________. — $|-12|$

39. The notation $|-15|$ means the ________ ________ of -15. — absolute value

40. So $|-7| =$ ____ since the distance between -7 and 0 is ____. — 7; 7

41. Notice that $|4| = 4$ and $|-4| = 4$, since both 4 and -4 are a distance of ____ units from 0. — 4

42. If $|n| = 20$, n can be ____ or ____ since these two numbers are both 20 unit lengths from 0. — 20; -20

43. The absolute value of n is the distance between n and ________. — 0 or the origin

44. Simplify: $|27| =$ ____ and $|-16| =$ ____. — 27; 16

45. If $|n| = 30$, we (do\do not) know if n is to the left or right of zero. — do not

46. The absolute value of n just tells the distance between ____ and the origin. — n

47. $|0| =$ ____ since the distance between 0 and itself is zero unit lengths. — 0

48. The absolute value of an integer n (is\is not) always positive or zero. — is

49. To do arithmetic with absolute values, we *compute the absolute values first* and then do the operations.

$$|-7| + |-5| = 7 + \_\_\_\_ = \_\_\_\_$$

— 5; 12

50. Simplify: $|25| + |-20| =$ ____ + ____ = ____. — 25; 20; 45

51. Simplify: $|-14| + |6| =$ ____. — 20

52. Simplify: $|-10| - |-7| =$ ____ − ____ = ____. — 10; 7; 3

53. Simplify: $|-12| \cdot |2| =$ ____ · ____ = ____. — 12; 2; 24

54. Simplify: $|-48| \div |8| =$ ____ ÷ ____ = ____. — 48; 8; 6

55. Simplify:

a. $|-6| \cdot |-5|$ *b.* $|-9| + |3|$ *c.* $|20| - |-10|$ *d.* $|14| \div |-7|$

— *a.* 30; *b.* 12; *c.* 10; *d.* 2

is; -8

56. The *additive inverse* of a number n is a number written $-n$. So the additive inverse of a number (is\is not) the negative of the number. The additive inverse of 8 is ____.

0; 0

57. The additive inverse of a number n satisfies the conditions that

$$n + (-n) = 0 \qquad \text{and} \qquad (-n) + n = 0$$

Since -5 is the additive inverse of 5, it is true that $5 + (-5) =$ ____ and $-5 + 5 =$ ____.

is; -9

58. Each number has exactly one additive inverse. Thus -10 (is\is not) the only additive inverse of 10. If z represents the additive inverse of 9, then z must equal ____.

-17; 0

59. The additive inverse of 17 is ____, so $17 + (-17) =$ ____.

-6; -6

60. The additive inverse of 6 is ____, so $6 + ($____$) = 0$.

43

61. Since $43 + (-43) = 0$, then -43 is the additive inverse of ____.

additive inverse

62. Since $-14 + 14 = 0$, then -14 is the ________ ________ of 14.

yes ($-n$ is the additive inverse of n)

63. Is each negative number the additive inverse of some counting number? (yes\no)

-18; 18; 18; yes

64. What is the additive inverse of 18? ____ What is the absolute value of 18? ____ What is the absolute value of the additive inverse of 18? ____ Are these absolute values equal? (yes\no)

$-(-6)$

65. Since the additive inverse of n is $-n$, the additive inverse of -3 is $-(-3)$. The additive inverse of -6 is written ____.

0; 4

66. The additive inverse of -10 is $-(-10)$. So $-10 + [-(-10)] =$ ____. But we also know that $-10 + 10 = 0$. Therefore $-(-10)$ must equal 10. What does $-(-4)$ equal? ____

6; n

67. The additive inverse of a negative number is the corresponding positive number. So $-(-6) =$ ____. In general, $-(-n) =$ ____.

-5; 5; -7; 7

68. Find the additive inverse of 5, -5, 7, and -7.

15; 33

69. Simplify: $-(-15) =$ ____ and $-(-33) =$ ____.

0

70. A number plus its additive inverse equals ____.

are

71. If the sum of two numbers equals 0, then the two numbers (are\are not) additive inverses of each other.

PRACTICE PROBLEMS

1. Draw a number line and label the points corresponding to 0, 1, 5, -3, and -8.
2. Indicate the elements in the following sets.
   a. set of negative integers *b.* set of integers

3. Use the appropriate $>$ or $<$ symbol.

a. 25 ____ 100 *b.* -15 ____ 40 *c.* -16 ____ -25 *d.* -18 ____ -17

e. -14 ____ 14 *f.* 0 ____ -3 *g.* 3 ____ -5 *h.* -7 ____ -4

4. Use the number line to determine the distance between the numbers in the following pairs.

a. $-3, 6$ *b.* $0, -10$ *c.* $25, 0$ *d.* $-7, 6$

5. What notation do we use to designate the absolute value of n?

6. Simplify:

a. $|16|$ *b.* $|-25|$ *c.* $|0|$ *d.* $|-3|$

e. $|5| + |-7|$ *f.* $|8| - |-5|$ *g.* $|-14| + |-6|$ *h.* $|-12| - |6|$

i. $|20| \cdot |-4|$ *j.* $|-15| \cdot |2|$ *k.* $|45| \div |-5|$ *l.* $|-30| \div |-6|$

7. A number plus its additive inverse always equals ____.

8. Find the additive inverse.

a. 7 *b.* -9 *c.* $|-25|$ *d.* 14

9. Simplify:

a. $-(-14)$ *b.* $7 + (-7)$ *c.* $-10 + 10$ *d.* $-(-a)$

SELFTEST

1. Locate $9, -3, -6, 0$, and 4 on the number line.

2. Label the numbers as integers, negative integers, both, or neither.

a. -5 *b.* 0 *c.* 18 *d.* 1 *e.* $-3/2$ *f.* -107

3. Use the appropriate $>$ or $<$ symbol in the blanks.

a. -6 ____ -3 *b.* -5 ____ 2 *c.* -2 ____ -8 *d.* $|-4|$ ____ 0

4. Simplify:

a. $|-3|$ *b.* $|-7| + |-9|$ *c.* $|25| \cdot |-4|$ *d.* $|-18| \div |3|$

5. What is the additive inverse of -9? What is the additive inverse of 9?

6. Simplify:

a. $18 + (-18)$ *b.* $-5 + 5$ *c.* $-(-7)$

SELFTEST ANSWERS

1. Number line with points at -6, -3, 0, 4, 9 (tick labels 0, 1)

2. *a.* both *b.* integer *c.* integer *d.* integer *e.* neither *f.* both

3. *a.* $-6 < -3$ *b.* $-5 < 2$ *c.* $-2 > -8$ *d.* $|-4| > 0$

4. *a.* 3 *b.* 16 *c.* 100 *d.* 6

5. additive inverse of -9 is 9; additive inverse of 9 is -9

6. *a.* 0 *b.* 0 *c.* 7

2.2 ADDITION OF INTEGERS

In elementary school we learned how to add two positive integers together. Now we shall describe techniques for computing sums that involve negative numbers.

We still want the commutative and the associative laws of addition to hold. Any rules given should yield results consistent with these laws and with the definition of negative numbers.

There are two cases to consider when we add integers. In one case, the numbers involved have the same sign. In the other case, the numbers have different signs. Let's look at an example from each case and see what happens when we use the commutative and associative laws and the fact that a number plus its additive inverse equals zero.

Example 1 Add $-5 + (-3)$: We know that

$$-5 + (-3) + 5 + 3 = 0$$

by the definition of additive inverse and applications of the commutative and associative laws. But this equation is the same as

$$[-5 + (-3)] + 8 = 0$$

Therefore $[-5 + (-3)]$ must be the additive inverse of 8. That is, $[-5 + (-3)]$ is -8. We conclude that $-5 + (-3) = -8$.

Example 2 Add $-5 + 9$: This time we are to add a positive number to a negative number. Let's replace 9 by the sum $5 + 4$. So

$$
\begin{aligned}
-5 + 9 &= -5 + (5 + 4) \\
&= (-5 + 5) + 4 && \text{by the associative law} \\
&= 0 + 4 && \text{since 5 and } -5 \text{ are additive inverses of each other} \\
&= 4
\end{aligned}
$$

We can follow one of these two patterns (depending on the case involved) whenever we wish to add two integers. But the patterns are long. The following rules give results that are consistent with those of the patterns.

Addition Rule 1: **To add two numbers that have the same sign, *add* the absolute values of the numbers and give the result the sign of the original numbers.**

Addition Rule 2: **To add two numbers with unlike signs (i.e., one number is positive and the other negative), *subtract* the smaller absolute value from the larger absolute value. Give the result the sign of the number with the larger absolute value.**

Let's look at our first two examples again. This time we will use the appropriate rule.

Example 3 Add $-5 + (-3)$: The numbers have the same sign, so we use rule **1** and *add* the absolute values of the numbers.

$$|-5| + |-3| = 5 + 3 = 8$$

The numbers in the original sum are negative, so the final result will be negative. Therefore

$$-5 + (-3) = -8$$

Example 4 Add $-5 + 9$: The numbers have different signs so we use rule **2**. We *subtract* the smaller absolute value from the larger absolute value.

$$|9| - |-5| = 9 - 5 = 4$$

The number with the larger absolute value is 9. Since 9 is positive the final result is positive. Therefore

$$-5 + 9 = 4$$

Example 5 Add $-8 + 6$: Again the numbers have different signs, so we *subtract* the absolute values.

$$|-8| - |6| = 8 - 6 = 2$$

To determine if the answer is 2 or -2, we look at the sign of the number with the larger absolute value. Since -8 has the larger absolute value, the answer is -2. Thus

$$-8 + 6 = -2$$

Example 6 Add $-10 + 12 + (-5) + (-2)$: We can add all the negative numbers together first (by the commutative law). So

$$\begin{aligned} -10 + 12 + (-5) + (-2) &= -10 + (-5) + (-2) + 12 \\ &= -17 + 12 && \text{by rule } \mathbf{1} \\ &= -5 && \text{by rule } \mathbf{2} \end{aligned}$$

Let's look at what happens on the number line when we add two integers.

Add two positive integers: $4 + 5$

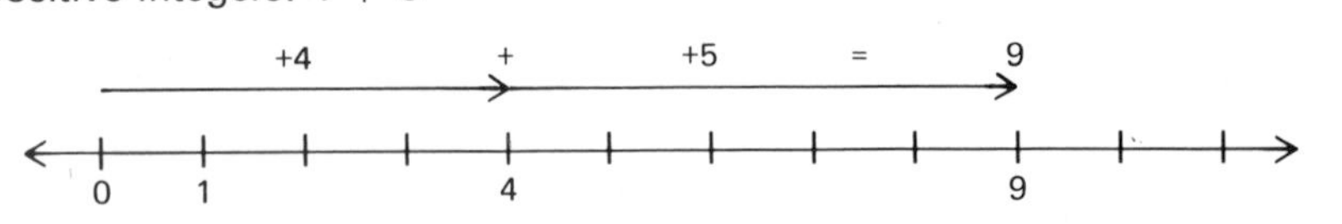

Add two negative integers: $-2 + (-4)$

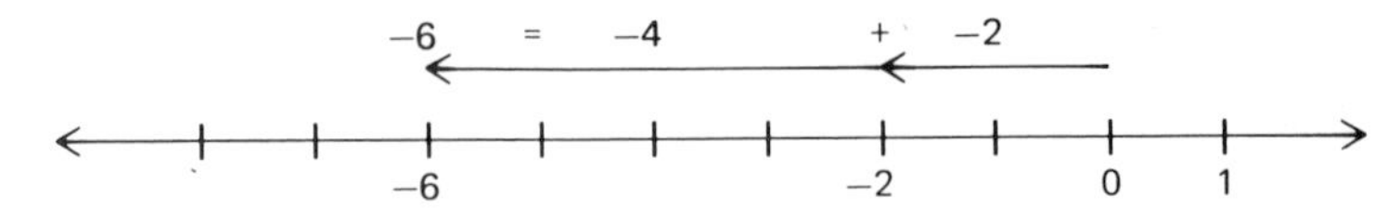

Add a positive and a negative integer: $-3 + 6$

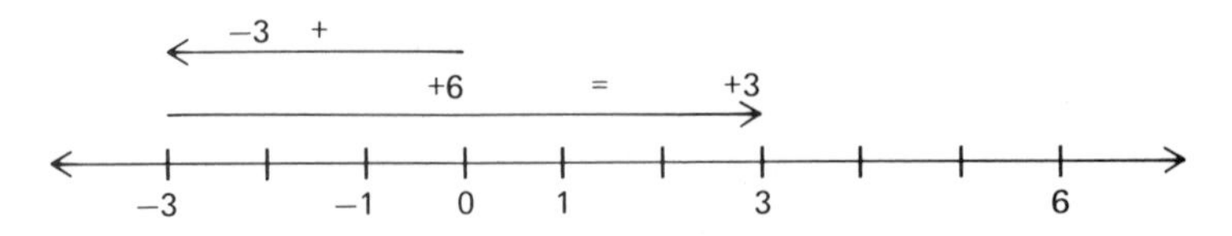

We observe that:

1. The sum of two positive numbers lies to the *right* of both.
2. The sum of two negative numbers lies to the *left* of both.
3. The sum of a positive and a negative is always *between* the two numbers.

PROGRAM 2.2

1. Recall that positive integers are the counting numbers. A number written without a sign is assumed to be positive. So 6 is a (positive\negative) number. — positive

2. We use the rules learned in elementary school to add two positive numbers together. So $25 + 5 =$ ____. — 30

3. Recall that 0 is the additive identity. So $0 + n = n + 0 = n$ for any integer n. Thus $-28 + 0 =$ ____ and $0 + 18 =$ ____. — -28; 18

negative

4. There are two other cases that occur when we add integers. In one case, we add two negative integers together. The sum $-6 + (-3)$ is a problem in this case since -6 and -3 are both (positive\negative) numbers.

negative
positive

5. In the other case, we add a positive integer to a negative integer. The sum $-7 + 3$ is in the second case since -7 is (positive\negative) but 3 is (positive\negative).

add
negative

6. ***Addition Rule 1:*** To add two negative numbers, *add* their absolute values and give the result a negative sign. To add $-7 + (-3)$ we (add\subtract) the absolute values and give the result a (positive\negative) sign.

absolute
7; 3; 10
-10

7. To evaluate $-7 + (-3)$, we add the ______ values of -7 and -3.

$$|-7| + |-3| = \_\_\_\_ + \_\_\_\_ = \_\_\_\_$$

But the sum of two negative numbers is always negative, so

$$-7 + (-3) = \_\_\_\_$$

add
22; 7; 29
negative
-29

8. To evaluate $-22 + (-7)$ we (add\subtract) the absolute values.

$$|-22| + |-7| = \_\_\_\_ + \_\_\_\_ = \_\_\_\_$$

But the sum of two negative numbers is always (positive\negative). Therefore

$$-22 + (-7) = \_\_\_\_$$

49; 10; 59
negative
-59

9. Evaluate $-49 + (-10)$:

$$|-49| + |-10| = \_\_\_\_ + \_\_\_\_ = \_\_\_\_$$

But the sum of $-49 + (-10)$ is (positive\negative) so

$$-49 + (-10) = \_\_\_\_$$

52
negative
-52

10. Evaluate $-14 + (-38)$:

$$|-14| + |-38| = \_\_\_\_$$

But the sum of two negative numbers is (positive\negative). Thus

$$-14 + (-38) = \_\_\_\_$$

$-$
-98

11. What is the sign of the sum $-72 + (-26)$? (+\−). Evaluate:

$$-72 + (-26) = \_\_\_\_$$

-65

12. Evaluate: $-15 + (-50) = \_\_\_\_$.

negative

13. The sum of two negative numbers is always (positive\negative).

subtract

14. ***Addition Rule 2:*** To add a positive number and a negative number together, *subtract* the smaller absolute value from the larger absolute value. Give the result the sign of the number with the larger absolute value. To add $-4 + 3$ we (add\subtract) the absolute values.

negative

15. The sign of the final result is the same as the number with the larger absolute value. So $-4 + 3$ is (positive\negative) since $|-4| > |3|$.

positive

16. The sign of $-7 + 9$ is (positive\negative) since $|9| > |-7|$.

17. In this frame, we will compute $-5 + 3$. In $-5 + 3$, the signs of the numbers are (alike\unlike) so we (add\subtract) their absolute values. — unlike; subtract

$$|-5| - |3| = \_\_\_\_ - \_\_\_\_ = \_\_\_\_$$ — 5; 3; 2

The final result is $(2\backslash -2)$ since $|-5| > |3|$. Thus $-5 + 3 = \_\_\_\_$. — -2; -2

18. In $-5 + 8$, the numbers have (like\unlike) signs. So we (add\subtract) their absolute values. — unlike; subtract

$$|8| - |-5| = \_\_\_\_ - \_\_\_\_ = \_\_\_\_$$ — 8; 5; 3

Since $|8| > |-5|$, the final result is $(3\backslash -3)$. Thus $-5 + 8 = \_\_\_\_$. — 3; 3

19. In $7 + (-12)$, the numbers have (like\unlike) signs, so we (add\subtract) the absolute values. — unlike; subtract

$$|-12| - |7| = \_\_\_\_ - \_\_\_\_ = \_\_\_\_$$ — 12; 7; 5

Which statement is true, $|-12| > |7|$ or $|7| > |-12|$? So the final result — $|-12| > |7|$

is (positive\negative). Thus $7 + (-12) = \_\_\_\_$. — negative; -5

20. To compute $-11 + 21$, we (add\subtract) the absolute values. Compute — subtract

the difference of the absolute values. ________ Which is larger, — $|21| - |-11| = 10$

$|-11|$ or $|21|$? ____ So $-11 + 21$ is (positive\negative). Thus — $|21|$; positive

$$-11 + 21 = \_\_\_\_$$ — 10

21. What is the sign of $-15 + 10$? $(+\backslash -)$ So $-15 + 10 = \_\_\_\_$. — $-$; -5

22. Compute $-20 + 27$: ____. — 7

23. To add two numbers with *like* signs, we *add* the absolute values. To add two numbers with *unlike* signs, we *subtract* the absolute values. To add $-4 + (-8)$, we (add\subtract) the absolute values. To add $(-10) + 15$, we (add\subtract) the absolute values. — add; subtract

24. To compute $-12 + 20$, we (add\subtract) the absolute values. — subtract

25. To compute $-15 + (-7)$, we (add\subtract) the absolute values. — add

26. The sum $-18 + (-25)$ is (positive\negative). — negative

27. The sum $-18 + 25$ is (positive\negative). — positive

28. The sum $18 + (-25)$ is (positive\negative). — negative

29. The sum of two negative numbers is (always\sometimes\never) negative. — always

30. The sum of a positive number and a negative number is (always\sometimes\never) negative. — sometimes

31. Add: $-18 + (-30) = \_\_\_\_$. — -48

32. Add: $75 + (-100) = \_\_\_\_$. — -25

33. $-5 + (-8) + (-7) = \_\_\_\_ + (-7) = \_\_\_\_$. — -13; -20

-37 **34.** Add: $-10 + (-12) + (-15) = $ ____.

35. To add several positive and negative numbers together, we find the sum of all the negative numbers and then the sum of all the positive numbers. Finally, we add these sums.

6; 20 $\quad 15 + (-4) + 5 + (-2) = (-\_\_\_\_) + \_\_\_\_$

14 $\quad = \_\_\_\_$

5 **36.** Add: $-3 + (-2) + 7 + (-5) + 8 = $ ____.

PRACTICE PROBLEMS

You should try to do this set of addition problems in less than 15 minutes.

1. $-18 + (-9)$	**2.** $42 + 17$	**3.** $38 + (-49)$
4. $-12 + (-10)$	**5.** $-25 + 30$	**6.** $-40 + 30$

7. $\begin{array}{r} -29 \\ +\quad 32 \\ \hline \end{array}$	**8.** $\begin{array}{r} 14 \\ +\quad -86 \\ \hline \end{array}$	**9.** $\begin{array}{r} -92 \\ +\quad -55 \\ \hline \end{array}$	**10.** $\begin{array}{r} -30 \\ +\quad 40 \\ \hline \end{array}$
11. $\begin{array}{r} -16 \\ +\quad 13 \\ \hline \end{array}$	**12.** $\begin{array}{r} 24 \\ +\quad -13 \\ \hline \end{array}$	**13.** $\begin{array}{r} -96 \\ +\quad -21 \\ \hline \end{array}$	**14.** $\begin{array}{r} -14 \\ +\quad 29 \\ \hline \end{array}$
15. $\begin{array}{r} 0 \\ +\quad 46 \\ \hline \end{array}$	**16.** $\begin{array}{r} 58 \\ +\quad -62 \\ \hline \end{array}$	**17.** $\begin{array}{r} 33 \\ +\quad -15 \\ \hline \end{array}$	**18.** $\begin{array}{r} -51 \\ +\quad 37 \\ \hline \end{array}$
19. $\begin{array}{r} -21 \\ +\quad 41 \\ \hline \end{array}$	**20.** $\begin{array}{r} -285 \\ +\quad 285 \\ \hline \end{array}$	**21.** $\begin{array}{r} -59 \\ +\quad 0 \\ \hline \end{array}$	**22.** $\begin{array}{r} 72 \\ +\quad -40 \\ \hline \end{array}$
23. $\begin{array}{r} 48 \\ +\quad -66 \\ \hline \end{array}$	**24.** $\begin{array}{r} -25 \\ +\quad -36 \\ \hline \end{array}$	**25.** $\begin{array}{r} 84 \\ +\quad 92 \\ \hline \end{array}$	**26.** $\begin{array}{r} -362 \\ +\quad 151 \\ \hline \end{array}$
27. $\begin{array}{r} -52 \\ +\quad 48 \\ \hline \end{array}$	**28.** $\begin{array}{r} -104 \\ +\quad -208 \\ \hline \end{array}$	**29.** $\begin{array}{r} 73 \\ +\quad -67 \\ \hline \end{array}$	**30.** $\begin{array}{r} -55 \\ +\quad -38 \\ \hline \end{array}$
31. $\begin{array}{r} -632 \\ +\quad -632 \\ \hline \end{array}$	**32.** $\begin{array}{r} 22 \\ +\quad -14 \\ \hline \end{array}$	**33.** $\begin{array}{r} 83 \\ +\quad -83 \\ \hline \end{array}$	**34.** $\begin{array}{r} -14 \\ +\quad 7 \\ \hline \end{array}$
35. $\begin{array}{r} 43 \\ +\quad -28 \\ \hline \end{array}$	**36.** $\begin{array}{r} -9 \\ +\quad -11 \\ \hline \end{array}$	**37.** $\begin{array}{r} -81 \\ +\quad 76 \\ \hline \end{array}$	**38.** $\begin{array}{r} -44 \\ +\quad -57 \\ \hline \end{array}$
39. $\begin{array}{r} -408 \\ +\quad 0 \\ \hline \end{array}$	**40.** $\begin{array}{r} -97 \\ +\quad 97 \\ \hline \end{array}$	**41.** $\begin{array}{r} -4836 \\ +\quad -5640 \\ \hline \end{array}$	

42. $33 + (-13) + (-10)$	**43.** $-63 + (-7) + (-10) + 12$
44. $14 + 46 + (-25)$	**45.** $-29 + (-1) + 33$
46. $-45 + 15 + 30$	**47.** $15 + (-18) + (-32) + 20$

48. $3 + (-3)$	**49.** $-8 + 8$	**50.** $-25 + (-25)$

SELFTEST

Simplify:

1. $-10 + (-16)$
2. $-38 + (-14)$
3. $-42 + 12$
4. $17 + -3$
5. $-8 + 208$
6. $-15 + (-20)$
7. $53 + 7$
8. $34 + (-6)$
9. $56 + (-81)$
10. $-95 + 60$
11. $-307 + (-43)$
12. $38 + 0$
13. $0 + (-67)$
14. $-34 + 34$
15. $-55 + 15 + (-5)$
16. $-30 + (-10) + 6$
17. $2 + (-8) + 4 + (-3)$
18. $-15 + 0 + 15$
19. $-62 + (-9) + 62 + 8$
20. $44 + (-16) + 2 + (-3)$
21. $14 + 13 + (-3) + (-4)$
22. $100 + (-23) + (-80) + 2$

SELFTEST ANSWERS

1. -26 **2.** -52 **3.** -30 **4.** 14 **5.** 200 **6.** -35 **7.** 60 **8.** 28
9. -25 **10.** -35 **11.** -350 **12.** 38 **13.** -67 **14.** 0 **15.** -45
16. -34 **17.** -5 **18.** 0 **19.** -1 **20.** 27 **21.** 20 **22.** -1

2.3 SUBTRACTION OF INTEGERS

Subtraction is closely related to addition, for to subtract b from a, we add the additive inverse of b to a.

Subtraction Rule: **If a and b are integers, then**

$$\mathbf{a - b = a + (-b) \qquad \text{and} \qquad a - (-b) = a + b}$$

Subtraction means "add the negative of." So $5 - 3$ means $5 + (-3)$ and $7 - (-2)$ means

$$7 + [-(-2)] = 7 + 2$$

The *subtrahend* is the quantity that is subtracted. So to compute a difference, we add the additive inverse of the subtrahend.

Example 1 $\qquad 4 - 6 = 4 + (-6) = -2$

Example 2 $\qquad -2 - 7 = -2 + (-7) = -9$

Example 3 $\qquad 8 - (-5) = 8 + 5 = 13$

The minus symbol, $-$, does double duty. It stands for the operation of subtraction as well as for the adjective "negative." The context determines the meaning of the minus symbol. In $5 - 8$, the minus symbol means subtract. In -3, it means negative.

In a problem that involves several subtractions, we convert them all to additions. So we must take the additive inverse of each subtrahend.

Example 4 $\qquad 8 - 3 - (-5) = 8 + (-3) + 5 = 10$

Example 5 $\qquad -6 - (-2) - 7 = -6 + 2 + (-7) = -11$

If a problem involves both additions and subtractions, we change all the subtractions to additions, but we leave the additions as additions. Then our problem is entirely an addition problem.

Example 6 $-3 + 10 - 12 - (-7) = -3 + 10 + (-12) + 7 = 2$

Again let's look at the number line; this time to see how subtraction works. We use subtraction to find the difference between two numbers. The difference between 4 and -5 is $4 - (-5)$ or $+9$.

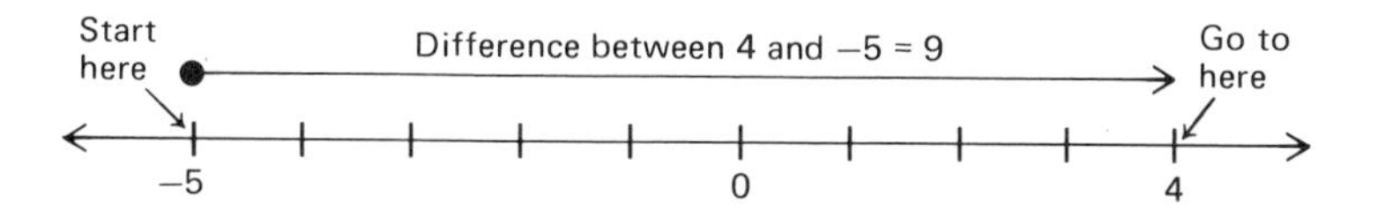

The difference is $+9$ since we must go 9 units in a positive direction to get from the subtrahend -5 to 4.

PROGRAM 2.3

subtract	**1.** The minus symbol has two meanings. It stands for the operation of subtraction and the descriptive adjective "negative." In $4 - 2$, the minus symbol stands for (subtract\negative).
negative	**2.** In -13, the minus symbol stands for ________.
subtract	**3.** The use determines the meaning of the symbol. If minus is the only symbol between two numbers, it tells us how to combine the numbers. So it means ________.
subtract	**4.** In $8 - 10$, the minus symbol means ________.
negative	**5.** If the minus symbol is not between two numbers, it tells us what kind of number we have. So it means ________.
negative	**6.** In $-28 + 2$, the minus symbol means ________.
negative	**7.** Sometimes the minus symbol describes the second number in a combination. If it tells us that the second number is negative, then there must be another symbol to tell us how to combine the numbers. In $8 + (-3)$, the minus symbol means ________.
subtract negative	**8.** In $16 - (-10)$, the first minus means ________ and the second one means ________.
−1; 7	**9.** The subtraction rule says that for integers a and b, $a - b = a + (-b)$ and $a - (-b) = a + b$ So $3 - 4 = 3 + (-4) =$ ____ and $3 - (-4) = 3 + 4 =$ ____.
6	**10.** To subtract b from a, we add the additive inverse of b to a. To compute $6 - 8$, we add the additive inverse of 8 to ____.

11. So $6 - 8 = 6 + (\underline{\quad})$ since the additive inverse of 8 is -8.

-8

12. $6 - 8 = 6 + (-8) = \underline{\quad}$.

-2

13. To compute $5 - 9$, we take the additive inverse of ____ and add it to ____.

9; 5

14. So $5 - 9 = 5 + (\underline{\quad}) = \underline{\quad}$.

-9; -4

15. The additive inverse of a number is the negative of that number. The additive inverse of 3 is ____.

-3

16. The additive inverse of $-6 = -(-6) = \underline{\quad}$.

6

17. The additive inverse of a negative number is always (positive\negative).

positive

18. The additive inverse of a positive number is always (positive\negative).

negative

19. So to take the additive inverse of a number, we change the sign of the number. The additive inverse of 8 is ____. (Remember 8 stands for $+8$.)

-8

20. The additive inverse of -12 is ____.

12

21. The additive inverse of 16 is ____.

-16

22. The additive inverse of -32 is ____.

32

23. To find the additive inverse of a number, we change the (sign\absolute value) of the number.

sign

24. To subtract, we change the sign of the subtrahend (the number we are subtracting) and add. So $15 - 10 = 15$ (+\−) (-10) = (5\−5).

+; 5

25. To subtract, we change two symbols. We change the subtraction symbol to the ________ symbol. We also change the ________ of the number we are subtracting.

addition; sign

26. $10 - (-25) = 10 + (+25) = \underline{\quad}$.

35

27. $-13 - (-12) = -13 + (+12) = \underline{\quad}$.

-1

28. $82 - (-100) = 82 + \underline{\quad} = \underline{\quad}$.

100; 182

29. $-63 - 70 = -63 + (\underline{\quad}) = \underline{\quad}$.

-70; -133

30. We did not change the -63 to 63 in the previous frame because we were not subtracting the -63. In $-18 - (-10)$, we (do\do not) change the -18 to 18.

do not

31. But to compute $-18 - (-10)$, we do change the -10 to ____ and the subtraction symbol between -18 and (-10) to (+\−).

$+10$
$+$

32. $-18 - (-10) = \underline{\quad} + \underline{\quad} = \underline{\quad}$.

-18; 10; -8

33. $15 - 20 = \underline{\quad} + (\underline{\quad}) = \underline{\quad}$.

15; -20; -5

34. $75 - (-20) = \underline{\quad} + \underline{\quad} = \underline{\quad}$.

75; 20; 95

Answer	Frame
+; −8	**35.** $-28 - (-20) = -28\ (+\backslash-)\ 20 =$ ____.
+; 115	**36.** $100 - (-15) = 100\ (+\backslash-)\ 15 =$ ____.
62; 8; 70	**37.** $62 - (-8) =$ ____ + ____ = ____.
−33; −9; −42	**38.** $-33 - 9 =$ ____ + (____) = ____.
−19	**39.** $-14 - 5 =$ ____.
sign add	**40.** To subtract, change the ________ of the subtrahend (number we are subtracting) and then ________.
addition	**41.** To subtract several numbers, we change each subtraction to (addition\multiplication) and we take the additive inverse of each subtrahend.
−3; 2; 5	**42.** $6 - 3 - (-2) = 6 +$ (____) + ____ = ____.
−5; −4; −17	**43.** $-8 - 5 - 4 = -8 +$ (____) + (____) = ____.
8; −9; 2; −15	**44.** $-16 - (-8) - 9 - (-2) = -16 +$ ____ + (____) + ____ = ____.
15; 16; −7; 24	**45.** $15 - (-16) - 7 =$ ____ + ____ + (____) = ____.
40	**46.** $30 - (-20) - 10 =$ ____.
3	**47.** $43 - 45 - (-5) =$ ____.
subtrahend	**48.** If a problem involves both subtraction and addition, we change each subtraction to addition and we take the additive inverse of each (number\subtrahend).
−8; 2	**49.** $3 + 7 - 8 = 3 + 7 +$ (____) = ____.
−2; −11	**50.** $-6 + (-3) - 2 = -6 + (-3) +$ (____) = ____.

PRACTICE PROBLEMS

You should try to do these problems in less than 15 minutes.

Simplify:

1. $4 - (-6)$ **2.** $-25 - (-75)$ **3.** $48 - (-22)$

4. $-750 - 50$ **5.** $-4 - (-4)$ **6.** $0 - (-94)$

7. $40 - (-40)$ **8.** $18 - 3$ **9.** $-20 - 9$

10. $-14 - (-7)$ **11.** $8 - 8$ **12.** $35 - (-5)$

13. $-64 - (-64)$ **14.** $-420 - (-20)$ **15.** $75 - 95$

16. $\begin{array}{r} 984 \\ - \quad -362 \\ \hline \end{array}$ **17.** $\begin{array}{r} 7 \\ - \quad 0 \\ \hline \end{array}$ **18.** $\begin{array}{r} 0 \\ - \quad 0 \\ \hline \end{array}$ **19.** $\begin{array}{r} -837 \\ - \quad 422 \\ \hline \end{array}$

20. $\begin{array}{r} 0 \\ - \quad -78 \\ \hline \end{array}$ **21.** $\begin{array}{r} -24 \\ - \quad -225 \\ \hline \end{array}$ **22.** $\begin{array}{r} 0 \\ - \quad 42 \\ \hline \end{array}$ **23.** $\begin{array}{r} 362 \\ - \quad 495 \\ \hline \end{array}$

24. $\begin{array}{r} -940 \\ -\ \ -936 \\ \hline \end{array}$ **25.** $\begin{array}{r} -22 \\ -\ \ 48 \\ \hline \end{array}$ **26.** $\begin{array}{r} 33 \\ -\ \ -12 \\ \hline \end{array}$ **27.** $\begin{array}{r} -10 \\ -\ \ -21 \\ \hline \end{array}$

28. $-95 - 10 - (-5)$ **29.** $48 - 50 - 6$ **30.** $38 - (-22) - 7$

31. $-5 - 82 - (-5)$ **32.** $-25 - 25 - 20$ **33.** $60 - (-60) - (-60)$

34. $406 - (-403) - 9$ **35.** $438 - (-73) - 438$ **36.** $3 - 5 - 3 - (-5)$

37. $-4 - 3 - (-7)$ **38.** $-605 - 25 + 30$ **39.** $9 + (-9) - 9$

40. $-7 - (-3) + 7$ **41.** $88 - 98 - (-15)$ **42.** $25 + (-10) - (-15)$

43. $27 - 50 + 3$ **44.** $-30 + (-50) - 30$ **45.** $18 - (-22) + (-1)$

46. $16 + (-8) - (-2)$ **47.** $-15 - (-15) + (-15)$

48. $-2 + (-4) - (-7) + 8$ **49.** $15 - 20 + (-5) - 10$

50. $-6 - (-6) + 8 - 8$

SELFTEST

Simplify:

1. $10 - 6$ **2.** $12 - (-8)$ **3.** $-14 - 15$

4. $-25 - (-5)$ **5.** $-5 - (-5)$ **6.** $9 - 9$

7. $0 - (-7)$ **8.** $0 - 3$ **9.** $5 - 9 + (-5)$

10. $16 - (-8) - 4$ **11.** $-12 + (-3) - 8$ **12.** $-37 - (-4) - 17$

13. $82 - (-5) + (-3)$ **14.** $63 - (-10) + (-4)$ **15.** $-57 - 12 + (-6)$

16. $-29 - 16 + (-4)$ **17.** $-3 - (-3)$ **18.** $-9 - 0$

19. $13 - (-3) + (-2)$ **20.** $46 - (-6) - 6$

SELFTEST ANSWERS

1. 4 **2.** 20 **3.** −29 **4.** −20 **5.** 0 **6.** 0 **7.** 7 **8.** −3
9. −9 **10.** 20 **11.** −23 **12.** −50 **13.** 84 **14.** 69 **15.** −75
16. −49 **17.** 0 **18.** −9 **19.** 14 **20.** 46

2.4 MULTIPLICATION OF INTEGERS

Recall that multiplication is a shortened form of addition. For example,

$$3 \cdot 8 = 8 + 8 + 8 = 24$$

Suppose we wish to multiply $3(-8)$. If we follow the same pattern, we get

$$3(-8) = (-8) + (-8) + (-8) = -24$$

Now let's try to multiply $(-4)5$. We cannot follow the pattern because we do not know how to write the number 5 a negative four times. However, by the commutative law of multiplication, $(-4)5 = 5(-4)$. Now we can follow the pattern to obtain

$$(-4)5 = 5(-4) = -4 + (-4) + (-4) + (-4) + (-4) = -20$$

When we try to multiply two negative numbers together the pattern evades us. However, we can use the distributive law to discover that the sign of the product should be positive. Thus, in order to evaluate $(-9)(-5)$, we start with the known fact that

$$-9(-5 + 5) = 9(0) = 0$$

By the distributive property, we see that

$$\begin{aligned} -9(-5 + 5) &= (-9)(-5) + (-9)(5) \\ &= (-9)(-5) + (-45) \end{aligned}$$

Therefore $(-9)(-5) + (-45) = 0$. Thus $(-9)(-5)$ is the additive inverse of -45. But 45 is also the additive inverse of -45. Since the additive inverse is unique, we conclude that

$$(-9)(-5) = 45$$

The following rules are easy to use and they give results that are consistent with the results of the previous examples.

Multiplication Rule 1: **The product of a negative number and a positive number is the negative of the product of the absolute values of the numbers. In particular, the product of a negative number and a positive number is negative.**

Multiplication Rule 2: **The product of two negative numbers is the product of their absolute values. In particular, the product of two negative numbers is positive.**

Example 1 Evaluate $(-25)(7)$: Since we are to multiply a negative number with a positive number, we use rule **1**.

$$(-25)(7) = -(|-25| \cdot |7|) = -(25 \cdot 7) = -175$$

Example 2 Evaluate $(-4)(-50)$: This time we use rule **2** since we are to multiply two negative numbers together.

$$(-4)(-50) = |-4| \cdot |-50| = (4)(50) = 200$$

Example 3 Evaluate $(-5)(-3)(2)$: Use the rules to multiply two numbers at a time.

$$(-5)(-3)(2) = 15 \cdot 2 = 30$$

Example 4 Evaluate $(-4)(-2)(3)(-1)(5)$:

$$(-4)(-2)(3)(-1)(5) = (8)(-3)(5) = (-24)(5) = -120$$

Example 5 Evaluate $(3)(-2) + (7 - 9)(4)$: Follow the usual order of operations.

$$(3)(-2) + (7 - 9)(4) = -6 + (-2)(4) = -6 + (-8) = -14$$

PROGRAM 2.4

1. Multiplication of integers breaks into two cases. In case one, one integer is positive and the other is negative. In case two, both integers have the same sign. So $(-1)(6)$ and $3(-2)$ are in case ____ but $(-4)(-6)$ and $3(5)$ are in case ____.

one

two

2. ***Multiplication Rule 1:*** The product of two numbers with unlike signs (i.e., one is positive and the other is negative) is the negative of the product of the absolute values of the numbers. So

$$(-2)(6) = -(|-2| \cdot |6|) = -(2 \cdot 6) = \_\_\_\_$$

−12

3. The product of a negative number and a positive number is always (positive\negative). — negative

4. $(3)(-4) = -(|3| \cdot |-4|) = -(3 \cdot 4) = \_\_\_\_.$ — −12

5. $(5)(-2) = -(|5| \cdot |-2|) = -(\_\_\_\_ \cdot \_\_\_\_) = \_\_\_\_.$ — 5; 2; −10

6. $(-10)(15) = -(|-10| \cdot |15|) = \_\_\_\_.$ — −150

7. $(-7)(2) = \_\_\_\_.$ — −14

8. Suppose a and b are integers. Then $a(-b) = -ab$ and $(-a)b = \_\_\_\_.$ — $-ab$

9. $3(-a) = \_\_\_\_ = (-3)a.$ — $-3a$

10. $-5(0) = \_\_\_\_.$ (Recall that 0 is neither positive nor negative.) — 0

11. The product of two numbers with unlike signs is (positive\negative). — negative

12. ***Multiplication Rule 2:*** The product of two numbers with like signs is the product of the absolute values of the numbers. So the product of two negative numbers is (positive\negative). — positive

13. $(-2)(-4) = |-2| \cdot |-4| = 2 \cdot 4 = \_\_\_\_.$ — 8

14. $(-7)(-3) = |-7| \cdot |-3| = \_\_\_\_ \cdot \_\_\_\_ = \_\_\_\_.$ — 7; 3; 21

15. $(-10)(-81) = |-10| \cdot |-81| = \_\_\_\_.$ — 810

16. $(-12)(-6) = \_\_\_\_ \cdot \_\_\_\_ = \_\_\_\_.$ — $|-12|$; $|-6|$; 72

17. $(-5)(-4) = \_\_\_\_.$ — 20

18. $(-3)(-8) = \_\_\_\_.$ — 24

19. Let m and n be integers. Then $(-m)(-n) = mn$ and $(-n)(-m) = \_\_\_\_.$ — nm

20. $(-9)(-m) = \_\_\_\_.$ — $9m$

21. The product of two negative numbers is (positive\negative). The product of a negative number and a positive number is (positive\negative). — positive; negative

22. $(-5)(-7) = \_\_\_\_.$ — 35

23. $(-5)(7) = \_\_\_\_.$ — −35

24. $(5)(-7) = \_\_\_\_.$ — −35

25. We use the same rules to multiply several numbers together. But we apply the rules to only two numbers at a time.

$$(-3)(-2)(-6) = (\_\_\_\_)(-6) = \_\_\_\_$$

6; −36

$-10; 70$	**26.** $(-5)(2)(-7) = (\_\_\_)(-7) = \_\_\_$.
$15; -30$	**27.** $(-3)(-5)(-2) = (\_\_\_)(-2) = \_\_\_$.
$-24; -72$	**28.** $(-4)(6)(3) = (\_\_\_)(3) = \_\_\_$.
80	**29.** $(-10)(-4)(2) = \_\_\_$.
-60	**30.** $(4)(3)(-5) = \_\_\_$.
	31. We can use the commutative law to place all the negative integers together and all the positive integers together.
160	$(-8)(2)(-10) = (-8)(-10)(2) = \_\_\_$
	32. $(-4)(2)(-1)(-3)(5) = (-4)(-1)(-3)(2)(5)$
-12	$= (\_\_\_)(10)$
-120	$= \_\_\_$
positive	**33.** If there are an even number of negative numbers, we pair each negative number with another negative number. The product of each pair of negative numbers is (positive\negative).
9	**34.** $(-2)(-2)(-3)(-3)(7) = (4)(\_\_\_)(7)$
252	$= \_\_\_$
positive	**35.** If there are an even number of negative numbers, the final result will be (positive\negative).
positive	**36.** The product $(8)(-9)(-5)(-3)(-6)(7)$ is (positive\negative) since there are an even number of negative factors.
positive	**37.** The product $(-1)(-2)(-3)(-4)(-5)(-7)$ is (positive\negative).
negative	**38.** If there are an odd number of negative factors, there will be one negative number left without a negative partner. So the entire product will be (positive\negative).
negative	**39.** $(-2)(-3)(-10)$ is (positive\negative).
negative	**40.** $(-4)(8)(-7)(-2)(-26)(-5)$ is (positive\negative).
-60	**41.** $(-2)(-3)(-10) = \_\_\_$.
-120	**42.** $(-2)(-5)(-4)(3) = \_\_\_$.
1	**43.** $(-1)(-1)(-1)(-1) = \_\_\_$.
	44. To do problems which involve multiplications, additions, and subtractions we follow the usual order of operations: *First* we compute expressions in parentheses. *Then* we do all multiplications and divisions from left to right. *Finally* we do additions and subtractions from left to right.
$-15; 14; -1$	$(-5)(3) + (-7)(-2) = \_\_\_ + \_\_\_ = \_\_\_$

45. $(-7)(3) - (-2)(5) =$ ____. -11

46. $(7 - 9)(3) = ($____$)(3) =$ ____. -2; -6

47. $(-9)(-3 + 6) =$ ____. -27

48. $(-2)(-5 + 8) + (-3) =$ ____ $+ (-3) =$ ____. -6; -9

49. $7(3 - 8) - (-2) =$ ____. -33

50. $(3)(-2)(-5) + (7 - 8)(-2) =$ ____. 32

PRACTICE PROBLEMS

Simplify:

1. $(-42)(-2)$
2. $(-33)(3)$
3. $(12)(-4)$
4. $0(-25)$
5. $(-8)(-6)$
6. $(-3)(7)$
7. $(-5)(-9)$
8. $(3)(8)$
9. $(-1)(-385)$
10. $(-55)(0)(96)$
11. $(-3)(-2)(-5)$
12. $(-1)(16)(-2)$
13. $(-3)(4)(5)$
14. $(-4)(-2)(-3)$
15. $(-1)(-8)(2)(-3)$
16. $(-2)(-2)(-2)(-2)$
17. $(-1)(-1)(-1)(5)$
18. $(3)(5)(-4)$
19. $(-1)(-85)(-2)$
20. $(-1)(-85)(2)$
21. $(-3)(-3)(3)$
22. $(-15)(-3) + (-20)$
23. $(72)(-2) + 50$
24. $38 + (-2)(40)$
25. $15 + (-16)(-5)$
26. $-12 + (-20)(-2)$
27. $-8 - (20)(-3)$
28. $18 - (-25)(-1)$
29. $-45 - (-9)(5)$
30. $84 - (-2)(6)$
31. $-5 - (-4)(5)$
32. $7 + (-5)(10)$
33. $(-2)(30) - 65$
34. $-6(4 + 3)$
35. $-8(4 - 8)$
36. $-2(-9) + 6$
37. $28(-3 + 3)$
38. $0(92 + 8)$
39. $-38(-2) - 1$
40. $(-36)(-2) + (-4)(3)$
41. $25(-3) - (5)(-4)$
42. $50(-2) + (100)(-1)$
43. $-6(-5 - 8) + (-4)(2)$
44. $8[3 - (-6)] - (6)(-2)$

SELFTEST

Simplify:

1. $(-15)(-3)$
2. $(-25)(3)$
3. $(8)(-10)$
4. $(-5)(-9)$
5. $0(-42)$
6. $(-37)(2)$
7. $(-8)(-2)(3)$
8. $(-8)(-3)(-2)$
9. $(-1)(-5)(10)(2)$

10. $(-1)(-5)(10)(-2)$ **11.** $(-4)(2) + (-8) - 12$ **12.** $(6)(-3) - 15 + (-9)$

13. $64 - (-9)(4) + (-2)(-5)$ **14.** $-15(8 - 12) + 3(-6)$

15. $[54 - (-6)](5 - 6)(3 - 3)$

SELFTEST ANSWERS

1. 45 **2.** −75 **3.** −80 **4.** 45 **5.** 0 **6.** −74 **7.** 48 **8.** −48
9. 100 **10.** −100 **11.** −28 **12.** −42 **13.** 110 **14.** 42 **15.** 0

2.5 DIVISION OF INTEGERS

Division is defined in terms of multiplication. So we might expect that the rules for deciding the sign of the quotient are related to the rules for determining the sign of the product. We will see that the rules are essentially the same.

Recall that the expression $a \div b = c$ means a divided by b equals c. We also write

$$\frac{a}{b} = c$$

to mean the same thing. In these expressions a is called the *dividend*, b is called the *divisor* and the result c is called the *quotient*. Remember that

$$\frac{a}{0} \quad \text{and} \quad a \div 0$$

are undefined. But both

$$\frac{0}{a} = 0 \quad \text{and} \quad 0 \div a = 0$$

if $a \neq 0$.

By definition, $a \div b = c$ if c is a number such that $bc = a$. We use this definition to find the quotient of two integers.

Example 1 Evaluate $-24 \div (-3)$: The quotient is 8 since $(-3)8 = -24$.

Example 2 Evaluate $50 \div (-25)$: This time the quotient is -2 since $(-25)(-2) = 50$.

The following rules conform to the results of the previous examples.

Division Rule 1: **To divide two integers, one of which is negative and one of which is positive, we take the negative of the quotient of their absolute values. So the sign of the quotient is negative.**

Division Rule 2: **To divide two integers which have the same sign, we divide their absolute values. So the sign of the quotient is positive.**

Example 3 Evaluate $70 \div (-7)$: Since we are to divide a positive number by a negative number, we use rule **1**.

$$70 \div (-7) = -(|70| \div |-7|) = -(70 \div 7) = -10$$

Example 4 Evaluate $-36 \div (-12)$: We use rule **2** this time since the dividend and the divisor are both negative.

$$-36 \div (-12) = |-36| \div |-12| = 3$$

Example 5 Evaluate $-95/5$: Again we use rule **1** since the numbers have different signs.

$$\frac{-95}{5} = -\left(\frac{|-95|}{|5|}\right) = -\left(\frac{95}{5}\right) = -19$$

Example 6 Evaluate $7/-7$: The quotient will be negative since the dividend and divisor have different signs.

$$\frac{7}{-7} = -\left(\frac{|7|}{|-7|}\right) = -\left(\frac{7}{7}\right) = -1$$

Example 7 Evaluate $-50/-10$: This time the quotient will be positive since both integers have the same sign.

$$\frac{-50}{-10} = \frac{|-50|}{|-10|} = \frac{50}{10} = 5$$

Example 8 Evaluate $36/-6 + (-14)/-2 - (-16)/8$: We follow the usual order of operations.

$$\frac{36}{-6} + \frac{-14}{-2} - \frac{-16}{8} = -6 + 7 - (-2) = -6 + 7 + 2 = 3$$

PROGRAM 2.5

1. Division is defined in terms of multiplication. Let a, b, and c be integers. By definition, $a \div b = c$ if c is an integer such that $b \cdot c = a$. So $10 \div 2 = 5$ since $2 \cdot$ ____ $= 10$. — 5

2. In the expression $a \div b = c$, c is called the *quotient*. In $20 \div 5 = 4$, the quotient is ____. — 4

3. The quotient of $30 \div 3$ is ____ since 3 times this value equals 30. — 10

4. The rules for determining the sign of the quotient of two integers are analogous to the rules for determining the sign of the product of two integers. Again there are two cases. In case one, one integer is positive and the other is negative. In case two, both integers have the same sign. The quotients $15 \div (-3)$ and $-60 \div 6$ are both in case ____. But the quotients $20 \div 4$ and $-8 \div (-2)$ are both in case ____. — one; two

5. ***Division Rule 1:*** The quotient of two integers with different signs (i.e., one is negative and the other positive) is the negative of the quotient of their absolute values. So $6 \div (-3)$ is (positive\negative). — negative

6. Evaluate $-28 \div 7$:

$$-28 \div 7 = -(|-28| \div |7|) = -(28 \div 7) = \_\_\_\_$$

— -4

7. Evaluate $50 \div (-10)$:

$$50 \div (-10) = -(|50| \div |-10|) = -(\_\_\_\_ \div \_\_\_\_)$$

$$= \_\_\_\_$$

— 50; 10; -5

$\lvert 14\rvert$; $\lvert -2\rvert$; -7	**8.** Evaluate: $14 \div (-2) = -(\_\_\_\_ \div \_\_\_\_) = \_\_\_\_$.
-3	**9.** Evaluate: $-75 \div 25 = \_\_\_\_$.
negative	**10.** The quotient of two numbers with different signs is (positive\negative).
positive	**11.** ***Division Rule 2:*** The quotient of two integers which have the same sign is the quotient of their absolute values. So the quotient $-4 \div (-2)$ is (positive\negative).
positive; 2	**12.** Evaluate $-16 \div (-8)$: Since -16 and -8 have the same signs, the quotient will be (positive\negative). $$-16 \div (-8) = \lvert -16\rvert \div \lvert 8\rvert = 16 \div 8 = \_\_\_\_$$
45; 9; 5	**13.** Evaluate $-45 \div (-9)$: $$-45 \div (-9) = \lvert -45\rvert \div \lvert -9\rvert = \_\_\_\_ \div \_\_\_\_ = \_\_\_\_$$
$\lvert -36\rvert$; $\lvert -6\rvert$; 6	**14.** $-36 \div (-6) = \_\_\_\_ \div \_\_\_\_ = \_\_\_\_$.
4	**15.** $-12 \div (-3) = \_\_\_\_$.
positive	**16.** The quotient of two numbers with the same sign is (positive\negative).
negative	**17.** The quotient of two numbers with different signs is (positive\negative).
12; 4	**18.** Fractions are another way to indicate quotients. The notation a/b means a divided by b. The fraction 12/4 means ____ divided by ____.
negative	**19.** So $\dfrac{-70}{5}$ is (positive\negative).
positive	**20.** $\dfrac{-100}{-4}$ is (positive\negative).
-2	**21.** $\dfrac{-12}{6} = -12 \div 6 = \_\_\_\_$.
-8; 2; -4	**22.** $\dfrac{-8}{2} = \_\_\_\_ \div \_\_\_\_ = \_\_\_\_$.
6	**23.** $\dfrac{-24}{-4} = \_\_\_\_$.
10	**24.** $\dfrac{-90}{-9} = \_\_\_\_$.
-6	**25.** $\dfrac{60}{-10} = \_\_\_\_$.
is	**26.** $\dfrac{-25}{5}$ (is\is not) equal to $\dfrac{25}{-5}$.
is not	**27.** $\dfrac{-25}{5}$ (is\is not) equal to $\dfrac{-25}{-5}$.

28. $\frac{-18}{-6}$ (is\is not) equal to $\frac{18}{6}$.

is

29. Remember that for all numbers n except 0, $n \div n = 1$ since $1 \cdot n = n$. So $7 \div 7 =$ ____.

1

30. $-10 \div (-10) =$ ____.

1

31. $\frac{-4}{4} =$ ____.

-1

32. $17 \div (-17) =$ ____.

-1

33. To simplify expressions which involve several operations, we follow the usual order of operations. First we compute expressions in parentheses. Then we do all multiplications and (additions\subtractions\divisions) from left to right as they occur. Finally, we do all additions and (subtractions\multiplications\divisions) from left to right as they occur.

divisions

subtractions

34. $16 \div (-2) + 8(-3) =$ ____ $+$ ____ $=$ ____.

-8; -24; -32

35. $[5 \div (-5)](-7) - 8 =$ ____ $- 8 =$ ____.

7; -1

36. $25 + \frac{-12}{4} =$ ____.

22

37. $\frac{35}{-7} + 6(-5) =$ ____.

-35

PRACTICE PROBLEMS

Simplify:

1. $24 \div (-6)$
2. $82 \div (-82)$
3. $-48 \div (-8)$
4. $0 \div (-23)$
5. $-65 \div (-5)$
6. $-44 \div 2$
7. $\frac{-26}{-13}$
8. $\frac{100}{-25}$
9. $\frac{0}{-84}$
10. $\frac{-14}{14}$
11. $\frac{6}{-6}$
12. $\frac{-3}{-3}$
13. $\frac{-56}{0}$
14. $\frac{-33}{-11}$
15. $\frac{-100}{4}$
16. $\frac{55}{-11}$
17. $\frac{-275}{-25}$
18. $\frac{101}{-101}$
19. $\frac{-400}{50}$
20. $\frac{0}{6}$
21. $\frac{63}{7}$
22. $(-300 \div 20) \div (-5)$
23. $[32 \div (-8)] \div (2)$
24. $60 \div (-24 \div 4)$
25. $75 \div [15 \div (-3)]$
26. $-45 + [-8 \div (-2)]$
27. $28 - (-50 \div 2)$
28. $63 \div (-7) - (-6)$
29. $(25)(-3) \div (-5)$
30. $\frac{-10 + 10}{-40}$
31. $\frac{(-12)(-5)}{15}$
32. $\frac{-25 + 10}{5}$
33. $\frac{100}{(-2)(-25)}$

34. $(-3)(20) + \left(\frac{-27}{3}\right)$

35. $(3)\left(\frac{-5}{5}\right) + 3$

36. $4 + (-4)\left(\frac{-6}{-6}\right)$

37. $\left(\frac{14}{-2}\right) - (5)(-4)$

38. $-\left(\frac{-90}{9}\right)$

39. $-\left(\frac{-20}{-4}\right)$

SELFTEST

Simplify:

1. $\frac{-36}{9}$

2. $40 \div (-5)$

3. $(-9) \div (-3)$

4. $\frac{8}{-1}$

5. $\frac{0}{-5}$

6. $\frac{-7}{7}$

7. $\frac{-8}{-8}$

8. $21 \div [(-3)(7)]$

9. $[(-4)(-2)] \div 4$

10. $(82 - 4) \div (-2)$

11. $(65 \div 5) \div (-13)$

12. $-28 \div (3 - 10)$

13. $(-24 \div 8) - 10$

14. $\frac{-25}{5} + \frac{12}{-3}$

15. $\left(\frac{3}{-3}\right)(4) + 4$

SELFTEST ANSWERS

1. -4 2. -8 3. 3 4. -8 5. 0 6. -1 7. 1 8. -1 9. 2
10. -39 11. -1 12. 4 13. -13 14. -9 15. 0

POSTTEST 2

1. *a.* Use set notation to indicate the set of integers and the set of whole numbers.
   b. Give an example of an integer that is not a whole number, and an example of a number that is not an integer.

2. Locate the integers -8, 6, 0, -2, 5 on the number line. Of these integers, which is the largest and which is the smallest?

3. Give the additive inverse of each of the following.
   a. x *b.* $-y$ *c.* 2 *d.* 0 *e.* -5 *f.* 1

4. Simplify:
   a. $|-16|$
   b. $|-8| + |-2|$
   c. $|-4| - |1|$
   d. $-9 + 3$
   e. $18 - 20$
   f. $-4 + 6 - 17$

5. Simplify:
   a. $7 - (-4)$
   b. $-14 - 6$
   c. $-2 + 1 - 16$
   d. $-6 - (-8) + 2$
   e. $4 - 0$
   f. $-(-7 + 3)$

6. Simplify:
   a. $8(-3)$
   b. $(-2)(-6)$
   c. $(-1)(-2)(-3)$
   d. $4(-2)(-3)$
   e. $(-1)5$
   f. $(18)(0)$

7. Simplify:
   a. $90 \div (-5)$
   b. $(15 - 30) \div (-3)$
   c. $-12 \div 4$
   d. $7 \div (-1)$
   e. $-4 \div (-2)$
   f. $0 \div (-1)$

Each problem above refers to a section in this chapter as shown in the table.

Problems	*Section*
1–3	2.1
4	2.1 and 2.2
5	2.3
6	2.4
7	2.4 and 2.5

POSTTEST 2 ANSWERS

1. *a*. Set of integers $= \{\cdots, -3, -2, -1, 0, 1, 2, 3, \cdots\}$; set of whole numbers $= \{0, 1, 2, 3, \cdots\}$.
b. -1 is an integer that is not a whole number; $^1/_2$ is not an integer.

2.

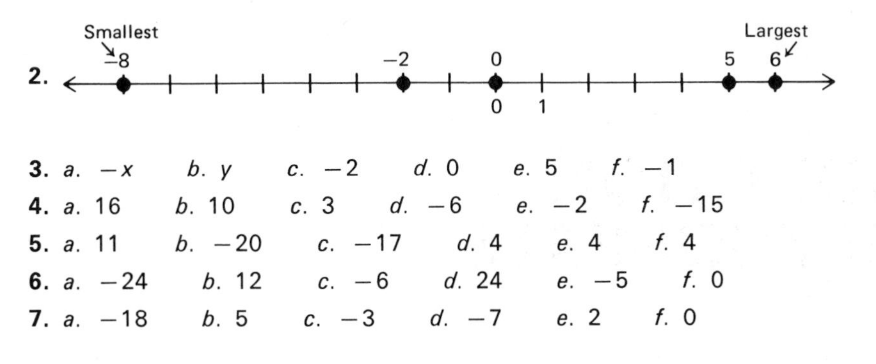

3. *a*. $-x$ *b*. y *c*. -2 *d*. 0 *e*. 5 *f*. -1

4. *a*. 16 *b*. 10 *c*. 3 *d*. -6 *e*. -2 *f*. -15

5. *a*. 11 *b*. -20 *c*. -17 *d*. 4 *e*. 4 *f*. 4

6. *a*. -24 *b*. 12 *c*. -6 *d*. 24 *e*. -5 *f*. 0

7. *a*. -18 *b*. 5 *c*. -3 *d*. -7 *e*. 2 *f*. 0

INTRODUCTION TO POLYNOMIALS

BASIC SKILLS

Upon completion of the appropriate section the student should be able to:

3.1 *a.* Identify the base or exponent in an expression.
b. Write an expression in factored form in the appropriate exponential form and vice versa for whole-number exponents.
c. Multiply and divide exponential expressions.

3.2 *a.* Recognize a polynomial expression in a given variable with integral coefficients.
b. Identify the terms, coefficients, and constants of a polynomial.
c. Give the degree of a polynomial in one or several variables.
d. Write a polynomial in standard form.
e. Evaluate a polynomial for a given value.

3.3 *a.* Identify like terms.
b. Add or subtract like terms in polynomials.
c. Simplify an expression of the form

(polynomial) + (monomial)(polynomial)

3.4 *a.* Give the additive inverse of a polynomial.
b. Subtract one polynomial from another.
c. Simplify an expression of the form

(monomial)(polynomial) − (monomial)(polynomial)

3.5 *a.* Multiply two binomials.
b. Multiply a binomial and a trinomial.

3.6 *a.* Divide a polynomial by a binomial and give the remainder.

PRETEST 3

1. *a.* What is the base of x^5? What is the exponent?
   b. What is the base of -6^2? What is the exponent?
   c. What is the base of $(-2)^3$? What is the exponent?

2. *a.* Write $xxxxxyyy$ in exponential form.
   b. Write x^3y^4 in factored form without exponents.

3. Simplify:

   a. $\dfrac{4x^6 5xy^2}{2x^2}$

   b. $(16x^3y)(2xy)$

   c. $5 + 3(2^3 \div 4)$

   d. $4^2 + 1$

4. Given the polynomial $3x + 1 - 2x^2$
   a. Write the polynomial in standard form.
   b. What is the degree of the polynomial?
   c. What is the coefficient of x^2 and what is the constant term?
   d. Evaluate the polynomial for $x = 3$.

5. Simplify:

   a. $6xy + 4xy + 7$

   b. $2x + 3(x^2 - 2x + 1)$

   c. $3(x + 2) + 6(x^2 + x)$

6. Simplify:

   a. $6x^2 - 9(2x^2 + 1)$

   b. $3(x + 2) - 5(4x - 1)$

   c. $-x^2 - 2 - (x^2 - x - 1)$

7. Multiply and simplify:

   a. $5x(x^2 + 2x - 1)$

   b. $(2x + 3)(x - 1)$

   c. $(x - 5)(x^2 + 3x + 2)$

8. Divide and check your answer: $(16x^2 + 2x - 3) \div (8x - 3)$.

Each problem above refers to a section in this chapter as shown in the table.

Problems	*Section*
1–3	3.1
4	3.2
5	3.3
6	3.4
7	3.5
8	3.6

PRETEST 3 ANSWERS

1. *a.* x; 5 *b.* 6; 2 *c.* -2; 3 **2.** *a.* x^5y^3 *b.* $xxxyyyy$

3. *a.* $10x^5y^2$ *b.* $32x^4y^2$ *c.* 11 *d.* 17

4. *a.* $-2x^2 + 3x + 1$ *b.* 2 *c.* -2; 1 *d.* -8

5. *a.* $10xy + 7$ *b.* $3x^2 - 4x + 3$ *c.* $6x^2 + 9x + 6$

6. *a.* $-12x^2 - 9$ *b.* $-17x + 11$ *c.* $-2x^2 + x - 1$

7. *a.* $5x^3 + 10x^2 - 5x$ *b.* $2x^2 + x - 3$ *c.* $x^3 - 2x^2 - 13x - 10$

8. $2x + 1$

3.1 EXPONENTIAL NOTATION

The word *factor* has a special meaning in mathematics. Expressions that are multiplied together are the *factors* of a product. In $3x(2 + y)$, the factors are 3, x, and $(2 + y)$. To write an expression in factored form means to write it as a product of factors. The number 12 has several factored forms. Among them are $12 \cdot 1$, $3 \cdot 4$, $2 \cdot 6$, and $2 \cdot 3 \cdot 3$.

It is appropriate to use exponential notation to indicate that the same factor occurs repeatedly in a product.

Definition: **For counting numbers n, the *exponential notation***

$$x^n \quad \text{means} \quad x \cdot x \cdot x \cdot \ldots \cdot x \quad \text{(use } x \text{ as a factor } n \text{ times)}$$

and

$$x^0 = 1 \qquad \text{for all } x \neq 0$$

The *base* of x^n is x. The *exponent* or *power* is n. We read x^n as "x to the nth power." A number which is represented as a factor may be represented in exponential form. For example,

$$32 = 2 \cdot 2 \cdot 2 \cdot 2 \cdot 2 = 2^5$$

The *number* is 32. The exponential form is 2^5. The base is 2 and the exponent is 5. The following examples show first the factored form and then the exponential form.

Example 1 $$x = x^1$$

Example 2 $$3 \cdot 3 \cdot 3 \cdot 3 = 3^4$$

Example 3 $$y \cdot y \cdot y \cdot y \cdot y \cdot y \cdot y \cdot y = y^8$$

Unless enclosed in parentheses, the base is the positive value immediately beneath the exponent.

Example 4 The base of -2^4 is 2.

$$-2^4 = -(2 \cdot 2 \cdot 2 \cdot 2) = -16$$

Example 5 The base of $(-2)^4$ is -2.

$$(-2)^4 = (-2)(-2)(-2)(-2) = 16$$

We can use exponential notation to simplify certain products. For example,

$$16 \cdot 32 = 2^4 \cdot 2^5$$

To multiply the numbers 16 and 32, we add the exponents 4 and 5 in the following way:

$$16 \cdot 32 = 2^4 \cdot 2^5 = 2^{4+5} = 2^9$$

Of course if the base were a variable number, such as x, then we wouldn't always know the value of the answer. But we would still have a correct answer.

We multiply two exponential expressions by multiplying the factored forms of each expression. It turns out that if the base of two expressions are the same, we can *multiply by adding the exponents.*

Example 6 $$x^5x^2 = (xxxxx)(xx) = xxxxxxx = x^7$$

Since the bases are the same, we can also simplify the problem by adding the exponents.

$$x^5x^2 = x^{5+2} = x^7$$

Example 7 $$3y^4x^2 2y^3x = 3(yyyy)(xx)(2)(yyy)(x)$$
$$= 6(yyyyyyy)(xxx)$$
$$= 6y^7x^3$$

Another way to multiply is to add the exponents of factors with similar bases.

$$3y^4x^2 2y^3x = 3 \cdot 2y^4y^3x^2x$$
$$= 6y^{4+3}x^{2+1} \qquad \text{(recall } x = x^1\text{)}$$
$$= 6y^7x^3$$

To divide exponential expressions, we use the fact that the divisor times the quotient must equal the dividend.

Example 8 $$x^5 \div x^2 = x^3 \qquad \text{since} \qquad x^2x^3 = x^5$$

Example 9 $$\frac{x^8}{x^2} = x^6 \qquad \text{since} \qquad x^2x^6 = x^8$$

In general we use the following equations to work with exponential expressions.

For whole numbers n and m and $x \neq 0$

$$x^nx^m = x^{n+m}$$

$$\frac{x^n}{x^m} = x^{n-m} \qquad \text{if } n \geq m$$

$$x^0 = 1$$

The relations follow from the basic definitions of exponential notation:

$$x^n \cdot x^m = \underbrace{(x \cdot x \cdot x \cdot \ldots \cdot x)}_{n \text{ factors}}\underbrace{(x \cdot x \cdot x \cdot \ldots \cdot x)}_{m \text{ factors}} = x^{n+m}$$

$$\frac{x^n}{x^m} = x^{n-m} \qquad \text{since} \qquad (x^m)(x^{n-m}) = x^{m+n-m} = x^n$$

$$x^0 = 1 \qquad \text{since} \qquad \frac{x^n}{x^n} = x^{n-n} = x^0 \qquad \text{and} \qquad \frac{x^n}{x^n} = 1$$

Example 10 $$4x^2y^3 3x^5y^7 = 12(x^{2+5})(y^{3+7}) = 12x^7y^{10}$$

Example 11 $$\frac{14x^6y^4}{2x^4y^3} = 7(x^{6-4})(y^{4-3}) = 7x^2y$$

Example 12 $$\frac{15x^4y^2}{5xy^2} = 3(x^{4-1})(y^{2-2}) = 3x^3y^0 = 3x^3 \qquad \text{since } y^0 = 1$$

Multiplication indicated by exponential notation takes priority over other multiplications and divisions. The order of operation for expressions containing exponents is:

1. Compute expressions in parentheses.
2. Evaluate exponential expressions.
3. Do all other multiplications and divisions from left to right.
4. Do all additions and subtractions from left to right.

Example 13 $$3 \cdot 7^2 = 3 \cdot 49 = 147$$

Example 14 $(3 \cdot 7)^2 = 21^2 = 441$

Example 15

$$\begin{aligned} 5 + 3(2^5 \div 4)^2 &= 5 + 3(32 \div 4)^2 \\ &= 5 + 3(8)^2 \\ &= 5 + 3(64) \\ &= 5 + 192 \\ &= 197 \end{aligned}$$

PROGRAM 3.1

Answer	Frame
x	**1.** The parts of a product that are multiplied together are the *factors* of the product. In $6 \cdot 5 \cdot x$, the factors are 6, 5, and ____.
$3 \cdot 3$	**2.** An expression is in *factored form* if it is written as a product of factors. The factored forms of 25 are $25 \cdot 1$ and $5 \cdot 5$. The factored forms of 9 are $9 \cdot 1$ and ____.
3 three	**3.** In $2 \cdot 2 \cdot 2$, we use 2 as a factor ____ times. Exponential notation for $2 \cdot 2 \cdot 2$ is 2^3. The exponent is 3. It tells us to use 2 as a factor ____ times.
4; 4; 2	**4.** The notation 4^2 means ____ · ____. The exponent is ____. It tells us to use 4 as a factor twice.
$x \cdot x \cdot x \cdot x$	**5.** In general, for counting numbers n, $x^n = x \cdot x \cdot x \cdot \ldots \cdot x$ (use x as a factor n times) So $x^4 =$ ____ in factored form.
$y \cdot y \cdot y$	**6.** Write y^3 in factored form. ____
$y \cdot y \cdot y \cdot y \cdot y \cdot y \cdot y \cdot y$	**7.** Write y^8 in factored form. ____
z^6	**8.** The exponential notation for $z \cdot z \cdot z \cdot z \cdot z \cdot z$ is ____.
x^9 $y \cdot y \cdot y \cdot y \cdot y$	**9.** *Factored form* / *Exponential form*: $x \cdot x \cdot x \cdot x \cdot x \cdot x \cdot x \cdot x \cdot x =$ ____ ; ____ $= y^5$
5; 2	**10.** In the exponential notation x^n, n is the *exponent* and x is the *base*. In 2^5, the exponent is ____ and the base is ____.
exponent; base	**11.** In 8^3, 3 is the (base\exponent) and 8 is the (base\exponent).
exponent; base	**12.** In 7^4, 4 is the ____ and 7 is the ____.
6; x; six	**13.** The exponent tells us how many times to use the base as a factor. In x^6, the exponent ____ tells us to use ____ as a factor ____ times.

14. The base is always the positive value immediately under the exponent, or the quantity enclosed by parentheses immediately under the exponent. So the base of -7^4 is 7. The base of $(-3)^2$ is -3. What is the base of $(-9)^2$?

____. The base of -9^2 is ____. | -9; 9

15. A minus symbol is included in the base only if it is enclosed in parentheses. The base of $-x^2$ is x since the negative sign (is\is not) in parentheses. The base of $(-x)^2$ is ____ since the negative sign is in the parentheses. | is not; $-x$

16. The base of $(-5)^2$ is ____. So $(-5)^2 = (-5)(-5) =$ ____. | -5; 25

17. The base of -5^2 is ____. Thus $-5^2 = -(5 \cdot 5) =$ ____. | 5; -25

18. $(-7)^2 =$ ____ but $-7^2 =$ ____. | 49; -49

19. We multiply two exponential expressions by first converting them to factored form.

$$x^3x^4 = (x \cdot x \cdot x)(x \cdot x \cdot x \cdot x) = x \cdot x \cdot x \cdot x \cdot x \cdot x \cdot x = x^{\_\_\_\_}$$

7

20. $2^4 \cdot 2^6 = (2 \cdot 2 \cdot 2 \cdot 2)($________$) = 2^{____}$. | $2 \cdot 2 \cdot 2 \cdot 2 \cdot 2 \cdot 2$; 10

21. $y^3y = (y \cdot y \cdot y)($____$) =$ ____. | y; y^4

22. The general rule for multiplication is $x^n \cdot x^m = x^{n+m}$. So

$$x^{12}x^7 = x^{\_\_\_\_+\_\_\_\_} = x^{\_\_\_\_}$$

12; 7; 19

23. $5^{10} \cdot 5^{11} = 5^{\_\_\_\_}$. | 21

24. $x^{25}x^6 =$ ____. | x^{31}

25. $3^5 \cdot 3^4 =$ ____ (exponential notation). | 3^9

26. $4x^3y^9x^5y^8 = 4x^{\_\_\_\_}y^{\_\_\_\_}$. | 8; 17

27. $7x^4y^9x^3y^2 =$ ________. | $7x^7y^{11}$

28. To multiply two exponential expressions *with the same base,* we (add\multiply) exponents. | add

29. It is necessary to distinguish between numbers used as exponents and numbers used as numerical factors. In $4x^35x^2$, the 4 and 5 are (exponents\numerical factors). The 3 and 2 are (exponents\numerical factors). | numerical factors; exponents

30. In $6x^47x^8$, the numerical factors are ________. The exponents are ________. | 6 and 7; 4 and 8

31. In a product, we add the exponents with the same base, but we multiply the numerical factors. In the expression $4x^35x^{11}$, we (add\multiply) the 4 and 5. We (add\multiply) the 3 and 11 since they are both exponents of x. | multiply; add

32. Thus $4x^35x^{11} = 4 \cdot 5x^{3+11} =$ ________. | $20x^{14}$

33. $6x^34x^2 =$ ________. | $24x^5$

$35x^{12}$ **34.** $5x^7 7x^5 =$ ________.

$-32x^{12}$ **35.** $-4x^3 8x^9 =$ ________.

$12x^5y^9$ **36.** $2x^2y^4 6x^3y^5 =$ ________.

$-15x^7y^6$ (recall $y = y^1$) **37.** $-3x^3y^5 5x^4y =$ ________.

38. To divide two exponential expressions, we use the fact that the divisor times the quotient equals the dividend. Thus $x^8 \div x^3 = x^5$ since $x^3 \cdot x^5 = x^8$.

x^2; x^2 Likewise, $x^5 \div x^3 =$ ____ since $x^3 \cdot$ ____ $= x^5$.

y^5; y^5 **39.** $y^9 \div y^4 =$ ____ since $y^4 \cdot$ ____ $= y^9$.

x^3; x^3 **40.** $x^5/x^2 =$ ____ since $x^2 \cdot$ ____ $= x^5$.

41. The general rule for division is

$$\frac{x^n}{x^m} = x^{n-m} \qquad \text{if } n \geq m \text{ and } x \neq 0$$

2 So $x^8/x^6 = x^{8-6} = x$——.

9; 2; 7 **42.** $\dfrac{x^9}{x^2} = x$——-—— $= x$——.

7 **43.** $\dfrac{x^{11}}{x^4} = x$——.

3 **44.** $\dfrac{3^7}{3^4} = 3$——.

x^{11} **45.** $\dfrac{x^{20}}{x^9} =$ ____.

subtract **46.** To divide exponential expressions which have *the same base*, we (subtract\divide) the exponents.

7; 3 (recall $y = y^1$) **47.** $\dfrac{x^{10}y^4}{x^3y} = x$——$y$——.

xy or x^1y^1 **48.** $\dfrac{x^5y^3}{x^4y^2} =$ ____.

numerical factors; exponents; subtract **49.** In $15x^7/5x^4$, the numbers 15 and 5 are (numerical factors\exponents). The 7 and 4 are both (numerical factors\exponents). So we divide the 5 into the 15, but we (divide\subtract) the 7 and 4.

7; 4; $3x^3$ **50.** $\dfrac{15x^7}{5x^4} = 3x$——-—— $=$ ____.

$5x^3$ **51.** $\dfrac{30x^5}{6x^2} =$ ____.

$5y^8$ **52.** $\dfrac{10y^{10}}{2y^2} =$ ____.

53. $\dfrac{48x^6y^5}{6x^3y^2} = \_\_\_\_ x^{\_\_\_\_} y^{\_\_\_\_}$. 8; 3; 3

54. $\dfrac{36x^9y^4}{9x^3y} = \_\_\_\_\_\_\_\_$. $4x^6y^3$

55. $x/x = 1$ since $x \cdot 1 = x$. But by the division rule,

$$\frac{x}{x} = x^{1-1} = x^{\_\_\_\_}.$$ 0

Therefore we conclude that $x^0 = \_\_\_\_$. 1

56. $y^0 = 1$. In fact, any nonzero number to the 0th power equals ____. 1

57. $5^0 = \_\_\_\_$. 1

58. $(-10)^0 = \_\_\_\_$. 1

59. $\dfrac{10x^5}{2x^5} = 5x^{5-5} = 5x^{\_\_\_\_} = \_\_\_\_$. 0; $5 \cdot 1$ or 5

60. $\dfrac{24y^6}{12y^6} = 2y^{\_\_\_\_} = \_\_\_\_$. 0; $2 \cdot 1$ or 2

61. $\dfrac{18x^5y^3}{9x^4y^3} = \_\_\_\_$. $2x$

62. Multiplication indicated by exponents takes priority over other multiplications. We do operations in the following order.

a. Compute all expressions in parentheses.
b. Evaluate all exponential expressions.
c. Do multiplications and divisions from left to right.
d. Do additions and subtractions from left to right.

In $3 \cdot 4^2$ we compute ____ first. Then, $3 \cdot 4^2 = 3 \cdot 16 = \_\_\_\_$. 4^2; 48

63. $3 \cdot 6^2 = 3 \cdot \_\_\_\_ = \_\_\_\_$. 36; 108

64. $(3 \cdot 6)^2 = (\_\_\_\_)^2 = \_\_\_\_$. 18; 324

65. $(4 \cdot 5)^2 = \_\_\_\_$. 400

66. $56 \div 2^3 = 56 \div \_\_\_\_ = \_\_\_\_$. 8; 7

67. $36 \div 3^2 = \_\_\_\_$. 4

68. $(33 - 3^2) \div 2^3 = \_\_\_\_ \div \_\_\_\_ = \_\_\_\_$. 24; 8; 3

69. $(14 - 2^3) \div 3 = \_\_\_\_$. 2

70. $(2^2 + 3^2)^2 = \_\_\_\_$. 169

PRACTICE PROBLEMS

1. Identify the exponent and the base in each.

a. 3^4 *b.* -2^5 *c.* $(-2)^3$ *d.* x^m

2. Write in factored form.

a. x^5 *b.* y^2 *c.* x *d.* $(-x)^4$
e. x^3y^8 *f.* 2^6 *g.* $(-3)^3$ *h.* -8^2

Simplify:

3. x^3x^4 **4.** yy^5 **5.** $2y^3y^6$

6. $-4x^2x$ **7.** $8x^2y4x^3$ **8.** $5xy^43x^5y^8$

9. $-4x^2y^47x^{10}y^8$ **10.** $-3z^4x^5(-2)x^9y^2$ **11.** $8xy(-5)y^4x^7$

12. $7z^43x^8zx7y^2$ **13.** $x^2y^3x^4y^5x^8y^9$ **14.** $3x^2y^42xy^34x^6y^2$

15. $\frac{x^9}{x^2}$ **16.** $\frac{y^4}{y}$ **17.** $\frac{z^3}{z^3}$ **18.** $\frac{10x^9}{2x^3}$

19. $\frac{28x^2y^4}{7x^2y^2}$ **20.** $\frac{45x^5y^6}{9x^4y^3}$ **21.** $\frac{30x^8z^4}{15x^4}$ **22.** $\frac{14x^7y^7}{2x^7y^7}$

23. $\frac{18x^9z^5y^3}{2x^3zy^2}$ **24.** $\frac{36x^{14}y^6z^8}{x^7yz^6}$ **25.** 4^2 **26.** 3^3

27. -2^4 **28.** $(-2)^4$ **29.** -3^4 **30.** $(-3)^4$

31. $(-8)^2$ **32.** -8^2 **33.** $\frac{6^2}{6}$ **34.** $\frac{2^5}{2^2}$

35. $\frac{10^2}{5^2}$ **36.** $4 \cdot 7^2$ **37.** $(3 \cdot 6)^2$ **38.** $7 + 2^4$

39. $(5 - 2)^2$ **40.** $5^2 - 2^2$ **41.** $2(8^2 + 2^2)$ **42.** $9^2 \div 3^3$

43. $6 \cdot 5^2 - 5^2$ **44.** $(32 \div 4^2) - 2^3$ **45.** $4 + 2(9^2 \div 3^2)^2$

46. $5(2^5 \div 2^4)^3 - 7^2$ **47.** $36x^2y \div 3^2x^2y$

48. *a.* What is the base in $(5y)^2$?

b. $(5y)^2 = (5y)(5y) =$ ____.

c. $5^2y^2 =$ ____.
d. Does $(5y)^2 = 5^2y^2$? (Compare the results of parts *b.* and *c.*)
e. In general is it true that $(ax)^2 = a^2x^2$? Why?

49. *a.* What is the base of 3 in $(x^2)^3$?

b. $(x^2)^3 = (x^2)(x^2)(x^2) =$ ________ in factored form.

c. $x^{2 \cdot 3} = x^6 =$ ________ in factored form.
d. Is $(x^2)^3 = x^{2 \cdot 3}$?

50. *a.* $2^4 =$ ____.

b. $2 \cdot 4 =$ ____.
c. Is 2^4 equal to $2 \cdot 4$?

SELFTEST

1. Identify the base and the exponent.
   a. 6^4 b. -2^7 c. $(-2)^3$
2. Write in factored form.
   a. $-x^6$ b. $(-x)^4$ c. y^8
3. Simplify:
   a. $4xy^5 6x^8 y^3$ b. $\dfrac{30x^2y^6z^4}{6x^2y^2z}$ c. $3 \cdot 8^2 + (7-5)^4$

SELFTEST ANSWERS

1. a. base is 6, exponent is 4 b. base is 2, exponent is 7 c. base is -2, exponent is 3
2. a. $-xxxxxx$ b. $(-x)(-x)(-x)(-x)$ c. $yyyyyyyy$
3. a. $24x^9y^8$ b. $5y^4z^3$ c. 208

3.2 POLYNOMIAL EXPRESSIONS

We form polynomial expressions by adding certain types of products together.

Definition: **A *polynomial in x with integer coefficients* is the sum of products of the form**

$$(\text{integer})x^{\text{whole number}}$$

The products that are added are the *terms* of the polynomial. The numerical factor of a term is the *coefficient* of the term. The highest power of x in the polynomial is the *degree* of the polynomial.

Consider the expression $4x^3 + 5x^2$. Is it a polynomial? First we identify the terms or products that are added. The first term is $4x^3$ and the last one is $5x^2$. Both terms are of the form $(\text{integer})x^{\text{whole number}}$. Consequently, the expression is a polynomial. The coefficient within the term $4x^3$ is 4. The coefficient within the term $5x^2$ is 5.

The expression x is also a polynomial. For the coefficient is understood to be 1 and the exponent of x is also understood to be 1. The number 8 is a polynomial because the x factor is x^0 which equals 1. Terms such as 8 are *constant terms* since they do not include a variable.

Polynomials with negative terms can be written as a difference rather than a sum of products. But the terms and coefficients of one form are the same as the terms and coefficients of the other form.

Example 1 $$6x^2 + (-3x) = 6x^2 - 3x$$

In both forms the terms are $6x^2$ and $-3x$. The coefficient in the x term is -3, no matter which form we use.

To write a polynomial in x in *standard form* we write it in descending powers of x. The polynomial $8x^3 + 12x^2 - 9x + 15$ is in standard form. But $8x^2 + 4x^5 - x^3$ is not in standard form since the powers of x are not in decreasing order.

Of course all of the above discussion remains true if we replace the variable x by a new variable.

Example 2 $$5y^6 + 2y - 4$$

is a polynomial in y. It is of degree 6. The terms are $5y^6$, $2y$, and -4.

Polynomials with just one term are *monomials*. Those with two terms are *binomials*. So $12x^3$, 10, and z are monomials. The polynomials $y + 5$ and $3x^2 - 4$ are binomials.

Some polynomials have several variables. A polynomial in x and t has terms of the form

$$(\text{integer})s^{\text{whole number}}t^{\text{whole number}}$$

So $7s^3t^2 + 4s^2t - s$ is a polynomial in s and t. The term $7s^3t^2$ has degree 3 in s and degree 2 in t. The degree of $7s^3t^2$ is the sum of the degree in s and the degree in t. Therefore $7s^3t^2$ has degree $3 + 2$ or 5. What is the degree of $8x^4y^2 + 10x^3y + x - 4$? The term with highest degree is $8x^4y^2$. The degree of $8x^4y^2$ is $4 + 2$ or 6. Therefore the degree of the polynomial is 6.

Example 3

$$xyz^2 + x^2 + y - z$$

is a polynomial in x, y, and z. The term of highest degree is xyz^2. Its degree is $1 + 1 + 2$, or 4. So the degree of the polynomial is 4.

Example 4

$$9x^5 - 17x^4y^3 + 4x - 1$$

is a polynomial in x and y. The term of highest degree is $-17x^4y^3$ since this term has degree $4 + 3$ or 7.

Example 5 $x + y + z$ has degree 1.

We can evaluate polynomials for specific values of the variables in the polynomial. We just replace the variables by the given value. In the following examples, evaluate the given polynomials for the given values.

Example 6 When $x = 5$, $3x + 4 = 3(5) + 4 = 15 + 4 = 19$.

Example 7 If $x = 2$, $3x + 4 = 3(2) + 4 = 6 + 4 = 10$.

Example 8 If $x = -2$, $3x + 4 = 3(-2) + 4 = -6 + 4 = -2$.

Example 9 When $y = -1$,

$$4y^2 + 6y - 3 = 4(-1)^2 + 6(-1) - 3 = 4(1) + (-6) - 3 = -5$$

Example 10 If $y = 0$,

$$4y^2 + 6y - 3 = 4(0)^2 + 6(0) - 3 = 4(0) + 6(0) - 3 = -3$$

Example 11 When $y = 2$,

$$4y^2 + 6y - 3 = 4(2)^2 + 6(2) - 3 = 4(4) + 12 - 3 = 25$$

Example 12 If $x = 3$ and $y = -2$, then

$$\begin{aligned} x^3y - x^2y^2 + x &= (3)^3(-2) - (3)^2(-2)^2 + (3) \\ &= 27(-2) - (9)(4) + 3 \\ &= -87 \end{aligned}$$

Example 13 If $x = 1$ and $y = -4$, then

$$\begin{aligned} x^3y - x^2y^2 + x &= (1)^3(-4) - (1)^2(-4)^2 + (1) \\ &= 1(-4) - 16 + 1 \\ &= -19 \end{aligned}$$

Example 14 If $x = 0$ and $y = 5$, then

$$\begin{aligned} x^3y - x^2y^2 + x &= (0)^3(5) - (0)^2(5)^2 + (0) \\ &= 0 + 0 + 0 \\ &= 0 \end{aligned}$$

PROGRAM 3.2

1. A *polynomial in x with integer coefficients* is the sum of products of the form $$(\text{integer})x^{\text{whole number}}$$ Is $3x^4 + 5x^2 + 8x$ a polynomial in x? (yes\no)	yes (each product is of the proper form)
2. Is $5x^3 + (-2x) + 8x^0$ a polynomial in x? (yes\no)	yes
3. $4y^3 + (-5y^2) + 3$ is a polynomial in ____.	y
4. The degree of a polynomial in x is the highest power of x that occurs in the expression. The degree of $5x^9 + 3x^4 + 8x^2$ is ____.	9
5. The highest power of y in $y^4 + 6y^2 + 3y$ is ____. So the degree of $y^4 + 6y^2 + 3y$ is ____.	4 4
6. The degree of $5x^3 + (-8x^4) + 9x^2$ is ____.	4
7. The products that are added together to form a polynomial are the *terms* of the polynomial. The terms of $7x^4 + 3x^3 + (-5x^2) + 4x$ are ____, ____, ____, and ____.	$7x^4$; $3x^3$ $-5x^2$; $4x$
8. The terms of $9x^3 + 4x^2 + (-2)$ are ____, ____, and ____.	$9x^3$; $4x^2$; -2
9. The polynomial $10x^4 + (-5x^3)$ can be written $10x^4 - 5x^3$. The terms of both forms are the same. The terms of $10x^4 + (-5x^3)$ are ____ and ____. So the terms of $10x^4 - 5x^3$ are ____ and ____.	$10x^4$; $-5x^3$ $10x^4$; $-5x^3$
10. The terms of $-8x^9 + 3x^4 - 6x^2$ are ____, ____, and ____.	$-8x^9$; $3x^4$; $-6x^2$
11. The terms of $3y^5 - 4y^3 + 9y^2 - 3y + 5$ are ____, ____, ____, ____, and ____.	$3y^5$; $-4y^3$; $9y^2$; $-3y$; 5
12. The degree of a term in one variable is the power of the variable in that term. In $3x^2 + 2x^3$, which term is of degree 3? ____.	$2x^3$
13. Which term in $4x^2 + 3x$ is of degree 1? ____.	$3x$ (recall $x = x^1$)
14. The number 5 is of degree 0 since $5 = 5x^0$. (Recall $x^0 = 1$.) Which term in $3x^2 + 2x + 4$ has degree 0? ____.	4
15. In $7y^4 + 2y^3 - 6y + 7$, the term 7 has degree ____. Which term has degree 1? ____.	0 $-6y$
16. Terms of degree 0 are *constant terms*. The constant term in $5x^2 + 3x^8 + 9$ is ____.	9
17. The numerical factor of a term is the *coefficient* of the term. The coefficient of $8x^2$ is 8. The coefficient of $-12x^5$ is ____.	-12

18. In the polynomial $15x^8 - 6x^5 + 2x$, the coefficient of the fifth-degree term is _____. The coefficient of the first degree term is _____.

-6; 2

19. List the coefficients in the polynomial $5x^2 - 3x$. _________.

5, -3

20. What is the coefficient of $1x^4$? _____.

1

21. If a term has coefficient 1, we usually omit the 1. The coefficient of x^5 is _____.

1

22. In y^7, the coefficient is understood to be _____. In $-x^2$ the coefficient is understood to be _____.

1

-1

23. In $x^5 - x^3$, the terms are _____ and _____. The coefficients are _____ and _____.

x^5; $-x^3$; 1

-1

24. To write a polynomial in x in *standard form,* we write it in descending powers of x. So $3x^5 + 2x^4 - x + 3$ is in standard form. Write $8x^2 + 5x^3 + 4$ in standard form. _______________.

$5x^3 + 8x^2 + 4$

25. Is $6y^2 + 2y^3 + 2 - 8y$ in standard form? (yes\no)

no

26. Write $5y^3 + 6y^4 - 3y^2 - 5$ in standard form. _______________.

$6y^4 + 5y^3 - 3y^2 - 5$

27. The expression $8y^3 + 2y^4 - 3 - y^2$ is a (polynomial\term) in y. It is of degree _____. The standard form is _______________. The terms are _______________. What is the constant term? _____. The coefficients of the nonconstant terms are _________.

polynomial

4; $2y^4 + 8y^3 - y^2 - 3$

$2y^4, 8y^3, -y^2, -3$; -3

2, 8, -1

28. A polynomial with only one term is a *monomial.* Those with two terms are *binomials.* The expression $4x^6$ is a (binomial\monomial).

monomial

29. How many terms does a binomial have? _____.

two

30. How many terms does a monomial have? _____.

one

31. $4x^2 - 2$ is a (binomial\monomial\neither).

binomial

32. The expression $3x^4y^2 + 5x^3y + x$ is a polynomial in x and y. The degree of $3x^4y^2$ is the sum of the exponents of x and y. So the degree of $3x^4y^2$ is $4 + 2$ or _____. The degree of $5x^3y$ is _____. The degree of x is _____. The term with highest degree is _____. It has degree 6, so the degree of the polynomial $3x^4y^2 + 5x^3y + x$ is _____.

6; 4; 1

$3x^4y^2$

6

33. The degree of $x^2y^3 + 1$ is _____.

5

34. The degree of $xy + x$ is _____.

2

35. To evaluate a polynomial for a specific value of the variable, we replace the variable by the given number and simplify. Evaluate $4x + 2$ for $x = 5$. We put 5 in place of x. The resulting expression is $4(5) + 2$ or _____.

22

36. Evaluate $4x + 2$ for $x = 3$. We put ____ in place of x. The resulting expression is $4(\_\_\_\_) + 2$ or ____. — 3; 3; 14

37. Evaluate $4x + 2$ for $x = -1$. This time we put ____ in place of x. The resulting expression is $4(\_\_\_\_) + 2$ or ____. — -1; -1; -2

38. Evaluate $4x + 2$ for $x = -2$. We replace ____ by -2. The new expression is $4(\_\_\_\_) + 2$ or ____. — x; -2; -6

39. Evaluate $5y^2 - y$ for $y = 3$. We put 3 in place of y and obtain

$$5y^2 - y = 5(\_\_\_\_)^2 - (\_\_\_\_) = \_\_\_\_ - \_\_\_\_ = \_\_\_\_$$

3; 3; 45; 3; 42

40. If $y = -2$, then $5y^2 - y = 5(\_\_\_\_)^2 - (\_\_\_\_) = \_\_\_\_$. — -2; -2; 22

41. When $y = -1$, $5y^2 - y = \_\_\_\_$. — 6

42. When $x = 0$, $3x^2 + 2x - 5 = \_\_\_\_ + \_\_\_\_ - \_\_\_\_ = \_\_\_\_$. — $3(0)^2$; $2(0)$; 5; -5

43. When $x = 4$, $3x^2 + 2x - 5 = \_\_\_\_$. — 51

44. When $x = -4$, $3x^2 + 2x - 5 = \_\_\_\_$. — 35

45. If $y = -3$, then $4y^2 - 2y = \_\_\_\_$. — 42

46. Evaluate $x^2 - 3x + 2$ for $x = 1$. ________________ — $1^2 - 3(1) + 2 = 0$

47. To evaluate $xy + 2x - y$ for $x = 3$ and $y = 4$, we use ____ in place of x and ____ in place of y. For $x = 3$ and $y = 4$, — 3; 4

$$xy + 2x - y = (3)(4) + 2(\_\_\_\_) - (\_\_\_\_) = \_\_\_\_$$

3; 4; 14

48. For $x = 5$ and $y = 1$,

$$xy + 2x - y = (\_\_\_\_)(\_\_\_\_) + 2(\_\_\_\_) - \_\_\_\_ = \_\_\_\_$$

5; 1; 5; 1; 14

49. For $x = 0$ and $y = -4$, $xy + 2x - y = $ ________________. — $0(-4) + 2(0) - (-4) = 4$

50. For $x = 8$ and $y = -2$, $3xy - 6y = 3(\_\_\_\_)(\_\_\_\_) - 6(\_\_\_\_) = \_\_\_\_$. — 8; -2; -2; -36

51. For $x = -2$ and $y = 6$, $3xy - 6y = \_\_\_\_$. — -72

52. For $x = 3$ and $y = 2$,

$$6x^2y^3 - 2xy^2 + x = 6(\_\_\_\_)^2(\_\_\_\_)^3 - 2(\_\_\_\_)(\_\_\_\_)^2 + (\_\_\_\_)$$
$$= \_\_\_\_$$

3; 2; 3; 2; 3; 411

53. For $x = 2$ and $y = 3$, $6x^2y^3 - 2xy^2 + x = \_\_\_\_$. — 614

54. For $x = -3$ and $y = 2$, $6x^2y^3 - 2xy + x = \_\_\_\_$. — 441

PRACTICE PROBLEMS

1. Identify the terms in each of the following polynomials.

a. $4x^2 + 8$ *b.* $9x^3 - 6x^2 + 5$ *c.* $3x^2$
d. $10x^5 - 7x^3 + x^8 - 2$ *e.* $-7xy^4 + 8x^3y - x^2$

2. What is the degree of each polynomial in Problem 1?

3. Which terms are constant in the polynomials in Problem 1?

4. What is the coefficient of the second-degree term in each of the polynomials in Problem 1?

5. Write each of the polynomials in Problem 1 in standard form with respect to x.

6. Identify the binomials and monomials in Problem 1.

7. Evaluate $6x - 5$ for

a. $x = 0$ *b.* $x = 3$ *c.* $x = -10$ *d.* $x = 1$

8. Evaluate $5x^2 - 5$ for

a. $x = 0$ *b.* $x = 2$ *c.* $x = -2$ *d.* $x = 1$

9. Evaluate $2x^2 - 3x + 4$ for

a. $x = 3$ *b.* $x = 0$ *c.* $x = -1$ *d.* $x = -2$

10. Evaluate $x^2y^2 - x + y$ for

a. $x = 0$ and $y = 1$ *b.* $x = -2$ and $y = 3$

SELFTEST

1. Consider the polynomial $4x^5 - 8x^3 + 2x^4 - x^2 - 4$.

a. Write it in standard form.
b. List the terms.
c. Give the coefficient of the fifth-degree term.
d. What is the coefficient of the second-degree term?
e. What is the degree of the polynomial?
f. Which term has degree 0?

2. Evaluate $4x^2 - 3x - 6$ for

a. $x = 0$ *b.* $x = 1$ *c.* $x = -1$ *d.* $x = -5$

3. Evaluate $2xy^2 - x + y$ for

a. $x = 1$ and $y = 0$ *b.* $x = 2$ and $y = -1$

SELFTEST ANSWERS

1. *a.* $4x^5 + 2x^4 - 8x^3 - x^2 - 4$ *b.* $4x^5, 2x^4, -8x^3, -x^2, -4$ *c.* 4 *d.* -1
e. 5 *f.* -4

2. *a.* -6 *b.* -5 *c.* 1 *d.* 109 **3.** *a.* -1 *b.* 1

3.3 ADDITION OF POLYNOMIALS

Terms of a polynomial that have exactly the same variables to the same powers are *like terms.* In $3x^2 + 2x + 5x^2$, the terms $3x^2$ and $5x^2$ are like terms because the variable factors in both are x^2. But $2x$ and $3x^2$ are not like terms, for the powers of the variables differ. In $2x$, x is to the first power. But in $3x^2$, x is to the second power.

The distributive law enables us to add or subtract like terms. So we can simplify polynomials that have like terms.

Example 1 Simplify $7xz + 4xz$: By the distributive law, $7xz + 4xz = (7 + 4)xz = 11xz$.

Example 2 Simplify $9xy^2 - 6xy^2$: By the distributive law, $9xy^2 - 6xy^2 = (9 - 6)xy^2 = 3xy^2$.

In general, the distributive law tells us that the coefficient of the sum of like terms is the sum of the coefficients of the terms. The variable factors of the sum are exactly the same as those in the original like terms.

The sum of two polynomials is again a polynomial. We usually simplify the resulting polynomial by combining like terms. Then we arrange the polynomial in standard form.

Example 3 Add $10x^3 + 3x^2$ and $5x^3 - 2x^2 + 4$.

$(10x^3 + 3x^2) + (5x^3 - 2x^2 + 4)$

$= 10x^3 + 3x^2 + 5x^3 - 2x^2 + 4$ by associative law
$= 10x^3 + 5x^3 + 3x^2 - 2x^2 + 4$ by commutative law
$= 15x^3 + x^2 + 4$ like terms combined

Example 4 Add $12x^2 - 3$ to $18x^2 + x$.

$$\begin{aligned} 12x^2 - 3 + (18x^2 + x) &= 12x^2 - 3 + 18x^2 + x \\ &= 12x^2 + 18x^2 + x - 3 \\ &= 30x^2 + x - 3 \end{aligned}$$

Example 5 Add $6x^3 + 2x^2 - 5x + 8$ to $-9x^2 + 4x - 7$.

$6x^3 + 2x^2 - 5x + 8 + (-9x^2 + 4x - 7)$

$$\begin{aligned} &= 6x^3 + 2x^2 - 5x + 8 + (-9x^2) + 4x - 7 \\ &= 6x^3 + 2x^2 - 9x^2 - 5x + 4x - 7 + 8 \\ &= 6x^3 - 7x^2 - x + 1 \end{aligned}$$

The next examples involve both multiplication and addition. The distributive law tells us how to multiply a polynomial by a monomial.

Example 6 Add $4x$ to $3(x^2 + 2x - 1)$. We do the multiplications first. Then we add like terms.

$$\begin{aligned} 4x + 3(x^2 + 2x - 1) &= 4x + 3x^2 + 6x - 3 \\ &= 3x^2 + 10x - 3 \end{aligned}$$

Example 7

$$\begin{aligned} 2(x + 5) + x(3x + 4) &= 2x + 10 + 3x^2 + 4x \\ &= 3x^2 + 6x + 10 \end{aligned}$$

Example 8

$$\begin{aligned} 3x^2 + (-2)(x^2 + 4) &= 3x^2 + (-2x^2) - 8 \\ &= x^2 - 8 \end{aligned}$$

PROGRAM 3.3

1. Terms in a polynomial that contain exactly the same variables to the same powers are *like terms.* In $4x + 3x - 2$, the terms $4x$ and $3x$ (do\do not) contain the same variable factor to the same power. Thus $4x$ and $3x$ (are\are not) like terms.

do
are

2. Are $5x^2$ and $2x$ like terms? (yes\no) (*Hint:* Compare the exponents of the variables.)

no

$3x^2$; $-4x^2$; x^2	**3.** The like terms in $3x^2 + 2x - 4x^2 + x^2$ are ____, ____, and ____.
$4xy$; $-2xy$	**4.** The like terms in $4xy + 4x^2y - 2xy$ are ____ and ____.
$2yx^2$; $3x^2y$	**5.** The like terms in $3x^2y + 2yx^2 + 4y^2x$ are ____ and ____. (*Hint:* By the commutative law, $x^2y = yx^2$.)
$8x^2y$; $-5yx^2$	**6.** The like terms in $8x^2y + 2xy^2 - 5yx^2$ are ____ and ____.
3; 4; 7	**7.** The distributive property says that $ax + bx = (a + b)x$. Thus $$3x + 4x = (\_\_\_\_ + \_\_\_\_)x = (\_\_\_\_)x$$
7; 3; 10	**8.** By the distributive property, we can add like terms since we can "factor out" the variable factors. $$7x + 3x = (\_\_\_\_ + \_\_\_\_)x = \_\_\_\_x$$
8 + 5; $13xy$	**9.** $8xy + 5xy = (\_\_\_\_\_\_\_\_)xy = \_\_\_\_\_\_\_\_$.
4 + 3; $7zy^2$	**10.** $4zy^2 + 3zy^2 = (\_\_\_\_\_\_\_\_)zy^2 = \_\_\_\_\_\_\_\_$.
$13x$	**11.** $8x + 5x = \_\_\_\_$.
7; 9; $-2x$	**12.** $7x - 9x = (\_\_\_\_ - \_\_\_\_)x = \_\_\_\_$.
$2xy$	**13.** $5xy - 3xy = \_\_\_\_$.
12; 4; 6; $10y$	**14.** $12y + 4y - 6y = (\_\_\_\_ + \_\_\_\_ - \_\_\_\_)y = \_\_\_\_$.
$2xy$ or $2yx$	**15.** $9xy - 10xy + 3yx = \_\_\_\_$.
are	**16.** To add like terms, we add the coefficients. The variable factors of the result (are\are not) the same as those in the original like terms.
xy^2z	**17.** $12xy^2z + 6xy^2z = 18\_\_\_\_\_\_\_\_$.
$13st$ or $13ts$	**18.** $4st + 9ts = \_\_\_\_$.
$3x + 2y$ no	**19.** What is the simplest form of $3x + 2y$? ________. Can we simplify the sum of unlike terms? (yes\no)
already in simplest form	**20.** Simplify: $3x + 2 = \_\_\_\_$.
$5x$	**21.** Simplify: $3x + 2x = \_\_\_\_$.
5; 2; 1	**22.** Simplify: $3x^2 + 2x + 5 + 2x^2 - 4 = \_\_\_\_x^2 + \_\_\_\_x + \_\_\_\_$.
$7x^2 + 8x - 9$	**23.** Let's add the polynomials $3x^2 + 8x$ and $4x^2 - 9$. By the associative law of addition, $$3x^2 + 8x + (4x^2 - 9) = 3x^2 + 8x + 4x^2 - 9 = \_\_\_\_\_\_\_\_\_\_\_\_$$
yes	**24.** Is the sum of the polynomials in the previous frame a polynomial? (yes\no)

25. The sum of two polynomials is again a polynomial. $x^3 - 5x + (x^4 + 7x) =$ ________________ (in standard form). — $x^4 + x^3 + 2x$

26. Add $-5x^2 + 3$ to $2x^2 + x$. Write the result in standard form. ________________ — $-3x^2 + x + 3$

27. Add $5x^4 + 3x^3 - 9x - 5$ to $6x^3 + 8$. The result in standard form is ________________. The result is a polynomial of degree ____. — $5x^4 + 9x^3 - 9x + 3$; 4

28. Simplify: $6x^3 + 8x - 5 + (4x^3 - 8x + 4) =$ ________________. — $10x^3 - 1$

29. Simplify: $2x^4 - 8x + (5x^2 + 9) + (-2x^4 + 3x^2) =$ ________________. — $8x^2 - 8x + 9$

30. Sometimes no like terms occur in a sum of polynomials. In such cases we simply write the polynomial in standard form.

$$10x^5 + 8 + (4x^3 - x) = \_\_\_\_\_\_\_\_$$

— $10x^5 + 4x^3 - x + 8$

31. Simplify: $3x^4 - 2x^2 + (8x^3 + 3x^2) =$ ________________. — $3x^4 + 8x^3 + x^2$

32. Simplify: $7x^5 + x^3 + (4x^3 - 8) =$ ________________. — $7x^5 + 5x^3 - 8$

33. Simplify and write in standard form in y.

$$4y^2 + 9y + (2 + 6y^3) = \_\_\_\_\_\_\_\_$$

— $6y^3 + 4y^2 + 9y + 2$

34. After we simplify expressions in parentheses, we do (additions\multiplications). — multiplications

35. By the distributive law, $3(x + 2) = 3x + 6$. So

$$4 + 3(x + 2) = 4 + \_\_\_\_\_\_ = \_\_\_\_\_\_$$

— $3x + 6$; $3x + 10$

36. By the distributive law, $5(x^2 + 2x - 4) =$ ________________. So — $5x^2 + 10x - 20$

$$6x^2 + 5(x^2 + 2x - 4) = 6x^2 + \_\_\_\_\_\_\_\_$$

— $5x^2 + 10x - 20$

$$= \_\_\_\_\_\_\_\_$$

— $11x^2 + 10x - 20$

37. Simplify: $5x^2 + 4(2x^2 - 3x + 4) =$ ________________. — $13x^2 - 12x + 16$

38. Simplify: $x^2 + 4x + 2(x^2 - 5) =$ ________________. — $3x^2 + 4x - 10$

39. Simplify: $5x^3 + 8 + 7(5 - x^3) =$ ________________. — $-2x^3 + 43$

40. By the distributive law, $x(4x^2 + 3) = 4x^3 +$ ____. — $3x$

41. Simplify: $5x + x(4x^2 + 3) =$ ________________. — $4x^3 + 8x$

42. Simplify: $2(x - 5) + x(3x + 4) =$ ________________. — $3x^2 + 6x - 10$

PRACTICE PROBLEMS

1. Identify the like terms in each polynomial.

   a. $3x + 2x$ *b.* $4x^2 + 3x - x^2$ *c.* $4x + 4y$

   d. $6xy + 8x^2y - 3yx^2$ *e.* $-3x^2 + 7x^2 - 7y^2$

Add the upper and lower polynomials.

2. $\begin{array}{r} 12x + 3 \\ -2x + 4 \\ \hline \end{array}$

3. $\begin{array}{r} -25x^2 + 15 \\ 10x^2 - 4 \\ \hline \end{array}$

4. $\begin{array}{r} 5x - 3 \\ 12x^2 - 3x \\ \hline \end{array}$

5. $\begin{array}{r} 3x^4 - 2x^3 + 3x - 5 \\ -2x^4 + 6x^3 \qquad + 3 \\ \hline \end{array}$

6. $\begin{array}{r} x^3 - 4x^2 + 8 \\ x^4 - 3x^3 - 2x^2 + 7 \\ \hline \end{array}$

7. $\begin{array}{r} 48x^2 + 33x + 9 \\ 5x^3 + 14x^2 - 22x \\ \hline \end{array}$

8. $\begin{array}{r} -15x^4 + 3x^3 - 16x + 2 \\ 120x^4 + 7x^3 + 19x - 8 \\ \hline \end{array}$

9. $\begin{array}{r} -52x^4 + 16x^2 - 5x - 45 \\ 10x^4 - 4x^2 \qquad - 205 \\ \hline \end{array}$

10. $\begin{array}{r} 83x^3 - 17x^2 - 98 \\ -6x^4 \qquad + 9x^2 - 18 \\ \hline \end{array}$

11. $\begin{array}{r} x^5 + 3x^4 - 2x + 8 \\ - 3x^4 \qquad - 9 \\ \hline \end{array}$

12. $\begin{array}{r} 4x^3 - x^2 \qquad - 16 \\ 9x^4 - 5x^3 + x^2 + 7x + 8 \\ \hline \end{array}$

Simplify and write in standard form:

13. $3x + 2x$

14. $7x - 5x$

15. $-6y - 6$

16. $2x^2 + 3x - 8x^2$

17. $5x + 3 - 2x$

18. $-4y^3 - 2y^2 + 7 - 8y^3 + 6y^2$

19. $6x^2 - 12x - 3 + (4x^2 - 6x + 7)$

20. $-24x^2 + 8x + 6 + (-4x^2 + 7x - 3)$

21. $10y^2 - y + 8 + (-10y^2 - 7y + 4)$

22. $3x^2 + x - 7 + (2x^4 - x^2 + 6)$

23. $-5x^4 + 3x^2 + (2x^4 - 3x - 5)$

24. $4(2x^2 - x) + (-3)(4x^2 + 2x)$

25. $(50x^3 + 7x^2) + (-26 + 30x^3)$

26. $-2(3x + 4) + (-7)(-x + 5)$

27. $6(x + 8) - x$

28. $4x + 5(2x - 1)$

29. $-5(x^2 + 1) + 3x + 8$

30. $x^3 - 2x + 2(-4x - x^3)$

31. $-1(x + 4) + 3(x - 2)$

32. $6(x^2 + x) + (-1)(2x^2 - x)$

33. $8x + (-1)(6x^2 + 3x - 2)$

34. $-5(3x + 1) + (-2)(x - 5)$

35. $-10x^4 + (-8x^2 + 20x) + (14x^4 + 3x - 8)$

36. $4x^2 + (3x + 2) + (5x^2 + 3) + (-4x^2 + 7x)$

SELFTEST

1. Is the sum of two polynomials a polynomial? yes

2. Add:

a. $\begin{array}{r} 4x^2 - 6x + 8 \\ -2x^2 - x + 9 \\ \hline 2x^2 - 7x + 17 \end{array}$

b. $\begin{array}{r} 5x^2 + 8x - 6 \\ x^3 - 5x^2 - 2x \\ \hline x^3 \qquad 6x - 6 \end{array}$

3. Simplify and write in standard form:

a. $-5x^3 + 9x^2 + (-4x^3 + x^2 - 8) + (3x^2 + 4)$
b. $4(x^2 + 1) + (-1)(6x^2 + 2)$
c. $-3(2x^2 - 2) + (-3)(-8x^2 + 7)$
d. $x(2x^2 - 3x + 1) + x^3 - 6x^2 + 8$

SELFTEST ANSWERS

1. yes
2. *a.* $2x^2 - 7x + 17$ *b.* $x^3 + 0x^2 + 6x - 6$ or $x^3 + 6x - 6$
3. *a.* $-9x^3 + 13x^2 - 4$ *b.* $-2x^2 + 2$ *c.* $18x^2 - 15$ *d.* $3x^3 - 9x^2 + x + 8$

3.4 SUBTRACTION OF POLYNOMIALS

To subtract one polynomial from another, we follow the subtraction rule for integers. We take the additive inverse of the subtrahend and then add.

We know how to find the additive inverse of a monomial. We simply change the sign. So the additive inverse of $5x$ is $-5x$, and the additive inverse of $-3x$ is $3x$. But how do we find the additive inverse of a polynomial with several terms? We change the sign of *each* term.

Example 1 The additive inverse of $-7x^4$ is $7x^4$.

Example 2 The additive inverse of $8x^2 + 3$ is $-8x^2 - 3$.

Example 3 The additive inverse of $-2x^4 + 4x^2 - 6x + 8$ is $2x^4 - 4x^2 + 6x - 8$.

Example 4 The additive inverse of $a + b$ is denoted by $-(a + b)$. So $-(-3x + 2)$ is the additive inverse of $-3x + 2$. Notice that $-(-3x + 2) = 3x - 2$. Likewise the additive inverse of $4x - 5$ is $-(4x - 5)$ or $-4x + 5$. The additive inverse of $-6x - 8$ is $-(-6x - 8)$ or $6x + 8$. The additive inverse of $7x + 2y$ is $-(7x + 2y)$ or $-7x - 2y$.

To subtract, we add the additive inverse of the subtrahend (the quantity we are subtracting).

Example 5 To subtract $4x^2 - 3x + 7$ from $3x^2 + 5x + 4$, we write

$$\begin{array}{r} 3x^2 + 5x + 4 \\ -\ (4x^2 - 3x + 7) \\ \hline \end{array}$$

which simplifies to the sum

$$\begin{array}{r} 3x^2 + 5x + 4 \\ -\ 4x^2 + 3x - 7 \\ \hline -x^2 + 8x - 3 \end{array}$$

Example 6 To subtract $-9x^4 + 3x - 2$ from $7x^4 - 5x^2 + 7x - 5$, we write

$$\begin{array}{r} 7x^4 - 5x^2 + 7x - 5 \\ -\ (-9x^4 \qquad + 3x - 2) \\ \hline \end{array}$$

which simplifies to the sum

$$\begin{array}{r} 7x^4 - 5x^2 + 7x - 5 \\ 9x^4 \qquad - 3x + 2 \\ \hline 16x^4 - 5x^2 + 4x - 3 \end{array}$$

When written horizontally the calculations appear as in Examples 7, 8, and 9.

Example 7 $4x^2 - (2x^2 - 8) = 4x^2 + (-2x^2 + 8)$

$$= 4x^2 + (-2x^2) + 8 = 2x^2 + 8$$

Example 8 $9x^2 - 4x - (-3x^2 + 6x - 5) = 9x^2 - 4x + (3x^2 - 6x + 5)$

$$= 9x^2 - 4x + 3x^2 - 6x + 5$$
$$= 12x^2 - 10x + 5$$

With some practice it is possible (and desirable) to do several steps mentally.

Example 9 $5x^3 + 3x - (4x^3 - x + 7) = 5x^3 + 3x - 4x^3 + x - 7$

$$= x^3 + 4x - 7$$

There is another way to view an expression of the type $-(a + b)$.

By the distributive law $-1(a + b) = -a - b$

By the definition of additive inverse $-(a + b) = -a - b$

Therefore $-(a + b) = -1(a + b)$.

Thus the additive inverse of an expression is -1 times the expression. Consequently, we can view subtraction as the process of adding -1 times the subtrahend.

Example 10 $-(5 + x) = -1(5 + x) = -5 - x$

Example 11 $-(-x^2 + 3x - 4) = -1(-x^2 + 3x - 4) = x^2 - 3x + 4$

Example 12 $4x^2 - (x - 8) = 4x^2 + (-1)(x - 8)$

$$= 4x^2 - x + 8$$

Example 13 $5x^2 - 2x - 3(x^2 + 2x) = 5x^2 - 2x + (-3)(x^2 + 2x)$

$$= 5x^2 - 2x - 3x^2 - 6x$$
$$= 2x^2 - 8x$$

Example 14 $x^2 + 8 - 2x(-x + 5) = x^2 + 8 + (-2x)(-x + 5)$

$$= x^2 + 8 + 2x^2 - 10x$$
$$= 3x^2 - 10x + 8$$

PROGRAM 3.4

1. If R and S are any numbers such that $R + S = 0$, then we say that R and S are the additive inverses of each other. For example $5 + (-5) = 0$, so the additive inverse of 5 is ____, and the additive inverse of -5 is ____. Are $4x$ and $-4x$ additive inverses of each other? (yes\no) Is it true that $4x + (-4x) = 0$? (yes\no)

   -5; 5
   yes
   yes

2. The additive inverse of $-3x^2$ is ____ since $-3x^2 +$ ____ $= 0$.

   $3x^2$; $3x^2$

3. What is the additive inverse of $7x - 3x^2$? It must be $-7x + 3x^2$ since

   $$7x - 3x^2 + (-7x + 3x^2) = \_\_\_\_.$$

   0

4. To find the additive inverse of a polynomial we change the sign of *each* term. The additive inverse of $4x^2 - 2x + 6$ is $-4x^2 + 2x - 6$. The additive inverse of $2x^2 - 4x + 8$ is ________.

   $-2x^2 + 4x - 8$

5. The additive inverse of $-3x^4 + 5x^2 - 2$ is ______. (Did you change the sign of each term?) — $3x^4 - 5x^2 + 2$

6. The additive inverse of $-x^3 - 6x^2 + 3x - 2$ is ______. — $x^3 + 6x^2 - 3x + 2$

7. The additive inverse of $a + b$ is denoted by $-(a + b)$. The additive inverse of $3x + 2$ is denoted by ______. — $-(3x + 2)$

8. Since $-(a + b)$ is the additive inverse of $a + b$, $-(a + b) = -a - b$. Likewise $-(3x + 4) =$ ______. — $-3x - 4$

9. We represent the additive inverse of $-2x + 5$ by $-(-2x + 5)$. So

$-(-2x + 5) =$ ______ — $2x - 5$

10. $-(x + 4) =$ ______. — $-x - 4$

11. $-(-x + 4) =$ ______. — $x - 4$

12. $-(x^2 - 5) =$ ______. — $-x^2 + 5$

13. $-(4x^2 + 2x - 8) =$ ______. — $-4x^2 - 2x + 8$

14. $-(-2x^3 - 4x + 6) =$ ______. — $2x^3 + 4x - 6$

15. To subtract, we add the additive inverse of the subtrahend. So

$2x$ subtract $5x$	is the same as	$2x$ add $-5x$	
		(answer) ______	$-3x$

16.

$3x + 5$ subtract $(7x + 8)$	is the same as	$3x + 5$ add $-7x - 8$	
		(answer) ______	$-4x - 3$

17.

$4x^2 - 2$ subtract $(-5x^2 - 4)$	is the same as	$4x^2 - 2$ add $(5x^2 + 4)$	
		(answer) ______	$9x^2 + 2$

18.

$-5x^2 + 2x - 4$ subtract $(2x^2 - 4x + 8)$	is the same as	$-5x^2 + 2x - 4$ add (______)	$-2x^2 + 4x - 8$
		(answer) ______	$-7x^2 + 6x - 12$

19.

$9x^3 - 8$ subtract $(-4x^3 + 6)$	is the same as	$9x^3 - 8$ (subtract\add) (______)	add; $4x^3 - 6$
		(answer) ______	$13x^3 - 14$

20. $\quad -2x^3 + 3x^2 - 4$

subtract $(\ 3x^3 - 4x^2 - 1)$

(answer)______________

$-5x^3 + 7x^2 - 3$

21. $\quad x^3 + 2x^2 - 7$

subtract $(2x^3 - 2x^2 - 7)$

(answer)______________

$-x^3 + 4x^2$

22. Subtract: $\quad -5x^2 - 8$

$\quad 2x^2 + 1$

____

$-7x^2 - 9$

(Remember to change each sign in the subtrahend or second polynomial.)

23. Subtract: $\quad 3x^4 - 2x^2 + 5$

$\quad 2x^4 \qquad\quad - 8$

____

$x^4 - 2x^2 + 13$

24. Subtract: $\quad 5x^3 + 3x^2 + 10$

$\quad -4x^3 + 3x^2 - \ 7$

____

$9x^3 + 17$

25. To subtract, we change the sign of each term in the subtrahend and then (add\subtract).

add

26. change the sign of each term in the subtrahend

$4x - (3x + 2) = 4x + (-3x - 2) =$ ________

change subtract to add

$x - 2$

27. $2x - (-5x + 1) = 2x + ($________$) =$ ________.

$5x - 1$; $7x - 1$

28. $-3x - (4x - 6) = -3x + ($________$) =$ ________.

$-4x + 6$; $-7x + 6$

29. $-4x - (10x - 8) = -4x + ($________$) =$ ________.

$-10x + 8$; $-14x + 8$

30. $25x + 3 - (6x + 8) = 25x + 3 + ($________$) =$ ________.

$-6x - 8$; $19x - 5$

31. $4x - 5 - (-3x + 2) = 4x - 5 + ($________$) =$ ________.

$3x - 2$; $7x - 7$

32. $12x^2 + 3x - (2x^2 + 4x - 8) = 12x^2 + 3x + ($________________$)$

$=$ ________________.

$-2x^2 - 4x + 8$

$10x^2 - x + 8$

33. $x^2 - 5x - (4x^2 + x - 9) =$ ________________.

$-3x^2 - 6x + 9$

34. In $3x^2 + 2x - (4x + 1)$ we (do\do not) change the + sign between $3x^2$ and $2x$ to a − sign. We (do\do not) change the + sign between $4x$ and 1 to minus. We only change a + sign to a − sign if it is in a subtrahend.

do not

do

35. In $4x^2 + 3x - (2x^2 + 8)$, we change the $+$ sign between $2x^2$ and _____ to a $-$ sign. Since $2x^2$ is part of the subtrahend, it changes to $-2x^2$. We (do\do not) change the $+$ sign between $4x^2$ and $(3x)$.

8

do not

36. $4x^2 + 2x - (2x^2 + 8) =$ _______________.

$2x^2 + 2x - 8$

37. We change a $-$ sign to a $+$ sign if the $-$ sign means subtract, or if it is in a subtrahend. In $3x^2 + (-5x) - (4x^2 - 2x)$, we change the $-$ sign between $4x^2$ and $2x$ to a $+$ sign because $4x^2 - 2x$ is a subtrahend. Do we change the $-$ sign in front of the $5x$? (yes\no) The minus sign between $(-5x)$ and $(4x^2 - 2x)$ means subtract. Do we change this $-$ sign to a $+$ sign? (yes\no) Do we change $4x^2$ to $-4x^2$? (yes\no)

no

yes; yes

38. $3x^2 + (-5x) - (4x^2 - 2x) =$ _______.

$-x^2 - 3x$

39. Make the proper changes of sign.

$$2x^2 - (-5x + 3)$$
$$\downarrow \quad \downarrow \quad \downarrow$$
$$2x^2 \_\_ (\_\_5x \_\_ 3)$$

$+; +; -$

40. Make the proper changes of sign.

$$-3x^2 - (\ 3x^2 - 4x + 2)$$
$$\downarrow \quad \downarrow \quad \downarrow \quad \downarrow \quad \downarrow$$
$$\_\_3x^2 \_\_ (\_\_3x^2 \_\_ 4x \_\_ 2)$$

$-; +; -; +; -$

41. Make the proper changes of sign.

$$2x + (-3x + 5) - (-6x + 8)$$

$$2x + (\_\_3x \_\_ 5) \_\_ (\_\_6x \_\_ 8)$$

$-; +; +; +; -$

42. $2x^2 + (-7x) - (-5x + 3) = 2x^2 + (-7x) + (5x - 3)$

$= $ _______________

$2x^2 - 2x - 3$

43. $-3x^2 - (-8x) - (3x^2 - 4x + 2) = -3x^2 + (8x) + (-3x^2 + 4x - 2)$

$= $ _______________

$-6x^2 + 12x - 2$

44. $2x + (-3x + 5) - (-6x + 8) = 2x + (-3x) + 5 + 6x - 8$

$= $ _______

$5x - 3$

45. $5y^4 - 3y^2 - (6y^4 + y^2 - 1) =$ _______________.

$-y^4 - 4y^2 + 1$

46. To take the additive inverse of a polynomial, we change the sign of each term. So $-(5x + 2) =$ _______. But by the distributive law,

$$-1(5x + 2) = -5x - 2$$

Therefore $-(5x + 2)$ (equals\does not equal) $-1(5x + 2)$.

$-5x - 2$

equals

47. $-(-5x^2 + 6x - 4) =$ _______________.

$-1(-5x^2 + 6x - 4) =$ _______________.

So $-(-5x^2 + 6x - 4)$ (equals\does not equal) $-1(-5x^2 + 6x - 4)$.

$5x^2 - 6x + 4$

$5x^2 - 6x + 4$

equals

48. Another way to find the additive inverse of a polynomial is to use the distributive law and multiply the polynomial by -1.

$$-(3x^2 - 5x + 8) = -1(3x^2 - 5x + 8) = \_\_\_\_\_\_\_\_\_\_$$

$-3x^2 + 5x - 8$

$x^2 - 3x + 5$

49. $-(-x^2 + 3x - 5) = -1(-x^2 + 3x - 5) =$ ____________.

$10x^2 + 8x + 6$

50. $-(-10x^2 - 8x - 6) = -1(-10x^2 - 8x - 6) =$ ____________.

(-1); $-7x^2 - 3x - 2$

51. $-(7x^2 + 3x + 2) =$ ____$(7x^2 + 3x + 2) =$ ____________.

52. Now we can view subtraction as the process of adding -1 times the subtrahend.

$$2x^2 - (5x^2 + 6x) = 2x^2 + (-1)(5x^2 + 6x)$$

$(-5x^2) - 6x$

$$= 2x^2 + \text{\_\_\_\_\_\_\_\_}$$

$-3x^2 - 6x$

$$= \text{\_\_\_\_\_\_\_\_}$$

(-1)

53. $8 - (-7x^2 + 3x - 2) = 8 +$ ____$(-7x^2 + 3x - 2)$

$7x^2 - 3x + 2$

$= 8 +$ ____________

$7x^2 - 3x + 10$

$=$ ____________

$-4x^2 + 2x - 7$

54. $x^2 - (-4x^2 + 2x - 7) = x^2 + (-1)($____________$)$

$5x^2 - 2x + 7$

$=$ ____________

12

55. $-3(x^2 + 4) = -3x^2 -$ ____.

56. $x^2 - 3(x^2 + 4) = x^2 + (-3)(x^2 + 4)$

$(-3x^2) - 12$

$= x^2 +$ ________

$-2x^2 - 12$

$=$ ________

-2

57. $8x - 2(x - 5) = 8x + ($____$)(x - 5)$

$(-2x) + 10$

$= 8x +$ ________

$6x + 10$

$=$ ________

$-4x^2 - 8$

58. $-4(x^2 + 2) =$ ________.

$(-4x^2) - 8$

59. $6 - 4(x^2 + 2) = 6 +$ ________

$-4x^2 - 2$

$=$ ________

$5x + 25$

60. $5(x + 5) =$ ________.

$2x - 6$

61. $-2(-x + 3) =$ ________.

25; $-x + 3$

62. $5(x + 5) - 2(-x + 3) = 5x +$ ____ $+ (-2)($________$)$

$7x + 19$

$=$ ________

PRACTICE PROBLEMS

Simplify:

1. $-(3x^2 + 2)$	**2.** $-(-5x^2 + x)$	**3.** $-(7x - 8)$
4. $-(-2x - 9)$	**5.** $-(-5x^2 + 3x - 2)$	**6.** $-(-2x^5 + 4x^2 + 11)$
7. $-(6x^3 - 2x - 5)$	**8.** $-3(4x^2 + x - 8)$	**9.** $-5(3x + 2)$

10. $-1(-4x^2 + x - 8)$ 11. $-6(10x - 7)$ 12. $-2(-x^2 - x)$

13. $-10(3x + 8)$ 14. $-9(-4x + 6)$

Subtract the bottom polynomial from the top one.

15. $\begin{array}{r} 4x^2 \\ -6x^2 \\ \hline \end{array}$ 16. $\begin{array}{r} -8x^2 \\ 9x^2 \\ \hline \end{array}$ 17. $\begin{array}{r} 6x^2 \\ -10x^2 \\ \hline \end{array}$ 18. $\begin{array}{r} 10x^2 \\ 12x^2 \\ \hline \end{array}$

19. $\begin{array}{r} -43x^3 \\ -10x^3 \\ \hline \end{array}$ 20. $\begin{array}{r} -18x^3 \\ 5x^3 \\ \hline \end{array}$ 21. $\begin{array}{r} 9x^2 \\ -9x^2 \\ \hline \end{array}$ 22. $\begin{array}{r} -15x \\ -15x \\ \hline \end{array}$

23. $\begin{array}{r} 25x^3 - 4x^2 + 18x - 50 \\ 5x^3 - 16x^2 - 8x + 30 \\ \hline \end{array}$ 24. $\begin{array}{r} -15x^4 + 3x^3 - 16x + 2 \\ 12x^4 \qquad\quad + 9x - 8 \\ \hline \end{array}$

25. $\begin{array}{r} - 17x^3 - 28x^2 + 15 \\ 3x^4 - 5x^3 - 30x^2 + 20 \\ \hline \end{array}$ 26. $\begin{array}{r} -14x^3 + 3x^2 - 5x - 6 \\ 10x^3 \qquad\quad + 5x - 7 \\ \hline \end{array}$

27. $\begin{array}{r} 10x^5 + 3x^4 - 9x^2 + 8 \\ 4x^5 \qquad\quad - 9x^2 - 3 \\ \hline \end{array}$ 28. $\begin{array}{r} 48x^2 + 33x + 9 \\ -x^3 + 12x^2 - 22x \\ \hline \end{array}$

Simplify and write in standard form:

29. $-5x - (x - 4)$ 30. $-7 - (-x + 8)$

31. $6x - (-5 - x)$ 32. $4x - (2x + x - 1)$

33. $2x + 8 - (3x + 7)$ 34. $5x - 6 - (4x - 9)$

35. $-2x^2 + 4 - (5x^2 - 3x + 7)$ 36. $-x^2 - 3 - (x^2 + 4x - 2)$

37. $4x^2 + (-7) - (-2x^2 - 8x + 5)$ 38. $-x^2 + (-10x) - (-7x^2 + 8x - 6)$

39. $-(6x^2 + 2x - 1) - (-8x + 5)$ 40. $-(-3x^2 - 4x + 6) - (2x^2 + 6x)$

41. $10x^4 - 8x^2 + 20x + (14x^4 + 3x - 8)$

42. $65x^3 + (19x^2 + 10) - (-30x^3 + 9x^2 - 72)$

43. $18x^4 + (35x^3 + 20x^2) - (-10x^3 + 40x^2 + 18)$

44. $-1(4x^5 + 3x^3 - 44) + (15x^6 - 8x^2 + 22)$

45. $-2(y - 3) + 4$ 46. $6y - 5(y^2 + 5)$

47. $3x^2 - 4(-2x^2 - 7)$ 48. $2(4x - 8) - 6(-x^2 + 5)$

49. $-8(-2y - 3) + 4(y + 1)$ 50. $13x^2 - 4x - 10(x^2 - 8)$

51. $-5y^2 + 8 - 4(-2y + 3)$ 52. $7(3x^2 - 2x + 1) - 4(2x^2 + x - 3)$

53. $x(-x^2 + 5x) - 3(x^3 + 4x^2)$ 54. $-(30x + 8) + (40x - 7)$

55. $-2(5 - x) + 8(x + 4)$

SELFTEST

Simplify:

1. $-(-5x + 3)$ 2. $-(7x^2 - 2x + 8)$

3. $-4(2x^2 + 8x - 2)$ 4. $3x^2 - (2x^2 + x - 7)$

5. $-4x^2 + 3 - (-6x^2 + 1)$

6. $x^2 + (5x - 1) - (2x + 1)$

7. $-4(x^2 + 2x - 1) + 3(x - 5)$

8. $10(x^2 + 8) - 5(-x^2 + 4x - 10)$

Subtract:

9.
$$\begin{array}{r} 3x^2 + 8x - 4 \\ -5x^2 + 7x - 4 \\ \hline \end{array}$$

10.
$$\begin{array}{rrrr} 10x^3 & & -6x & +5 \\ 3x^3 & +2x^2 & & -4 \\ \hline \end{array}$$

SELFTEST ANSWERS

1. $5x - 3$ 2. $-7x^2 + 2x - 8$ 3. $-8x^2 - 32x + 8$ 4. $x^2 - x + 7$
5. $2x^2 + 2$ 6. $x^2 + 3x - 2$ 7. $-4x^2 - 5x - 11$ 8. $15x^2 - 20x + 130$
9. $8x^2 + x$ 10. $7x^3 - 2x^2 - 6x + 9$

3.5 PRODUCT OF POLYNOMIALS

The product of two polynomials is again a polynomial. The distributive law allows us to multiply a polynomial by a monomial. We multiply each term of the polynomial by the monomial and add the results.

Example 1 $4x^2(2x^2 + 3x + 5) = 8x^4 + 12x^3 + 20x^2$

To multiply two polynomials together, we multiply one polynomial by each term of the other and add the results. Column notation aids in keeping track of the terms.

Example 2 Multiply $5x + 4$ by $3x + 2$.

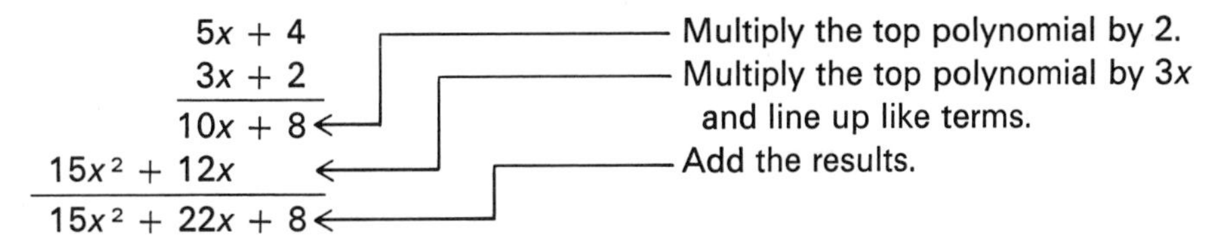

5x + 4
3x + 2
10x + 8
15x² + 12x
15x² + 22x + 8
Multiply the top polynomial by 2.
Multiply the top polynomial by 3x and line up like terms.
Add the results.

Example 3 Multiply $6x^3 - 2x + 4$ by $-3x^2 + x$.

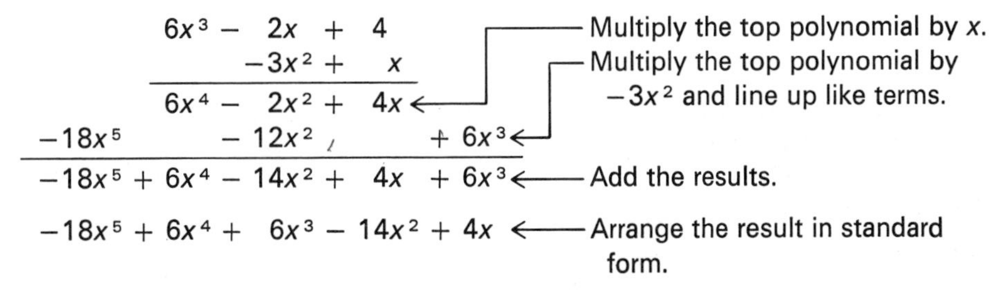

We often use a horizontal scheme also based on the distributive law to multiply binomials. The distributive law says

$$q(r + s) = qr + qs$$

If we let q represent $(a + b)$, r represent c, and s represent d, we discover that

$$(a + b)(c + d) = (a + b)c + (a + b)d$$

Two more applications of the distributive law yield

$(a + b)c + (a + b)d = ac + bc + ad + bd$ So $(a + b)(c + d) = ac + bc + ad + bd$

Again, as in the vertical scheme, we multiply the first polynomial by each term of the second polynomial and add the results. The diagram indicates the multiplications to be done. The numbers *suggest* an order for the multiplications, but we can do the multiplications in any order as long as we do all four of them.

$$(a + b)(c + d) = ac + bc + ad + bd$$

Example 4

$$\begin{aligned}(x + 2)(x + 3) &= x^2 + 2x + 3x + 6\\ &= x^2 + 5x + 6\end{aligned}$$

Example 5

$$\begin{aligned}(y - 4)^2 = (y - 4)(y - 4) &= y^2 - 4y - 4y + 16\\ &= y^2 - 8y + 16\end{aligned}$$

Example 6

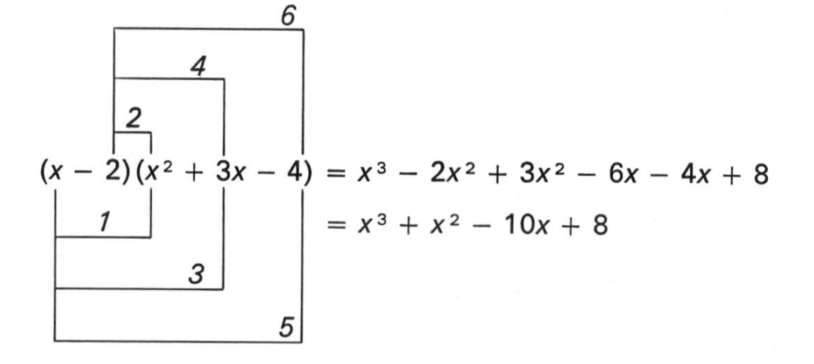

$$\begin{aligned}(x - 2)(x^2 + 3x - 4) &= x^3 - 2x^2 + 3x^2 - 6x - 4x + 8\\ &= x^3 + x^2 - 10x + 8\end{aligned}$$

Example 7

$$\begin{aligned}(2x - 3)^3 &= (2x - 3)(2x - 3)(2x - 3)\\ &= (2x - 3)(4x^2 - 6x - 6x + 9)\\ &= (2x - 3)(4x^2 - 12x + 9)\\ &= 8x^3 - 12x^2 - 24x^2 + 36x + 18x - 27\\ &= 8x^3 - 36x^2 + 54x - 27\end{aligned}$$

PROGRAM 3.5

1. By the distributive law, $3x(4x^2 + 8) =$ ________.

   $12x^3 + 24x$

2. The (distributive\transitive) law enables us to multiply a polynomial by a monomial.

   distributive

3. We multiply each term of the polynomial by the monomial and add or subtract the results.

   $5x^2(3x^3 - 4x^2 - x + 2) =$ ________________

   $15x^5 - 20x^4 - 5x^3 + 10x^2$

   $-2x(3x - 4x^2 - x + 2) =$ ________________

   $-6x^2 + 8x^3 + 2x^2 - 4x$

4. To multiply two polynomials together, we multiply the first one by each term of the second one and add the results. To multiply $2x + 4$ by $x + 3$ we

1. Multiply the first polynomial by 3.
2. Multiply the first polynomial by x.
3. Add the results.

$$\begin{array}{rrrl} & 2x + & 4 & \\ & x + & 3 & \\ \hline & 6x + & 12 & \text{(step 1)} \\ 2x^2 + & 4x & & \text{(step 2)} \\ \hline \_\_x^2 + & \_\_x + & \_\_ & \text{(step 3)} \end{array}$$

2; 10; 12

5. Multiply:

$$\begin{array}{r} 3x + 2 \\ 2x + 4 \\ \hline \end{array}$$

________ 1. Multiply the top polynomial by 4.

________ 2. Multiply the top polynomial by $2x$ and line up like terms.

________ 3. Add the results.

$12x + 8$

$6x^2 + 4x$

$6x^2 + 16x + 8$

6. Multiply:

$$\begin{array}{r} 5x - 3 \\ x^2 + 2x \\ \hline \end{array}$$

________ 1. Multiply the top polynomial by ____.

________ 2. Multiply the top polynomial by x^2.

________ 3. Add the results.

$10x^2 - 6x$; $2x$

$5x^3 - 3x^2$

$5x^3 + 7x^2 - 6x$

7. Multiply:

$$\begin{array}{r} 2x^2 + 3x + 3 \\ x - 1 \\ \hline \end{array}$$

____

____

________ (Add the results)

________ (Arrange the results)

$-2x^2 - 3x - 3$

$2x^3 + 3x^2 + 3x$

$2x^3 + x^2 + 0x - 3$

$2x^3 + x^2 - 3$

8. Multiply:

$$\begin{array}{l} x + 1 \\ x^2 + 3x - 1 \\ \hline \end{array}$$

____

____

____

____

$-x - 1$

$3x^2 + 3x$

$x^3 + x^2$

$x^3 + 4x^2 + 2x - 1$

9. Multiply $(-x^3 - 4x^2 + 2)(x - 1)$:

$$\begin{array}{r} -x^3 - 4x^2 + 2 \\ x \quad - 1 \\ \hline x^3 + 4x^2 - 2 \quad\quad \\ -x^4 - 4x^3 \quad\quad\quad\quad + 2x \\ \hline -x^4 - 3x^3 + 4x^2 - 2 + 2x \\ -x^4 - 3x^3 + 4x^2 + 2x - 2 \end{array}$$

10. When we multiply two binomials together, we can use a horizontal notational scheme instead of the previous vertical scheme. The scheme follows from three applications of the distributive law. By the first application,

$$(a + b)(c + d) = (a + b)c + (a + b)d$$

By two more applications of the distributive law,

$(a + b)c + (a + b)d =$ ____ $+$ ____ $+$ ____ $+$ ____ — ac; bc; ad; bd

11. We just derived the following scheme

(brackets: 1 joins a and c; 2 joins b and c; 3 joins a and d; 4 joins b and d)

$(a + b)(c + d) =$ ____ $+$ ____ $+$ ____ $+$ ____ (1, 2, 3, 4) — ac; bc; ad; bd

(Do the multiplications in the given order.)

12. Multiply:

(brackets 1, 2, 3, 4 as above)

$(x + 2)(x + 1) =$ ____ $+$ ____ $+$ ____ $+$ ____ — x^2; $2x$; x; 2

$=$ ____ $+$ ____ $+$ ____ — x^2; $3x$; 2

13. Multiply: $(y + 3)(y - 2) =$ ____ $+$ ____ $-$ ____ $-$ ____ — y^2; $3y$; $2y$; 6

$=$ ____ $+$ ____ $-$ ____ — y^2; y; 6

14. Multiply: $(x - 5)(x - 1) =$ ________________ — $x^2 - 5x - x + 5$

$=$ ________________ — $x^2 - 6x + 5$

15. Multiply: $(x + 3)^2 = (x + 3)(x + 3) =$ ________________. — $x^2 + 6x + 9$

16. Multiply: $(y - 1)^2 =$ ________________. — $y^2 - 2y + 1$

17. Multiply: $(-2x + 4)^2 =$ ________________. — $4x^2 - 16x + 16$

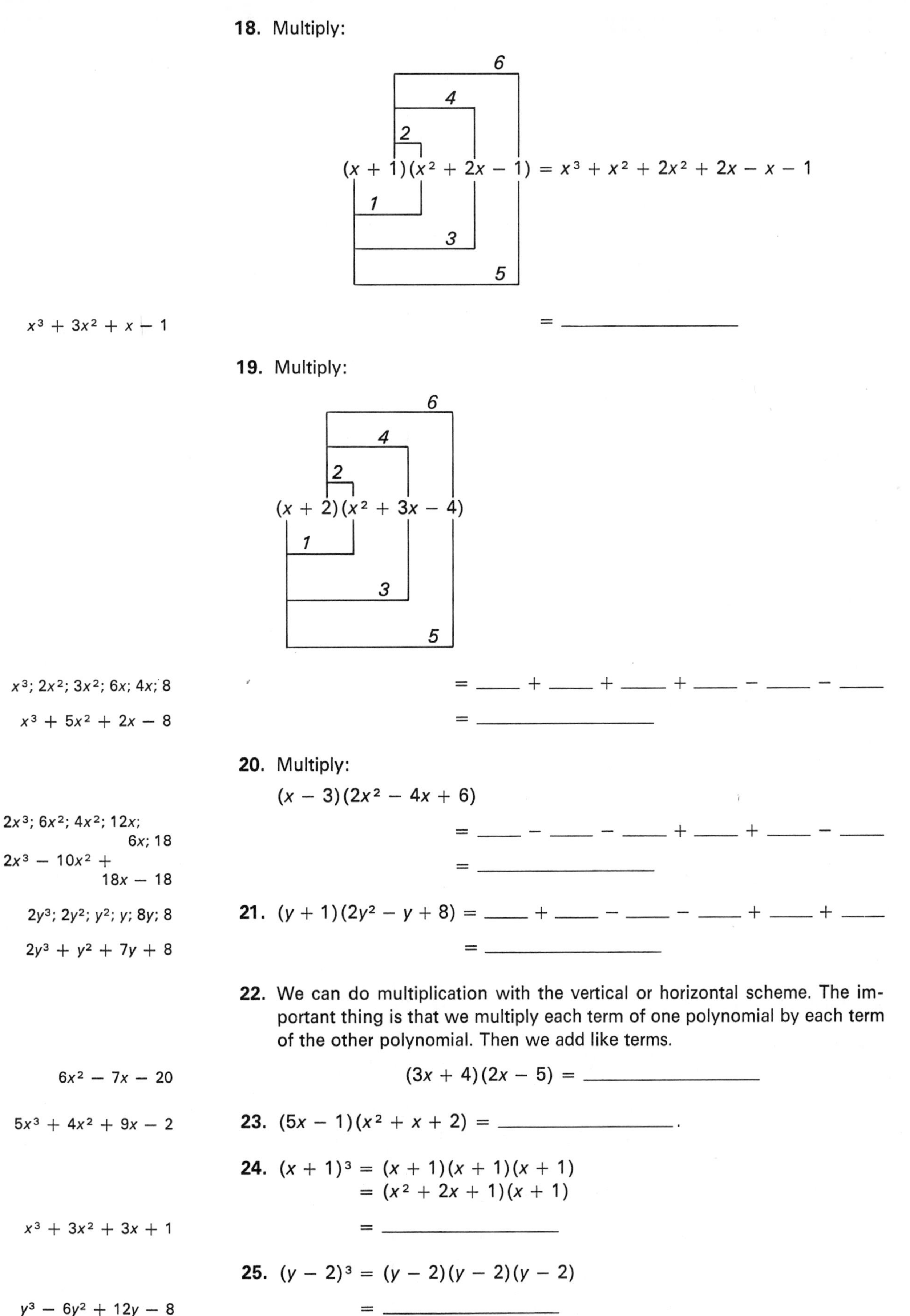

18. Multiply:

$$(x + 1)(x^2 + 2x - 1) = x^3 + x^2 + 2x^2 + 2x - x - 1$$

$= \_\_\_\_\_\_\_\_\_\_$

$x^3 + 3x^2 + x - 1$

19. Multiply:

$$(x + 2)(x^2 + 3x - 4)$$

$= \_\_\_ + \_\_\_ + \_\_\_ + \_\_\_ - \_\_\_ - \_\_\_$

x^3; $2x^2$; $3x^2$; $6x$; $4x$; 8

$= \_\_\_\_\_\_\_\_\_\_$

$x^3 + 5x^2 + 2x - 8$

20. Multiply:

$(x - 3)(2x^2 - 4x + 6)$

$= \_\_\_ - \_\_\_ - \_\_\_ + \_\_\_ + \_\_\_ - \_\_\_$

$2x^3$; $6x^2$; $4x^2$; $12x$; $6x$; 18

$= \_\_\_\_\_\_\_\_\_\_$

$2x^3 - 10x^2 + 18x - 18$

21. $(y + 1)(2y^2 - y + 8) = \_\_\_ + \_\_\_ - \_\_\_ - \_\_\_ + \_\_\_ + \_\_\_$

$2y^3$; $2y^2$; y^2; y; $8y$; 8

$= \_\_\_\_\_\_\_\_\_\_$

$2y^3 + y^2 + 7y + 8$

22. We can do multiplication with the vertical or horizontal scheme. The important thing is that we multiply each term of one polynomial by each term of the other polynomial. Then we add like terms.

$(3x + 4)(2x - 5) = \_\_\_\_\_\_\_\_\_\_$

$6x^2 - 7x - 20$

23. $(5x - 1)(x^2 + x + 2) = \_\_\_\_\_\_\_\_\_\_.$

$5x^3 + 4x^2 + 9x - 2$

24. $(x + 1)^3 = (x + 1)(x + 1)(x + 1)$

$= (x^2 + 2x + 1)(x + 1)$

$= \_\_\_\_\_\_\_\_\_\_$

$x^3 + 3x^2 + 3x + 1$

25. $(y - 2)^3 = (y - 2)(y - 2)(y - 2)$

$= \_\_\_\_\_\_\_\_\_\_$

$y^3 - 6y^2 + 12y - 8$

PRACTICE PROBLEMS

Multiply and simplify:

1. $3x(4x^2 - 2x + 2)$
2. $2x^2y(3x^3 - 5)$
3. $-4x(8x^2 - 3x + 6)$
4. $(x^2 + 5x - 2)(-10x)$
5. $(3x^2 + 2xy - 4y^2)(2xy)$
6. $-xy(8y^2 - 20xy + 5x^2)$
7. $(3x + 2)(2x^2 + 4x + 3)$
8. $(5y + 1)(6y^2 + 8y + 7)$
9. $(x - 5)(4x^2 + 3x + 9)$
10. $(2x - 3)(x^2 + 4x + 8)$
11. $(4x - 6)(x^2 - 2x - 4)$
12. $(-x + 4)(2x^2 - x - 1)$
13. $(2x + 1)(x + 4)$
14. $(x + 5)(3x + 2)$
15. $(x - 3)(x + 2)$
16. $(x - 5)(x + 10)$
17. $(x - 7)(4x - 3)$
18. $(6x - 9)(x - 2)$
19. $(2x + y)(3x - y)$
20. $(4x + 2y)(x - 5y)$
21. $(x^2 + y)(x^2 - y)$
22. $(x - 3y)(x^2 + 2y)$
23. $(x - 20)^2$
24. $(x + 2)^2$
25. $(3x + 1)^2$
26. $(5x - 4)^2$
27. $(2y + 1)^3$
28. $(x - 2)^3$

SELFTEST

1. The ________________ property enables us to multiply polynomials together.

Multiply and simplify:

2. $2x(-4x^2 + 3x - 1)$
3. $(5x - 2)(x^2 + 3x - 10)$
4. $(2x + 5)(3x - 4)$
5. $(7x - 2)^2$
6. $(y + 1)^3$

SELFTEST ANSWERS

1. distributive 2. $-8x^3 + 6x^2 - 2x$ 3. $5x^3 + 13x^2 - 56x + 20$
4. $6x^2 + 7x - 20$ 5. $49x^2 - 28x + 4$ 6. $y^3 + 3y^2 + 3y + 1$

3.6 DIVISION OF POLYNOMIALS

In this section we will divide one polynomial by another. The division pattern is very reminiscent of long division of whole numbers. We use the following steps.

1. Arrange the polynomials in standard form.
2. Divide the first term of the divisor into the first term of the dividend. The result is a term in the quotient.
3. Multiply the most recent entry in the quotient by the entire divisor.
4. Subtract that result from the dividend.
5. If the remainder has lower degree than the divisor, stop. If not, repeat the process from Step 2 and divide the divisor into the remainder.

Example 1 Divide $x^2 - 48 - 2x$ by $6 + x$. First we arrange the polynomials in standard form.

$$
\begin{array}{lrl}
 & x - 8 & \\
x + 6 & \overline{)\, x^2 - 2x - 48} & \\
 & \underline{x^2 + 6x} & \longleftarrow \text{Step 3. Multiply the divisor by } x. \\
 & -8x - 48 & \longleftarrow \text{Step 4. Subtract.} \\
\text{Step 3.} & \underline{-8x - 48} & \\
\text{Step 4.} & 0 + 0 &
\end{array}
$$

By step 2, the first term in the quotient is x.

Step 3. Multiply the divisor by x.

Step 4. Subtract.

Step 5. The degree of the remainder is not lower than that of the divisor. So, repeating Step 2, we divide the divisor into the remainder. The second term in the quotient is -8.

Example 2 Quotients of polynomials are similar to quotients of integers. Consider $x^2 + 6x + 9$ divided by $x + 3$.

$$
\begin{array}{rl}
 & x + 3 \\
x + 3 & \overline{)\, x^2 + 6x + 9} \\
 & \underline{x^2 + 3x} \\
 & 3x + 9 \\
 & \underline{3x + 9} \\
 & 0 \text{ remainder}
\end{array}
$$

If we let $x = 10$ we get

$x + 3 = 10 + 3 = 13$ and $x^2 + 6x + 9 = 10^2 + 6 \cdot 10 + 9 = 169$

Comparing the calculations with $x = 10$ we get

$$
\begin{array}{rl}
 & x + 3 \\
x + 3 & \overline{)\, x^2 + 6x + 9} \\
 & \underline{x^2 + 3x} \\
 & 3x + 9 \\
 & \underline{3x + 9} \\
 & 0
\end{array}
,\qquad
\begin{array}{rl}
 & 10 + 3 \\
10 + 3 & \overline{)\, 10^2 + 6 \cdot 10 + 9} \\
 & \underline{10^2 + 3 \cdot 10} \\
 & 3 \cdot 10 + 9 \\
 & \underline{3 \cdot 10 + 9} \\
 & 0
\end{array}
,\qquad \text{or} \qquad
\begin{array}{rr}
 & 13 \\
13 & \overline{)\, 169} \\
 & \underline{13} \\
 & 39 \\
 & \underline{39} \\
 & 0
\end{array}
$$

If a polynomial has degree n, but is missing some terms of degree less than n, we use placeholders for the missing terms. The polynomial $6x^5 + 3x^2 + 8$ is missing the x^4, x^3, and x terms. So we use placeholders to write the polynomial as $6x^5 + 0x^4 + 0x^3 + 3x^2 + 0x + 8$. The use of placeholders makes the division process work more smoothly.

Example 3 Divide $5x + 4x^3 + 3$ by $x^2 - x$. We arrange the polynomials in standard form and use placeholders in the dividend.

$$
\begin{array}{rl}
 & 4x + 4 \text{ with remainder } 9x + 3 \\
x^2 - x & \overline{)\, 4x^3 + 0x^2 + 5x + 3} \\
 & \underline{4x^3 - 4x^2} \\
 & 4x^2 + 5x + 3 \\
 & \underline{4x^2 - 4x} \\
 & 9x + 3
\end{array}
$$

(The remainder has degree lower than the divisor, so we stop the division process.)

To check our division, we simply multiply the quotient by the divisor and add the remainder. This result should equal the dividend.

Example 4 Check the division in Example 3.

$$
\begin{aligned}
(x^2 - x)(4x + 4) + 9x + 3 &= 4x^3 - 4x + 9x + 3 \\
&= 4x^3 + 5x + 3
\end{aligned}
$$

which is the dividend of Example 3.

PROGRAM 3.6

1. Polynomial division is very much like long division of numbers. But we must first arrange the polynomials in standard form before we do the division. Are the polynomials in $(4x + 2x^2) \div (x + 2)$ in standard form? (yes\no)

   no

   If not, rewrite the problem. ________ $\div$ ________

   $(2x^2 + 4x)$; $(x + 2)$

2. We then use long division notation.

   $(2x^2 + 4x) \div (x + 2)$ is the same as ________ $\overline{)\quad\quad}$

   $x + 2 \overline{)2x^2 + 4x}$

3. To obtain the first term of the quotient, we divide the first term of the divisor into the first term of the dividend. In the answer to Frame 2, therefore, we would divide ____ into ____, and place the result here.

   $x + 2 \overline{)2x^2 + 4x}$

   x; $2x^2$

   $2x$

4. Our next step is to multiply the entire divisor by the most recent entry in the quotient. We multiply ________ by ____ and place the result here.

   $x + 2$; $2x$

   $$\begin{array}{r} 2x \\ x + 2 \overline{)2x^2 + 4x} \\ \underline{\quad\quad\quad} \end{array}$$

   $2x^2 + 4x$

5. As in long division, our last step is to subtract the result of the previous step from the dividend.

   $$\begin{array}{r} 2x \\ x + 2 \overline{)2x^2 + 4x} \\ \underline{2x^2 + 4x} \\ \underline{\quad\quad\quad} \end{array}$$ Subtract

   $0 + 0$

   The last step gives us a *remainder*. The remainder in this example is ____.

   0

6. Divide $8x + 2x^2$ by $x + 4$. First we use the long division notation with the polynomials in standard form.

   ____ ← Divide the first term of the dividend by the first term of the divisor.

   $2x$

   $x + 4 \overline{)2x^2 + 8x}$

   ________ ← Multiply the entire divisor by the first term in the quotient.

   $2x^2 + 8x$

   ________ ← Subtract to find the remainder.

   $0 \; + 0$

7. Divide $5x^2 + 25x$ by $5 + x$: The divisor is ________. So the long division notation with the polynomials in standard form is ________________.

   $x + 5$

   $x + 5 \overline{)5x^2 + 25x}$

   First, divide ____ into $5x^2$, with the result ________. Then multiply ________ by $5x$, so that

   x; $\overline{)\;\;}$ with $5x$ above

   $x + 5$

   $$\begin{array}{r} 5x \\ x + 5 \overline{)5x^2 + 25x} \\ \underline{\quad\quad\quad} \end{array}$$

   $5x^2 + 25x$

   ________ ← Subtract to find the remainder

   $0 + 0$

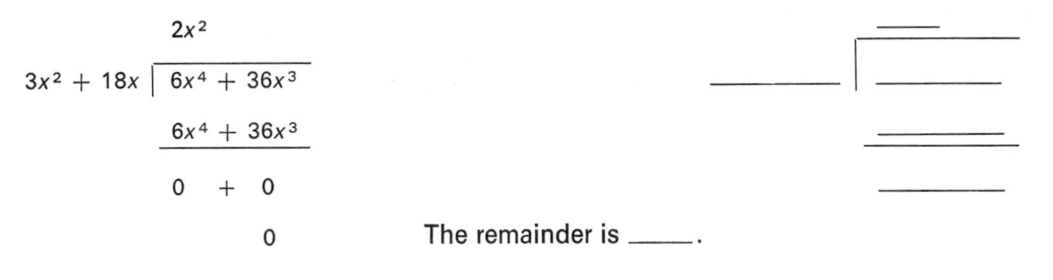

8. Divide $6x^4 + 36x^3$ by $18x + 3x^2$.

$$
\begin{array}{r|l}
 & 2x^2 \\
3x^2 + 18x & 6x^4 + 36x^3 \\
 & 6x^4 + 36x^3 \\ \hline
 & 0 \;+\; 0 \\
 & 0
\end{array}
$$

The remainder is ____.

9. Divide $2x^3 - 4x^2$ by $2x - 4$.

$$
\begin{array}{r|l}
 & x^2 \\
2x - 4 & 2x^3 - 4x^2 \\
 & 2x^3 - 4x^2 \\ \hline
 & 0
\end{array}
$$

yes

10. If the remainder has degree equal to or higher than the divisor, then we repeat the division process. Only this time we divide the divisor into the remainder. Suppose the divisor is $2x + 1$ and the remainder is $4x^2 + 2x + 8$. Do we divide the divisor into the remainder? (yes\no)

11. Complete the division:

$$
\begin{array}{r|l}
 & x + \_\_\_\_ \\
x + 3 & x^2 + 5x + 6 \\
 & x^2 + 3x \\ \hline
 & \quad 2x + 6 \\
 & \quad \_\_\_\_ \\ \hline
 & \quad \_\_\_\_
\end{array}
$$

2

$$
\begin{array}{l}
2x + 6 \\ \hline
0 + 0
\end{array}
$$

Is the degree of the remainder higher than or equal to that of the divisor? (yes\no)

yes

12. Divide $x + 5$ into $x^2 + 3x - 10$. (Watch the sign of the second term in the quotient.)

$$
\begin{array}{r|l}
 & x \;-\; 2 \\
x + 5 & x^2 + 3x - 10 \\
 & x^2 + 5x \\ \hline
 & \quad - 2x - 10 \\
 & \quad - 2x - 10 \\ \hline
 & \quad\; 0 \;+\; 0
\end{array}
$$

13. Divide $3x + 8$ into $3x^2 + 5x - 12$ and show the remainder.

$$
\begin{array}{r|l}
 & x \;-\; 1 \text{ remainder } -4 \\
3x + 8 & 3x^2 + 5x - 12 \\
 & 3x^2 + 8x \\ \hline
 & \quad - 3x - 12 \\
 & \quad - 3x - \;\;8 \\ \hline
 & \qquad\quad - \;\;4
\end{array}
$$

14. The polynomial $10x^5 + 5x^2 + 2x - 1$ does not have terms of degree ____ and ____. In such cases, we use placeholders for the missing terms. We write the polynomial as

4

3

$$10x^5 + 0x^4 + \_\_\_\_ + 5x^2 + 2x - 1$$

$0x^3$

15. The use of placeholders helps us to line up like terms in the division process. When we divide $x^2 - 4$ by $x - 2$, we use a placeholder.

$$x - 2 \overline{) x^2 + \_\_\_\_ - 4}$$

Complete the division.

$0x$

$$\begin{array}{r} x + 2 \\ x - 2 \overline{) x^2 + 0x - 4} \\ \underline{x^2 - 2x} \\ 2x - 4 \\ \underline{2x - 4} \\ 0 + 0 \end{array}$$

16. Divide $6x^3 - 3x^2 + 40$ by $3x + 6$.

$$\begin{array}{r} 2x^2 - 5x + 10 \\ 3x + 6 \overline{) 6x^3 - 3x^2 + 0x + 40} \\ \underline{6x^3 + 12x^2} \\ -15x^2 + 0x + 40 \\ \underline{-15x^2 - 30x} \\ 30x + 40 \\ \underline{30x + 60} \\ \text{remainder} \quad -20 \end{array}$$

17. To check division, we multiply the quotient by the divisor and add the remainder. The result should equal the dividend. Divide $x^2 - 13$ by $x + 4$.

$$\begin{array}{r} x - 4 \\ x + 4 \overline{) x^2 + 0x - 13} \\ \underline{x^2 + 4x} \\ -4x - 13 \\ \underline{-4x - 16} \\ \text{remainder} \quad 3 \end{array}$$

Check your results.

$$(x - 4)(x + 4) + 3 = x^2 - 16 + 3 = x^2 - 13$$

18. With these techniques we can divide any two polynomials as long as the degree of the divisor is (larger than or equal to\smaller than or equal to) the degree of the dividend.

smaller than or equal to

PRACTICE PROBLEMS

Divide and check your results.

1. $(5x + x^2 + 6) \div (x + 2)$
2. $(x^2 - 8 - 2x) \div (2 + x)$
3. $(2x^2 - 20 + 8x) \div (x + 4)$
4. $(16x^2 + 2x - 3) \div (8x - 3)$
5. $(x^3 + 3x - x^2 - 3) \div (x - 1)$
6. $(x^2 - 36) \div (x - 6)$
7. $(4x^3 - 15x^2 + 4x - 32) \div (x - 4)$
8. $(8x^3 + 16x^2 - 2x - 22) \div (2x + 4)$
9. $(8x^3 + 16) \div (2x + 4)$
10. $(12x^4 - 2x^3 - 10) \div (2x - 2)$
11. $(2x^6 - x^5 + 2x^3 - x^2 - 2x + 1) \div (2x - 1)$

SELFTEST

Divide and check your results.

1. $(3x^2 - 12 + 16x) \div (x + 6)$
2. $(15x^4 + 3x + 10x^3 + 2) \div (3x + 2)$
3. $(6x^4 + 13x^3 - 8x^2 - 7) \div (2x - 1)$

SELFTEST ANSWERS

1. $3x - 2$ 2. $5x^3 + 1$ 3. $3x^3 + 8x^2$, with remainder -7

POSTTEST 3

1. *a.* What is the base of 16^4? What is the exponent?
   b. What is the base of $-x^3$? What is the exponent?
   c. What does x^0 equal?

2. *a.* Write *xxxyy* in exponential form.
   b. Write x^5y^2 in factored form without exponents.

3. Simplify:

   a. $\dfrac{6x^4y^2}{2x^3y}$

   b. $(3x^2yz)(2xy)$
   c. $2 + (3^3 \div 9)$
   d. $5^2 - 1$

4. Given the polynomial $3 - 2x + x^2$:
   a. Write the polynomial in standard form.
   b. What is the degree of the polynomial?
   c. What is the coefficient of x and what is the constant term?
   d. Evaluate the polynomial when $x = 2$.

5. Simplify:
   a. $4x^2 + 6x + 2x^2 - 1$
   b. $2x + 4(5x + 2)$
   c. $5(x + 1) + 3(x^2 + x)$

6. Simplify:
   a. $10x^2 - 2(5x + 1)$
   b. $4(x - 1) - (6x + 2)$
   c. $-x^2 - 5 - (x^2 - 3x + 1)$

7. Multiply and simplify:
   a. $2x(x^2 - 3x)$
   b. $(x + 1)(3x - 2)$
   c. $(x^2 + x - 1)(x + 3)$

8. Divide and check your answer: $(x^2 + 2x + 1) \div (x + 1)$

Each problem above refers to a section in this chapter as shown in the table.

Problems	*Section*
1–3	3.1
4	3.2
5	3.3
6	3.4
7	3.5
8	3.6

POSTTEST 3 ANSWERS

1. *a.* 16; 4 *b.* x; 3 *c.* $x^0 = 1$ if $x \neq 0$ **2.** *a.* x^3y^2 *b.* $xxxxxyy$

3. *a.* $3xy$ *b.* $6x^3y^2z$ *c.* 5 *d.* 24 **4.** *a.* $x^2 - 2x + 3$ *b.* 2 *c.* -2; 3 *d.* 3

5. *a.* $6x^2 + 6x - 1$ *b.* $22x + 8$ *c.* $3x^2 + 8x + 5$

6. *a.* $10x^2 - 10x - 2$ *b.* $-2x - 6$ *c.* $-2x^2 + 3x - 6$

7. *a.* $2x^3 - 6x^2$ *b.* $3x^2 + x - 2$ *c.* $x^3 + 4x^2 + 2x - 3$

8. $x + 1$

4

FACTORING POLYNOMIALS

BASIC SKILLS

Upon completion of the indicated section the student should be able to:

4.1 *a.* Use the distributive law to factor out the greatest monomial factor common to all terms in a polynomial. (Factor out the greatest common factor.)

b. Factor out -1 from a polynomial.

4.2 *a.* Recognize polynomials (up to the fourth degree) that are perfect squares or the difference of perfect squares.

b. Factor polynomials that are either perfect squares or the difference of perfect squares by using the formulas

$$a^2 + 2ab + b^2 = (a + b)^2$$
$$a^2 - 2ab + b^2 = (a - b)^2$$
$$a^2 - b^2 = (a + b)(a - b)$$

4.3 Factor completely many polynomials of the form

$$x^2 + bx + c \qquad (b \text{ and } c \text{ integers})$$

4.4 Factor completely many polynomials of the form

$$ax^2 + bx + c \qquad (a, b, \text{ and } c \text{ integers})$$

4.5 Find the least common multiple of two or three polynomials.

PRETEST 4

1. Factor out the largest possible monomial factor.

a. $8x^2y^2 + 4xy$

b. $45x^2 + 15x - 30$

c. $14x^2y - 28x^2y + 14xy^2$

2. Factor completely:

a. $x^2 - 36$

b. $64x^2 - 49$

c. $25x^2 + 20x + 4$

3. Factor completely:

a. $x^2 + 2x - 15$

b. $x^2 - 4x + 3$

c. $x^2 - 9x + 8$

4. Factor completely:

a. $9x^2 - x - 8$

b. $3x^2 + 28x - 20$

c. $2x^2 + 9x - 5$

5. Find the least common multiple.

a. $2x$, $x + 5$

b. $2x + 1$, $4x^2 + 4x + 1$

c. x, $10x^2$, $8x^2 - x$

Each problem above refers to a section in this chapter as shown in the table.

Problems	*Section*
1	4.1
2	4.2
3	4.3
4	4.4
5	4.5

PRETEST 4 ANSWERS

1. *a.* $4xy(2xy + 1)$ *b.* $15(3x^2 + x - 2)$ *c.* $14xy(y - x)$
2. *a.* $(x + 6)(x - 6)$ *b.* $(8x + 7)(8x - 7)$ *c.* $(5x + 2)(5x + 2)$
3. *a.* $(x + 5)(x - 3)$ *b.* $(x - 3)(x - 1)$ *c.* $(x - 1)(x - 8)$
4. *a.* $(9x + 8)(x - 1)$ *b.* $(3x - 2)(x + 10)$ *c.* $(2x - 1)(x + 5)$
5. *a.* $2x(x + 5)$ *b.* $4x^2 + 4x + 1$ *c.* $10x^2(8x - 1)$

4.1 COMMON MONOMIAL FACTORS

To factor a polynomial means to write it as a product of other polynomials. In this chapter we will study techniques of factoring polynomials with *integer* coefficients into other polynomials with *integer* coefficients. Just as we have prime numbers which factor no further with whole-number factors, we also have *prime polynomials.* These are polynomials which admit no further factorization into polynomials with integer coefficients. Our goal will be to factor certain polynomials *completely* (i.e., into a product of prime polynomials with integer coefficients).

But for convenience we will not factor integer factors into primes. So we will consider $8(x + 1)$ as the complete factorization of $8x + 8$.

We have already used the distributive law to partially factor some polynomials. For if each term of a polynomial has a common factor, we can factor out that common factor. We always want to factor out the largest possible factor common to all terms.

Distributive Law: $\boldsymbol{ab + ac = a(b + c)}$.

Example 1 $$5x^2 + 20 = 5(x^2 + 4)$$

Example 2 $$24x^2 - 16x = 8x(3x - 2)$$

Example 3 $$25x^3 + 50x^2 - 25x + 75 = 25(x^3 + 2x^2 - x + 3)$$

In each of the above examples, we factored a polynomial into a product of a *monomial* and a polynomial. Factoring out a monomial is one of the easier factoring techniques. We generally attempt it first since we can usually do it by inspection.

Again, we always wish to factor out the largest possible monomial factor. Usually a second look at the remaining polynomial will reveal any additional common factors that we might have missed the first time. If we discover that we did miss some common factors, we simply continue factoring until all the polynomial factors are prime.

Example 4 $$24x^3 - 36x^2 + 72x = 6(4x^3 - 6x^2 + 12x)$$

Each term of the second polynomial still has a factor of $2x$.

$$\begin{aligned} 6(4x^3 - 6x^2 + 12x) &= 6(2x)(2x^2 - 3x + 6) \\ &= 12x(2x^2 - 3x + 6) \end{aligned}$$

We can change the sign of each term in a polynomial by factoring out -1 or any negative value.

Example 5 $$3x - 2 = -1(-3x + 2)$$

Example 6 $$-5x^2 + 15x = -5x(x - 3)$$

The commutative law applies to polynomial multiplication so the order of the polynomial factors does not matter. Usually we write monomial factors first.

Complete factorization is unique except for orders of factors and changes of sign caused by negative factors.

Example 7 The complete factorization of $2x^2 - 10x$ is given by any one of the following:

a. $2x(x - 5)$ *b.* $(x - 5)2x$ *c.* $-2x(-x + 5)$ *d.* $(-x + 5)(-2x)$

PROGRAM 4.1

1. Recall that a counting number $p \neq 1$ is a prime number if the only counting number factors are 1 and p itself. The first three prime numbers are 2, 3, 5.

   The next five are ____, ____, ____, ____, ____.

   7, 11, 13, 17, 19

2. A polynomial is *prime* if it cannot be written as a product of other polynomials with *integer* coefficients. Since $x + 2$ cannot be written as a product of other polynomials with integer coefficients, $x + 2$ is ______.

   prime

   $$4x^2 + 2x = 2x(2x + 1)$$

   so $4x^2 + 2x$ (is\is not) prime.

   is not

3. The process of writing a polynomial as a product of other polynomials is called *factoring*. To *factor completely* means to factor into a product of prime polynomials. Is $3(x + 4)$ the complete factorization of $3x + 12$? (yes\no). Is $2(4x + 6)$ the complete factorization of $8x + 12$? (yes\no).

   yes; no [the factor $4x + 6$ factors as $2(2x + 3)$]

4. For convenience, we will not factor integer factors into primes. So we will consider $4(x + 1)$ as the complete factorization of $4x + 4$ even though 4 factors as $2 \cdot 2$. Will we consider $10(x + 3)$ as the complete factorization of $10x + 30$? (yes\no).

   yes

5. A simple application of the distributive law gives us one factoring technique. The distributive law says

   $$ab + ac = a(b + c)$$

   By the distributive law, $4x + 3x = x($____ + ____$)$.

   4; 3

6. A factor that is common to all terms of a given polynomial is called a *common factor*. In $3x + 9$, ____ is a common factor since it is a factor of $3x$ and also of 9.

   3

7. Consider the polynomial $8x^2 + 6x - 10xy$. Is $2x$ a factor of $8x^2$? (yes\no). Is $2x$ a factor of $6x$? (yes\no). Is $2x$ a factor of $10xy$? (yes\no). So $2x$ (is\is not) a common factor in $8x^2 + 6x - 10xy$.

   yes
   yes; yes
   is

8. Is 3 a common factor of $12x^2 - 9x + 14$? (Yes\No) because 3 is not a factor of ____.

   No
   14

9. The greatest common factor in $4x^2 + 8x - 6$ is ____.

   2

10. Since 5 is a common factor of $10x^2 - 15$, we can write $10x^2 - 15$ as $5($____$) - 5($____$)$. By the distributive law,

    $2x^2$; 3

    $$5(2x^2) - 5(3) = 5(\_\_\_\_ - \_\_\_\_)$$

    $2x^2$; 3

11. If there is a common factor in a polynomial, we can factor out that factor by the __________ law.

    distributive

12. What is the greatest common factor in $12x^2 + 9x - 6$? ____. So

    3

    $$12x^2 + 9x - 6 = \_\_\_\_(4x^2) + \_\_\_\_(3x) - \_\_\_\_(2)$$

    3; 3; 3

    By the distributive law, $3(4x^2) + 3(3x) - 3(2) =$ ____$(4x^2 + 3x - 2)$.

    3

$5x$	**13.** What is the greatest common factor in $25x^3 - 15x^2 + 10x$? ____
$5x$; $5x$; $5x$ $5x^2$; $3x$; 2	**14.** So $25x^3 - 15x^2 + 10x =$ ____$(5x^2)$ $-$ ____$(3x)$ $+$ ____(2). By the distributive law, $25x^3 + 15x^2 - 10x = 5x($____ $-$ ____ $+$ ____).
$7x$	**15.** What is the greatest common factor of $35x^4 + 7x$?____.
  $7x$; $5x^3$; 1	**16.** So $35x^4 + 7x = 7x(5x^3) + 7x(1)$. By the distributive law, $35x^4 + 7x =$ ____(____ $+$ ____)
4 4; $7x$; 1	**17.** What is the greatest common factor in $28x + 4$? ____. So, by the distributive law, $28x + 4 =$ ____(____ $+$ ____).
$15x$ $15x(2x^2 - x + 3)$	**18.** What is the greatest common factor in $30x^3 - 15x^2 + 45x$? ____. So $30x^3 - 15x^2 + 45x =$ ____________
  one a.; b.; d.	**19.** An example of a monomial is $4x^5$. How many terms does a monomial have? ____. Which of the following are monomials? *a.* $3x$ *b.* 7 *c.* $4x + 2$ *d.* $6x^4$
  $2x + 1$ $3x$	**20.** We use the distributive law to factor out monomial factors. In the product $3x(2x + 1)$, the factors are $3x$ and ________. Which factor is a monomial? ____
$3x + 2$ $4x$	**21.** By the distributive law, $12x^2 + 8x = 4x($________). The monomial factor is ____.
$8x$ $8x$	**22.** By the distributive law, $24x^3 + 16x^2 - 32x =$ ____$(3x^2 + 2x - 4)$. The monomial factor is ____.
  4; $6x^2 - 2x + 1$ 4	**23.** Use the distributive law to factor out the greatest common monomial factor. $24x^2 - 8x + 4 =$ ____ (____________) The monomial factor is ____.
  $7xy$; $6x^2 - 2y + 3y^2$ $7xy$	**24.** Use the distributive law to factor out the greatest monomial factor common to all the terms. $42x^3y - 14xy^2 + 21xy^3 =$ ____ (____________) The monomial factor is ____.
Yes equals	**25.** To check a factorization, we simply multiply the factors. Is $4(x + 3)$ the factorization of $4x + 12$? (Yes\No) since $4(x + 3)$ (equals\does not equal) $4x + 12$.
  $22x^2 - 11x$	**26.** The factorization of $22x^2 - 11x$ is $11x(2x - 1)$ since $11x(2x - 1) =$ ________

27. Factor out the greatest monomial factor.

$16x^2 + 8x - 4 =$ ________________ (factored form)

Check your work.

$4(4x^2 + 2x - 1)$
$= 16x^2 + 8x - 4$

28. Factor out the greatest monomial factor.

$48x^3y + 16x^2y^2 - 8xy =$ ________________

Check your work.

$8xy(6x^2 + 2xy - 1)$
$= 48x^3y +$
$16y^2y^2 - 8xy$

29. If we wish to change the sign of each term in a polynomial, we can factor out -1.

$$-3x^2 + 2x - 5 = -1(3x^2 - 2x + 5)$$

$$-2x + 80 = -1(\_\_\_\_\_\_\_\_)$$

$2x - 80$

30. Factor out -1 in $-10x^2 + 3 =$ ____(________).

$-1(10x^2 - 3)$

31. Factor out -4 in $16x - 8 =$ ________.

$-4(-4x + 2)$

32. Complete factorization of polynomials into prime polynomials with integer coefficients is unique except for orders of factors and change of sign of pairs of factors. Which of the following are complete factorizations of $3x^2 + 3x$?

a. $3x(x + 1)$ *b.* $-3x(-x - 1)$ *c.* $(x + 1)3x$ *d.* $3(x^2 + x)$

a.; b.; c.

33. In general, we write the monomial factors first. Factor $2x^2 + 8x$ completely and write the monomial factor first. ________

$2x(x + 4)$

34. Are $-5(x + 3)$ and $5(-x - 3)$ both complete factorizations of $-5x - 15$? (yes\no)

yes

35. Are $-2(-x + 6)$ and $2(-x + 6)$ both complete factorizations of $2x - 12$? (yes\no) Why? ________________

no; $2(-x + 6)$
$= -2x + 12$

36. Which of the following are factorizations of $12x^2 + 8x - 4$?

a. $4(3x^2 + 2x - 1)$ *b.* $-4(-3x^2 - 2x - 1)$
c. $-4(3x^2 + 2x - 1)$ *d.* $-4(-3x^2 + 2x - 1)$

a.; b.

37. Factor: $45x^2 + 36x - 27 =$ ________________.

$9(5x^2 + 4x - 3)$ or
$-9(-5x^2 - 4x + 3)$

PRACTICE PROBLEMS

Factor out the largest possible monomial factor.

1. $3x + 9$	**2.** $12x + 16$	**3.** $14x^2 + 12x - 7$
4. $18x^2 + 27x + 9$	**5.** $10x^2 + 15x + 30$	**6.** $40x^2 + 20x - 8$
7. $26x^2 - 12x + 39$	**8.** $22x^2 - 44x + 33$	**9.** $45x^2 + 55x - 35$
10. $25x^2 - 85x + 65$	**11.** $12x^2 - 48x + 36$	**12.** $45x^2 - 15x + 30$
13. $8x^2y^2 + 4xy$	**14.** $16x^2y^2 - 12xy$	**15.** $26x^2y^2 + 2xy + 4$

16. $18xy + 12x - 2$ **17.** $75xy + 25x - 50$ **18.** $33x^2 + 66x - 99$

19. $105x^2 - 100x + 5$ **20.** $84x^2 - 12x - 12$ **21.** $xyz + x^2yz - x^3yz$

22. $x^2y^2z - xy^2z + xy^2z^2$ **23.** $14x^2y - 28x^2y^3 + 14xy^2$

24. $64x^2y - 128x^2y^2 + 16xy^2$

Factor out -1.

25. $3x^2 - 2x + 1$ **26.** $-8x^2 + x - 5$ **27.** $-14x + 1$

28. $17x - 8$ **29.** $-3x + 5$ **30.** $-2x - 5$

31. $-x - 1$ **32.** $x + 1$ **33.** $10x^2 + 6x + 5$

34. $12x^2 + 10x + 6$ **35.** $-14x^2 - 7x - 1$ **36.** $-25x^2 - 5x - 3$

37. $-16x^2 + 3x - 5$ **38.** $16x^2 - 8x + 4$

SELFTEST

1. Which of the following are factorizations of $25x^2 + 15x - 5$?

a. $5(5x^2 + 3x - 1)$ *b.* $-5(5x^2 + 3x - 1)$
c. $-5(-5x^2 - 3x - 1)$ *d.* $-1(-25x^2 + 15x + 5)$

2. Factor out the largest possible monomial factor.

a. $18x^2 + 27x - 36$ *b.* $30x^2y - 15xy + 60y$
c. $42x^2 + 12x + 4$ *d.* $50x + 30$

3. Factor out -1.

a. $-x - 8$ *b.* $-14x^2 + 3x - 7$ *c.* $12x^2 - 2x + 5$

SELFTEST ANSWERS

1. *a.*
2. *a.* $9(2x^2 + 3x - 4)$ *b.* $15y(2x^2 - x + 4)$ *c.* $2(21x^2 + 6x + 2)$ *d.* $10(5x + 3)$
3. *a.* $-1(x + 8)$ *b.* $-1(14x^2 - 3x + 7)$ *c.* $-1(-12x^2 + 2x - 5)$

4.2 SPECIAL POLYNOMIALS

Some polynomials factor into the product of two binomials. In this section we will see two such special types (perfect squares and the difference of two squares). Once we recognize these types we can factor them almost immediately.

Definition: **An algebraic expression is a *perfect square* if it equals the square of another algebraic expression.**

Thus 25 and $16x^2$ are perfect squares since $25 = 5^2$ and $16x^2 = (4x)^2$. But 30 is not a perfect square, for the only ways to factor 30 into the product of two integers are $1 \cdot 30, 15 \cdot 2, 10 \cdot 3, 6 \cdot 5$, and the corresponding negative factors. Likewise, a polynomial is a perfect square if it equals the square of another polynomial. Let's examine some squares of binomials.

$$(x + 4)^2 = x^2 + 8x + 16$$
$$(2y^2 - 3)^2 = 4y^4 - 12y^2 + 9$$

So $x^2 + 8x + 16$ and $4y^4 - 12y^2 + 9$ are perfect squares. How can we recognize perfect squares if we do not know their factorizations? Let's look at two general cases. By the multiplication rule,

$$(a + b)^2 = (a + b)(a + b) = a^2 + ba + ab + b^2 = a^2 + 2ab + b^2$$
$$(a - b)^2 = (a - b)(a - b) = a^2 - ba - ab + b^2 = a^2 - 2ab + b^2$$

Thus a polynomial is a perfect square if the first and last terms are perfect squares (say a^2 and b^2) and the middle term is $2ab$ or $-2ab$. Once we recognize a perfect square of either type we can factor it by inspection, because

$$a^2 + 2ab + b^2 = (a + b)^2 \qquad \text{and} \qquad a^2 - 2ab + b^2 = (a - b)^2$$

Example 1 $x^2 + 6x + 9$

The first term, x^2, is clearly a perfect square. The last term is also a perfect square since $9 = 3^2$.

$$\begin{array}{ccccc} x^2 & + & 6x & + & 9 \\ \downarrow & & \downarrow & & \downarrow \\ x^2 & + & 2(x)(3) & + & (3)^2 \\ \downarrow & & \downarrow\downarrow\downarrow & & \downarrow \\ a^2 & + & 2\ a\ \ b & + & b^2 \end{array}$$

The middle term corresponds to $2ab$. Thus $x^2 + 6x + 9$ is a perfect square. It factors as $(a + b)^2$ or $(x + 3)^2$.

Example 2

$$\begin{array}{ccccc} 4x^2 & + & 20x & + & 25 \\ \downarrow & & \downarrow & & \downarrow \\ (2x)^2 & + & 2(2x)(5) & + & (5)^2 \\ \downarrow & & \downarrow\downarrow\downarrow & & \downarrow \\ a^2 & + & 2\ a\ \ b & + & b^2 \end{array}$$

Again we see a perfect square. This time $a = 2x$, $b = 5$, and $2ab = 20x$. Thus $4x^2 + 20x + 25$ factors as $(2x + 5)^2$.

Example 3 $16x^2 + 9x + 4$

In this example, $16x^2$ corresponds to a^2 and 4 corresponds to b^2. So $a = 4x$ and $b = 2$ and $2ab = 2(4x)(2) = 16x$. But the middle term of $16x^2 + 9x + 4$ is $9x$. It is not of the form $2ab$. Consequently, $16x^2 + 9x + 4$ is not a perfect square.

Example 4 $9x^2 - 24x + 16$

Here $a^2 = 9x^2$ and $b^2 = 16$. So $a = 3x$, $b = 4$, and $2ab = 2(3x)(4) = 24x$. The middle term of $9x^2 - 24x + 16$ is of the form $-2ab$, By the equation,

$$a^2 - 2ab + b^2 = (a - b)^2$$

we see that $9x^2 - 24x + 16$ factors as $(3x - 4)^2$.

Example 5 $5x^2 - 20x + 20$

Since neither the first nor last term is a perfect square, the polynomial is not a perfect square. However,

$$5x^2 - 20x + 20 = 5(x^2 - 4x + 4)$$

The polynomial $x^2 - 4x + 4$ is a perfect square since $a^2 = x^2$, $b^2 = 4$, and $-2ab = -4x$. Therefore the complete factorization of $5x^2 - 20x + 20$ is $5(x - 2)^2$.

Example 6 $9x^4 + 6x^2 + 1$

In this example, $a^2 = 9x^4$ so $a = 3x^2$, and $b^2 = 1$ so $b = 1$. Thus

$$2ab = 2(3x^2)(1) = 6x^2$$

Therefore $9x^4 + 6x^2 + 1$ is a perfect square and $9x^4 + 6x^2 + 1 = (3x^2 + 1)^2$.

Another special polynomial is the *difference of perfect squares*. This type of polynomial also factors into the product of two binomials. By the multiplication, we see that

$$(a + b)(a - b) = a^2 + ba - ab - b^2 = a^2 - b^2$$

Therefore

$$a^2 - b^2 = (a + b)(a - b)$$

We use this formula to factor the difference of two squares.

Example 7 $x^2 - 16$

The first term, x^2, is clearly a perfect square. Since $16 = 4^2$, the second term is also a perfect square. So x corresponds to a and 4 corresponds to b in the formula $a^2 - b^2 = (a + b)(a - b)$. Therefore,

$$x^2 - 16 = (x + 4)(x - 4)$$

Example 8 $36x^2 - 49$

The two terms are perfect squares since $36x^2 = (6x)^2$ and $49 = 7^2$. Now $6x$ corresponds to a and 7 corresponds to b in $a^2 - b^2 = (a + b)(a - b)$. Thus

$$36x^2 - 49 = (6x + 7)(6x - 7)$$

Example 9 $11x^2 - 4$

The last term is a perfect square but $11x^2$ is not. So $11x^2 - 4$ is not an example of the difference of two squares.

Example 10 $9x^2 + 16$

This polynomial is the sum, not the difference of two perfect squares, so we cannot use the formula for the difference of two squares to factor this polynomial.

Example 11 $12x^2 - 75$

Neither $12x^2$ nor 75 is a perfect square. However, we can factor out 3.

$$12x^2 - 75 = 3(4x^2 - 25)$$

The polynomial $4x^2 - 25$ is the difference of two squares since $4x^2 = (2x)^2$ and $25 = 5^2$. Therefore the factored form of $12x^2 - 75$ is $3(2x + 5)(2x - 5)$.

Example 12 $64x^4 - 81$

Here, both terms are perfect squares since $64x^4 = (8x^2)^2$ and $81 = 9^2$. So $8x^2$ corresponds to a, and 9 corresponds to b in $a^2 - b^2 = (a + b)(a - b)$. Therefore

$$64x^4 - 81 = (8x^2 + 9)(8x^2 - 9)$$

PROGRAM 4.2

Part I: Perfect Squares

1. An algebraic expression is a *perfect square* if it equals the square of another algebraic expression. Thus 16 is a perfect square since $16 = 4^2$; and 25 is a perfect square since $25 = \_\_\_\_^2$.

5

2. $4x^2y^2$ is a perfect square since $4x^2y^2 = (2 \cdot \_\_ \cdot \_\_)^2$. — x; y

3. Is $36x^4$ a perfect square? (Yes\No) because $36x^4 = \_\_\_\_\_\_\_\_$. — Yes; $(6x^2)^2$

4. Is $15x^2$ a perfect square? (yes\no) — no

5. Some polynomials are perfect squares. Since $(x + 3)^2 = x^2 + 6x + 9$, $x^2 + 6x + 9$ (is\is not) a perfect square. — is

6. Since $(x - 5)^2 = x^2 - 10x + 25$, is $x^2 - 10x + 25$ a perfect square? (yes\no) — yes

7. Since $x^4 - 4x^2 + 4 = (x^2 - 2)^2$, $x^4 - 4x^2 + 4$ is a ________ ________. — perfect square

8. Recall that, by the distributive law,

$$(a + b)^2 = (a + b)(a + b) = a^2 + ab + ba + b^2$$
$$= a^2 + 2ab + b^2$$

$$(x + 6)^2 = (x + 6)(x + 6) = \_\_\_\_\_\_\_\_\_\_$$ — $x^2 + 12x + 36$

9. And also that

$$(a - b)^2 = (a - b)(a - b) = a^2 - ba - ab + b^2 = a^2 - 2ab + b^2$$

Therefore $(x - 3)^2 = (x - 3)(x - 3) = \_\_\_\_\_\_\_\_\_\_$. — $x^2 - 6x + 9$

10. So $a^2 + 2ab + b^2$ is a perfect square since it equals $(a + b)^2$. Also $a^2 - 2ab + b^2$ is a perfect square since it equals $(\_\_\_\_\_\_)^2$. — $a - b$

11. Let's examine the form of polynomials that are perfect squares. The two general cases for expressions with three terms are

$$a^2 + 2ab + b^2 = (a + b)^2 \quad \text{and} \quad a^2 - 2ab + b^2 = (a - b)^2$$

In both cases the first and last terms are the perfect squares a^2 and ____ respectively. The middle terms are either $2ab$ or ____ respectively. — b^2; $-2ab$

12. Let's see if $x^2 + 8x + 16$ fits the pattern $a^2 + 2ab + b^2$.

x^2 corresponds to a^2 so x corresponds to ____ — a

16 corresponds to b^2 so 4 corresponds to ____ — b

Thus $2ab = 2(\_\_\_\_)(\_\_\_\_) = \_\_\_\_$. So the middle term corresponds to $2ab$. Therefore $x^2 + 8x + 16$ (is\is not) of the form $a^2 + 2ab + b^2$. Consequently, $x^2 + 8x + 16$ (is\is not) a perfect square. — x; 4; $8x$; is; is

13. Since $x^2 + 8x + 16$ is a perfect square of the form $a^2 + 2ab + b^2$, we can use the formula $a^2 + 2ab + b^2 = (a + b)^2$ to factor $x^2 + 8x + 16$.

Again x corresponds to ____ in the formula and 4 corresponds to ____. — a; b

So $x^2 + 8x + 16 = (x + \_\_\_\_)^2$. — 4

14. Is $x^2 + 12x + 36$ a perfect square of the form $a^2 + 2ab + b^2$? The answer is yes, since x^2 corresponds to a^2 so x corresponds to ____; 36 corresponds to b^2, so ____ corresponds to b; and therefore $2ab = 2 \cdot \_\_\_\_ \cdot \_\_\_\_ =$ ____. So the middle term of $x^2 + 12x + 36$ corresponds to $2ab$. — a; 6; x; 6; $12x$

x; 6

15. Since $x^2 + 12x + 36$ is a perfect square of the form $a^2 + 2ab + b^2$, we use the formula $a^2 + 2ab + b^2 = (a + b)^2$ to factor it. So

$$x^2 + 12x + 36 = (\_\_\_\_ + \_\_\_\_)^2$$

a²; b²

x; 3; −6x

16. $x^2 - 6x + 9$ is a perfect square of the form $a^2 - 2ab + b^2$ since x^2 corresponds to ____, 9 corresponds to ____, and

$$-2ab = -2(\_\_\_\_)(\_\_\_\_) = \_\_\_\_$$

x − 3

17. Since $x^2 - 6x + 9$ is a perfect square of the form $a^2 - 2ab + b^2$, we use the formula $a^2 - 2ab + b^2 = (a - b)^2$ to factor it. Thus $x^2 - 6x + 9$ factors as $(\_\_\_\_\_\_\_\_)^2$.

x²; x

81; 9

−18x

x − 9

18. Determine if $x^2 - 18x + 81$ is a perfect square.

$a^2 = \_\_\_\_$ so $a = \_\_\_\_$

$b^2 = \_\_\_\_$ so $b = \_\_\_\_$

$-2ab = \_\_\_\_$

So $x^2 - 18x + 81$ factors as $(\_\_\_\_\_\_\_\_)^2$.

2x

1; 1

4x

2x + 1

19. Determine if $4x^2 + 4x + 1$ is a perfect square of the form $a^2 + 2ab + b^2$.

$a^2 = 4x^2$ so $a = \_\_\_\_$

$b^2 = \_\_\_\_$ so $b = \_\_\_\_$

$2ab = \_\_\_\_$

So $4x^2 + 4x + 1 = (\_\_\_\_\_\_\_\_)^2$ in factored form.

9x²; 3x

25; 5

−30x

3x − 5

20. Determine if $9x^2 - 30x + 25$ is a perfect square.

$a^2 = \_\_\_\_$ so $a = \_\_\_\_$

$b^2 = \_\_\_\_$ so $b = \_\_\_\_$

$-2ab = \_\_\_\_$

Thus $9x^2 - 30x + 25$ factors as $(\_\_\_\_\_\_\_\_)^2$.

yes

21. Is $25x^2 + 20x + 4$ a perfect square? (yes\no)

(5x + 2)²

22. $25x^2 + 20x + 4 = (\_\_\_\_\_\_\_\_)^2$ in factored form.

no (6 is not a perfect square)

23. Is $x^2 + 3x + 6$ a perfect square? (yes\no)

no (2x² is not a perfect square)

24. Is $2x^2 + 6x + 9$ a perfect square? (yes\no)

9x²; 3x

4; 2

12x

no

no

25. Determine if $9x^2 + 6x + 4$ is a perfect square.

$a^2 = \_\_\_\_$ so $a = \_\_\_\_$

$b^2 = \_\_\_\_$ so $b = \_\_\_\_$

$2ab = \_\_\_\_$

Is the middle term of $9x^2 + 6x + 4$ in the form $2ab$? (yes\no). Is $9x^2 + 6x + 4$ a perfect square? (yes\no).

yes; (x + 6)²

26. Is $x^2 + 12x + 36$ a perfect square? (yes\no). If it is, factor it. ________.

Part II: Almost Perfect Squares

27. Some polynomials are not perfect squares but they contain a perfect square as a factor. $6x^2 + 24x + 24$ (is\is not) a perfect square. But

is not

$$6x^2 + 24x + 24 = 6(x^2 + 4x + 4)$$

Is $x^2 + 4x + 4$ a perfect square? (yes\no). So $6x^2 + 24x + 24$ factors as $6(\_\_\_\_\_)^2$.

yes

$x + 2$

28. Is $5x^2 - 50x + 125$ a perfect square? (yes\no). But

no

$$5x^2 - 50x + 125 = 5(\_\_\_\_\_\_\_\_)$$

$x^2 - 10x + 25$

29. Is $x^2 - 10x + 25$ a perfect square? (yes\no). So

yes

$$5x^2 - 50x + 125 = 5(x^2 - 10x + 25) = 5(\_\_\_\_\_)^2$$

$x - 5$

30. Is $8x^2 + 16x + 8$ a perfect square? (yes\no). But

no

$$8x^2 + 16x + 8 = \_\_(x^2 + 2x + 1)$$

8

So $8x^2 + 16x + 8$ factors as $\_\_(\_\_\_\_\_)^2$.

8; $x + 1$

31. To factor completely, it is usually easier to factor out the greatest common factor first.

$$3x^2 + 18x + 27 = \_\_(x^2 + 6x + 9) = \_\_(x + 3)^2$$

3; 3

32. Factor $7x^2 + 42x + 63$ completely.

$$7x^2 + 42x + 63 = 7(\_\_\_\_\_\_\_\_) = 7(\_\_\_\_\_)^2$$

$x^2 + 6x + 9$; $7(x + 3)^2$

33. Factor $4x^2 + 40x + 100$ completely.

$$4x^2 + 40x + 100 = \_\_(x^2 + 10x + 25) = \_\_(\_\_\_\_\_)^2$$

4; 4; $x + 5$

34. If the leading coefficient is negative, it is usually easier to factor out -1 first.

$$-x^2 + 8x - 16 = -1(\_\_\_\_\_\_\_\_)$$

$x^2 - 8x + 16$

$$= -1(\_\_\_\_\_)^2$$

$x - 4$

35. Factor completely:

$$-25x^2 - 60x - 36 = \_\_(25x^2 + 60x + 36)$$

-1

$$= \_\_(\_\_\_\_\_)^2$$

-1; $5x + 6$

Part III: Difference of Perfect Squares

36. Another type of polynomial that we can factor immediately is the *difference of perfect squares*. We follow the pattern

$$\begin{array}{c} a^2 - b^2 = (a + b)(a - b) \\ \downarrow \quad \downarrow \quad\quad \downarrow \quad \downarrow \; \downarrow \quad \downarrow \\ x^2 - 9 = x^2 - 3^2 = (x + 3)(\_\_\_\_) \end{array}$$

$x - 3$

37. Determine if $x^2 - 25$ fits the pattern $a^2 - b^2$. Since $a^2 = \_\_\_\_$ and $b^2 = \_\_\_\_$, $x^2 - 25 = (x + \_\_\_\_)(x - \_\_\_\_)$ because

x^2

25; 5; 5

$$a^2 - b^2 = (a + b)(a - b)$$

yes x; x	**38.** Does $x^2 - 49$ fit the pattern $a^2 - b^2$? (yes\no). So $x^2 - 49$ factors as (___ + 7)(___ − 7).
x; x; is	**39.** $x^2 - 64$ factors as (___ + 8)(___ − 8) since $x^2 - 64$ (is\is not) the difference of two perfect squares.
$(x + 9)(x - 9)$	**40.** Factor: $x^2 - 81$ = ______________.
$(x + 1)(x - 1)$	**41.** Factor: $x^2 - 1$ = ______________.
yes $2x$; 9; 3 $2x - 3$	**42.** Does $4x^2 - 9$ fit the pattern $a^2 - b^2$? (yes\no). $a^2 = 4x^2$ so a = ____ and b^2 = ____ so b = ____ Then $4x^2 - 9$ factors as $(2x + 3)$(________).
yes $(3x + 1)(3x - 1)$	**43.** Does $9x^2 - 1$ fit the pattern $a^2 - b^2$? (yes\no). Factor: $9x^2 - 1$ = ______________
$(5x + 9)(5x - 9)$	**44.** Factor: $25x^2 - 81$ = ______________.
no It is a sum not a difference	**45.** Does $x^2 + 4$ fit the pattern $a^2 - b^2$? (yes\no). Explain your answer. ______________.
no 20 is not a perfect square	**46.** Does $x^2 - 20$ fit the pattern $a^2 - b^2$? (yes\no). Explain your answer. ______________.
no $x + 5$; $x - 5$	**47.** Is $5x^2 - 125$ the difference of two perfect squares? (yes\no). But we can still factor $5x^2 - 125$. $5x^2 - 125 = 5(x^2 - 25) = 5$(________)(________)
$x + 3$; $x - 3$	**48.** $10x^2 - 90 = 10(x^2 - 9) = 10$(________)(________).
$x^2 - 25$; $3(x + 5)(x - 5)$	**49.** Factor $3x^2 - 75$ completely. 3(________) = ______________
$6(x^2 - 1)$; $6(x + 1)(x - 1)$	**50.** Factor $6x^2 - 6$ completely. ______________ = ______________

PRACTICE PROBLEMS

Which of the following are perfect squares? Factor such polynomials completely.

1. $x^2 + 2x + 1$	**2.** $x^2 + 12x + 36$	**3.** $x^2 - 2x + 1$
4. $x^2 + 4x + 4$	**5.** $x^2 + 14x + 49$	**6.** $x^2 + 16x + 64$
7. $x^2 + 15x + 4$	**8.** $x^2 + 20x + 100$	**9.** $4x^2 + 12x + 25$
10. $x^2 + 12x + 9$	**11.** $4x^2 + 12x + 9$	**12.** $9x^2 - 12x + 4$
13. $81x^2 + 126x + 49$	**14.** $x^2 + 8x - 16$	**15.** $-64x^2 - 96x - 36$
16. $-4x^2 - 16x - 64$	**17.** $x^4 + 8x^2 + 16$	**18.** $x^4 + 18x^2 + 81$

Which of the following are the difference of squares? Factor such polynomials completely.

19. $x^2 - 9$ **20.** $x^2 - 1$ **21.** $x^2 - 81$ **22.** $x^2 - 100$

23. $4x^2 - 9$ **24.** $16x^2 - 25$ **25.** $25x^2 - 81$ **26.** $64x^2 - 1$

27. $x^4 - 16$ **28.** $x^4 - 49$ **29.** $x^2 + 9$ **30.** $x^2 + 4$

SELFTEST

Factor completely.

1. $x^2 - 6x + 9$ **2.** $25x^2 + 50x + 25$ **3.** $-x^2 - 4x - 4$

4. $3x^2 - 3$ **5.** $x^2 - 4$ **6.** $25x^2 - 1$

7. $x^2 + 8x + 16$ **8.** $2x^2 - 50$

SELFTEST ANSWERS

1. $(x - 3)^2$ **2.** $25(x + 1)^2$ **3.** $-1(x + 2)^2$ **4.** $3(x + 1)(x - 1)$

5. $(x + 2)(x - 2)$ **6.** $(5x + 1)(5x - 1)$ **7.** $(x + 4)^2$ **8.** $2(x + 5)(x - 5)$

4.3 FACTORING

In the previous section we studied polynomials that are from products of the type $(x + a)^2$ or $(x - a)(x + a)$, where a is an integer.

Let's examine polynomials that come from a more general product of the form $(x + a)(x + b)$, where a and b are integers. Several applications of the distributive law yield

$$\begin{aligned}(x + a)(x + b) &= (x + a)x + (x + a)b \\ &= x^2 + ax + bx + ab \\ &= x^2 + (a + b)x + ab\end{aligned}$$

The product of the first term in each of the factors $x + a$ and $x + b$ is x^2 while the product of the last term is ab.

first terms

$$(x + a)(x + b) = x^2 + (a + b)x + ab$$

last terms

The middle term of the product is the sum of the inner and outer products.

inner product

$$(x + a)(x + b) = x^2 + (a + b)x + ab$$

outer product

Example 1 Multiply $(x + 3)(x - 5)$:

inner product — product of first terms

$$(x + 3)(x - 5) = x^2 - 2x - 15$$

— sum of inner and outer products

outer product — product of last terms

Does $x^2 + 5x + 6$ factor into the product of two binomials with integer coefficients? If we let the first term of each binomial factor be x, we are sure to get x^2 in the product. So we begin by writing

$$x^2 + 5x + 6 = (x + ?)(x + ?)$$

The product of the missing last terms should be 6. The only such pairs are 6,1; $-6, -1$; 3, 2; $-3, -2$. We try all four pairs to discover which if any combination gives the middle term $5x$.

inner product
$$(x + 6)(x + 1) = x^2 \underline{+ 7x} + 6$$
outer product

inner product
$$(x - 6)(x - 1) = x^2 \underline{- 7x} + 6$$
outer product

inner product
$$(x + 3)(x + 2) = x^2 \underline{+ 5x} + 6$$
outer product

inner product
$$(x - 3)(x - 2) = x^2 \underline{- 5x} + 6$$
outer product

We see that the third pair gives us the factorization of $x^2 + 5x + 6$.

General Factor Technique: **To factor polynomials of the form $x^2 + bx + c$, try all combinations of the form**

inner product
$$(x + q)(x + r)$$
outer product

where q and r are integers such that $qr = c$. The factorization is the one in which the sum of the inner and outer products gives the middle term bx.

Example 2 Factor $x^2 + 9x + 8$: Since 8 factors as $8 \cdot 1$, $(-8)(-1)$, $4 \cdot 2$, $(-4)(-2)$, the possible binomial factors are

inner product	inner product	inner product	inner product
$(x + 8)(x + 1)$,	$(x - 8)(x - 1)$,	$(x + 4)(x + 2)$,	$(x - 4)(x - 2)$
outer product	outer product	outer product	outer product

The only factorization in which the sum of the inner and outer products is $9x$ is the first one. So the factorization is $(x + 8)(x + 1)$ since

$$(x + 8)(x + 1) = x^2 + 9x + 8$$

Example 3 Factor $x^2 + 4x - 12$: The factorizations of -12 that must be considered are $(12)(-1)$, $(-12)(1)$, $(6)(-2)$, $(-6)(2)$, $(3)(-4)$, $(-3)(4)$. So the possible factorizations of $x^2 + 4x - 12$ are

a. $(x + 12)(x - 1)$ *b.* $(x - 12)(x + 1)$ *c.* $(x + 6)(x - 2)$
d. $(x - 6)(x + 2)$ *e.* $(x + 3)(x - 4)$ *f.* $(x - 3)(x + 4)$

The sum of the inner and outer products must be $4x$. The only such factorization is

$$c.\ (x + 6)(x - 2) = x^2 + 4x - 12$$

Example 4 Factor $x^2 - 6x + 9$: The pairs whose product is 9 are 9, 1; −9, −1; 3, 3; −3, −3. The possible factorizations are

$$(x + 9)(x + 1), \quad (x - 9)(x - 1), \quad (x + 3)(x + 3), \quad (x - 3)(x - 3)$$

and only $(x - 3)(x - 3) = x^2 - 6x + 9$.

Example 5 Factor $y^2 - 6y - 7$: Only pairs 7, −1 and −7, 1 have product −7. So the possible factorizations are

$$(y + 7)(y - 1) \quad \text{and} \quad (y - 7)(y + 1)$$

Since $(y - 7)(y + 1) = y^2 - 6y - 7$, $(y - 7)(y + 1)$ is the factorization. Is $(y + 1)(y - 7)$ also correct? *Hint:* Apply the commutative law.

Some polynomials of the form $x^2 + bx + c$ do not factor into the product of two binomials with integer coefficients.

Example 6 Factor $x^2 - 10x - 10$: Since the only pairs of integers with product −10 are 10, −1; −10, 1; 5, −2; −5, 2 the only possible factorizations are

$$(x + 10)(x - 1) = x^2 + 9x - 10$$
$$(x - 10)(x + 1) = x^2 - 9x - 10$$
$$(x + 5)(x - 2) = x^2 + 3x - 10$$
$$(x - 5)(x + 2) = x^2 - 3x - 10$$

None of the factorizations yield the proper middle term, $-10x$. Therefore, $x^2 - 10x - 10$ does not factor into binomials with integer coefficients.

Factorization does involve a bit of trial and error technique. Experience will enable you to eliminate some of the possibilities immediately. Efficiency is desirable since factoring is often an intermediate step in a longer process.

PROGRAM 4.3

1. Some polynomials are not perfect squares or the difference of two perfect squares. Match the following.

1. $x^2 - 36$ — *a.* perfect square — 1.*b.*
2. $4x^2 + 20x + 25$ — *b.* difference of perfect squares — 2.*a.*
3. $x^2 + 8x + 15$ — *c.* neither *a.* nor *b.* — 3.*c.*

2. Polynomials which do not fit previous patterns might come from products of the form $(x + a)(x + b)$ where a, b are integers.

$$(x + a)(x + b) = (x + a)x + (x + a)b$$

$= \_\_\_\_ + \_\_\_\_ + \_\_\_\_ + \_\_\_\_$ — x^2; ax; xb; ab

$= x^2 + (a + b)x + \_\_\_\_$ — ab

3. In the product of the two binomials

first terms

$$(x + a)(x + b) = x^2 + (a + b)x + ab$$

last terms

The product of the first terms gives the (first\middle\last) term of the product while the product of the last term gives the (first\middle\last) term of the product. — first; last

4. In the product

$$\overbrace{(x + 3)(x}^{\text{first terms}} + \underbrace{4)}_{\text{last terms}}$$

x^2; 12

the product of the first terms is ____. The product of the last terms is ____.

x^2

-20

5. In $(x - 4)(x + 5)$, the product of the first terms is ____. The product of the last terms is ____.

6. The middle term of the product

$$(x + a)(x + b) = x^2 + \underbrace{(a + b)x}_{\text{middle term}} + ab$$

comes from the sum of the *inner* and *outer* products

inner product

$(x + a)(x + b)$

outer product

$10x$

$10x$

In $(x + 4)(x + 6)$, the sum of inner and outer products is ____. The middle term in the product of $(x + 4)(x + 6)$ is ____.

7. In

inner product

$(x + 5)(x - 6)$

outer product

$5x$; $-6x$

$-x$

$-x$

the inner product is ____. The outer product is ____. The sum of the inner and outer products is ____. The middle term in the product $(x + 5)(x - 6)$ is ____.

$-4x$; $7x$

$3x$

$3x$

8. In $(x - 4)(x + 7)$, the inner product is ____. The outer product is ____. The sum of these products is ____. What is the middle term in the product $(x - 4)(x + 7)$? ____.

9. Multiply:

x^2; $12x$; 32

$(x + 4)(x + 8) =$ ____ + ____ + ____

product of first terms → (first blank)

sum of inner and outer products → (second blank)

product of last terms → (third blank)

10. Multiply:

x^2; $(-10x)$; 21

$(x - 7)(x - 3) =$ ____ + ____ + ____

product of first terms → (first blank)

sum of inner and outer products → (second blank)

product of last terms → (third blank)

11. To factor the polynomial $x^2 + 6x + 5$ into the product of two binomials we choose the first terms so that their product is x^2.

$$(x + ?)(x + ?)$$

We choose the last terms so that their product is 5. The possibilities are

$(x + 5)(x + 1)$ and $(x - 5)(x - \_\_)$ — 1

The actual factorization is the one in which the sum of the inner and outer products is the middle term of $x^2 + 6x + 5$. So the factorization $(x + 5)(x + 1)$ is correct since the middle term of the product is ____. — $6x$

12. To factor $x^2 - 4x + 3$ into the product of two binomials, we begin by writing $(x + ?)(x + ?)$. The product of the missing last terms should be ____. — 3

13. The two possible factorizations of $x^2 - 4x + 3$ are

$(x + 3)(x + \_\_)$ and $(x - 3)(x - \_\_)$ — 1; 1

The sum of the inner and outer products of $(x + 3)(x + 1)$ is ____. — $4x$

The sum of the inner and outer products of $(x - 3)(x - 1)$ is ____. — $-4x$

So the actual factorization is

$x^2 - 4x + 3 = (\_\_\_\_\_\_)(\_\_\_\_\_\_)$ — $x - 3$; $x - 1$

14. The general technique for factoring a polynomial of the form $x^2 + bx + c$ is

1. Write the polynomial in standard form.
2. Try all combinations of the form $(x + ?)(x + ?)$ where the product of the missing last terms is c.
3. The actual factorization is the one in which the sum of the inner and outer products is the middle term of $x^2 + bx + c$.

To factor $x^2 + 4x - 5$, we try the combinations

$$(x + 5)(x - 1) \quad \text{and} \quad (x - 5)(x + 1)$$

Which factorization satisfies condition 3? ________. — $(x + 5)(x - 1)$

Which is the correct factorization? ________. — $(x + 5)(x - 1)$

15. To factor $8x + x^2 + 7$, we first write the polynomial in standard form. ________. In standard form, the last term is ____ — $x^2 + 8x + 7$; 7

and the middle term is ____. The pairs of integers with product 7 are 7, ___, — $8x$; 1

and -1, ___. — -7

So the possible factorizations are

$(x + 7)(x + 1)$ and $(x - \_\_)(x - \_\_)$ — 7; 1

In which factorization is the sum of the inner and outer products $8x$? ________. — $(x + 7)(x + 1)$

Which is the correct factorization? ________. — $(x + 7)(x + 1)$

16. Is $x^2 + 5x + 6$ in standard form? (yes\no). The last term is ____ — yes; 6

and the middle term is ____. The pairs with product 6 are 6, ___; -6, ___; 3, ___; — $5x$; 1; -1; 2

-3	and __, -2. So the possible factorizations are
1; 1	*a.* $(x + 6)(x + \_\_)$ *b.* $(x - 6)(x - \_\_)$
2; 3; 2	*c.* $(x + 3)(x + \_\_)$ *d.* $(x - \_\_)(x - \_\_)$
c.	The correct factorization is (*a.**b.**c.**d.*) since the sum of the inner and outer products is the required middle term $5x$.

no	**17.** Factor $21 + x^2 + 10x$: Is the polynomial in standard form? (yes\\no). The
$x^2 + 10x + 21$; 21	standard form is __________. The last term is ____. The four
3; -3; 1; -1	pairs with product 21 are 7, __; -7, __; 21, __; -21, __. So the possible factorizations are
$x + 7$; $x + 3$	(______)(______)
$x - 7$; $x - 3$	(______)(______)
$x + 21$; $x + 1$	(______)(______)
$x - 21$; $x - 1$	(______)(______)
$x + 7$; $x + 3$	The correct factorization is (______)(______) since the sum of the inner and outer products is the required middle term $10x$.

	18. Factor $x^2 - 5x - 14$; The four pairs with product -14 are -14, 1;
14; -7; 7	__, -1; 2, __; -2, __. The trial factorizations are
$x - 14$; $x + 1$	(______)(______)
$x + 14$; $x - 1$	(______)(______)
$x + 2$; $x - 7$	(______)(______)
$x - 2$; $x + 7$	(______)(______)
$x + 2$; $x - 7$	The actual factorization is (______)(______) since its product is $x^2 - 5x - 14$.

-1	**19.** Factor $x^2 - 10x + 24$: The eight pairs with product 24 are 24, 1; -24, __;
2; -2; 3; -3 4; -4	12, __; -12, __; 8, __; -8, __; 6, __; -6, __. The eight trial factorizations are
$x + 24$; $x + 1$; $x - 24$; $x - 1$	(______)(______) (______)(______)
$x + 6$; $x + 4$; $x - 6$; $x - 4$	(______)(______) (______)(______)
$x + 12$; $x + 2$; $x - 12$; $x - 2$	(______)(______) (______)(______)
$x + 8$; $x + 3$; $x - 8$; $x - 3$	(______)(______) (______)(______)
$(x - 6)(x - 4)$	The actual factorization is __________.

20. The faster we can factor the better, since factoring is often an intermediate technique. Consequently we should do as many steps mentally as possible and write down only enough to keep track of trial factorizations. One pattern is the following: To factor $x^2 + 5x - 50$, we find the pairs with product -50 to be

$$-1, 50; \quad 1, -50; \quad -2, 25; \quad 2, -25; \quad -5, 10; \quad 5, -10$$

Mentally try the various pairs in the format.

$$(x + ?)(x + ?)$$

$(x - 5)(x + 10)$	The factorization is __________.

21. Factor $x^2 - 6x + 9$: The pairs with product 9 are ______; ______; ______; ______. Now, mentally try the pairs in the format $(x + ?)(x + ?)$. The factorization is ______.

9, 1; −9, −1; 3, 3; −3, −3

$(x - 3)(x - 3)$

The next frames will show that we can automatically eliminate some of the pairs for the second term in the format $(x + ?)(x + ?)$ without even trying them mentally.

22. Multiply and write the results in standard form.

a. $(x - 10)(x - 2) =$ ______.

b. $(x + 10)(x + 2) =$ ______.

The terms that are the same in the two products are the (first\middle\last) terms. The terms that differ are the (first\middle\last) terms. In product *a.* the middle term is (positive\negative). In product *b.* the middle term is (positive\negative).

$x^2 - 12x + 20$

$x^2 + 12x + 20$

first and last

middle

negative

positive

23. Multiply and write the results in standard form.

a. $(x + 7)(x - 3) =$ ______.

b. $(x - 7)(x + 3) =$ ______.

The first terms of the products (are\are not) the same. The corresponding last terms (are\are not) the same. The middle terms (are\are not) alike except for the sign.

$x^2 + 4x - 21$

$x^2 - 4x - 21$

are

are; are

24. Write the products in standard form.

a. $(x - 5)(x - 3) =$ ______.

b. $(x + 5)(x + 3) =$ ______.

What is the difference in the products *a.* and *b.*?

$x^2 - 8x + 15$

$x^2 + 8x + 15$

the sign of the middle term

25. The middle term of the product $(x + 4)(x - 6)$ is ____. Which of the following products have the same middle term except for the sign?

a. $(x - 6)(x + 4)$ b. $(x - 4)(x + 6)$ c. $(x - 4)(x - 6)$

$-2x$

a. and *b.*

26. The middle term of the product $(x + 4)(x + 5)$ is ____. Which of the following products has a middle term with absolute value $9x$? ____. (i.e., the coefficient of the term is the same or it is the same except for the sign.)

a. $(x - 4)(x - 5)$ b. $(x - 4)(x + 5)$ c. $(x + 4)(x - 5)$

$9x$

a.

27. If we change both the circled signs in $(x \oplus 4)(x \oplus 3)$, we (change\do not change) the sign of the middle term. Do the first and last terms change? (yes\no).

change

no

28. If we change both the circled signs in $(x \oplus 8)(x \ominus 6)$, we change only the sign of the (middle term\last term) in the product.

middle term

29. The products $(x - 3)(x + 6)$ and $(x + 3)(x - 6)$ (differ\do not differ) only in the sign of the middle term.

differ

30. The products $(x - 7)(x + 2)$ and ______ are the same except for the sign of the middle term.

$(x + 7)(x - 2)$

31. We use the fact that the products

$$(x + a)(x + b) \qquad \text{and} \qquad (x - a)(x - b)$$

are the same except for the sign of the middle term to speed the factoring process.

Factor $x^2 + 11x + 10$: The pairs with product 10 are

$-1; -2$

$$10, 1; \quad -10, \_\_; \quad 5, 2; \quad -5, \_\_$$

is not

Since the middle term of $(x + 5)(x + 2)$ does not have absolute value $11x$, it (is\is not) necessary to try the factorization $(x - 5)(x - 2)$. The middle term of $(x - 10)(x - 1)$ is $-11x$. But we need $+11x$. So we change both

$x + 10; x + 1$

signs to get the proper factorization (________)(________).

32. Factor $x^2 - 6x + 8$: The pairs with product 8 are

$1; -1; 2; -4$

$$8, \_\_; \quad -8, \_\_; \quad 4, \_\_; \quad \_\_, -2$$

$x - 4; x - 2$

The factorizations $(x + 4)(x + 2)$ and (________)(________) both have a middle term with absolute value $6x$. Which factorization has $-6x$ as a middle term? The actual factorization of $x^2 - 6x + 8$ is

$(x - 4)(x - 2)$

____

33. Factor $x^2 + 3x - 18$: The pairs with product -18 are

$-1; 1; \quad 9; -9; \quad -3; 3$

a. 18, __; -18, __ *b.* __, -2; __, 2 *c.* 6, __; -6, __

c.

a. and *b.*

The pairs in (*a.**b.**c.*) can be used in $(x + ?)(x + ?)$ to give a middle term with absolute value $3x$. We can eliminate all the pairs in (*a.**b.**c.*). The

$(x+6)(x-3)$

factorization is ________________.

34. Factor $x^2 - 2x - 24$: The pairs with product -24 are

$-1; 1$

a. 24, ____; -24, ____

$-2; 2$

b. 12, ____; -12, ____

$-3; 3$

c. 8, ____; -8, ____

$-4; 4$

d. 6, ____; -6, ____

The middle term of $(x + ?)(x + ?)$ has absolute value $2x$ when we use

d.; $(x - 6)(x + 4)$

pairs from (*a.**b.**c.**d.*). The factorization is ________________.

35. Some polynomials do not factor. The only possible factorizations of

$$x^2 + 4x - 3$$

$x - 1; x + 1$

no

are $(x + 3)$(________) and $(x - 3)$(________). Are either of these factorizations of $x^2 + 4x - 3$? (yes\no).

36. To conclude that a polynomial does not factor, we must try all the possibilities. If none of the trial factorizations work, then the polynomial (does\ does not) factor.

does not

no

37. Does $x^2 + 8x + 10$ factor?

yes; $(x + 4)(x + 4)$

38. Does $x^2 + 8x + 16$ factor? (yes\no). If so, how? ________________.

PRACTICE PROBLEMS

Factor as much as possible.

1. $x^2 + 3x + 2$
2. $x^2 + 4x + 4$
3. $x^2 + 5x + 6$
4. $x^2 + 6x + 8$
5. $x^2 - 2x - 8$
6. $x^2 + x - 6$
7. $x^2 + 2x - 15$
8. $x^2 - 3x - 10$
9. $x^2 - x - 12$
10. $x^2 + 4x - 12$
11. $x^2 + 11x - 12$
12. $x^2 + 7x + 12$
13. $x^2 + 11x + 18$
14. $x^2 - 19x + 18$
15. $x^2 + x - 20$
16. $x^2 - 8x - 20$
17. $x^2 + 10x + 9$
18. $x^2 - 6x + 5$
19. $x^2 + 8x + 16$
20. $x^2 + 14x + 48$
21. $x^2 - 16x + 48$
22. $x^2 + 5x - 36$
23. $x^2 + 12x + 36$
24. $x^2 - 10x + 25$
25. $x^2 + 25x + 25$
26. $x^2 + 4x + 6$
27. $x^2 + 28x - 60$
28. $x^2 - 13x + 40$
29. $x^2 + 22x + 40$
30. $x^2 + 3x + 5$
31. $x^2 - 4x + 5$
32. $x^2 - 4x - 5$
33. $x^2 - 6x$
34. $x^2 + 9x$
35. $x^2 - 6x - 16$
36. $x^2 + 15x + 36$
37. $x^2 - 15x + 54$
38. $x^2 + x$
39. $x^2 + 100x$
40. $x^2 - 48x - 100$
41. $x^2 - 10x + 24$
42. $x^2 - 18x + 81$
43. $x^2 - 49$
44. $x^2 - 16x - 36$
45. $x^2 - 16x + 28$
46. $x^2 + 63x - 64$
47. $x^2 - 4x + 4$
48. $x^2 + 21x$
49. $x^2 - 2x - 48$
50. $x^2 + 14x + 33$

SELFTEST

Factor completely:

1. $x^2 - 7x - 8$
2. $x^2 - 5x - 6$
3. $x^2 + 5x + 6$
4. $x^2 + 10x - 24$
5. $x^2 + x - 30$
6. $x^2 + 4x + 6$
7. $x^2 + 2x - 15$
8. $x^2 + x + 1$

SELFTEST ANSWERS

1. $(x - 8)(x + 1)$
2. $(x - 6)(x + 1)$
3. $(x + 2)(x + 3)$
4. $(x + 12)(x - 2)$
5. $(x + 6)(x - 5)$
6. does not factor
7. $(x - 3)(x + 5)$
8. does not factor

4.4 FACTORING $ax^2 + bx + c$

To factor polynomials of the type

$$ax^2 + bx + c$$

we use the same trial and error technique as before. But for this type of polynomial there are more possibilities to consider.

Example 1 Factor $6x^2 + 7x + 2$ into the product of two binomials: The general format is now

$$(?x + ?)(?x + ?)$$

We must supply not only the last terms, but also the coefficients of the first terms.

Since the product of the first terms should be $6x^2$, we select the coefficients from the pairs 6, 1 and 3, 2. We need not consider the pairs $-6, -1$ and $-3, -2$. For these pairs do not give us any new factorizations. Note that

$$(-6x + 2)(-x + 1) = (-1)(6x - 2)(-1)(x - 1) = (6x - 2)(x - 1)$$

The last factorization is one we obtain by using the pair 6, 1. Consequently, we need to select last terms to fit in the two patterns

$$(6x + ?)(x + ?) \qquad \text{or} \qquad (3x + ?)(2x + ?)$$

The last terms are either 2 and 1 or -2 and -1 since these are the only pairs with product 2.

$$\begin{array}{ll} (6x + 2)(x + 1) & (6x - 2)(x - 1) \\ (6x + 1)(x + 2) & (6x - 1)(x - 2) \\ (3x + 2)(2x + 1) & (3x - 2)(2x - 1) \\ (3x + 1)(2x + 2) & (3x - 1)(2x - 1) \end{array}$$

The actual factorization is $(3x + 2)(2x + 1)$ since this product yields the original polynomial $6x^2 + 7x + 2$.

General Factor Technique: **To factor polynomials of the form**

$$ax^2 + bx + c$$

try all combinations of the form

$$\underbrace{(qx + \overbrace{r)(sx}^{\text{inner product}} + t)}_{\text{outer product}} = qsx^2 + (qt + rs)x + rt$$

where q and s are positive numbers with product a, and r and t are integers with product c. The factorization is the combination in which the sum of the inner and outer products gives the middle term bx.

Example 2 Factor $8x^2 + 14x + 6$ completely: First we check for common factors and factor out these.

$$8x^2 + 14x + 6 = 2(4x^2 + 7x + 3)$$

Now we factor $4x^2 + 7x + 3$. The positive pairs with product 4 are 4, 1 and 2, 2. The pairs with product 3 are 3, 1 and $-3, -1$.

$$\begin{array}{ll} (4x + 3)(x + 1) & (4x - 3)(x - 1) \\ (4x + 1)(x + 3) & (4x - 1)(x - 3) \\ (2x + 3)(2x + 1) & (2x - 3)(2x - 1) \end{array}$$

The factorization of $4x^2 + 7x + 3$ is $(4x + 3)(x + 1)$. So the complete factorization of $8x^2 + 14x + 6$ is $2(4x + 3)(x + 1)$.

Some polynomials of the form $ax^2 + bx + c$ might actually be perfect squares. If the coefficients are large, it is especially worthwhile to see if the polynomial is a perfect square. If it is, it factors immediately as in Section 4.2. Otherwise, we might have to consider many possibilities.

Example 3 Factor $81x^2 + 126x + 49$ completely: The polynomial is a perfect square of the form $a^2 + 2ab + b^2$. In this case, $a = 9x$, $b = 7$, and $2ab = 126x$. Consequently it factors as $(a + b)^2$ or $(9x + 7)^2$.

Example 4 Factor $16x^2 - 46x + 15$ completely: Unfortunately, this polynomial is not a perfect square. Nor does it have any common factors. So we list the possibilities and hope one of the early tries works. Positive pairs with product 16 are 16, 1; 8, 2; 4, 4. Pairs with product 15 are 15, 1; 5, 3; −15, −1; −5, −3.

$(16x + 15)(x + 1)$	$(8x + 15)(2x + 1)$	$(4x + 15)(4x + 1)$
$(16x + 1)(x + 15)$	$(8x + 1)(2x + 15)$	$(4x + 5)(4x + 3)$
$(16x + 5)(x + 3)$	$(8x + 5)(2x + 3)$	
$(16x + 3)(x + 5)$	$(8x + 3)(2x + 5)$	

The other possible factorizations use −15, −1 and −5, −3. The factorizations using these values are almost the same as those above which use the corresponding positive pairs. The only difference is that the signs of the middle terms are opposites. The factorization $(8x + 3)(2x + 5)$ has middle term $+46x$. We need $-46x$. So, we change the signs of both last terms.

$$16x^2 - 46x + 15 = (8x - 3)(2x - 5)$$

If the leading term of a polynomial is negative, we simply factor out a negative one from the entire polynomial.

Example 5 Factor $-3x^2 + 14x + 5$ completely:

$$\begin{aligned} -3x^2 + 14x + 5 &= -1(3x^2 - 14x - 5) \\ &= -1(3x + 1)(x - 5) \end{aligned}$$

But $(-3x - 1)(x - 5)$ or $(3x + 1)(-x + 5)$ is also correct.

PROGRAM 4.4

1. In the previous section, we factored polynomials of the type $x^2 + bx + c$.

The coefficient of the x^2 term is ____.

1

2. We use the same techniques to factor polynomials of the form $ax^2 + bx + c$, where the coefficient a is not necessarily one. The coefficient of the x^2 term in $3x^2 + 2x - 5$ is ____.

3

3. To factor $2x^2 + 15x + 7$, we use the general format

$$(?x + ?)(?x + ?)$$

We provide the coefficients of the first terms and also the last terms.

$$2x^2 + 15x + 7 = (2x + 1)(\_\_x + \_\_)$$

1; 7

Remember the product of the first terms should be $2x^2$ and the sum of the inner and outer products should be $15x$.

4. In the product

inner product

$$(2x + 3)(4x - 5)$$

outer product

the inner product is ____. The outer product is ____.

$12x$; $-10x$

5. In $(6x + 5)(3x - 4)$, the inner product is ____. The outer product is ____.

$15x$; $-24x$

6. The general factor technique for polynomials of the form $ax^2 + bx + c$ is:

   1. Write the polynomial in standard form.
   2. Try all combinations of the form $(?x + ?)(?x + ?)$, where the coefficients of the first terms are positive and have product a. The product of the last terms should be c.
   3. The factorization is the combination in which the sum of the inner and outer products is the middle term ____.

bx

7. Which combination is the factorization of $15x^2 - x - 6$?

   a. $(15x + 1)(x - 6)$ *b.* $(15x + 3)(x - 2)$ *c.* $(15x + 1)(x - 6)$
   d. $(5x - 2)(3x + 3)$ *e.* $(5x + 1)(3x - 6)$ *f.* $(5x + 3)(3x - 2)$

f.

8. Which of the following is the factorization of $4x^2 + 11x + 7$?

   a. $(4x + 7)(x + 1)$ *b.* $(4x + 1)(x + 7)$ *c.* $(2x + 7)(2x + 1)$

a.

9. To factor $3x^2 + 8x + 5$, we use the format $(?x + ?)(?x + ?)$. The product of the first terms should be ____. The product of the last terms should be ____.

$3x^2$

5

10. The two possibilities for factoring $3x^2 + 8x + 5$ which use positive numbers are $(3x + 5)(x + \_\_)$ and $(3x + 1)(x + \_\_)$. It (does\does not) matter if the 5 goes with the $3x$ or with the x. The correct factorization of $3x^2 + 8x + 5$ is ________.

1; 5; does

$(3x + 5)(x + 1)$

11. The two possibilities for factoring $3x^2 + x - 2$ which have -2 and 1 as the last terms are

    $(3x - 2)(\_\_\_\_)$ and $(3x + 1)(\_\_\_\_)$

    The correct factorization is ________.

$x + 1$; $x - 2$

$(3x - 2)(x + 1)$

12. The factorization of $8x^2 + 2x - 15$ is $(2x + 3)(\_\_x - \_\_)$.

4; 5

13. The factorization of $8x^2 + 14x - 15$ is $(2x + 5)(\_\_x - \_\_)$.

4; 3

14. The factorization of $14x^2 + 15x + 4$ is $(7x + \_\_)(2x + \_\_)$.

4; 1

15. The factorization of $10x^2 + 17x + 3$ is $(\_\_\_ + 1)(\_\_\_ + 3)$.

$5x$; $2x$

16. The possible combinations for the factorization of $6x^2 + 11x + 5$ are

    $(6x + 1)$____ $(3x + 1)$____
    $(6x + 5)$____ $(3x + 5)$____

    and the corresponding combinations with last terms of the opposite signs.

    The correct factorization is ________.

$(x + 5)$; $(2x + 5)$

$(x + 1)$; $(2x + 1)$

$(6x + 5)(x + 1)$

17. To factor $4x^2 - 16x - 9$ we try the combinations

    $(4x + 9)$____ $(4x + 1)$____ $(4x + 3)$____
    $(2x + 9)$____ $(2x + 3)$____

    and the corresponding combinations with opposite signs on the last terms.

    The correct factorization is ________.

$(x-1)$; $(x-9)$; $(x-3)$

$(2x - 1)$; $(2x - 3)$

$(2x - 9)(2x + 1)$

18. The positive pairs with product 4 are ____, ____ and ____, ____. So to factor $6x^2 + 11x + 4$, we consider the combinations

$(6x + \_)(x + \_)$ $(3x + \_)(2x + \_)$

$(6x + \_)(x + \_)$ $(3x + \_)(2x + \_)$

$(6x + \_)(x + \_)$ $(3x + \_)(2x + \_)$

The correct factorization is ________________.

4, 1; 2, 2

4; 1; 4; 1

1; 4; 1; 4

2; 2; 2; 2

$(3x + 4)(2x + 1)$

19. The possible combinations for factoring $3x^2 - 2x - 5$ are

$(3x + _)$________ $(3x - _)$________

$(3x + _)$________ $(3x - _)$________

The correct factorization is ________________.

5; $(x - 1)$; 5; $(x + 1)$

1; $(x - 5)$; 1; $(x + 5)$

$(3x - 5)(x + 1)$

20. The positive pair with product 7 is __, __. The pairs with product 5 are __, __ and __, __. Factor $7x^2 - 12x + 5$: ________________.

7, 1

5, 1 and −5, −1; $(7x - 5)(x - 1)$

21. The positive pair with product 3 is __, __. The pairs with product 7 are __, __; __, __. Factor $3x^2 - 4x - 7$: ________________.

3, 1

7, 1; −7, −1; $(3x - 7)(x + 1)$

22. Many times the larger the coefficients in a polynomial, the more possibilities there are to be tried in the factoring process. So we first look for common monomial factors. To factor $12x^2 - 16x - 28$, we observe that ____ is the greatest common factor.

$$12x^2 - 16x - 28 = \_\_\_\_(3x^2 - 4x - 7)$$

In the previous frame, we factored $3x^2 - 4x - 7$. So the complete factorization of $12x^2 - 16x - 28$ is ____$(3x - 7)(x + 1)$.

4

4

4

23. Factor: $6x^2 + 16x + 10 = 2(\_\_\_\_\_\_\_\_\_\_\_\_)$

$= 2(\_\_\_\_\_\_)(\_\_\_\_\_\_)$

$3x^2 + 8x + 5$

$3x + 5$; $x + 1$

24. Factor $25x^2 + 50x - 75$ completely:

$$25x^2 + 50x - 75 = \_\_(x^2 + 2x - 3)$$

$$= \_\_(\_\_\_\_\_\_)(\_\_\_\_\_\_)$$

25

25; $x + 3$; $x - 1$

25. Polynomials of the form $ax^2 + bx + c$ might actually be perfect squares. It is worthwhile to make a quick check. The polynomial $9x^2 + 30x + 25$ is a perfect square of the form $a^2 + 2ab + b^2$. In this case, $a^2 = 9x^2$ and $b^2 =$ ____. Since $a^2 + 2ab + b^2 = (a + b)^2$,

$$9x^2 + 30x + 25 = (\_\_\_\_\_\_)^2$$

25

$3x + 5$

26. Is $36x^2 + 84x + 49$ a perfect square? (yes\no). Factor $36x^2 + 84x + 49$ completely. ________.

yes

$(6x + 7)^2$

27. $9x^2 + 6x - 3$ (is\is not) a perfect square. Factor it completely.

is not

$3(3x - 1)(x + 1)$. The order of the factors does not matter.

28. If the x^2 term is negative, we can factor out -1 and change the signs of all the terms.

$3x^2 - 16x + 5$

$$-3x^2 + 16x - 5 = -1(\_\_\_\_\_\_\_\_)$$

$3x - 1;\ x - 5$

$$= -1(\_\_\_\_)(\_\_\_\_)$$

29. Factor $-2x^2 - x + 3$ completely:

$2x^2 + x - 3$

$$-2x^2 - x + 3 = -1(\_\_\_\_\_\_\_\_)$$

$2x + 3;\ x - 1$

$$= -1(\_\_\_\_)(\_\_\_\_)$$

30. It is also acceptable not to factor out -1. In which case, we must worry about the signs of the coefficients of the first terms in $(?x + ?)(?x + ?)$. However,

$$-2x^2 - x + 3 = (-2x - 3)(x - 1)$$

is also correct. The complete factorization of $-x^2 + 6x + 7$ is any one of the following:

$(x - 7)(x + 1)$

$$-1\_\_\_\_\ \ \_\_\_\_$$

$x - 7$

$$(-x - 1)(\_\_\_\_)$$

$-x + 7$

$$(x + 1)(\_\_\_\_)$$

31. Factor $-10x^2 - 4x + 14$ completely.

$-2(5x + 7)(x - 1)$
or $2(-5x - 7)(x - 1)$
or $2(5x + 7)(-x + 1)$

PRACTICE PROBLEMS

Factor completely:

1. $2x^2 + 5x + 3$
2. $2x^2 + 7x + 3$
3. $3x^2 + 2x - 5$
4. $3x^2 + 14x - 5$
5. $2x^2 - 7x + 5$
6. $2x^2 - 7x + 3$
7. $4x^2 + 9x + 5$
8. $4x^2 + 12x + 5$
9. $4x^2 - 8x - 5$
10. $6x^2 - 3x - 3$
11. $6x^2 - 17x - 3$
12. $6x^2 + 9x + 3$
13. $6x^2 + 11x + 3$
14. $3x^2 + 2x - 16$
15. $3x^2 + 8x - 16$
16. $8x^2 + 10x - 7$
17. $8x^2 + x - 7$
18. $9x^2 + 6x - 8$
19. $9x^2 - x - 8$
20. $9x^2 - 14x - 8$
21. $6x^2 + x - 15$
22. $6x^2 + 9x - 15$
23. $6x^2 - 27x - 15$
24. $6x^2 - 9x - 15$
25. $3x^2 - 19x + 20$
26. $3x^2 - 17x + 20$
27. $-3x^2 + 28x + 20$
28. $-3x^2 + 23x - 20$
29. $2x^2 + 8x + 6$
30. $-2x^2 + 4x + 6$
31. $-6x^2 - 10x + 4$
32. $6x^2 - 2x - 4$

SELFTEST

Factor completely:

1. $3x^2 + 5x + 2$
2. $4x^2 - 12x + 5$
3. $4x^2 + 11x + 6$
4. $6x^2 - 4x - 10$
5. $10x^2 + 18x - 4$
6. $-3x^2 + x + 2$

SELFTEST ANSWERS

1. $(3x + 2)(x + 1)$ 2. $(2x - 5)(2x - 1)$ 3. $(4x + 3)(x + 2)$
4. $2(3x - 5)(x + 1)$ 5. $2(5x - 1)(x + 2)$ 6. $-1(3x + 2)(x - 1)$

4.5 LEAST COMMON MULTIPLES OF POLYNOMIALS

In Chapter 1 we discussed multiples of whole numbers. The number m is a *multiple* of n if n is a factor of m. Some multiples of 6 are $6 = 6 \cdot 1$, $12 = 6 \cdot 2$, $42 = 6 \cdot 7$, and $600 = 6 \cdot 100$.

Likewise, one polynomial P is a multiple of another polynomial R if R is a factor of P. The polynomials

$$3x + 3, \quad -2x - 2, \quad x^2 - 1, \quad \text{and} \quad x^2 + 3x + 2$$

are multiples of $x + 1$. For $x + 1$ is a factor of each of them.

$$3x + 3 = 3\underbrace{(x + 1)} \qquad -2x - 2 = -2\underbrace{(x + 1)}$$
$$x^2 - 1 = \underbrace{(x - 1)}(x + 1) \qquad x^2 + 3x + 2 = (x + 2)\underbrace{(x + 1)}$$

Any polynomial that has $x + 1$ as one of its factors is a multiple of $x + 1$.

Given several polynomials, is there a *common multiple*? That is, is there a polynomial which has each of the original polynomials as factors? The answer is yes. Some common multiples of

$$5, \quad x + 3, \quad \text{and} \quad x + 4$$

are

$$5(x + 3)(x + 4) \qquad 5 \cdot 2(x + 3)(x + 4) \qquad 5(x + 1)(x + 2)(x + 3)(x + 4)$$

We could continue forming common multiples. The criteria is that 5, $x + 3$, and $x + 4$ should all be factors of the new common multiple. But $5(x + 4)(x + 2)$ is *not* a common multiple of 5, $x + 3$, and $x + 4$ because $x + 3$ is not included as a factor.

The least common multiple of 5, $x + 3$, and $x + 4$ is $5(x + 3)(x + 4)$. All the necessary factors are present, but no factor is extra.

Let's find the least common multiple of 2, $x^2 - 9$, and $2x + 6$. At first glance it might appear that $2(x^2 - 9)(2x + 6)$ is the least common multiple. But there is a multiple that has fewer prime factors. To find it, we factor each of the original polynomials completely and use only the necessary factors.

$$2 = 2 \qquad x^2 - 9 = (x + 3)(x - 3) \qquad 2x + 6 = 2(x + 3)$$

The product

$$2(x + 3)(x - 3)$$

contains all the factors of the original polynomials. But some factors do double duty.

$$2 \cdot (x + 3) \cdot (x - 3)$$

(2 ← 2; $(x + 3) \cdot (x - 3)$ → $x^2 - 9$; $2 \cdot (x + 3)$ → $2x + 6$)

Technique to Form the Least Common Multiple (LCM): **First factor all the polynomials into primes. Then form a product (call it the LCM product) which employs every prime in each of the factorizations. The number of times any given prime should occur in the LCM product is the *maximal* number of times it occurred in *any one* of the factorizations. Then the LCM product will be the least common multiple.**

Example 1 Find the least common multiple of $2x$, $9y$, and $18xy$. In factored form

$$2x = 2x \qquad 9y = 3 \cdot 3y \qquad 18xy = 2 \cdot 3 \cdot 3xy$$

We form the LCM product which employs every prime in each of the factorizations. So the LCM product will employ the primes 2, 3, x, and y. But how many times should each of these primes occur in the LCM product? To answer this, we look at the maximal number of times that prime occurred in any of the factorizations.

What is the maximal number of times 2 was used in any factorization? It was used only once in $2x$, not at all in $9y$, and only once in $18y$. So the maximal number of times it was used is one.

What is the maximal number of times 3 was used in any factorization? It was not used in $2x$, it was used twice in $9y$, and twice in $18y$. So the maximal number of times is two.

Likewise the maximal number of times x was used in any factorization is one. The same is true for y. Now we form the LCM product which employs every prime 2, 3, x, and y the maximal number of times it was found in any factorization. This product is the least common multiple.

$$\text{LCM} = 2 \cdot 3 \cdot 3xy = 18xy$$

Example 2 Find the least common multiple of $8x$, $4x^2$, and $28xy$. The factored forms are

$$8x = 2 \cdot 2 \cdot 2x \qquad 4x^2 = 2 \cdot 2xx \qquad 28xy = 2 \cdot 2 \cdot 7xy$$

Factor	*Maximal number of times it is used in any factorization*
2	3
7	1
x	2
y	1

Thus the LCM product is made up of the factors listed on the left. We use each factor the indicated number of times.

$$\text{LCM} = 2 \cdot 2 \cdot 2 \cdot 7 \cdot xxy = 56x^2y$$

Example 3 What is the least common multiple of $5x$, $2x^3 + 4x^2$, and $x^2 + 4x + 4$? Again we look at the factored forms of the polynomials.

$$5x = 5x \qquad 2x^3 + 4x^2 = 2xx(x + 2) \qquad x^2 + 4x + 4 = (x + 2)(x + 2)$$

We use $x + 2$ twice, x twice, 5 once, and 2 once.

$$\text{LCM} = 5 \cdot 2 \cdot x \cdot x(x + 2)(x + 2) = 10x^2(x + 2)^2$$

PROGRAM 4.5

multiple; is

1. A number m is a *multiple* of n if n is a factor of m. Since $10 = 5 \cdot 2$, 10 is a ________________ of 5. It (is\is not) also a multiple of 2.

a.; *b.*; *d.*

2. Which of the following are multiples of 7?
   a. 14   *b.* 49   *c.* 20   *d.* 28

3. 18 is a multiple of 9 but 3 is not. To form multiples of 9, we (multiply\divide) other numbers by 9. | multiply

4. Is 4 a multiple of 8? (yes\no). Is 16 a multiple of 8? (yes\no). Why? | no; yes

________________. | 8 is a factor of 16

5. We can also talk about multiples of polynomials. The polynomial P is a multiple of R if R is a factor of P. Is $3(x + 2)$ a multiple of $x + 2$? (yes\no). | yes

Is $2(x + 3)$ a multiple of 3? (yes\no). Why? ________________. | no; 3 is not a factor of $2(x + 3)$

6. We form multiples of $x^2 + 4$ by using $x^2 + 4$ as a factor in a product.

$(x - 3)(x^2 + 4)$ is a ________________ of $x^2 + 4$. Another multiple of | multiple

$x^2 + 4$ is ________________. (There are many correct answers.) | (any polynomial) $\cdot\ (x^2 + 4)$

7. Which of the following are multiples of $x - 5$?

a. $(x + 6)(x - 5)$ *b.* $5 \cdot 6(x + 5)(x - 5)$ | *a.*; *b.*
c. $-1(x - 5)$ *d.* $x - 5$ | *c.*; *d.*

8. To create multiples of a polynomial P, we (multiply\divide) the polynomial P by other polynomials. | multiply

9. A *common multiple* of several polynomials is one that contains each of the original polynomials as factors. Therefore $2(x + 5)(x - 4)$ is a common multiple of 2, $x + 5$, and $x - 4$. A common multiple of 3, $x - 2$, and $x + 3$

is $3(x - 2)($________$)$. | $x + 3$

10. Which of the following are common multiples of 7, $x + 3$, and $x + 2$?

a. $7(x + 3)(x + 2)$ *b.* $7(x + 3)(x - 4)$ | *a.*
c. $7 \cdot 3(x + 2)(x + 3)$ *d.* $7(x + 1)(x + 2)(x + 3)$ | *c.*; *d.*

11. Factor $12x^2 + 24x$ completely:

$$12x^2 + 24x = \_\_\_\_\_\_\_\_$$ | $12x(x + 2)$

Is $12x^2 + 24x$ a common multiple of 12 and $x + 2$? (yes\no). Why? | yes

________________. | 12 and $x + 2$ are both factors

12. Is $3x + 9$ a common multiple of 3 and $x + 9$? (yes\no). Because $x + 9$ is | no

not ________________ of $3x + 9$. | a factor

13. The *least common multiple* (LCM) of several polynomials is the common multiple with fewest prime factors. Which is the least common multiple of 5 and $x + 8$?

a. $5(x + 8)$ *b.* $5(-2)(x + 8)$ *c.* $3(x + 8)$ | *a.*

14. The least common multiple is a product formed by using the minimal number of prime factors. The least common multiple of $2x + 2$ and $x^2 - 1$ is

$$\text{LCM} = 2(x + 1)(x - 1)$$

Which factors of the LCM make $2x + 2$? ________________. Which | 2 and $(x + 1)$

factors make $x^2 - 1$? ________________. | $(x + 1)$ and $(x - 1)$

15. ***Rule to Form the Least Common Multiple (LCM):***

1. Factor the polynomials into primes.
2. Form a product (called the LCM product) which uses every prime in each of the factorizations. The number of times any given prime should occur in the LCM product is the maximal number of times it occurs in any of the factorizations. The LCM product is the least common multiple.

a product

The least common multiple is (a product\a sum\a difference).

16. To find the LCM of $6x^2$ and $3xy$ we factor $6x^2$ and $3xy$ into primes.

x

$$6x^2 = 2 \cdot 3x\underline{\quad} \qquad 3xy = 3xy$$

The maximal number of times 2 occurs in any one of the factorizations is once.

once

The maximal number of times 3 occurs in any one of the factorizations is (once\twice\no times).

twice; once
twice

The factor x occurs (once\twice) in $2 \cdot 3xx$ and (once\twice) in $3xy$ so the maximal number of times it occurs in one factorization is (once\twice\three times).

The factor y occurs only once. So

x; x; y

$$\text{LCM} = 2 \cdot 3 \cdot \underline{\quad\quad} \cdot \underline{\quad\quad} \cdot \underline{\quad\quad}$$

17. What is the least common multiple of $4x^2y$; $3xy^2$; $6x^3yz$? First we look at the prime factorizations.

y; 2; y; z

$$4x^2y = 2 \cdot 2xxy \qquad 3xy^2 = 3xy\underline{\quad} \qquad 6x^3yz = 3 \cdot \underline{\quad} \cdot xxx\underline{\quad} \cdot \underline{\quad}$$

Factor	*Maximal number of times it occurs in a factorization*
2	2
3	____
x	3
y	____
z	____

1
2
1

x; y; y; z
^{3}; ^{2}

$$\text{LCM} = 2 \cdot 2 \cdot 3 \cdot xx \cdot \underline{\quad} \cdot \underline{\quad} \cdot \underline{\quad} \cdot \underline{\quad}$$
$$= 12x\text{—}y\text{—}z$$

18. Let's find the LCM of $x^2 + 5x + 4$, $x^2 - 16$, and $x^2 + 8x + 16$. The prime factorizations are

$x + 1$

$$x^2 + 5x + 4 = (x + 4)(\underline{\quad\quad\quad})$$

$x - 4$

$$x^2 - 16 = (x + 4)(\underline{\quad\quad\quad})$$

$x + 4$

$$x^2 + 8x + 16 = (x + 4)(\underline{\quad\quad\quad})$$

twice
once
once

The maximal number of times $(x + 4)$ is used as a factor in any one factorization is (once\twice\three times\four times). The maximal number of times for $(x - 4)$ is (once\twice\not at all). For $(x + 1)$ the maximal number of times is (once\twice\not at all). The least common multiple is given by

$x + 4$; $x - 4$; $x + 1$

$$\text{LCM} = (x + 4)(\underline{\quad\quad\quad})(\underline{\quad\quad\quad})(\underline{\quad\quad\quad})$$

19. Find the LCM of $5x - 10$, $x^2 + x - 6$, and $5x^2$. The prime factorizations are

$$5x - 10 = 5(\_\_\_\_\_\_\_)$$ $x - 2$

$$x^2 + x - 6 = (x + 3)(\_\_\_\_\_\_\_)$$ $x - 2$

$$5x^2 = 5x\_\_$$ x

In the LCM, we use 5 (once\twice\not at all). We use $(x - 2)$ (once\twice\three times) and $(x + 3)$ (once\twice\not at all). How many times do we use x? ________.

once; once
once

twice

$$\text{LCM} = 5 \cdot \_\_ \cdot \_\_ \cdot (x + 3)\_\_\_\_\_\_\_$$ x; x; $(x - 2)$

PRACTICE PROBLEMS

Find the least common multiple.

1. $3x$, $5y$
2. $2x$, $3y$
3. $2x$, $6y$
4. $4x$, $5x$
5. $2x$, $4y$, $8xy$
6. $3x$, $6y$, $12xy$
7. $4y$, $8x^2$, $3x$
8. $7y^2$, $14y$, $5x$
9. $2x$, $x + 3$
10. $8x$, $x - 4$
11. $3x$, $6x + 3$
12. $5x$, $10x - 5$
13. $x + 1$, $x^2 + 2x + 1$
14. $x + 2$, $x^2 + 3x + 2$
15. $2x + 4$, $2x^2 + 6x + 4$
16. $5x - 5$, $5x^2 + 15x + 10$
17. $x^2 + 9x + 8$, $x^2 + 5x + 4$
18. $x^2 - 4$, $x^2 - 4x + 4$

SELFTEST

Find the least common multiple.

1. $10x$, $5y$, $2xy$
2. $x + 4$, $x^2 - 16$
3. $2x$, $x^2 - 1$, $2x + 2$
4. x, $x - 1$, x^2
5. x^2, $3x^2 - 9x$

SELFTEST ANSWERS

1. $10xy$ 2. $(x + 4)(x - 4)$ or $x^2 - 16$ 3. $2x(x + 1)(x - 1)$
4. $x^2(x - 1)$ 5. $3x^2(x - 3)$

1. Factor out the largest possible monomial factor.
   a. $90x^2 - 65x + 15$
   b. $18x^2y + 6xy^2$
   c. $12x^2y^2 + 8x^2$

2. Factor completely:
   a. $x^2 - 9$
   b. $25x^2 - 16$
   c. $4x^2 + 12x + 9$

3. Factor completely:
   a. $x^2 + x - 2$
   b. $x^2 - 3x + 2$
   c. $x^2 + 6x - 7$

4. Factor completely:
   a. $4x^2 + 4x + 1$
   b. $6x^2 + 9x + 3$
   c. $2x^2 + 5x - 3$

5. Find the least common multiple.
   a. $5, 3x$
   b. $4x, 2x + 2$
   c. $6x^2, 5x, x - 1$

Each problem above refers to a section in this chapter as shown in the table.

Problems	*Section*
1	4.1
2	4.2
3	4.3
4	4.4
5	4.5

POSTTEST 4 ANSWERS

1. *a.* $5(18x^2 - 13x + 3)$ *b.* $6xy(3x + y)$ *c.* $4x^2(3y^2 + 2)$

2. *a.* $(x + 3)(x - 3)$ *b.* $(5x + 4)(5x - 4)$ *c.* $(2x + 3)(2x + 3)$

3. *a.* $(x + 2)(x - 1)$ *b.* $(x - 2)(x - 1)$ *c.* $(x + 7)(x - 1)$

4. *a.* $(2x + 1)(2x + 1)$ *b.* $3(2x + 1)(x + 1)$ *c.* $(2x - 1)(x + 3)$

5. *a.* $15x$ *b.* $4x(x + 1)$ *c.* $30x^2(x - 1)$

RATIONAL NUMBERS

BASIC SKILLS

Upon completion of the appropriate sections the student should be able to:

5.1 *a.* Recognize a rational number and locate it on the number line.
b. Determine if two fractions are equal.
c. Reduce a fraction to lowest terms.
d. Identify positive and negative fractions, proper and improper fractions.

5.2 *a.* Add two fractions.
b. Find the additive inverse of a rational number.
c. Subtract one fraction from another.

5.3 *a.* Multiply two fractions.
b. Divide two fractions.
c. Find the reciprocal or multiplicative inverse of a rational number.
d. Simplify expressions that involve addition, subtraction, multiplication, or division of rational numbers.

5.4 *a.* Identify a complex fraction.
b. Simplify a complex fraction.

PRETEST 5

1. *a.* Is every integer a rational number?
   b. Are the numbers 0.5, -2, 2/3, $-1/3$, 11/9 all rational numbers?
   c. Which are integers?
   d. Locate the above numbers on the number line.
   e. Which is the smallest?
   f. Which is the largest?

2. Which of the following fractions are equal?

$$\frac{-3}{2} \qquad \frac{-3}{4} \qquad \frac{-6}{4} \qquad \frac{6}{-8} \qquad \frac{-4}{-10} \qquad \frac{2}{5}$$

3. Simplify and leave your answer in reduced form.

   a. $\dfrac{x}{5} + \dfrac{2x}{5}$

   b. $\dfrac{1}{3x} + \dfrac{1}{2}$

   c. $\dfrac{2}{x} + \dfrac{1}{4}$

   d. $\dfrac{4}{21} - \dfrac{7}{3}$

   e. $\dfrac{6}{x} - \dfrac{2}{x}$

   f. $\dfrac{6}{y} - \dfrac{4}{x}$

4. Simplify and leave your answers in reduced form.

   a. $\left(\dfrac{7}{5}\right)\left(\dfrac{15}{21}\right)$

   b. $\left(\dfrac{6x}{y}\right)\left(\dfrac{-3y}{12x}\right)$

   c. $\left(\dfrac{1}{2} - \dfrac{4}{3}\right)\left(\dfrac{2}{5}\right)$

   d. $\dfrac{2x}{y} \div \dfrac{4}{y}$

   e. $\dfrac{5x}{y} \div \dfrac{20x}{y}$

   f. $\dfrac{7x}{2y} \div 14x$

5. Simplify and leave your answer in reduced form.

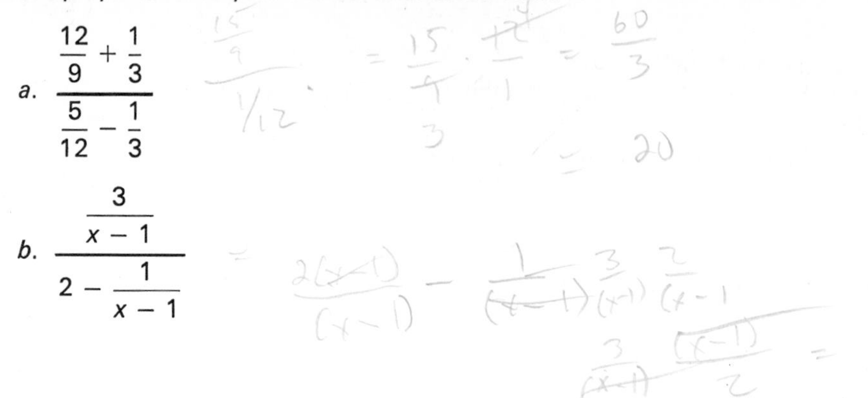

a. $\dfrac{\dfrac{12}{9}+\dfrac{1}{3}}{\dfrac{5}{12}-\dfrac{1}{3}}$

b. $\dfrac{\dfrac{3}{x-1}}{2-\dfrac{1}{x-1}}$

Each problem above refers to a section in this chapter as shown in the table.

Problems	*Section*
1 and 2	5.1
3	5.2
4	5.3
5	5.4

PRETEST 5 ANSWERS

1. *a.* Yes. *b.* Yes. *c.* The only integer is -2.
d.

Smallest -2; $-\frac{1}{3}$; 0.5; $\frac{2}{3}$; $\frac{11}{9}$ Largest

(number line marked -2, -1, 0, 1, 2)

e. The smallest is -3. *f.* The largest is $11/9$.

2. $-3/2 = -6/4$; $-3/4 = 6/-8$; $-4/-10 = 2/5$

3. *a.* $3x/5$ *b.* $(2 + 3x)/6x$ *c.* $(8 + x)/4x$ *d.* $-15/7$ *e.* $4/x$ *f.* $(6x - 4y)/xy$

4. *a.* 1 *b.* $\dfrac{-3}{2}$ *c.* $\dfrac{-1}{3}$ *d.* $\dfrac{x}{2}$ *e.* $\dfrac{1}{4}$ *f.* $\dfrac{1}{4y}$ **5.** *a.* 20 *b.* $\dfrac{3}{2x-3}$

5.1 RATIONAL NUMBERS AND THE NUMBER LINE

Let a and b be integers and suppose that b is not zero. Then a number of the form a/b is called a fraction or a rational number. We call a the numerator and b the denominator of the fraction a/b. Some examples of rational numbers are 1/3, 18/17, −5/4, 0/10, 8/−16, and 0.31. The numerator of the fraction 81/17 is 81. The denominator is 17.

The fractions 5/8 and 5/8 are identical and so they are of course equal. But there are fractions which are equal and not identical. For example, 8/16 and 2/4 are equal even though they are not identical. We say that 8/16 and 2/4 represent the same rational number. The following rule may easily be applied to determine when two fractions are equal.

Rule for Equality of Fractions: **Let a/b and c/d be two fractions. Then**

$$\frac{a}{b} = \frac{c}{d}$$

if and only if $ad = bc$.

Example 1 $\quad \dfrac{7}{14} = \dfrac{3}{6} \quad$ since $\quad 7 \cdot 6 = 14 \cdot 3$

Example 2 $\quad \dfrac{-1}{3} = \dfrac{1}{-3} \quad$ since $\quad (-1)(-3) = 3 \cdot 1$

Example 3 $\quad \dfrac{-5x}{-6} = \dfrac{10x}{12} \quad$ since $\quad (-5x)(12) = (-6)(10x)$

A general application of the rule for equality of fractions allows us to reduce fractions.

Rule for Reducing Fractions: **If k is any nonzero integer, then**

$$\frac{a \cdot k}{b \cdot k} = \frac{a}{b}$$

This relation is true since $(ak)b = (bk)a$.

So if the numerator and denominator have a factor in common, we can cancel or omit that factor. To "reduce to lowest terms" means to cancel all the factors which the numerator and denominator have in common.

Example 4 Reduce 4/24:

$$\frac{4}{24} = \frac{1 \cdot \cancel{2} \cdot \cancel{2}}{2 \cdot \cancel{2} \cdot \cancel{2} \cdot 3} \qquad \text{by factorization}$$

$$= \frac{1}{6} \qquad \text{by the reducing rule}$$

Example 5 $\quad \dfrac{-25}{35} = \dfrac{\cancel{5}(-5)}{\cancel{5} \cdot 7} = \dfrac{-5}{7}$

Example 6 $\quad \dfrac{-6x^2}{-24x} = \dfrac{\cancel{2}\cancel{(-3)}\cancel{x}x}{\cancel{2}(2)(2)\cancel{(-3)}\cancel{x}} = \dfrac{x}{4}$

Example 7 $\quad \dfrac{6x + 8}{10x^2 + 4x} = \dfrac{\cancel{2}(3x + 4)}{\cancel{2}(5x^2 + 2x)} = \dfrac{3x + 4}{5x^2 + 2x}$

Rule on Positive and Negative Fractions: **We say the fraction a/b is positive if the numerator and denominator are both positive or both negative. If the numerator and denominator have opposite signs, then the fraction is negative. If a/b is positive then $a/b > 0$. If a/b is negative, then $a/b < 0$. Thus $-5/-4$ is positive since the numerator and denominator are both negative. But $4/-3$ is negative and $4/-3 < 0$ since the numerator, 4, and the denominator, -3, have opposite signs.**

If a fraction is negative, we usually write the fraction with the negative sign in the numerator. For

$$\frac{a}{-b} = \frac{-a}{b}$$

since $ab = (-a)(-b)$. Thus we write $-2/3$ instead of $2/-3$ and $-x/4$ instead of $x/-4$.

We adopt the notational convention that if a is any integer then

$$a = \frac{a}{1}$$

Using this convention, we may think of any integer as being a fraction. Therefore the integers -17, 1, 0, and x are also fractions since $-17 = -17/1$, $1 = 1/1$, $0 = 0/1$, and $x = x/1$. The fraction 8/4 may be identified with the integer 2 since $8/4 = 2/1 = 2$.

The fraction a/b is *proper* if $|a| < |b|$. But if $|a| \geq |b|$, we say that a/b is *improper*. Proper fractions are all less than 1 but larger than -1.

Example 8 $\quad \frac{-23}{21}$ is improper since $|-23| \geq |21|$

Example 9 $\quad \frac{-3}{-5}$ is proper since $|-3| < |-5|$

Example 10 $\quad \frac{6}{-6}$ is improper since $|6| = |-6|$

Now that we have seen some algebraic properties of fractions, let us look at a geometric interpretation of them. The fraction 2/3 has a simple geometric meaning. Take the unit length from 0 to 1 on the number line. Divide this unit length into three equal parts. Starting at zero and proceeding to the right, place two of these parts end to end. The new length so created corresponds to the fraction 2/3.

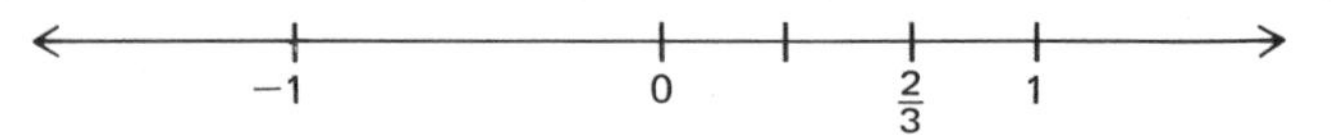

We can count by parts of a unit as well as by whole units. For example, we can count by thirds beginning with 0/3.

Whole units	0			1			2			3
Parts of units	$\frac{0}{3}$	$\frac{1}{3}$	$\frac{2}{3}$	$\frac{3}{3}$	$\frac{4}{3}$	$\frac{5}{3}$	$\frac{6}{3}$	$\frac{7}{3}$	$\frac{8}{3}$	$\frac{9}{3}$

The numerator tells *how many* parts we have counted, the denominator tells *which* part.

The fraction $-3/4$ is negative. It will be located on the negative part of the number line. This time we divide the unit length between 0 and -1 into four equal parts. Three of these parts placed end to end (starting at 0 and going left) position us at the rational number $-3/4$.

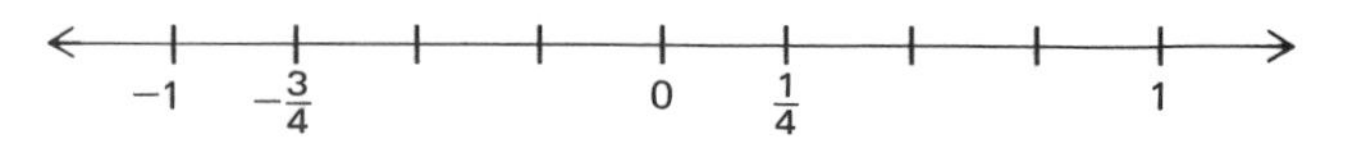

To locate the improper fraction $9/-5$, we look at the equal fraction $-9/5$. We then divide *each* unit length into five equal pieces. Then beginning from zero, we proceed to the left until we have passed 9 of these pieces.

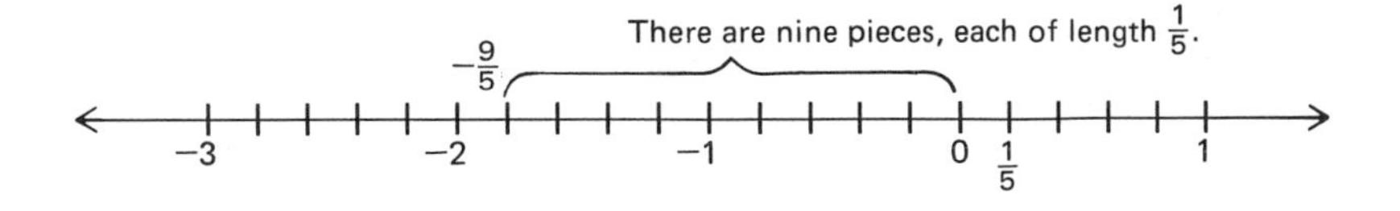

We can compare two fractions by looking at their relative positions on the number line. If they both correspond to the same point, they are equal. If a/b is to the right of c/d then $a/b > c/d$. We can determine if a fraction is proper. For if it lies to the right of -1 and to the left of 1, it is proper. Otherwise it is improper.

Example 11 Compare the fractions $3/2$, $-4/2$, $7/4$, $6/4$, and $-2/1$. First we graph the fractions with denominator 2.

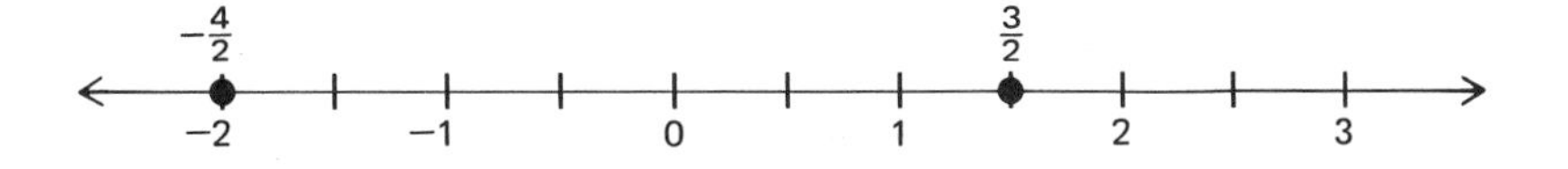

Then we graph the fractions with denominator 4 on a line with the same unit length.

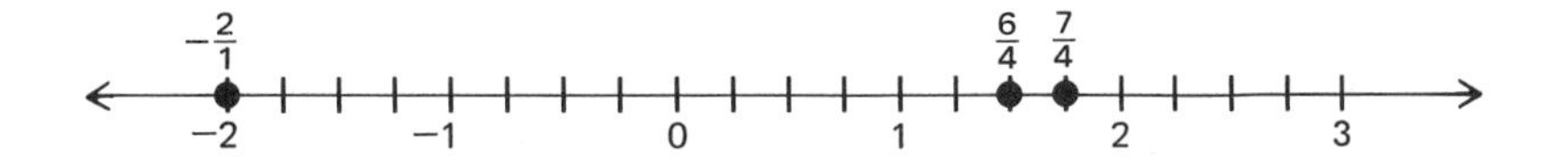

If we mentally superimpose the second graph on the first we see that $7/4$ is the largest fraction and

$$\frac{3}{2} = \frac{6}{4}, \quad \frac{-2}{1} = \frac{-4}{2}, \quad \frac{-4}{2} < \frac{3}{2}, \quad \text{and} \quad \frac{3}{2} > \frac{-2}{1}$$

Also we can see that all the fractions are improper.

PROGRAM 5.1

1. Let a be an integer and b be a nonzero integer. Then a number of the form a/b is called a rational number or fraction. We call a the numerator and b the denominator. Therefore 18/31 is an example of a __________ number or fraction; and 18 is the __________, while 31 is the __________.

   rational
   numerator
   denominator

2. The fraction 6/11 is a __________ number with numerator _____ and denominator _____.

   rational; 6
   11

3. The fractions 1/2 and 3/6 are not identical (the numerators are not the same and the denominators are not the same). But they are equal. The next rule allows us to determine whether two fractions are equal.

Rule for Equality of Fractions:

$$\frac{a}{b} = \frac{c}{d}$$

if and only if $ad = bc$. So $1/2 = 3/6$ because $1 \cdot 6 =$ ____ $\cdot$ ____. — 2; 3

4. $2/3 = -4/-6$ because (____)(____) = (____)(____). — 2; −6; 3; −4

5. 3/8 (=\≠) 2/5 because 3 · 5 (=\≠) 8 · 2. — ≠, ≠

6. −4/8 (=\≠) 5/−10 because (−4)(−10) (=\≠) (8)(5). — =, =

7. $4/2x$ (=\≠) 2/4. — ≠

8. $16x^2/48x$ (=\≠) $x/3$. — =

9. $ak/bk = a/b$ since (ak)(____) = (bk)(____). — b; a

10. The previous frame gives the basis for reducing fractions.

Rule for Reducing Fractions: For any nonzero integer k,

$$\frac{ak}{bk} = \frac{a}{b}$$

So $\dfrac{8 \cdot 4}{3 \cdot 4} =$ ____. — $\frac{8}{3}$

11. To reduce a fraction to lowest terms means to factor the numerator and denominator and cancel all the factors common to both parts of the fraction. Reduce to lowest terms:

$$\frac{25}{75} = \frac{1 \cdot \cancel{5} \cdot \cancel{5}}{3 \cdot \cancel{5} \cdot \cancel{5}} = \frac{1}{3}$$

$$\frac{9}{18} = \frac{1 \cdot \cancel{3} \cdot \cancel{3}}{2 \cdot \cancel{3} \cdot \cancel{3}} = \_\_\_\_$$

— $\frac{1}{2}$

12. $\dfrac{-40}{-100} = \dfrac{-1 \cdot 2 \cdot 2 \cdot 2 \cdot 5}{-1 \cdot 2 \cdot 2 \cdot 5 \cdot 5} =$ ____. — $\frac{2}{5}$

13. $\dfrac{15}{95} = \dfrac{5 \cdot \_\_}{5 \cdot 19} =$ ____. — 3; $\frac{3}{19}$

14. $\dfrac{24xy}{-3x^2} = \dfrac{3 \cdot \_\_ \cdot \_\_ \cdot \_\_}{-1 \cdot 3 \cdot x \cdot x} =$ ____. — 8; x; y; $\frac{8y}{-x}$ or $\frac{-8y}{x}$

15. $\dfrac{5(x+3)}{10(x+3)} = \dfrac{1 \cdot 5 \cdot (x+3)}{2 \cdot 5 \cdot (x+3)} =$ ____. — $\frac{1}{2}$

16. $\dfrac{-27x^4(x+2)}{9x^3(x+2)} =$ ____ (in lowest terms) — $\frac{-3x}{1}$ or $-3x$

17. To reduce a fraction, we cancel (factors\terms) which the numerator and denominator have in common. — factors

18. Does $(3 + x)/(4 + x)$ reduce to 3/4? (yes\no). (Remember, we can cancel only factors.) — no

yes

19. Does $\dfrac{3x}{4x}$ reduce to $\dfrac{3}{4}$? (yes\no).

20. We must carefully distinguish the fractions in the previous two frames. We cannot cancel terms. Which of the following reduce?

b.

a. $\dfrac{5 + x}{6 + x}$ *b.* $\dfrac{5x}{6x}$

$\dfrac{5}{6}$

21. $\dfrac{5x}{6x} =$ ____.

22. Which of the following reduce?

a.

a. $\dfrac{4(x + 2)}{3(x + 2)}$ *b.* $\dfrac{4x + 2}{3x + 2}$

$\dfrac{4}{3}$

23. $\dfrac{4(x + 2)}{3(x + 2)} =$ ____.

24. We use the distributive property to factor polynomials.

$x + 1; \dfrac{x^2 + 1}{x + 1}$

$$\frac{3x^2 + 3}{3x + 3} = \frac{3(x^2 + 1)}{3(\_\_\_\_\_\_)} = \_\_\_\_\_\_$$

$5; \dfrac{x^2 + 2}{3x^2 - 1}$

25. Reduce: $\dfrac{5x^2 + 10}{15x^2 - 5} = \dfrac{\_\_(x^2 + 2)}{5(3x^2 - 1)} =$ ______.

$x^2 + 3; \dfrac{x^2 + 3}{6x^2 + 1}$

26. Reduce: $\dfrac{3x^2 + 9}{18x^2 + 3} = \dfrac{3(\_\_\_\_\_\_)}{3(6x^2 + 1)} =$ ______.

is

27. For every integer a, $a = a/1$. So every integer (is\is not) also a rational number.

4

28. The integer 4 is the same as the rational number ___/1.

$x/1$

29. The integer x is the rational number ____.

0

30. The rational number 0/1 is the same as the integer ____.

$-42/1$

31. The integer -42 is equal to the rational number ____.

positive

negative

32. The fraction a/b is positive if the numerator and denominator both have the same sign. Otherwise, a/b is negative. So $-3/-4$ is (positive\negative) and $-8/9$ is (positive\negative).

positive

33. $-5/-9 > 0$ since $-5/-9$ is (positive\negative).

$<$

34. $4/-7$ ($>$\\$<$) 0 since $4/-7$ is negative.

35. Fill in the appropriate $<$, $>$, or $=$ symbol.

a. $>$; *b.* $<$

a. $\dfrac{2}{3}$ ____ 0 *b.* $\dfrac{-4}{5}$ ____ 0

c. $>$; *d.* $=$; *e.* $<$

c. $\dfrac{-3}{-4}$ ____ 0 *d.* $\dfrac{0}{-2}$ ____ 0 *e.* $\dfrac{7}{-11}$ ____ 0

36. We can write any negative fraction with the minus sign in the numerator. For

$$\frac{a}{-b} = \frac{-a}{b} \quad \text{since} \quad ab = (-b)(-a)$$

So $4/-5 =$ __/5. — -4

37. $21/-52 = -21/$____. — 52

38. $x/-2 =$ ____. — $-x/2$

39. The fraction a/b is *proper* if $|a| < |b|$. It is *improper* if $|a| \geq |b|$. So $-4/2$ is (improper\proper) since $|-4| \geq |2|$. — improper

40. $3/-8$ is (proper\improper). — proper

41. (Proper\Improper) fractions are larger than -1 but smaller than 1. — Proper

42. To locate the proper fraction 3/5 on a number line, we divide the interval from 0 to 1 into 5 equal pieces. We proceed from 0 and go to the right until we have passed 3 of these pieces. This point corresponds to 3/5. Locate 4/5 on this number line.

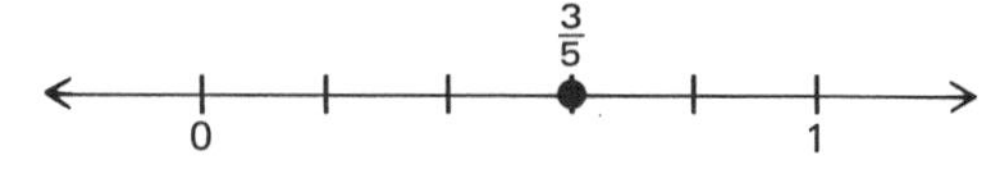

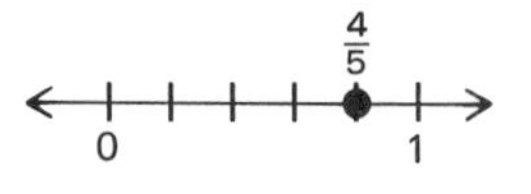

43. To locate $-3/4$ on a number line, we divide the interval from 0 to -1 into ____ equal pieces. Then beginning from 0 we proceed to the (left\right) until we have passed 3 of these pieces. Locate $-3/4$ on this number line. — 4; left

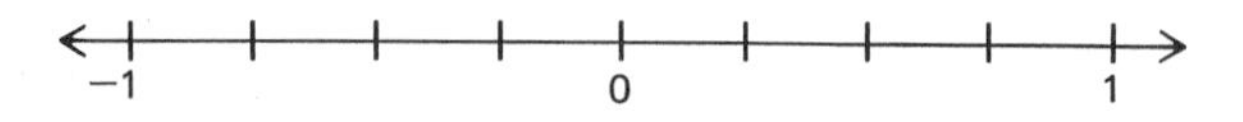

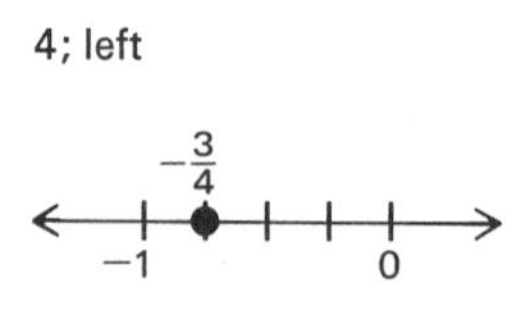

44. Locate $5/-6$ and $5/6$ on a number line.

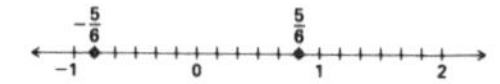

45. To locate an improper fraction on a number line, we divide *each* unit length into the number of equal pieces designated by the denominator. The position of 15/7 is shown here. Locate 10/7, 6/7, and 7/7.

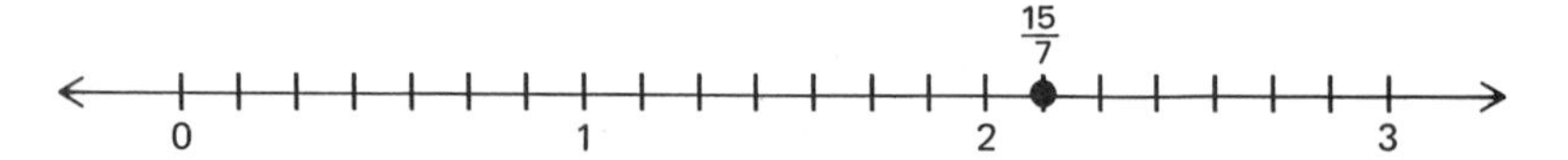

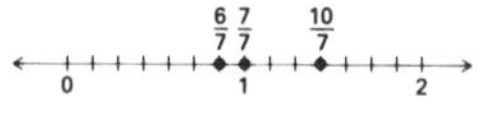

46. Locate 7/10, $-10/10$, $-8/10$, $-17/10$ and 12/10 on the number line below.

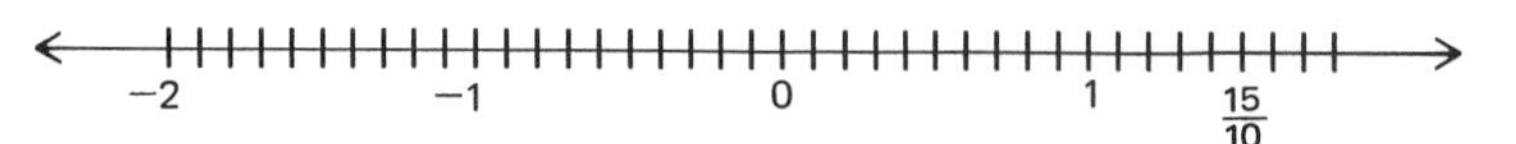

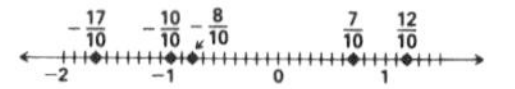

47. Two fractions are equal if they correspond to the same point on the number line. Locate 0.5, 2/2, 2/4, and 4/4 on the same number line.

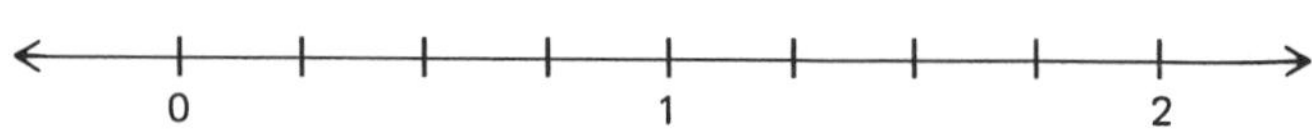

0.5 = 2/4, 2/2 = 4/4;
2/2 = 4/4 = 1

Which fractions are equal? ________. Which fractions equal 1? ________.

2/2

48. The fraction a/b is larger than c/d if a/b lies to the right of c/d on the number line. Which is larger? 2/4 or 2/2? ____. (Refer to the previous frame for the number line.)

right

49. 4/4 > 0.5 since 4/4 lies to the (right\left) of 1/2 on the number line.

50. Locate 3/9, 4/9, −10/9, and 13/9 on the number line. On the same line place 1/3 and −6/3.

$-\frac{6}{3}$ $-\frac{10}{9}$ $\frac{1}{3}, \frac{3}{9}\frac{4}{9}$ $\frac{13}{9}$
−2 −1 0 1 2

−2 −1 0 1 2

>; >

51. Using your answers to the above number line, choose the appropriate > or < symbol.

−10/9 (<\>) −6/3 13/9 (<\>) 1/3

PRACTICE PROBLEMS

Reduce to lowest terms:

1. $\frac{6 \cdot 3}{5 \cdot 3}$ **2.** $\frac{18}{7}$ **3.** $\frac{9}{21}$ **4.** $\frac{4}{-12}$

5. $\frac{-6}{-9}$ **6.** $\frac{8}{13}$ **7.** $\frac{4k^2}{20k}$ **8.** $\frac{42k}{7}$

9. $\frac{4x^2}{8xy}$ **10.** $\frac{25x^4}{30x^6}$ **11.** $\frac{6xy}{3x^2}$ **12.** $\frac{18xy}{12xy^4}$

13. $\frac{(x+4)^2}{(x+4)}$ **14.** $\frac{(x-2)^3}{(x-2)}$ **15.** $\frac{x(x+3)}{x}$ **16.** $\frac{x(x-4)}{(x-4)}$

17. $\frac{3x^2-6x}{3x}$ **18.** $\frac{7x^2+3x}{(7x+3)}$ **19.** $\frac{2x^2-5x}{x}$ **20.** $\frac{4x^2+8x}{2x}$

21. $\frac{6x+4}{3x+2}$ **22.** $\frac{8x-4}{2x-1}$ **23.** $\frac{2x-2}{x}$ **24.** $\frac{2x^2-1}{2x}$

25. $\frac{x^2+x}{3x}$ **26.** $\frac{4x^2-x}{6x}$ **27.** $\frac{5x^2+10}{5x}$ **28.** $\frac{20x+4}{10x}$

29. Which of the following are positive and which are negative?

a. $\frac{1}{5}$ *b.* $\frac{-4}{7}$ *c.* $\frac{14}{-3}$ *d.* $\frac{-9}{-7}$ *e.* $\frac{6}{-12}$ *f.* $\frac{25}{5}$

30. Which of the following are improper fractions?

a. $\frac{-9}{1}$ *b.* $\frac{7}{8}$ *c.* $\frac{12}{-15}$ *d.* $\frac{4}{-3}$ *e.* $\frac{-1}{2}$ *f.* $\frac{-6}{-5}$

31. Locate each of the following fractions on the number line. Which of the fractions is largest? Which is smallest? 1/5, −4/3, −5/2, 12/4.

32. *a.* For what integer values of x will the number $x/2$ be an integer?
b. For what integer values of x will $x/2$ not be an integer?

33. *a.* For what integer values of x will $x/10$ be a proper fraction?
b. For what integer values of x will $x/10$ be an improper fraction?

34. *a.* For what integer values of x will $12/x$ be a proper fraction?
b. For what integer values of x will $12/x$ be an improper fraction?

SELFTEST

1. Give two reasons why $4/6 = 2/3$.

2. *a.* If x is any nonzero integer, why is it true that $2x/3x = 2/3$?
b. Why must x be nonzero?

3. Reduce to lowest terms:

a. $\frac{5x}{8x}$ *b.* $\frac{-33}{22}$ *c.* $\frac{14}{42}$ *d.* $\frac{-7}{-35}$ *e.* $\frac{9x+3}{12x+6}$ *f.* $\frac{6x+2}{8x+2}$

4. *a.* Give examples of rational numbers that are not integers.
b. Why is every integer a rational number?

5. Label each of the following fractions as positive or negative, proper or improper.

a. $\frac{-3}{2}$ *b.* $\frac{1}{2}$ *c.* $\frac{3}{1}$ *d.* $\frac{2}{-1}$

6. *a.* Locate all the fractions in Problem 5 on a single number line.
b. Which fraction is the largest? *c.* Which is the smallest?

SELFTEST ANSWERS

1. $4/6 = 2/3$ since $4 \cdot 3 = 6 \cdot 2$. Also $4/6 = (2 \cdot 2)/(2 \cdot 3) = 2/3$ by the cancellation rule.

2. *a.* Because $(2x)3 = 2(3x)$. It is also true by the cancellation rule.
b. Because we divided by x in $2x/3x$ and we cannot divide by zero.

3. *a.* $\frac{5}{8}$ *b.* $\frac{-3}{2}$ *c.* $\frac{1}{3}$ *d.* $\frac{1}{5}$ *e.* $\frac{3x+1}{4x+2}$ *f.* $\frac{3x+1}{4x+1}$

4. *a.* 1/2, 3/7, 11/10, etc. *b.* If n is an integer then $n/1$ is a rational number and $n/1 = n$.

5. *a.* negative, improper *b.* positive, proper *c.* positive, improper *d.* negative, improper

6. *a.*

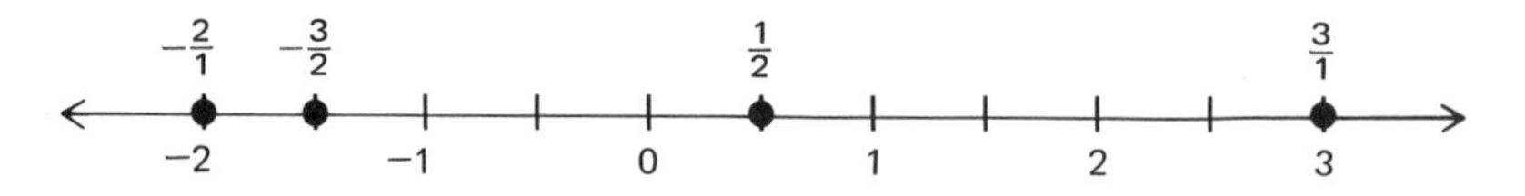

b. 3/1. *c.* 2/−1.

5.2 ADDITION AND SUBTRACTION OF RATIONAL NUMBERS

How can we add the fractions 9/16 and 3/16? Let us find the fraction $9/16 + 3/16$ on the number line. We divide each unit length into 16 equal pieces. Then we locate 9/16. Next we start at 9/16 and mark off 3 more pieces to the right. The stopping point represents $9/16 + 3/16$.

We have marked off 12 pieces in all, so the stopping point is also 12/16. Thus 9/16 plus 3/16 and 12/16 correspond to the same point. Therefore $9/16 + 3/16 = 12/16$.

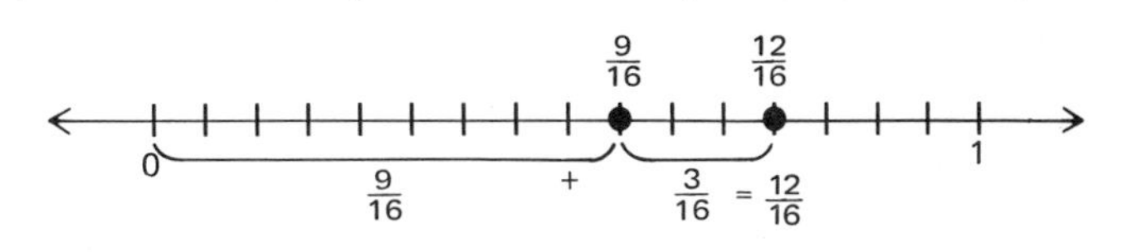

We notice that 9/16 and 3/16 have 16 as a common denominator. The above discussion could be carried through for any two fractions with common denominators. We state this result as a rule.

Rule for Adding Fractions with Common Denominators: **Let a/b and c/b be fractions with common denominator b. Then**

$$\frac{a}{b} + \frac{c}{b} = \frac{a + c}{b}$$

We add only the *numerators* of fractions with like denominators.

Example 1

$$\frac{9}{5} + \frac{12}{5} = \frac{9 + 12}{5} = \frac{21}{5}$$

Example 2

$$\frac{817}{41} + \frac{-63}{41} = \frac{817 + (-63)}{41} = \frac{754}{41}$$

Example 3

$$\frac{4x}{-7} + \frac{6x}{-7} = \frac{4x + 6x}{-7} = \frac{10x}{-7} = \frac{-10x}{7}$$

How do we add fractions that do not have common denominators? We simply rewrite the fractions so that they do have common denominators. Then we apply the above rule. Now the problem is to rewrite fractions so they have a common denominator. What shall we use for a common denominator?

For example, here are two ways to compute $1/2 + 1/3$.

1. First, find a common multiple of 2 and 3: $12 = 2 \cdot 6$ and $12 = 3 \cdot 4$ so 12 is a common multiple of 2 and 3. Then, by an application of the reducing rule,

$$\frac{1}{2} = \frac{6 \cdot 1}{6 \cdot 2} = \frac{6}{12} \quad \text{and} \quad \frac{1}{3} = \frac{4 \cdot 1}{4 \cdot 3} = \frac{4}{12}$$

Therefore

$$\frac{1}{2} + \frac{1}{3} = \frac{6}{12} + \frac{4}{12}$$

Finally, by the addition rule,

$$\frac{6}{12} + \frac{4}{12} = \frac{6 + 4}{12} = \frac{10}{12}$$

2. First, again find a common multiple of 2 and 3: $18 = 2 \cdot 9$ and $18 = 3 \cdot 6$ so 18 is a common multiple of 2 and 3. Then, by the reducing rule,

$$\frac{1}{2} = \frac{9 \cdot 1}{9 \cdot 2} = \frac{9}{18} \quad \text{and} \quad \frac{1}{3} = \frac{6 \cdot 1}{6 \cdot 3} = \frac{6}{18}$$

And then, again from the addition rule,

$$\frac{1}{2} + \frac{1}{3} = \frac{9}{18} + \frac{6}{18} = \frac{9 + 6}{18} = \frac{15}{18}$$

Thus we see that as a common denominator for the fractions 1/2 and 1/3 we could use any common multiple of the denominators 2 and 3. Of course it is most convenient to deal with numbers that are as small as possible; and the least common multiple of 2 and 3 is 6.

$$\frac{1}{2} = \frac{3 \cdot 1}{3 \cdot 2} = \frac{3}{6} \quad \text{and} \quad \frac{1}{3} = \frac{2 \cdot 1}{2 \cdot 3} = \frac{2}{6}$$

$$\frac{1}{2} + \frac{1}{3} = \frac{3}{6} + \frac{2}{6} = \frac{3+2}{6} = \frac{5}{6}$$

All of the computations give us the same value for $1/2 + 1/3$. For the result 10/12 from the first computation reduces to 5/6. The result 15/18 from the second computation also reduces to 5/6. And the result of computation with the least common multiple *is* 5/6.

It is usually preferable to use the least common multiple of the denominators for the common denominator, because then the final result will not need to be reduced very much. However, any common multiple will work for the common denominator. We can always reduce the final result if we need to.

Given the fractions a/b and c/d, is there a general formula for $a/b + c/d$? The answer is yes. For bd is certainly a common multiple of b and d. The following calculations give us the general formula.

$$\frac{a}{b} = \frac{ad}{bd} \quad \text{and} \quad \frac{c}{d} = \frac{bc}{bd}$$

so

$$\frac{a}{b} + \frac{c}{d} = \frac{ad}{bd} + \frac{bc}{bd} = \frac{ad+bc}{bd}$$

General Formula for Adding Fractions:

$$\boldsymbol{\frac{a}{b} + \frac{c}{d} = \frac{ad+bc}{bd}}$$

Example 4

$$\frac{1}{2} + \frac{1}{3} = \frac{1 \cdot 3 + 2 \cdot 1}{2 \cdot 3} = \frac{3+2}{6} = \frac{5}{6}$$

Example 5

$$\frac{2}{x+3} + \frac{4}{x} = \frac{2x + 4(x+3)}{(x+3)x} = \frac{6x+12}{x(x+3)}$$

Example 6

$$\frac{5}{7} + \frac{-3}{2} = \frac{5 \cdot 2 + 7(-3)}{7 \cdot 2} = \frac{10 + (-21)}{14} = \frac{-11}{14}$$

Example 7

$$\frac{3x}{2} + \frac{5x}{4} = \frac{3x(4) + 2(5x)}{2 \cdot 4} = \frac{12x + 10x}{8} = \frac{22x}{8} = \frac{11x}{4}$$

Let us look at a specific example to discover information about the additive inverse of a fraction. The additive inverse of 3/5 is denoted by

$$-\left(\frac{3}{5}\right) \quad \text{or} \quad -\frac{3}{5}$$

So $-3/5$ is the unique fraction such that

$$\frac{3}{5} + \left(-\frac{3}{5}\right) = 0$$

But

$$\frac{3}{5} + \frac{-3}{5} = 0$$

as well. So

$$-\frac{3}{5} = \frac{-3}{5}$$

Rule of Additive Inverse of a Fraction: **The additive inverse of a/b is denoted by $-(a/b)$ and**

$$-\left(\frac{a}{b}\right) = -\frac{a}{b} = \frac{-a}{b} \quad \text{or} \quad \frac{a}{-b}$$

Now we can use the subtraction rule to find the difference of two fractions. For to subtract a/b is the same as to add the additive inverse of a/b.

$$\frac{a}{b} - \frac{c}{d} = \frac{a}{b} + \left(-\frac{c}{d}\right) \quad \text{by the subtraction rule}$$

$$= \frac{a}{b} + \frac{-c}{d} \quad \text{since } -\frac{c}{d} = \frac{-c}{d}$$

$$= \frac{ad + b(-c)}{bd} \quad \text{by the addition rule}$$

$$= \frac{ad - bc}{bd}$$

Rule for Subtracting Fractions:

$$\frac{a}{b} - \frac{c}{d} = \frac{ad - bc}{bd}$$

Example 8

$$\frac{1}{2} - \frac{1}{3} = \frac{1 \cdot 3 - 2 \cdot 1}{2 \cdot 3} = \frac{1}{6}$$

Example 9

$$\frac{2}{5x} - \frac{3}{4} = \frac{2 \cdot 4 - 3(5x)}{(5x)4} = \frac{8 - 15x}{20x}$$

Example 10

$$\frac{-7x}{9} - \frac{4x}{3} = \frac{(-7x)3 - 9(4x)}{9 \cdot 3} = \frac{-21x - 36x}{27} = \frac{-57x}{27} = \frac{-19x}{9}$$

Example 11

$$\frac{2}{5x + 1} - \frac{3}{4x} = \frac{2(4x) - 3(5x + 1)}{(5x + 1)4x} = \frac{-7x - 3}{4x(5x + 1)}$$

Again, we normally reduce a fraction to lowest terms.

In many problems, we can use the addition or subtraction formulas quite effectively. But if we generate polynomials of degree three or higher, there might be difficulty with reducing the result. If the original fractions contain polynomials of degree two or more, it is usually best to use the *least* common denominator. The *least common denominator* (LCD) is just the least common multiple of the denominators (see Section 4.5).*

Example 12 Add:

$$\frac{4}{x^2 - 9} + \frac{x}{x^2 + 5x + 6}$$

First we factor the denominators.

$$x^2 - 9 = (x + 3)(x - 3) \qquad x^2 + 5x + 6 = (x + 3)(x + 2)$$

$$\text{LCM} = (x + 3)(x - 3)(x + 2)$$

* Omit examples 12 and 13 if you have not completed Sections 4.2–4.5.

Next we convert the fractions of the original problem into equal fractions with $(x+3)(x-3)(x+2)$ as denominator.

$$\frac{4}{x^2-9}+\frac{x}{x^2+5x+6}=\frac{4(x+2)}{(x+3)(x-3)(x+2)}+\frac{x(x-3)}{(x+3)(x+2)(x-3)}$$

$$=\frac{4(x+2)+x(x-3)}{(x+3)(x-3)(x+2)}$$

$$=\frac{4x+8+x^2-3x}{(x+3)(x-3)(x+2)}$$

$$=\frac{x^2+x+8}{(x+3)(x-3)(x+2)}$$

Since x^2+x+8 does not factor, the result is in reduced form. We may leave the denominator in factored form.

Example 13 Subtract: $\dfrac{x}{x^2+3x+2}-\dfrac{5}{x^2-3x-4}$

We factor the denominators and construct the least common denominator.

$$x^2+3x+2=(x+1)(x+2) \qquad x^2-3x-4=(x+1)(x-4)$$

$$\text{LCM}=(x+1)(x+2)(x-4)$$

Now we change the fractions to equal fractions with denominators $(x+1)(x+2)(x-4)$.

$$\frac{x}{x^2+3x+2}-\frac{5}{x^2-3x-4}=\frac{x(x-4)}{(x+1)(x+2)(x-4)}-\frac{5(x+2)}{(x+1)(x-4)(x+2)}$$

$$=\frac{x(x-4)-5(x+2)}{(x+1)(x+2)(x-4)}$$

$$=\frac{x^2-4x-5x-10}{(x+1)(x+2)(x-4)}$$

$$=\frac{x^2-9x-10}{(x+1)(x+2)(x-4)}$$

$$=\frac{(x-10)\cancel{(x+1)}}{\cancel{(x+1)}(x+2)(x-4)}$$

$$=\frac{x-10}{(x+2)(x-4)} \quad \text{or} \quad \frac{x-10}{x^2-2x-8}$$

PROGRAM 5.2

1. The fractions a/b and c/b have common denominators. Their common denominator is ____. The sum of a/b and c/b has numerator $a+c$ and denominator b. Therefore,

$$\frac{a}{b}+\frac{c}{b}= \_\_\_\_\_$$

b

$\dfrac{a+c}{b}$

19

2. The fractions 3/19 and 9/19 have a common denominator. It is ____. Since they have a common denominator, we can use the rule in frame 1 to add them. When we add 3/19 and 9/19 the resulting fraction has numerator

$3 + 9$ or 12; 19

________ and denominator ____. So

$\frac{12}{19}$

$$\frac{3}{19} + \frac{9}{19} = \_\_\_\_$$

3. Compute each of the following and then reduce your answer to lowest terms.

$20; \frac{4}{1}$ or 4; $-3; \frac{-1}{2}$

$$\frac{9}{5} + \frac{11}{5} = \frac{\_\_\_\_}{5} = \_\_\_\_ \qquad \frac{-7}{6} + \frac{4}{6} = \frac{\_\_\_\_}{6} = \_\_\_\_$$

$2x; \frac{x}{7}$; $\frac{-10}{11x}$; $\frac{-10}{11x}$

$$\frac{8x}{14} + \frac{-6x}{14} = \frac{\_\_\_\_}{14} = \_\_\_\_ \qquad \frac{-3}{11x} + \frac{-7}{11x} = \_\_\_\_\_\_ = \_\_\_\_\_\_$$

4. All of the following fractions are equal.

$$-\left(\frac{a}{b}\right) = -\frac{a}{b} = \frac{-a}{b} = \frac{a}{-b}$$

equals

So we see that the additive inverse of a/b, which is $-(a/b)$, (equals\does not equal) $-a/b$.

5. By the subtraction rule,

$$\frac{a}{b} - \frac{c}{b} = \frac{a}{b} + \left(-\frac{c}{b}\right)$$

$$= \frac{a}{b} + \frac{-c}{b} \qquad \text{since } -\frac{c}{b} = \frac{-c}{b}$$

$a + (-c)$ or $a - c$

$$= \frac{\_\_\_\_\_\_}{b} \qquad \text{by the addition rule}$$

equals

-9

6. The additive inverse of 9/13 (equals\does not equal) $-(9/13)$. But $-(9/13) = \_\_/13$. So by the subtraction rule,

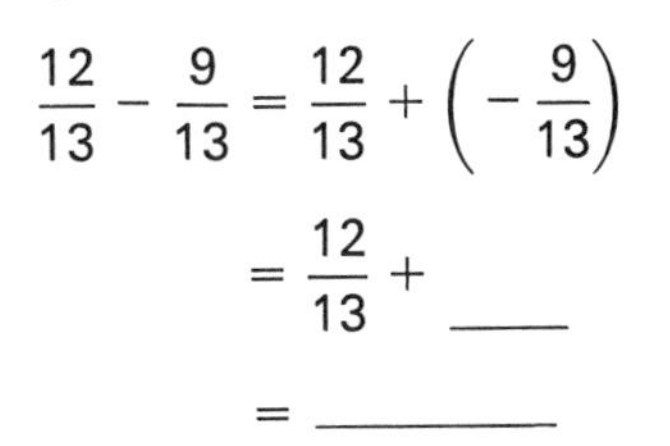

$$\frac{12}{13} - \frac{9}{13} = \frac{12}{13} + \left(-\frac{9}{13}\right)$$

$\frac{-9}{13}$

$$= \frac{12}{13} + \_\_\_\_$$

$\frac{12-9}{13}$ or $\frac{3}{13}$

$$= \_\_\_\_\_\_$$

7. Compute each of the following and then write your answers in lowest terms.

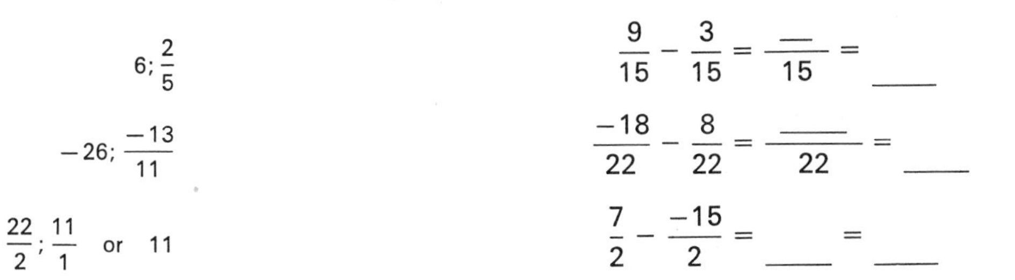

$6; \frac{2}{5}$

$$\frac{9}{15} - \frac{3}{15} = \frac{\_\_}{15} = \_\_\_\_$$

$-26; \frac{-13}{11}$

$$\frac{-18}{22} - \frac{8}{22} = \frac{\_\_\_\_}{22} = \_\_\_\_$$

$\frac{22}{2}; \frac{11}{1}$ or 11

$$\frac{7}{2} - \frac{-15}{2} = \_\_\_\_ = \_\_\_\_$$

do not

8. The fractions 2/15 and 5/6 (do\do not) have a common denominator. To add these fractions we must first write them so they do have a common denominator. Any common multiple of 15 and 6 will serve as a common denominator. The least common multiple is preferred. The least common multiple of 15 and 6 is ____ By the reducing rule,

30

$$\frac{2}{15} = \frac{2 \cdot 2}{2 \cdot 15} = \_\_\_\_ \qquad \text{and} \qquad \frac{5}{6} = \frac{5 \cdot 5}{5 \cdot 6} = \_\_\_\_$$

$\frac{4}{30}$; $\frac{25}{30}$

So

$$\frac{2}{15} + \frac{5}{6} = \frac{4}{30} + \frac{25}{30} = \_\_\_\_$$

$\frac{29}{30}$

9. We can use the above process to derive a general addition rule. Let a/b and c/d be fractions which do not have a common denominator. The number bd is a multiple of both ____ and ____. The number bd may not be the least common multiple but it is a common multiple. By the reducing rule,

b; d

$$\frac{a}{b} = \frac{ad}{bd} \qquad \text{and} \qquad \frac{c}{d} = \frac{bc}{bd}$$

Therefore

$$\frac{a}{b} + \frac{c}{d} = \frac{ad}{bd} + \frac{bc}{bd} = \frac{\_\_\_\_\_\_}{bd}$$

$ad + bc$

Because both fractions ad/bd and bc/bd have a common denominator (namely, bd) we can just add their (numerators\denominators\numerators and denominators). So the general formula for adding fractions is

numerators

$$\frac{a}{b} + \frac{c}{d} = \_\_\_\_\_\_$$

$\frac{ad + bc}{bd}$

10. Use the general formula to add 2/15 and 5/6.

$$\frac{2}{15} + \frac{5}{6} = \frac{2 \cdot 6 + 15 \cdot 5}{15 \cdot 6} = \_\_\_\_$$

$\frac{87}{90}$

When reduced to lowest terms, 87/90 is____. This result is the same answer obtained in frame 8 where we used the least common denominator. No matter how two fractions are added the answers must always be the same when written in ______ ______.

$\frac{29}{30}$

lowest terms or reduced form

11. In this frame, we will compute the same sum in two different ways and then compare the answers.

(1). Use the general addition formula to add

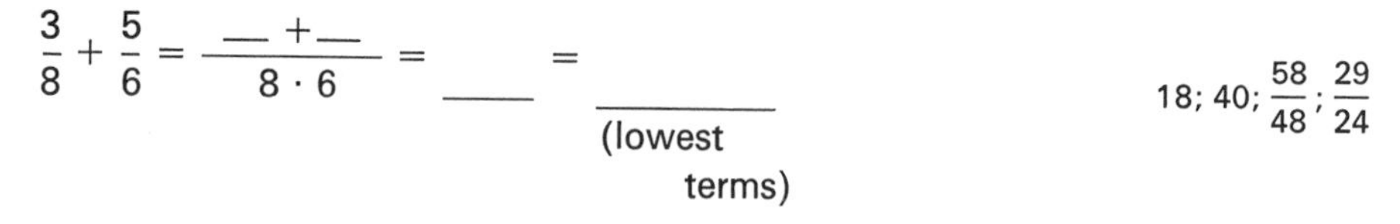

$$\frac{3}{8} + \frac{5}{6} = \frac{\_\_ + \_\_}{8 \cdot 6} = \_\_\_\_ = \underset{\text{(lowest terms)}}{\_\_\_\_\_\_}$$

18; 40; $\frac{58}{48}$; $\frac{29}{24}$

(2). Use the least common multiple of 8 and 6 as the common denominator. The least common multiple of 6 and 8 is____.

24

$$\frac{3}{8} = \frac{3 \cdot 3}{3 \cdot 8} = \_\_\_\_ \qquad \text{and} \qquad \frac{5}{6} = \frac{(\_\_)5}{4 \cdot 6} = \_\_\_\_$$

$\frac{9}{24}$; 4; $\frac{20}{24}$

So

$$\frac{3}{8} + \frac{5}{6} = \frac{\_\_\_}{24} + \frac{\_\_\_}{24} = \_\_\_\_$$

9; 20; $\frac{29}{24}$

equals

When we reduce the sum obtained by using the general formula, it (equals\does not equal) the sum obtained by using any other common denominator.

12. We may use either of the previous two techniques to add fractions. The general formula gives us the sum quickly even if the sum is not in reduced form. But in some cases (especially if there are polynomials of degree two or more in the fractions) the result is so difficult to reduce that it is better to use the *least* common denominator. First we will practice use of the general formula. Use the general formula

$3; 4; \frac{11}{12}$

$$\frac{1}{4} + \frac{2}{3} = \frac{1(\_\_) + 2(\_\_)}{4 \cdot 3} = \_\_\_\_$$

13. Do the following problems by using the addition formula.

$15; 8; 23; \quad 3; \frac{53}{21}$

$\frac{6 + 50}{15}; \frac{56}{15}$

$\frac{9 + 10}{18}; \frac{19}{18}$

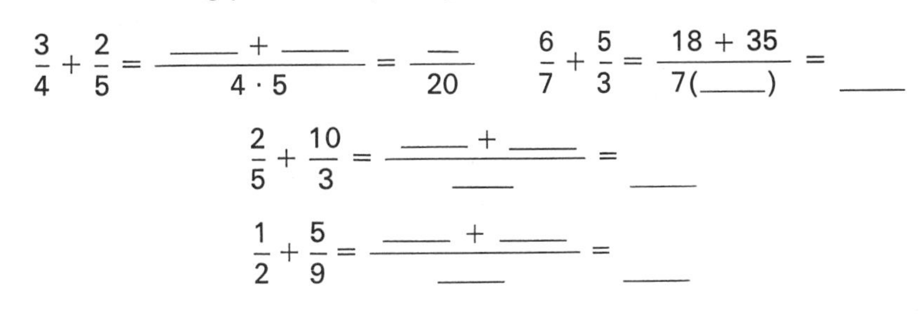

14. Add the following by using the general addition formula. Then reduce the results as far as possible.

$\frac{-35 + 135}{75}; \frac{100}{75}; \frac{4}{3}$

$\frac{72 + (-40)}{48}; \frac{32}{48}; \frac{2}{3}$

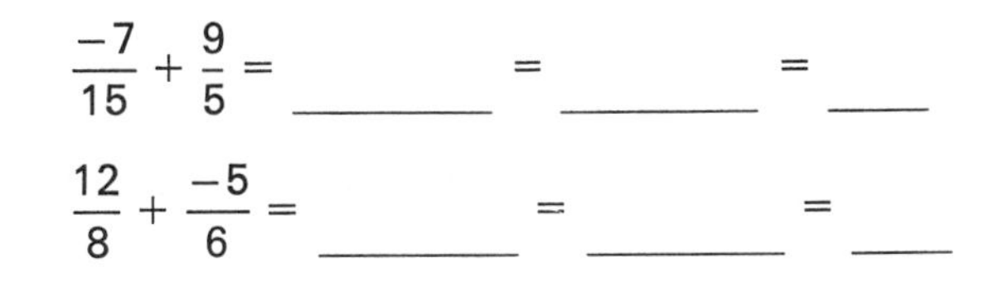

15. Use the addition formula to add

$3y; 2x; \frac{2y + 9x}{xy}$

$$\frac{3}{x} + \frac{2}{y} = \frac{\_\_\_\_ + \_\_\_\_}{xy} \qquad \frac{2}{x} + \frac{9}{y} = \_\_\_\_\_\_$$

$\frac{3z + 5x}{xz}$

16. Add: $\frac{3}{x} + \frac{5}{z} = \_\_\_\_\_\_$

no

Does the numerator simplify? (yes\no).

17. When we add two fractions with polynomials in the numerator or denominator, we find that the numerator of the sum sometimes simplifies. So we must do two steps.

1. Use the addition formula.
2. Simplify the numerator.

(We may leave the denominator in factored form.)

1. $\frac{2}{x} + \frac{4}{x + 3} = \frac{2(x + 3) + 4x}{x(x + 3)}$

$$= \frac{2x + 6 + 4x}{x(x + 3)}$$

$6x + 6 \quad \text{or} \quad 6(x + 1)$

2. $= \frac{\_\_\_\_\_\_}{x(x + 3)}$

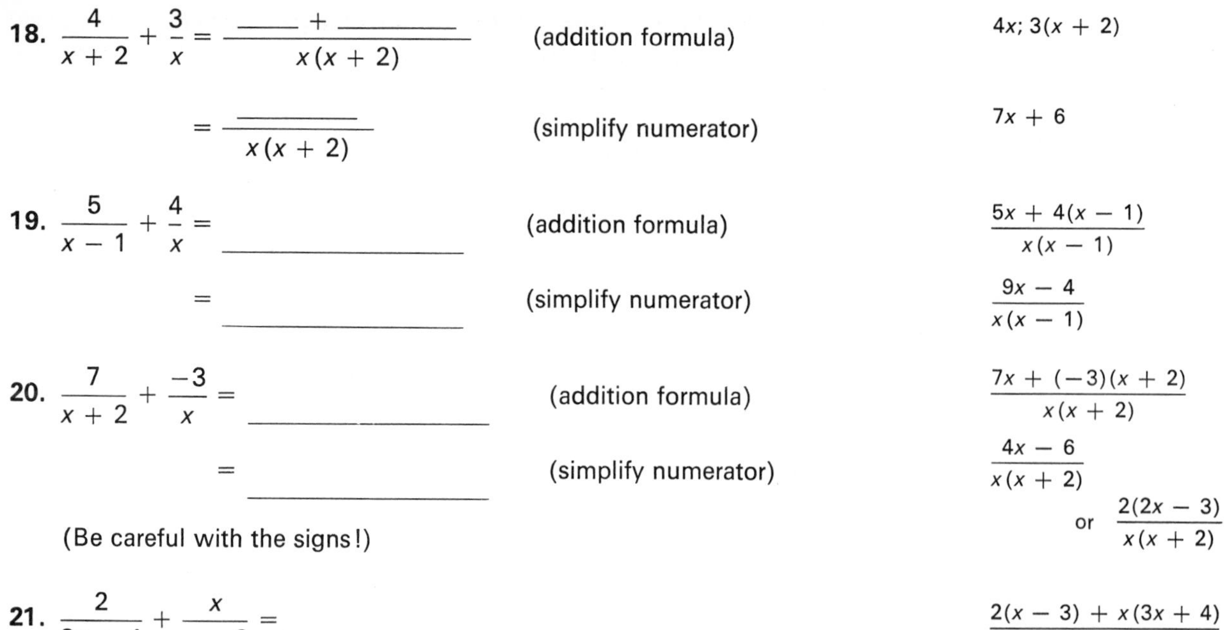

18. $\frac{4}{x+2} + \frac{3}{x} = \frac{\_\_\_\_ + \_\_\_\_\_\_}{x(x+2)}$ (addition formula)

$4x;\ 3(x+2)$

$= \frac{\_\_\_\_\_\_}{x(x+2)}$ (simplify numerator)

$7x + 6$

19. $\frac{5}{x-1} + \frac{4}{x} =$ ____________ (addition formula)

$\frac{5x + 4(x-1)}{x(x-1)}$

$=$ ____________ (simplify numerator)

$\frac{9x - 4}{x(x-1)}$

20. $\frac{7}{x+2} + \frac{-3}{x} =$ ____________ (addition formula)

$\frac{7x + (-3)(x+2)}{x(x+2)}$

$=$ ____________ (simplify numerator)

$\frac{4x - 6}{x(x+2)}$ or $\frac{2(2x-3)}{x(x+2)}$

(Be careful with the signs!)

21. $\frac{2}{3x+4} + \frac{x}{x-3} =$ ____________

$\frac{2(x-3) + x(3x+4)}{(3x+4)(x-3)}$

$=$ ____________

$\frac{3x^2 + 6x - 6}{(3x+4)(x-3)}$

22. Add and simplify the numerator:

$$\frac{6}{2x-3} + \frac{x}{x+1} = \_\_\_\_\_\_\_\_\_\_\_\_$$

$\frac{2x^2 + 3x + 6}{(2x-3)(x+1)}$

23. To reduce a sum of fractions we must apply two more steps.

1. Factor the numerator and denominator.
2. Reduce the fraction by canceling the factors which the numerator and denominator have in common.

$$\frac{2x+4}{6x} = \frac{\not{2}(x+2)}{\not{2} \cdot 3x} = \frac{x+2}{3x}$$

$$\frac{2x+6}{4x} = \frac{2(\_\_\_\_\_\_)}{2 \cdot 2x} = \_\_\_\_\_\_$$

$(x+3);\ \frac{x+3}{2x}$

24. Add and reduce the answer:

$$\frac{3}{x} + \frac{1}{4x} = \frac{12x + x}{x(4x)} = \underset{\text{factored form}}{\_\_\_\_\_\_} = \underset{\text{reduced form}}{\_\_\_\_\_\_}$$

$\frac{13x}{4xx};\ \frac{13}{4x}$

25. Add and reduce:

$\frac{x+1}{3x} + \frac{2}{3} =$ ____________ (use addition formula)

$\frac{3(x+1) + 6x}{(3x)(3)}$

$=$ ____________ (simplify numerator)

$\frac{9x + 3}{(3x)(3)}$

$=$ ____________ (factored form)

$\frac{3(3x+1)}{3 \cdot 3x}$

$=$ ____________ (reduced form)

$\frac{3x+1}{3x}$

Answers	Frames
	26. Add and reduce: (Follow the steps in the previous frame.)
$\frac{2x + 4(x + 1)}{(x + 1)2x}$	$\frac{1}{x+1} + \frac{4}{2x} =$ ____
$\frac{6x + 4}{(x + 1)2x}$	$=$ ____
$\frac{2(3x + 2)}{2x(x + 1)}$	$=$ ____
$\frac{3x + 2}{x(x + 1)}$	$=$ ____
	27. To subtract fractions with unlike denominators, we change the problem to that of adding the additive inverse of the subtrahend and proceed as with addition
$\frac{1}{15}$	$\frac{2}{5} - \frac{1}{3} = \frac{2}{5} + \frac{-1}{3} =$ ____
$\frac{-1}{2}$; $\frac{3}{10}$	**28.** $\frac{12}{15} - \frac{1}{2} = \frac{12}{15} +$ ____ $=$ ____.
	29. We can use the same technique to derive the general subtraction formula.
$\frac{-c}{d}$	$\frac{a}{b} - \frac{c}{d} = \frac{a}{b} +$ ____
	$= \frac{ad + b(-c)}{bd}$
	$= \frac{ad - bc}{bd}$
	So the general formula is
$\frac{ad - bc}{bd}$	$\frac{a}{b} - \frac{c}{d} =$ ____
	30. Use the subtraction formula and simplify your answers. Reduce if possible.
$\frac{1}{12}$	$\frac{3}{4} - \frac{8}{12} =$ ____
$\frac{3y - 4x}{xy}$	$\frac{3}{x} - \frac{4}{y} =$ ____
	[Omit frames 31–33 if you have not completed Sections 4.2–4.5.]
is	**31.** We will work the next problems by using the least common denominator. This technique is especially useful if the fractions contain polynomials of degree two or more. Recall that the least common denominator (is\is not) the least common multiple of the denominators.
	32. Add: $\frac{4}{25x^2y} + \frac{7}{100xy^3}$
	First we factor the denominators.
	$25x^2y = 5 \cdot 5xxy$
$\cdot 2$; y	$100xy^3 = 5 \cdot 5 \cdot 2$ ____xyy____
y; y; y	least common denominator = LCM $= 5 \cdot 5 \cdot 2 \cdot 2xx \cdot$ ____ $\cdot$ ____ $\cdot$ ____

Next we convert the fractions to equal fractions with the least common denominator.

$$\frac{4}{25x^2y} = \frac{4}{5 \cdot 5xxy} = \frac{4(\_\_\_\_)}{5 \cdot 5 \cdot 2 \cdot 2xxyyy}$$

$4y^2$ or $4yy$

$$\frac{7}{100xy^3} = \frac{7}{5 \cdot 5 \cdot 2 \cdot 2xyyy} = \frac{7(\_\_)}{5 \cdot 5 \cdot 2 \cdot 2xxyyy}$$

x

Thus the sum is

$$\frac{\_\_\_\_ + \_\_\_\_}{5 \cdot 5 \cdot 2 \cdot 2xxyyy} \quad \text{or} \quad \frac{\_\_\_\_\_\_}{100x^2y^3}$$

$16y^2$; $7x$; $16y^2 + 7x$

33. Add: $\dfrac{3}{x^2 - 4} + \dfrac{2}{x^2 - x - 6}$

First factor the denominators.

$x^2 - 4 = (x + 2)(\_\_\_\_\_\_)$ $\quad x^2 - x - 6 = (x + 2)(\_\_\_\_\_\_)$

$x - 2$; $x - 3$

least common denominator = LCM = $(x + 2)(\_\_\_\_\_\_)(\_\_\_\_\_\_)$

$x - 2$; $x - 3$

Next change each of the original fractions to equal fractions which have the common denominator.

$$\frac{3}{x^2 - 4} = \frac{3(\_\_\_\_\_\_)}{(x + 2)(x - 2)(x - 3)}$$

$x - 3$

$$\frac{2}{x^2 - x - 6} = \frac{2(\_\_\_\_\_\_)}{(x + 2)(x - 2)(x - 3)}$$

$x - 2$

So the numerator becomes $3(\_\_\_\_\_\_) + 2(\_\_\_\_\_\_)$, and the sum is

$x - 3$; $x - 2$

$$\frac{\_\_\_\_\_\_}{} $$

$$\frac{3(x - 3) + 2(x - 2)}{(x + 2)(x - 2)(x - 3)}$$

or $\quad \dfrac{\_\_\_\_\_\_}{\_\_\_\_\_\_}$ $\quad$ {simplify numerator and leave denominator in factored form

$$\frac{5x - 13}{(x + 2)(x - 2)(x - 3)}$$

PRACTICE PROBLEMS

1. Add:

a. $\dfrac{x}{3} + \dfrac{-x}{3}$ $\quad$ *b.* $\dfrac{7}{6} + \dfrac{9}{6}$ $\quad$ *c.* $\dfrac{4}{-17} + \dfrac{6}{-17}$

d. $\dfrac{1}{10x} + \dfrac{9}{10x}$ $\quad$ *e.* $\dfrac{-6}{14} + \dfrac{-8}{14}$ $\quad$ *f.* $\dfrac{y}{5} + \dfrac{7y}{5}$

2. Write each pair of fractions with a common denominator.

a. $\dfrac{1}{3}, \dfrac{1}{2}$ $\quad$ *b.* $\dfrac{4}{7x}, \dfrac{5}{4x}$ $\quad$ *c.* $\dfrac{2}{10}, \dfrac{3}{5}$

d. $\dfrac{19}{7}, \dfrac{3}{14}$ $\quad$ *e.* $\dfrac{-4}{9x}, \dfrac{5}{6y}$ $\quad$ *f.* $\dfrac{4}{10}, \dfrac{-2}{15}$

3. Add each of the pairs of fractions given in Problem 2.

4. The general formula for adding the fractions a/b and c/d is $a/b + c/d = (ad + bc)/bd$. Use this formula to add each of the following. Simplify and reduce the result.

a. $\frac{1}{3x} + \frac{1}{2}$ *b.* $\frac{2}{5} + \frac{-3}{4}$ *c.* $\frac{1}{7x} + \frac{2}{-5y}$ *d.* $\frac{4}{1} + \frac{2}{1}$

e. $\frac{-1}{3} + \frac{-2}{5}$ *f.* $\frac{x}{y} + \frac{z}{w}$ *g.* $\frac{2}{x} + \frac{3}{y}$ *h.* $\frac{4}{y} + \frac{5}{x}$

i. $\frac{6}{5x} + \frac{2}{3x}$ *j.* $\frac{2}{5y} + \frac{3}{7x}$ *k.* $\frac{1}{x} + \frac{2}{x+3}$ *l.* $\frac{5}{y-2} + \frac{6}{y}$

5. Use the method of finding a least common denominator to subtract:

a. $\frac{x}{2} - \frac{1}{4}$ *b.* $\frac{1}{3x} - \frac{1}{2}$ *c.* $\frac{4}{21} - \frac{7}{3}$

d. $\frac{-5}{6} - \frac{-3}{2}$ *e.* $\frac{8}{6} - \frac{1}{8}$ *f.* $\frac{-3x}{4} - \frac{1}{5y}$

6. The general formula for subtracting fractions is $a/b - c/d = (ad - bc)/bd$. Use this formula to do these computations. Simplify and reduce the result.

a. $\frac{2}{3} - \frac{4}{9}$ *b.* $\frac{6}{x} - \frac{5}{4}$ *c.* $\frac{6}{x} - \frac{5}{x}$ *d.* $\frac{8}{5} - \frac{7}{x}$

e. $\frac{2}{3x} - \frac{5}{2y}$ *f.* $\frac{6}{5x} - \frac{1}{2y}$ *g.* $\frac{5}{x} - \frac{1}{x+2}$ *h.* $\frac{2}{x+3} - \frac{4}{x}$

7. Answer each of the following as true or false.

a. $\frac{-1}{3} = \frac{1}{-3}$ *b.* $-\frac{a}{b} = \frac{-a}{-b}$ *c.* $\frac{2}{-5} = -\frac{2}{5}$

d. $\frac{a}{b} = \frac{-a}{-b}$ *e.* $-\left(\frac{a}{b}\right) = \frac{a}{b}$ *f.* $-\left(\frac{-a}{b}\right) = \frac{a}{-b}$

***8.** Use the least common denominator to do these problems. Reduce your answers.

a. $\frac{3}{9x^2y^3} + \frac{2}{27x^4y}$ *b.* $\frac{5}{14x^3y} - \frac{3}{7xy^4}$ *c.* $\frac{x}{x^2-9} + \frac{2}{x+3}$

d. $\frac{4}{x^2+2x+1} + \frac{3}{x+1}$ *e.* $\frac{2}{x^2+3x+2} + \frac{1}{x^2-4}$

SELFTEST

Add or subtract and reduce:

1. $\frac{5}{8} + \frac{7}{8}$ **2.** $\frac{5}{9} + \frac{-7}{9}$ **3.** $\frac{7}{x} + \frac{18}{x}$ **4.** $\frac{33}{y} - \frac{25}{y}$

5. $\frac{4}{9} + \frac{3}{5}$ **6.** $\frac{15}{7} - \frac{2}{9}$ **7.** $\frac{4}{x} - \frac{3}{y}$ **8.** $\frac{1}{x} + \frac{1}{y}$

9. $\frac{3}{x+2} + \frac{4}{x}$ **10.** $\frac{7}{x+3} + \frac{5}{x-2}$ ***11.** $\frac{3}{x^2-16} + \frac{1}{x+4}$

* Omit if you have omitted other starred parts of this section.

SELFTEST ANSWERS

1. $\frac{3}{2}$ 2. $\frac{-2}{9}$ 3. $\frac{25}{x}$ 4. $\frac{8}{y}$ 5. $\frac{47}{45}$ 6. $\frac{121}{63}$ 7. $\frac{4y - 3x}{xy}$ 8. $\frac{y + x}{xy}$

9. $\frac{7x + 8}{x(x + 2)}$ 10. $\frac{12x + 1}{(x + 3)(x - 2)}$ 11. $\frac{x - 1}{(x + 4)(x - 4)}$

5.3 MULTIPLICATION AND DIVISION OF RATIONAL NUMBERS

Rule on Multiplication of Fractions: **If a/b and c/d are any two fractions then**

$$\frac{a}{b} \cdot \frac{c}{d} = \frac{a \cdot c}{b \cdot d}$$

So to multiply fractions we multiply the numerators together and we multiply the denominators together.

Example 1 $$\frac{7}{2} \cdot \frac{5}{3} = \frac{7 \cdot 5}{2 \cdot 3} = \frac{35}{6}$$

Example 2 $$\frac{1}{4} \cdot x = \frac{1}{4} \cdot \frac{x}{1} = \frac{1 \cdot x}{4 \cdot 1} = \frac{x}{4}$$

Example 3 $$\frac{4}{7} \cdot \frac{7}{4} = \frac{4 \cdot 7}{7 \cdot 4} = \frac{28}{28} = 1$$

Example 4 $$\frac{-2y}{9} \cdot \frac{5x}{4y} = \frac{(-2y)(5x)}{9(4y)} = \frac{-10yx}{36y} = \frac{-5x}{18}$$

Multiplication of rational numbers satisfies the usual commutative, associative, and distributive laws. That is,

$$\frac{a}{b} \cdot \frac{c}{d} = \frac{c}{d} \cdot \frac{a}{b}$$

$$\left(\frac{a}{b} \cdot \frac{c}{d}\right)\frac{e}{f} = \frac{a}{b}\left(\frac{c}{d} \cdot \frac{e}{f}\right)$$

$$\frac{a}{b}\left(\frac{c}{d} + \frac{e}{f}\right) = \frac{ac}{bd} + \frac{ae}{bf}$$

We can extend the reducing rule so that

$$\frac{a}{b} = \frac{ak}{bk}$$

whenever a and b are rational numbers and k is a nonzero rational number.

If A and B are numbers such that $A \cdot B = 1$ then A and B are *multiplicative inverses* or *reciprocals* of each other. In other words, two numbers are *reciprocals of each other* if their product is 1. So 3/4 and 4/3 are multiplicative inverses of each other since $(3/4)(4/3) = 1$. If a/b is not zero then the multiplicative inverse of a/b is b/a since $(a/b)(b/a) = ab/ba = 1$.

We will use the reducing rule, the multiplication rule, and the multiplicative inverse to discover how to divide rational numbers. Let's look at a specific example.

$$\frac{2}{3} \div \frac{5}{7} = \frac{\frac{2}{3}}{\frac{5}{7}}$$

By the reducing rule, we can multiply the numerator and denominator by the same value. We chose to multiply by the multiplicative inverse of 5/7. For this is the only value that will give a 1 in the denominator.

$$\frac{2}{3} \div \frac{5}{7} = \frac{\frac{2}{3}}{\frac{5}{7}} = \frac{\frac{2}{3} \cdot \frac{7}{5}}{\frac{5}{7} \cdot \frac{7}{5}} = \frac{\frac{2}{3} \cdot \frac{7}{5}}{1}$$

$$= \frac{2}{3} \cdot \frac{7}{5} = \frac{14}{15}$$

We can follow the same process to obtain the general rule for division. Again we use the reducing rule to multiply the numerator and denominator by the multiplicative inverse of the denominator.

$$\frac{a}{b} \div \frac{c}{d} = \frac{\frac{a}{b}}{\frac{c}{d}} = \frac{\frac{a}{b} \cdot \frac{d}{c}}{\frac{c}{d} \cdot \frac{d}{c}} = \frac{\frac{a}{b} \cdot \frac{d}{c}}{1}$$

$$= \frac{a}{b} \cdot \frac{d}{c}$$

$$= \frac{ad}{bc}$$

Rule for Division of Fractions: **If a/b and c/d are fractions and $c/d \neq 0$,**

$$\frac{a}{b} \div \frac{c}{d} = \frac{a}{b} \cdot \frac{d}{c} = \frac{ad}{bc}$$

In words, a/b divided by c/d equals a/b times the reciprocal of c/d.

Example 5

$$\frac{4}{x} \div \frac{2}{x} = \frac{4}{x} \cdot \frac{x}{2} = 2$$

Example 6

$$\frac{9}{15} \div \frac{6}{5y} = \frac{9}{15} \cdot \frac{5y}{6} = \frac{y}{2}$$

Example 7

$$\frac{-8}{7} \div \frac{17}{35} = \frac{-8}{7} \cdot \frac{35}{17} = \frac{-40}{17}$$

Example 8

$$\frac{4}{3} \div \left(1 - \frac{17}{9}\right)$$

We first evaluate the value inside the parentheses.

$$\left(1 - \frac{17}{9}\right) = \frac{9}{9} - \frac{17}{9} = \frac{-8}{9}$$

So

$$\frac{4}{3} \div \left(1 - \frac{17}{9}\right) = \frac{4}{3} \div \frac{-8}{9}$$

$$= \frac{4}{3} \cdot \frac{9}{-8}$$

$$= \frac{-3}{2}$$

Example 9 $$\frac{-7}{156}\left[\left(\frac{12}{7} + 2\right) \div \left(\frac{1}{3} - \frac{1}{2}\right)\right]$$

We evaluate the values in the innermost parentheses and work outward.

$$\frac{12}{7} + 2 = \frac{12}{7} + \frac{14}{7} = \frac{26}{7}$$

$$\frac{1}{3} - \frac{1}{2} = \frac{2-3}{6} = \frac{-1}{6}$$

So

$$\left(\frac{12}{7} + 2\right) \div \left(\frac{1}{3} - \frac{1}{2}\right) = \frac{26}{7} \div \frac{-1}{6} = \frac{26}{7} \cdot \frac{6}{-1} = \frac{-156}{7}$$

Finally,

$$\frac{-7}{156}\left[\left(\frac{12}{7} + \frac{2}{1}\right) \div \left(\frac{1}{3} - \frac{1}{2}\right)\right] = \frac{-7}{156} \cdot \frac{-156}{7} = 1$$

PROGRAM 5.3

1. The product

$$\frac{a}{b} \cdot \frac{c}{d} = \frac{ac}{bd} \quad \text{so} \quad \frac{9}{7} \cdot \frac{4}{5} = \frac{9 \cdot 4}{7 \cdot 5} = \_\_\_\_$$

$\frac{36}{35}$

2. $\frac{5}{6} \cdot \frac{1}{2} = \frac{5 \cdot 1}{\_\_\_\_\_} = \_\_\_\_$

$6 \cdot 2; \frac{5}{12}$

To multiply we multiply the numerators of the fractions. Do we also multiply the denominators? (yes\no).

yes

3. Multiply: $\frac{8}{3} \cdot \frac{2}{5} = \_\_\_\_\_\_$.

$\frac{8 \cdot 2}{3 \cdot 5}$ or $\frac{16}{15}$

4. Look at the following example.

$$\frac{2}{3} \cdot 4 = \frac{2}{3} \cdot \frac{4}{1} = \frac{2 \cdot 4}{3 \cdot 1} = \frac{8}{3}$$

When we multiply a fraction by an integer we multiply the integer with the (numerator\denominator).

numerator

5. Multiply:

$7; \frac{28}{9}$

$$\frac{4}{9} \cdot 7 = \frac{4 \cdot \_\_\_}{9} = \_\_\_$$

6. The same procedure applies when we multiply a fraction by a variable.

$4y$

$$\frac{3}{7}x = \frac{3x}{7} \quad \text{also} \quad \frac{4}{5}y = \frac{\_\_\_}{5}$$

7. Multiply and reduce:

$3x$

$$\frac{3x}{4z} \cdot \frac{4}{5} = \frac{3x(4)}{4z(5)} = \frac{\_\_\_}{5z}$$

second fraction
first fraction

We canceled the 4's. The 4 in the numerator was in the (first fraction\second fraction). The 4 in the denominator was from the (first fraction\second fraction). It appears that in multiplication we can cancel factors from either denominator with factors from either numerator. This is indeed the case. It is often easier to *cancel before we multiply.*

8. Cancel like factors then multiply:

2

$$\frac{\cancel{3x}}{\cancel{5z}} \cdot \frac{\overset{2}{\cancel{10z}}}{\underset{3x}{\cancel{9x^2}}} = \frac{\_\_\_}{3x}$$

9. Cancel like factors then multiply:

$2x; 2x$

$$\frac{x(x+3)}{z} \cdot \frac{4z}{2(x+3)} = \frac{\_\_\_}{1} = \_\_\_$$

10. Cancel like factors and multiply:

$4x^2$

$$\frac{14xy^3}{25z} \cdot \frac{10x}{21y^4} = \frac{\_\_\_}{15yz}$$

11. Cancel and multiply:

$8x$

$$\frac{4x^3y}{5} \cdot \frac{10}{x^2y} = \_\_\_$$

12. The numbers A and B are called *multiplicative inverses* of each other if $A \cdot B = 1$. Since $(3/8) \cdot (8/3) = 1$, then 3/8 and 8/3 are __________ __________ of each other.

multiplicative inverses

multiplicative inverse

13. A number times its __________ __________ equals 1.

14. The multiplicative inverse of 4/5 is 5/4 since

$$\frac{4}{5} \cdot \frac{5}{4} = \frac{20}{20} = 1$$

The multiplicative inverse of 4/11 is 11/4 since

1

$$\frac{4}{11} \cdot \frac{11}{4} = \_\_\_$$

15. The multiplicative inverse of 3/4 is ____ since (3/4) · (____) = 1. — 4/3; 4/3

16. The multiplicative inverse of 2/5 is ____. — 5/2

17. Notice that we just invert the fraction to find its multiplicative inverse.

The multiplicative inverse of 8/1 is ____. — 1/8

The multiplicative inverse of $1/x$ is ____. — $x/1$ or x

The multiplicative inverse of $\frac{a}{b}$ is ____. — $\frac{b}{a}$

18. Another name for the multiplicative inverse is the reciprocal. The reciprocal of 9/10 is ____. The reciprocal of −8/17 is ____. — 10/9; 17/−8 or −17/8

19. A number times its reciprocal equals ____. — 1

20. To find the reciprocal of 5, we write 5 as the fraction ____. Then we take the reciprocal of 5/1. The reciprocal is ____, so the reciprocal of 5 is ____. — 5/1; 1/5; 1/5

21. The multiplicative inverse of x is ____ ($x \neq 0$). — $1/x$

22. The reducing rule says that $a/b = ak/bk$. This rule is true whenever a, b, and k are fractions and b, k are not zero. Therefore

$$\frac{\frac{1}{2}}{\frac{3}{4}} = \frac{\frac{1}{2}\cdot\frac{7}{8}}{\frac{3}{4}\cdot\frac{7}{8}} \quad \text{and} \quad \frac{\frac{2}{3}}{\frac{7}{16}} = \frac{\frac{2}{3}\cdot\frac{4}{5}}{\frac{7}{16}\cdot\_\_\_\_}$$

— $\frac{4}{5}$

23. We will use the reducing rule, the concept of reciprocal, and the multiplication rule to discover how to divide fractions. To compute (1/4) ÷ (2/3) we write

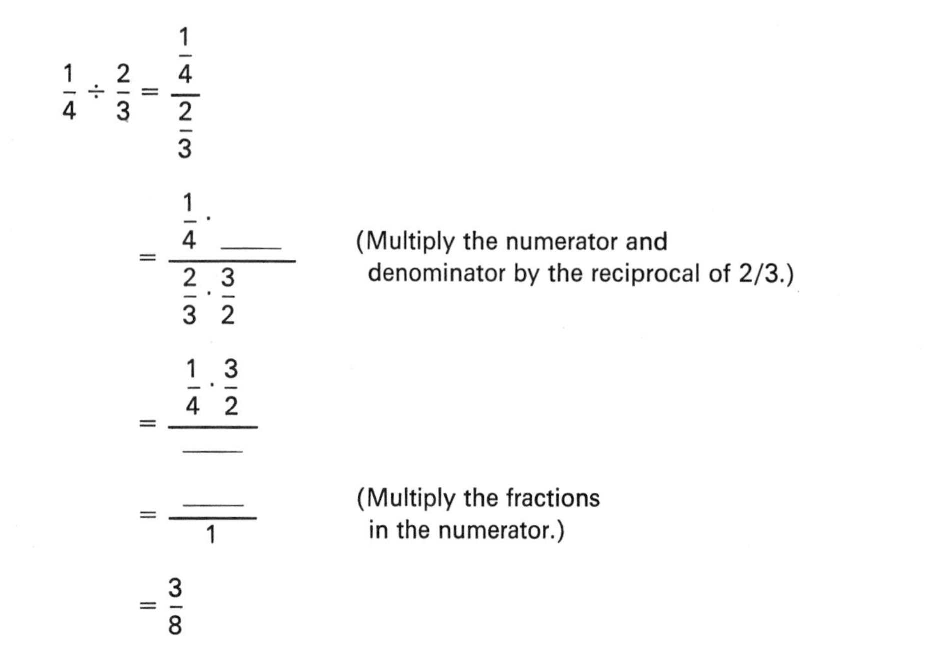

$$\frac{1}{4} \div \frac{2}{3} = \frac{\frac{1}{4}}{\frac{2}{3}}$$

$$= \frac{\frac{1}{4}\cdot\_\_\_\_}{\frac{2}{3}\cdot\frac{3}{2}}$$ (Multiply the numerator and denominator by the reciprocal of 2/3.) — $\frac{3}{2}$

$$= \frac{\frac{1}{4}\cdot\frac{3}{2}}{\_\_\_\_}$$ — 1

$$= \frac{\_\_\_\_}{1}$$ (Multiply the fractions in the numerator.) — $\frac{3}{8}$

$$= \frac{3}{8}$$

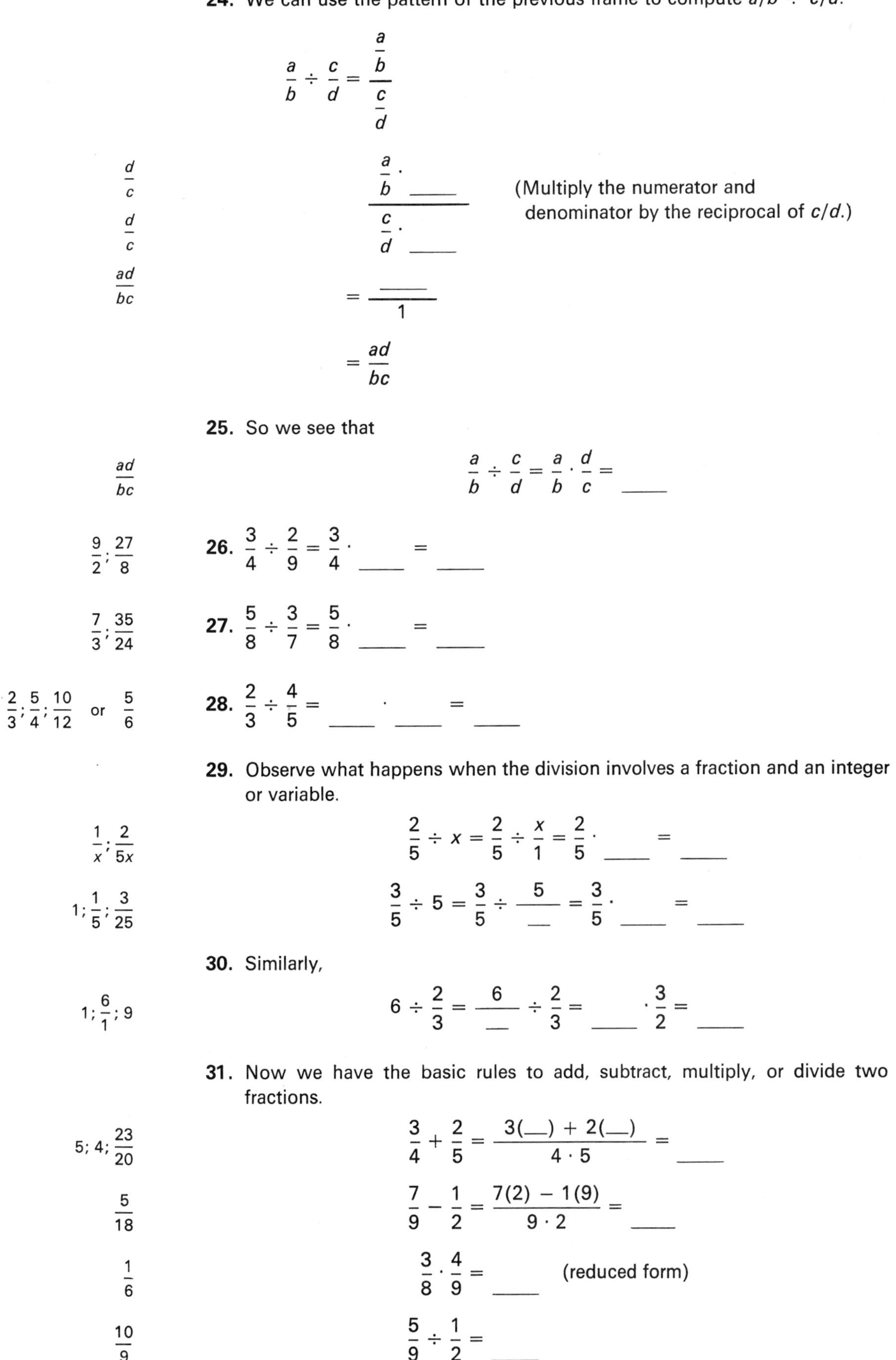

24. We can use the pattern of the previous frame to compute $a/b \div c/d$.

$$\frac{a}{b} \div \frac{c}{d} = \frac{\frac{a}{b}}{\frac{c}{d}}$$

$$\frac{\frac{a}{b} \cdot \_\_\_\_}{\frac{c}{d} \cdot \_\_\_\_}$$ (Multiply the numerator and denominator by the reciprocal of c/d.)

$$= \frac{\_\_\_\_}{1}$$

$$= \frac{ad}{bc}$$

$\frac{d}{c}$

$\frac{d}{c}$

$\frac{ad}{bc}$

25. So we see that

$$\frac{a}{b} \div \frac{c}{d} = \frac{a}{b} \cdot \frac{d}{c} = \_\_\_\_$$

$\frac{ad}{bc}$

26. $\frac{3}{4} \div \frac{2}{9} = \frac{3}{4} \cdot \_\_\_\_ = \_\_\_\_$

$\frac{9}{2}; \frac{27}{8}$

27. $\frac{5}{8} \div \frac{3}{7} = \frac{5}{8} \cdot \_\_\_\_ = \_\_\_\_$

$\frac{7}{3}; \frac{35}{24}$

28. $\frac{2}{3} \div \frac{4}{5} = \_\_\_\_ \cdot \_\_\_\_ = \_\_\_\_$

$\frac{2}{3}; \frac{5}{4}; \frac{10}{12}$ or $\frac{5}{6}$

29. Observe what happens when the division involves a fraction and an integer or variable.

$$\frac{2}{5} \div x = \frac{2}{5} \div \frac{x}{1} = \frac{2}{5} \cdot \_\_\_\_ = \_\_\_\_$$

$\frac{1}{x}; \frac{2}{5x}$

$$\frac{3}{5} \div 5 = \frac{3}{5} \div \frac{5}{\_\_\_\_} = \frac{3}{5} \cdot \_\_\_\_ = \_\_\_\_$$

$1; \frac{1}{5}; \frac{3}{25}$

30. Similarly,

$$6 \div \frac{2}{3} = \frac{6}{\_\_\_\_} \div \frac{2}{3} = \_\_\_\_ \cdot \frac{3}{2} = \_\_\_\_$$

$1; \frac{6}{1}; 9$

31. Now we have the basic rules to add, subtract, multiply, or divide two fractions.

$$\frac{3}{4} + \frac{2}{5} = \frac{3(\_\_) + 2(\_\_)}{4 \cdot 5} = \_\_\_\_$$

$5; 4; \frac{23}{20}$

$$\frac{7}{9} - \frac{1}{2} = \frac{7(2) - 1(9)}{9 \cdot 2} = \_\_\_\_$$

$\frac{5}{18}$

$$\frac{3}{8} \cdot \frac{4}{9} = \_\_\_\_$$ (reduced form)

$\frac{1}{6}$

$$\frac{5}{9} \div \frac{1}{2} = \_\_\_\_$$

$\frac{10}{9}$

32. The next series of frames show how to compute expressions that combine addition, subtraction, multiplication, and division of rational numbers. To compute

$$\left(5 + \frac{1}{3}\right) \div \left(\frac{6}{8} \cdot \frac{2}{3}\right)$$

first compute the values inside the parentheses. Using the addition formula,

$$\left(5 + \frac{1}{3}\right) = \frac{5}{1} + \frac{1}{3} = \frac{\_\_ + \_\_}{1 \cdot 3}$$ 15; 1

$$= \underline{\quad\quad} \text{ in lowest terms}$$ $\frac{16}{3}$

Using the rule for multiplication for fractions,

$$\left(\frac{6}{8} \cdot \frac{2}{3}\right) = \frac{6 \cdot 2}{\_\_ \cdot 3} = \underline{\quad\quad} \text{ in lowest terms}$$ 8; $\frac{1}{2}$

33. From the previous frame, it follows that

$$\left(5 + \frac{1}{3}\right) \div \left(\frac{6}{8} \cdot \frac{2}{3}\right) = \left(\underline{\quad\quad}\right) \div \left(\underline{\quad\quad}\right)$$ $\frac{16}{3}$; $\frac{1}{2}$

Using the rule for division,

$$\frac{16}{3} \div \frac{1}{2} = \frac{16}{3} \cdot \underline{\quad\quad} = \underline{\quad\quad} \text{ in lowest terms}$$ $\frac{2}{1}$; $\frac{32}{3}$

Thus the final result is that

$$\left(5 + \frac{1}{3}\right) \div \left(\frac{6}{8} \cdot \frac{2}{3}\right) = \underline{\quad\quad}$$ $\frac{32}{3}$

34. To compute

$$\frac{3}{2}\left[\left(\frac{7}{9} \div \frac{5}{3}\right) \div \left(\frac{1}{2} - \frac{4}{5}\right)\right]$$

first compute the values inside the innermost sets of parentheses. Then compute the value inside the next set of parentheses.

$$\left(\frac{7}{9} \div \frac{5}{3}\right) = \underline{\quad\quad} \text{ in lowest terms}$$ $\frac{7}{15}$

$$\left(\frac{1}{2} - \frac{4}{5}\right) = \frac{\underline{\quad\quad} - \underline{\quad\quad}}{2 \cdot \underline{\quad\quad}} = \underline{\quad\quad} \text{ in lowest terms}$$ $\frac{1 \cdot 5 - 2 \cdot 4}{2 \cdot 5}$; $\frac{-3}{10}$

35. Referring to the previous frame, we see that the next step is to compute

$$\left(\frac{7}{9} \div \frac{5}{3}\right) \div \left(\frac{1}{2} - \frac{4}{5}\right) = \underline{\quad\quad} \div \underline{\quad\quad}$$ $\frac{7}{15}$; $\frac{-3}{10}$

Now

$$\frac{7}{15} \div \frac{-3}{10} = \frac{7}{15} \cdot \underline{\quad\quad} = \underline{\quad\quad} \text{ in lowest terms}$$ $\frac{10}{-3}$; $\frac{-14}{9}$

Therefore

$$\left(\frac{7}{9} \div \frac{5}{3}\right) \div \left(\frac{1}{2} - \frac{4}{5}\right) = \underline{\quad\quad}$$ $\frac{-14}{9}$

36. Continuing with the previous two frames, we compute the last value.

$$\frac{3}{2}\left[\left(\frac{7}{9} \div \frac{5}{3}\right) \div \left(\frac{1}{2} - \frac{4}{5}\right)\right] = \frac{3}{2}\left(\frac{-14}{9}\right)$$

By the multiplication rule,

$\frac{-7}{3}$

$$\frac{3}{2}\left(\frac{-14}{9}\right) = \_\_\_\_ \quad \text{in lowest terms}$$

$\frac{-7}{3}$

Therefore the final result is ____.

PRACTICE PROBLEMS

1. Multiply (write your answer in reduced form):

a. $\frac{1}{2} \cdot \frac{2}{3}$ *b.* $\frac{7}{5} \cdot \frac{15}{21}$ *c.* $\frac{-x}{3} \cdot \frac{2y}{x}$ *d.* $\frac{6}{15} \cdot \frac{9}{2}$

e. $\frac{-2}{5} \cdot \frac{3}{-4}$ *f.* $\frac{2x}{3y} \cdot \frac{-x}{4y}$ *g.* $\frac{2x}{y} \cdot \frac{-3y}{8}$ *h.* $\frac{7x}{3y} \cdot \frac{9y}{21x}$

i. $\frac{2x}{y} \cdot \frac{3x}{8}$ *j.* $\frac{x+2}{3} \cdot \frac{4}{x-2}$ *k.* $\frac{x+1}{5} \cdot \frac{x}{x+1}$ *l.* $\frac{2x+3}{x} \cdot \frac{x+4}{6x+9}$

2. Compute and write your answer in reduced form.

a. $\frac{7}{3} \div (-9)$ *b.* $\frac{1}{5} \div \frac{6}{4}$ *c.* $\frac{-4}{9} \div \frac{-2}{7}$ *d.* $\frac{2x}{y} \div \frac{-4}{y}$

e. $\frac{3x}{3y} \div \frac{2x}{2y}$ *f.* $7xy \div \frac{-21x}{y}$ *g.* $\frac{3x}{2} \div x^2$ *h.* $\frac{4x}{5} \div \frac{10}{x^2}$

i. $\frac{5x}{y} \div \frac{3x}{y}$ *j.* $\frac{7x}{2y} \div \frac{4y}{x}$ *k.* $\frac{x+1}{3} \div \frac{2}{x+1}$ *l.* $\frac{3x-1}{4} \div \frac{x}{2}$

3. Compute and write your answer in reduced form.

a. $\left(\frac{1}{2} - \frac{4}{3}\right)\frac{2}{5}$ *b.* $\left(\frac{4x}{5} + 2\right)\left(\frac{1}{3} - 1\right)$

c. $\left(\frac{-6}{9} + \frac{x}{2}\right)\left(\frac{2}{3} - \frac{4}{5}\right)$ *d.* $\frac{4}{7} \cdot \frac{x}{3} \cdot \frac{1}{12x}$

4. Compute and write your answer in reduced form.

a. $\left(1 - \frac{4}{5}\right) \div \frac{6}{7}$ *b.* $\left(\frac{x}{3} - 2x\right) \div \frac{5x}{6}$

c. $\left(\frac{xy}{4} + 3xy\right) \div 12x$ *d.* $\left(\frac{9}{12} \div \frac{4}{7}\right) \div \frac{3}{48}$

5. Compute and write your answer in reduced form.

a. $\left[\left(\frac{8}{3} + 1\right) \div \left(\frac{2}{3} - \frac{1}{4}\right)\right]\frac{1}{4}$

b. $\left[\left(\frac{6}{5} \div \frac{-3}{4}\right)\frac{20}{7}\right] \div \frac{3}{14}$

SELFTEST

Evaluate and reduce:

1. $\frac{5}{8} \cdot \frac{3}{11}$
2. $\frac{9}{x} \cdot \frac{x}{3}$
3. $\frac{-4}{x} \cdot \frac{5}{2y}$
4. $\frac{(x+1)}{2} \cdot \frac{3}{x}$
5. $\frac{15}{2}x$
6. $8 \cdot \frac{9}{-4}$
7. $\frac{2}{3} \div \frac{1}{2}$
8. $\frac{-15}{4} \div \frac{5}{-2}$
9. $\frac{12x}{5} \div \frac{9x}{2}$
10. $\frac{7y}{25} \div \frac{5}{6y}$
11. $\frac{38}{7} \div 18$
12. $\frac{3}{4} \div \left(\frac{2}{3} + \frac{4}{5}\right)$

SELFTEST ANSWERS

1. $\frac{15}{88}$
2. $\frac{3}{1}$ or 3
3. $\frac{-10}{xy}$
4. $\frac{3(x+1)}{2x}$
5. $\frac{15x}{2}$
6. $\frac{-18}{1}$ or -18
7. $\frac{4}{3}$
8. $\frac{3}{2}$
9. $\frac{8}{15}$
10. $\frac{42y^2}{125}$
11. $\frac{19}{63}$
12. $\frac{45}{88}$

5.4 COMPLEX FRACTIONS

Fractions in which the numerator or denominator or both contain other fractions are called *complex fractions*. Each of the following fractions are complex.

$$\frac{2}{\frac{4}{5}} \qquad \frac{\frac{x}{2}+3}{\frac{5}{8}} \qquad \frac{\frac{1}{2}+\frac{4}{5}}{\frac{5}{9}-\frac{4}{5}} \qquad \frac{\frac{2}{x-1}}{1-\frac{3}{x+1}}$$

In this section we will see several ways to simplify complex fractions.

One way to simplify a complex fraction is to write both numerator and denominator as a single fraction. Then apply the rule on division of fractions.

Example 1

$$\frac{2}{\frac{4}{5}} = 2 \div \frac{4}{5}$$

$$= \frac{2}{1} \cdot \frac{5}{4} \qquad \text{by the division rule}$$

$$= \frac{5}{2} \qquad \text{in lowest terms}$$

Example 2 Simplify:

$$\frac{\frac{x}{2}+3}{\frac{5}{8}}$$

First we write the numerator as a single fraction.

$$\frac{x}{2} + 3 = \frac{x}{2} + \frac{3}{1} = \frac{x+6}{2}$$

$$\frac{\frac{x}{2} + 3}{\frac{5}{8}} = \frac{\frac{x+6}{2}}{\frac{5}{8}} = \frac{x+6}{2} \div \frac{5}{8}$$

$$= \frac{x+6}{2} \cdot \frac{8}{5} \qquad \text{by the division rule}$$

$$= \frac{4(x+6)}{5} \qquad \text{in lowest terms}$$

Example 3 Simplify:

$$\frac{\frac{1}{2} + \frac{4}{5}}{\frac{5}{9} - \frac{1}{4}}$$

First we write the numerator as a single fraction.

$$\frac{1}{2} + \frac{4}{5} = \frac{5+8}{10} = \frac{13}{10}$$

Then we write the denominator as a single fraction.

$$\frac{5}{9} - \frac{1}{4} = \frac{20-9}{36} = \frac{11}{36}$$

Therefore

$$\frac{\frac{1}{2} + \frac{4}{5}}{\frac{5}{9} - \frac{1}{4}} = \frac{\frac{13}{10}}{\frac{11}{36}} = \frac{13}{10} \div \frac{11}{36}$$

$$= \frac{13}{10} \cdot \frac{36}{11}$$

$$= \frac{234}{55} \qquad \text{in lowest terms}$$

Example 4 Simplify:

$$\frac{\frac{2}{x+1}}{1 - \frac{3}{x+1}}$$

First we write the denominator as a single fraction. Then we divide. So

$$\frac{\frac{2}{x+1}}{1 - \frac{3}{x+1}} = \frac{\frac{2}{x+1}}{\frac{x-2}{x+1}} = \frac{2}{x+1} \div \frac{x-2}{x+1}$$

$$= \frac{2}{x+1} \cdot \frac{x+1}{x-2}$$

$$= \frac{2}{x-2} \qquad \text{in reduced form}$$

Another way to handle complex fractions is to multiply both numerator and denominator of the complex fraction by the least common denominator of all fractions involved in the numerator and denominator.

Example 5 Simplify (3/7)/(5/4): The least common denominator of 3/7 and 5/4 is $7 \cdot 4$. So we multiply the numerator and denominator by $7 \cdot 4$.

$$\frac{\frac{3}{7}}{\frac{5}{4}} = \frac{\frac{3}{7} \cdot 7 \cdot 4}{\frac{5}{4} \cdot 7 \cdot 4} = \frac{\frac{3 \cdot \not{7} \cdot 4}{\not{7}}}{\frac{5 \cdot 7 \cdot \not{4}}{\not{4}}} = \frac{12}{35}$$

Example 6 Simplify:

$$\frac{\frac{2}{x-1}}{\frac{6}{x-5}}$$

The least common multiple of $x - 1$ and $x - 5$ is $(x - 1)(x - 5)$. So we multiply numerator and denominator by $(x - 1)(x - 5)$.

$$\frac{\frac{2}{x-1}}{\frac{6}{x-5}} = \frac{\frac{2}{x-1} \cdot (x-1)(x-5)}{\frac{6}{x-5} \cdot (x-1)(x-5)} = \frac{\frac{2\cancel{(x-1)}(x-5)}{\cancel{(x-1)}}}{\frac{6(x-1)\cancel{(x-5)}}{\cancel{(x-5)}}}$$

$$= \frac{2(x-5)}{6(x-1)}$$

$$= \frac{x-5}{3(x-1)}$$

Example 7 Simplify:

$$\frac{\frac{1}{5} - \frac{2}{3x}}{\frac{4}{2x} - \frac{x}{2}}$$

The least common denominator of $1/5$, $2/3x$, $4/2x$, and $x/2$ is $30x$ since $30x$ is the least common multiple of 5, $3x$, $2x$, and 2. So

$$\frac{\frac{1}{5} - \frac{2}{3x}}{\frac{4}{2x} - \frac{x}{2}} = \frac{\left(\frac{1}{5} - \frac{2}{3x}\right)30x}{\left(\frac{4}{2x} - \frac{x}{2}\right)30x} = \frac{\frac{1 \cdot 30x}{5} - \frac{2 \cdot 30x}{3x}}{\frac{4 \cdot 30x}{2x} - \frac{x \cdot 30x}{2}}$$

$$= \frac{6x - 20}{60 - 15x^2}$$

PROGRAM 5.4

1. Fractions in which the numerator or denominator or both contain fractions are complex fractions. Which fractions are complex?

a. $\frac{1 + \frac{3}{8}}{\frac{7}{6}}$ *b.* $\frac{x}{\frac{4}{1+4}}$ *c.* $\frac{3(x+1)}{5x^2 + 8}$

a. and *b.*

2. A complex fraction can be simplified by carrying out the indicated division. Thus

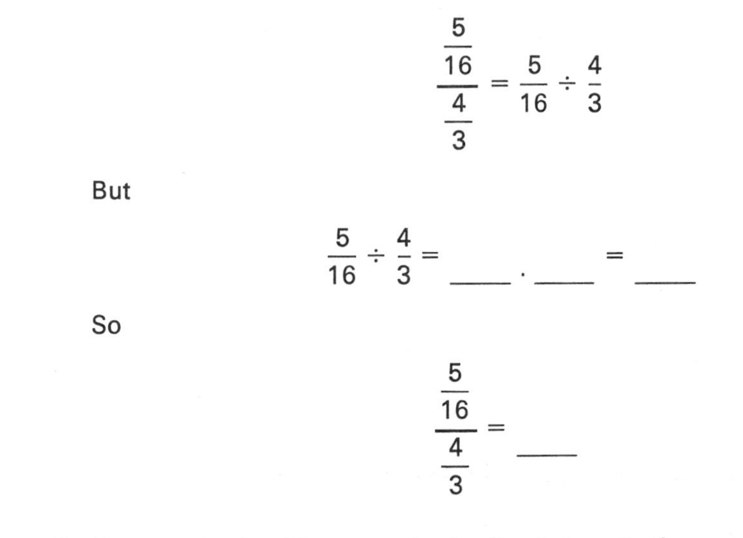

$$\frac{\frac{5}{16}}{\frac{4}{3}} = \frac{5}{16} \div \frac{4}{3}$$

But

$$\frac{5}{16} \div \frac{4}{3} = \_\_\_\_ \cdot \_\_\_\_ = \_\_\_\_$$

So

$$\frac{\frac{5}{16}}{\frac{4}{3}} = \_\_\_\_$$

Answers: $\frac{5}{16}; \frac{3}{4}; \frac{15}{64}$ — $\frac{15}{64}$

3. One way to simplify a complex fraction is to write the numerator as a single fraction; write the denominator as a single fraction; and then divide the numerator by the denominator as we did in the previous frame. To compute

$$\frac{1 + \frac{2}{3}}{\frac{2}{5} - \frac{1}{4}}$$

we first simplify the numerator. The numerator is ________. Then we simplify the denominator. The denominator is ________. Finally we (divide\multiply\add\subtract) the new numerator by the new denominator.

Answers: $1 + \frac{2}{3}$ — $\frac{2}{5} - \frac{1}{4}$ — divide

4. Simplify:

$$\frac{1 + \frac{2}{3}}{\frac{2}{5} - \frac{1}{4}}$$

First simplify the numerator.

$$1 + \frac{2}{3} = \frac{3 + 2}{3} = \_\_\_\_$$

Now simplify the denominator.

$$\frac{2}{5} - \frac{1}{4} = \frac{8 - 5}{\_\_\_\_} = \_\_\_\_ = \_\_\_\_ \quad \text{in lowest terms}$$

Now divide the simplified numerator by the simplified denominator.

$$\frac{5}{3} \div \_\_\_\_ = \_\_\_\_$$

So

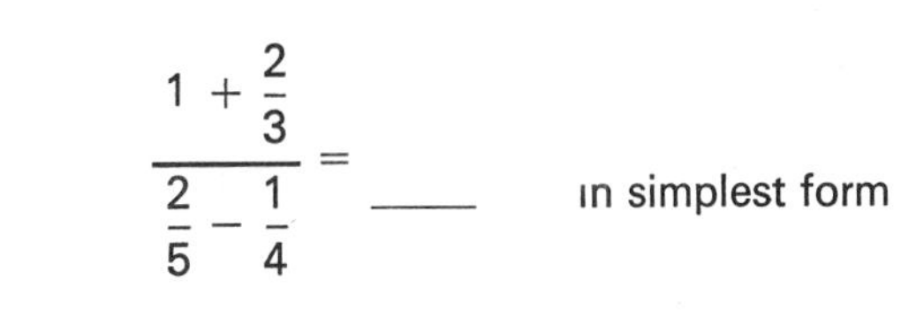

$$\frac{1 + \frac{2}{3}}{\frac{2}{5} - \frac{1}{4}} = \_\_\_\_ \quad \text{in simplest form}$$

Answers: $\frac{5}{3}$ — $20; \frac{3}{20}; \frac{3}{20}$ — $\frac{3}{20}; \frac{100}{9}$ — $\frac{100}{9}$

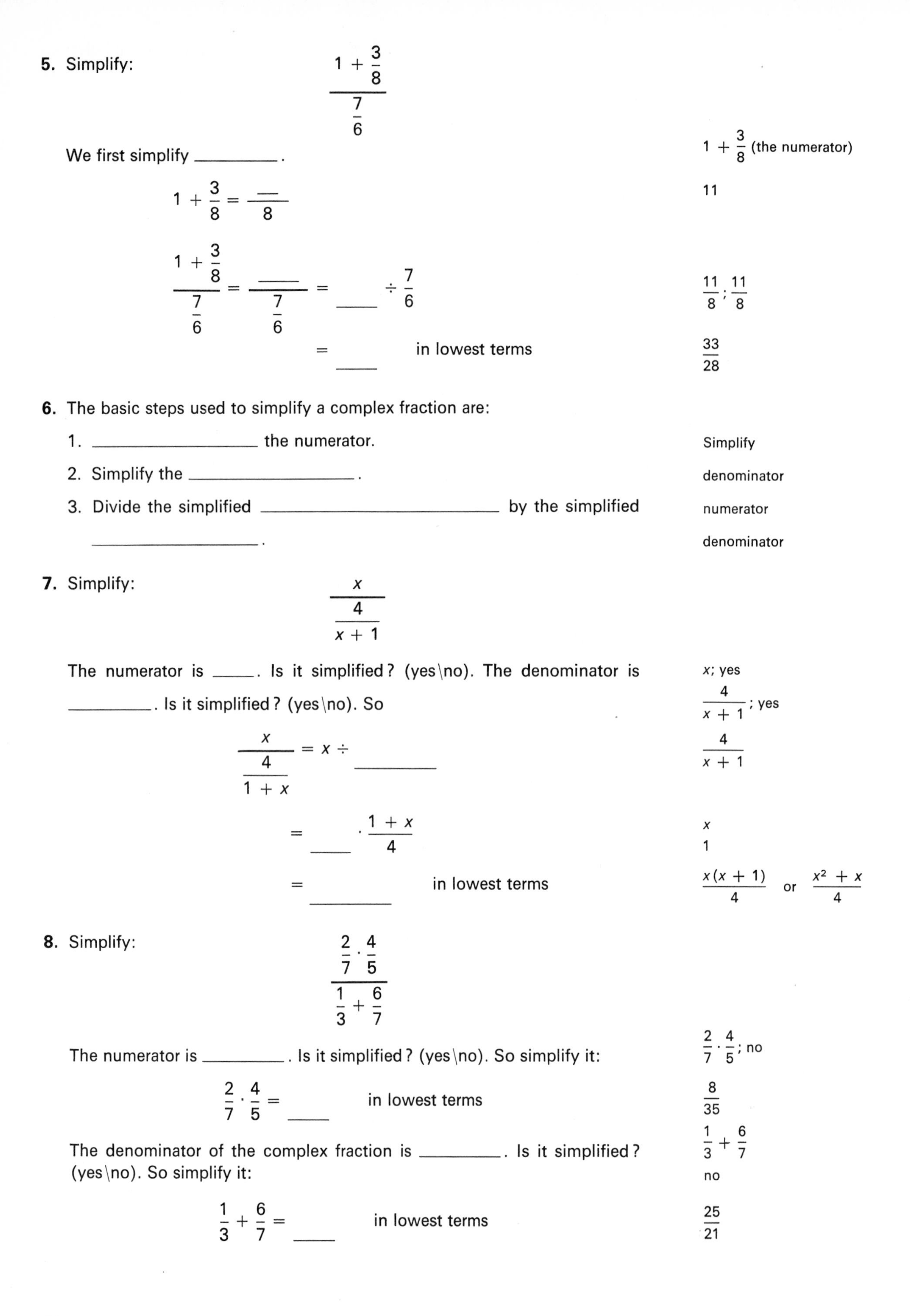

5. Simplify: $\dfrac{1 + \frac{3}{8}}{\frac{7}{6}}$

We first simplify ________.

$1 + \frac{3}{8}$ (the numerator)

$1 + \frac{3}{8} = \frac{\_\_}{8}$

11

$\dfrac{1 + \frac{3}{8}}{\frac{7}{6}} = \dfrac{\_\_\_}{\frac{7}{6}} = \_\_\_ \div \frac{7}{6}$

$\frac{11}{8}$; $\frac{11}{8}$

$= \_\_\_$ in lowest terms

$\frac{33}{28}$

6. The basic steps used to simplify a complex fraction are:

1. ________________ the numerator.
2. Simplify the ________________.
3. Divide the simplified ________________ by the simplified ________________.

Simplify

denominator

numerator

denominator

7. Simplify: $\dfrac{x}{\frac{4}{x + 1}}$

The numerator is ____. Is it simplified? (yes\no). The denominator is ________. Is it simplified? (yes\no). So

x; yes

$\frac{4}{x + 1}$; yes

$\dfrac{x}{\frac{4}{1 + x}} = x \div \_\_\_\_\_$

$\frac{4}{x + 1}$

$= \dfrac{\_\_\_}{\_\_\_} \cdot \dfrac{1 + x}{4}$

x

1

$= \_\_\_\_\_$ in lowest terms

$\frac{x(x + 1)}{4}$ or $\frac{x^2 + x}{4}$

8. Simplify: $\dfrac{\frac{2}{7} \cdot \frac{4}{5}}{\frac{1}{3} + \frac{6}{7}}$

The numerator is ________. Is it simplified? (yes\no). So simplify it:

$\frac{2}{7} \cdot \frac{4}{5}$; no

$\frac{2}{7} \cdot \frac{4}{5} = \_\_\_$ in lowest terms

$\frac{8}{35}$

The denominator of the complex fraction is ________. Is it simplified? (yes\no). So simplify it:

$\frac{1}{3} + \frac{6}{7}$

no

$\frac{1}{3} + \frac{6}{7} = \_\_\_$ in lowest terms

$\frac{25}{21}$

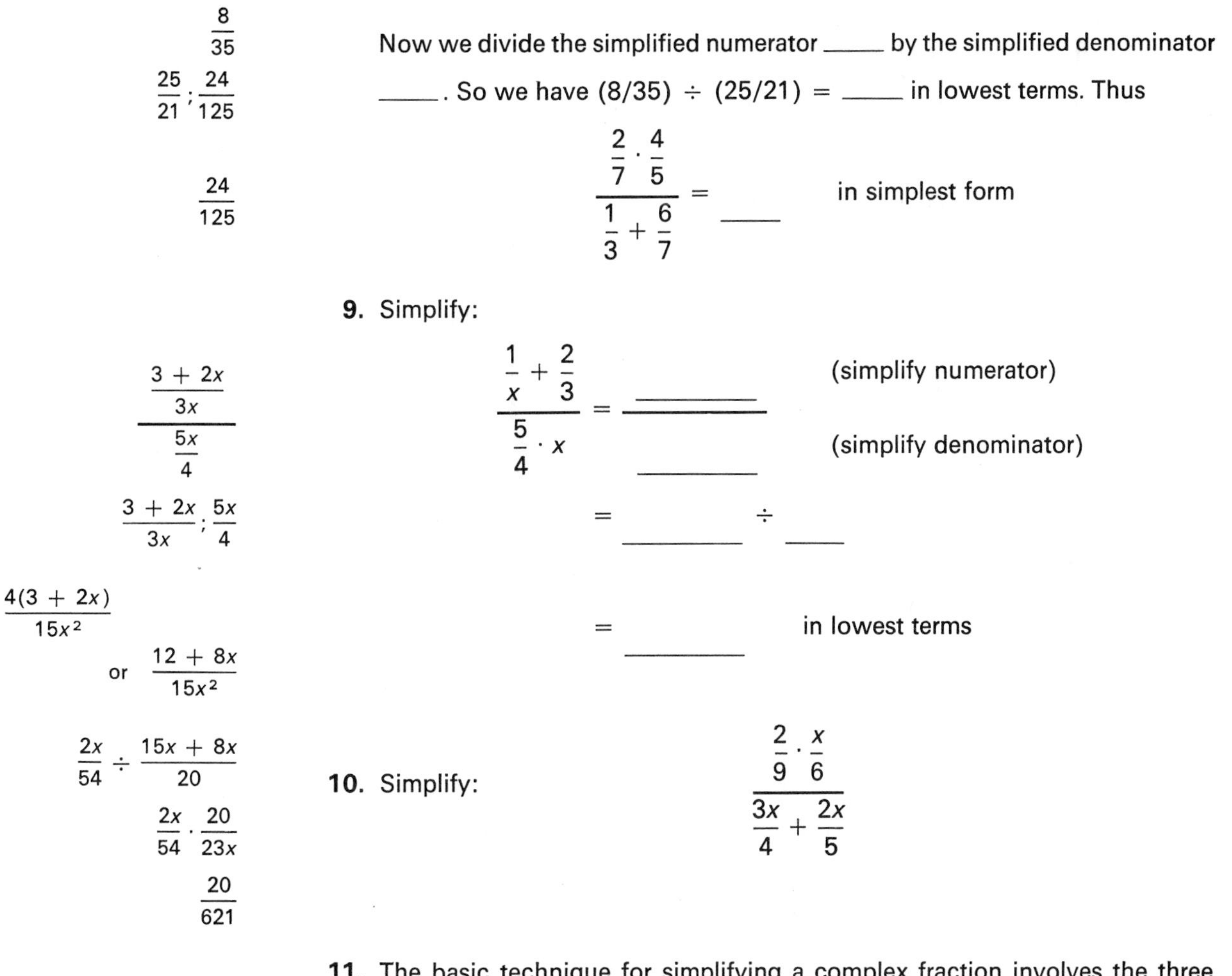

$\frac{8}{35}$

$\frac{25}{21}$; $\frac{24}{125}$

$\frac{24}{125}$

Now we divide the simplified numerator _____ by the simplified denominator _____. So we have $(8/35) \div (25/21) =$ _____ in lowest terms. Thus

$$\frac{\frac{2}{7} \cdot \frac{4}{5}}{\frac{1}{3} + \frac{6}{7}} = \_\_\_\_ \quad \text{in simplest form}$$

$\dfrac{\frac{3 + 2x}{3x}}{\frac{5x}{4}}$

$\frac{3 + 2x}{3x}$; $\frac{5x}{4}$

$\frac{4(3 + 2x)}{15x^2}$ or $\frac{12 + 8x}{15x^2}$

9. Simplify:

$$\frac{\frac{1}{x} + \frac{2}{3}}{\frac{5}{4} \cdot x} = \frac{\_\_\_\_}{\_\_\_\_} \quad \begin{matrix}\text{(simplify numerator)} \\ \text{(simplify denominator)}\end{matrix}$$

$$= \frac{\_\_\_\_}{\_\_\_\_} \div \frac{}{\_\_\_\_}$$

$$= \frac{}{\_\_\_\_} \quad \text{in lowest terms}$$

$\frac{2x}{54} \div \frac{15x + 8x}{20}$

$\frac{2x}{54} \cdot \frac{20}{23x}$

$\frac{20}{621}$

10. Simplify:

$$\frac{\frac{2}{9} \cdot \frac{x}{6}}{\frac{3x}{4} + \frac{2x}{5}}$$

simplify numerator

simplify denominator

divide the new numerator by the new denominator

11. The basic technique for simplifying a complex fraction involves the three steps:

1. ____________________
2. ____________________
3. ____________________

15

12. Another way to simplify a complex fraction is to multiply the numerator and denominator of the complex fraction by a common multiple of the denominators of all fractions occurring in the numerator or denominator of the complex fraction. By this process, we can cancel each denominator in the numerator in the complex fraction. We can also cancel each denominator in the denominator of the complex fraction. For example

$$\frac{\frac{4}{5}}{\frac{3}{8}} = \frac{\frac{4}{5} \cdot 40}{\frac{3}{8} \cdot 40} = \frac{32}{\_\_\_\_}$$

The final result is in simplified form.

13. Simplify:

$$\frac{1 + \frac{3}{8}}{\frac{7}{6}}$$

The fraction occurring in the numerator is ____. The fraction occurring in the denominator is ____. The denominators of 3/8 and 7/6 are ____ and ____. The least common multiple of 6 and 8 is ____. Now we multiply the numerator and denominator of the complex fraction by 24. Note the use of parentheses to insure that we multiply the entire numerator by 24.

$\frac{3}{8}$

$\frac{7}{6}$; 8

6; 24

$$\frac{1+\frac{3}{8}}{\frac{7}{6}} = \frac{\left(1+\frac{3}{8}\right)24}{\frac{7}{6}\cdot 24} = \frac{24+\frac{3\cdot 24}{8}}{\_\_\_\_}$$

$$= \frac{24 + \_\_}{28}$$

$$= \_\_\_\_$$

28

9

$\frac{33}{28}$

14. Simplify:

$$\frac{\frac{3}{2x}-\frac{1}{7}}{\frac{4}{x-1}}$$

The fractions in the numerator are ____ and ____. The fraction in the denominator is ________. The denominators of all the fractions involved in the complex fraction are $2x$, 7, and ______. A common multiple of $2x$, 7, and $x-1$ is $2x \cdot 7 \cdot (x-1) =$ ____$x(x-1)$. So now we will multiply the numerator and denominator of the complex fraction by the common multiple ________.

$(3/2x)$; $1/7$

$4/(x-1)$

$x-1$

14

$14x(x-1)$

15. (Continuation of problem in previous frame.)

$$\frac{\frac{3}{2x}-\frac{1}{7}}{\frac{4}{x-1}} = \frac{\left(\frac{3}{2x}-\frac{1}{7}\right)14x(x-1)}{\frac{4}{x-1}\cdot 14x(x-1)}$$

$$= \frac{\frac{3\cdot 14x(x-1)}{2x} - \frac{1\cdot 14x(x-1)}{7}}{\frac{\_\_\_\_}{x-1}}$$

$$= \frac{3\cdot 7(x-1) - \_\_\_\_}{\_\_\_\_}$$

$$= \frac{21(x-1) - 2x(x-1)}{56x}$$

$$= \frac{\_\_\_\_}{56x}$$

$56x(x-1)$

$2x(x-1)$

$56x$

$-2x^2 + 23x - 21$

16. Simplify:

$$\frac{3+\frac{1}{x}}{\frac{2}{x}\cdot\frac{1}{5}}$$

$\frac{1}{x}; \frac{2}{x}; \frac{1}{5}$

$5x$

multiply

$5x$

The fractions occurring in the numerator and denominator are _____. The least common multiple of the denominators of these fractions is _____. We (add\multiply) numerator and denominator of the complex fraction by _____. So

$5x$

5

2

$\frac{15x+5}{2}$

$$\frac{3+\frac{1}{x}}{\frac{2}{x}\cdot\frac{1}{5}} = \frac{\left(3+\frac{1}{x}\right)5x}{\frac{2}{x}\cdot\frac{1}{5}\cdot\underline{\quad}}$$

$$= \frac{15x + \underline{\quad}}{\underline{\quad}}$$

$$= \frac{\underline{\qquad}}{\underline{\qquad}} \quad \text{in lowest terms}$$

fractions

multiply

17. To simplify fractions by the second technique we first find the least common denominator of all the _____ in the numerator and denominator. We then (multiply\divide) numerator and denominator of the complex fraction by this number.

18. Use the second technique to simplify

$$\frac{\frac{1}{3}+\frac{1}{x}}{\frac{1}{5}-\frac{1}{3}}$$

$15x$

$15x$

$15x$

$\frac{-(5x+15)}{2x}$ or $\frac{-5x-15}{2x}$

We multiply the numerator and denominator of the complex fraction by the value _____. So

$$\frac{\frac{1}{3}+\frac{1}{x}}{\frac{1}{5}-\frac{1}{3}} = \frac{\left(\frac{1}{3}+\frac{1}{x}\right)\underline{\quad}}{\left(\frac{1}{5}-\frac{1}{3}\right)\underline{\quad}}$$

$$= \frac{\underline{\qquad}}{\underline{\qquad}} \quad \text{in lowest terms}$$

$\frac{\left(4+\frac{1}{x}\right)xy}{\left(3+\frac{1}{y}\right)xy} = \frac{4xy+y}{3xy+x}$

19. Use the second technique to simplify:

$$\frac{4+\frac{1}{x}}{3+\frac{1}{y}}$$

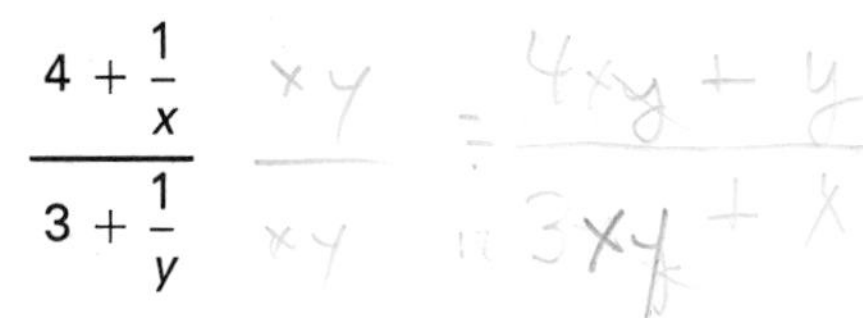

$\frac{20x+2x^2}{3}$

20. Use either technique to simplify:

$$\frac{5+\frac{x}{2}}{\frac{1}{x}\cdot\frac{3}{4}}$$

$\frac{6}{5}$

21. Use either technique to simplify $\frac{x/5}{x/6}$.

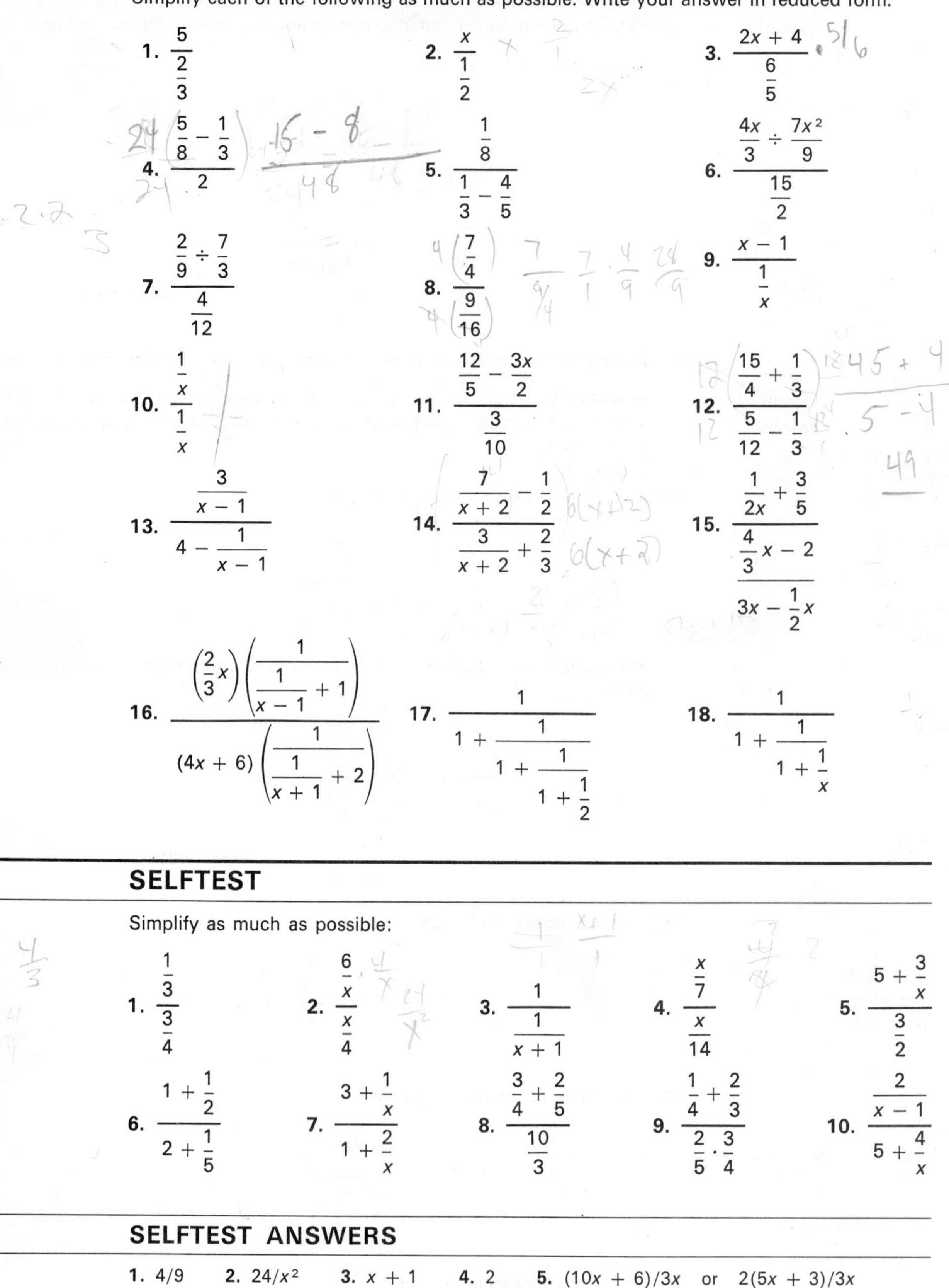

PRACTICE PROBLEMS

Simplify each of the following as much as possible. Write your answer in reduced form.

1. $\dfrac{\frac{5}{2}}{3}$

2. $\dfrac{x}{\frac{1}{2}}$

3. $\dfrac{2x+4}{\frac{6}{5}}$

4. $\dfrac{\frac{5}{8}-\frac{1}{3}}{2}$

5. $\dfrac{\frac{1}{8}}{\frac{1}{3}-\frac{4}{5}}$

6. $\dfrac{\frac{4x}{3}\div\frac{7x^2}{9}}{\frac{15}{2}}$

7. $\dfrac{\frac{2}{9}\div\frac{7}{3}}{\frac{4}{12}}$

8. $\dfrac{\frac{7}{4}}{\frac{9}{16}}$

9. $\dfrac{x-1}{\frac{1}{x}}$

10. $\dfrac{\frac{1}{x}}{\frac{1}{x}}$

11. $\dfrac{\frac{12}{5}-\frac{3x}{2}}{\frac{3}{10}}$

12. $\dfrac{\frac{15}{4}+\frac{1}{3}}{\frac{5}{12}-\frac{1}{3}}$

13. $\dfrac{\frac{3}{x-1}}{4-\frac{1}{x-1}}$

14. $\dfrac{\frac{7}{x+2}-\frac{1}{2}}{\frac{3}{x+2}+\frac{2}{3}}$

15. $\dfrac{\frac{1}{2x}+\frac{3}{5}}{\dfrac{\frac{4}{3}x-2}{3x-\frac{1}{2}x}}$

16. $\dfrac{\left(\frac{2}{3}x\right)\left(\dfrac{1}{\frac{1}{x-1}+1}\right)}{(4x+6)\left(\dfrac{1}{\frac{1}{x+1}+2}\right)}$

17. $\dfrac{1}{1+\dfrac{1}{1+\dfrac{1}{1+\frac{1}{2}}}}$

18. $\dfrac{1}{1+\dfrac{1}{1+\frac{1}{x}}}$

SELFTEST

Simplify as much as possible:

1. $\dfrac{\frac{1}{3}}{\frac{3}{4}}$

2. $\dfrac{\frac{6}{x}}{\frac{x}{4}}$

3. $\dfrac{1}{\frac{1}{x+1}}$

4. $\dfrac{\frac{x}{7}}{\frac{x}{14}}$

5. $\dfrac{5+\frac{3}{x}}{\frac{3}{2}}$

6. $\dfrac{1+\frac{1}{2}}{2+\frac{1}{5}}$

7. $\dfrac{3+\frac{1}{x}}{1+\frac{2}{x}}$

8. $\dfrac{\frac{3}{4}+\frac{2}{5}}{\frac{10}{3}}$

9. $\dfrac{\frac{1}{4}+\frac{2}{3}}{\frac{2}{5}\cdot\frac{3}{4}}$

10. $\dfrac{\frac{2}{x-1}}{5+\frac{4}{x}}$

SELFTEST ANSWERS

1. $4/9$ **2.** $24/x^2$ **3.** $x+1$ **4.** 2 **5.** $(10x+6)/3x$ or $2(5x+3)/3x$
6. $15/22$ **7.** $(3x+1)/(x+2)$ **8.** $69/200$ **9.** $55/18$ **10.** $2x/[(x-1)(5x+4)]$

POSTTEST 5

1. *a.* Is every rational number an integer?
   b. Which of the numbers 12/3, 5, 0.5, −3/2, 0 are rational numbers?
   c. Which are integers?
   d. Locate each of these numbers on the number line.
   e. Which is the largest and which is the smallest?

2. Which of the following fractions equal −3/5?

   a. $\frac{3}{-5}$

   b. $-\frac{15}{25}$

   c. $\frac{-3x}{5x}$

   d. $\frac{-6}{-10}$

   e. $\frac{5}{-3}$

   f. $\frac{6}{-10}$

3. Simplify and leave your answer in reduced form.

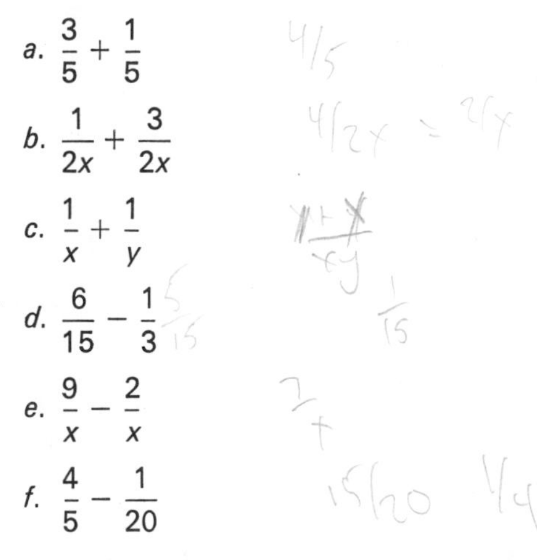

   a. $\frac{3}{5}+\frac{1}{5}$

   b. $\frac{1}{2x}+\frac{3}{2x}$

   c. $\frac{1}{x}+\frac{1}{y}$

   d. $\frac{6}{15}-\frac{1}{3}$

   e. $\frac{9}{x}-\frac{2}{x}$

   f. $\frac{4}{5}-\frac{1}{20}$

4. Simplify and leave your answer in reduced form.

   a. $\left(\frac{12}{5}\right)\left(\frac{5}{10}\right)$

   b. $\left(\frac{4x}{3}\right)\left(\frac{6}{8x}\right)$

   c. $\left(\frac{1}{3}+\frac{1}{2}\right)(6)$

   d. $\frac{4x}{y}\div\frac{x}{y}$

   e. $\frac{21y}{5x}\div\frac{7y}{10x}$

   f. $\frac{10x}{y}\div\frac{2}{y}$

5. Simplify and leave your answer in reduced form.

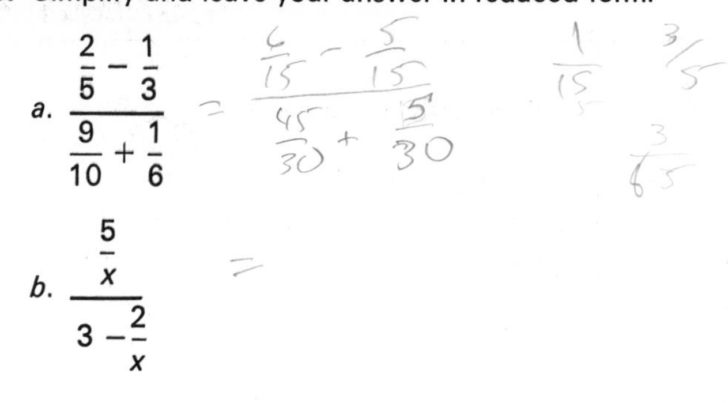

a. $\dfrac{\frac{2}{5} - \frac{1}{3}}{\frac{9}{10} + \frac{1}{6}}$

b. $\dfrac{\frac{5}{x}}{3 - \frac{2}{x}}$

Each problem above refers to a section in this chapter as shown in the table.

Problems	*Section*
1 and 2	5.1
3	5.2
4	5.3
5	5.4

1. *a.* no *b.* They are all rational.
   c. $12/3 = 4$, 5, and 0. *e.* The smallest is $-3/2$ and the largest is 5.
   d.

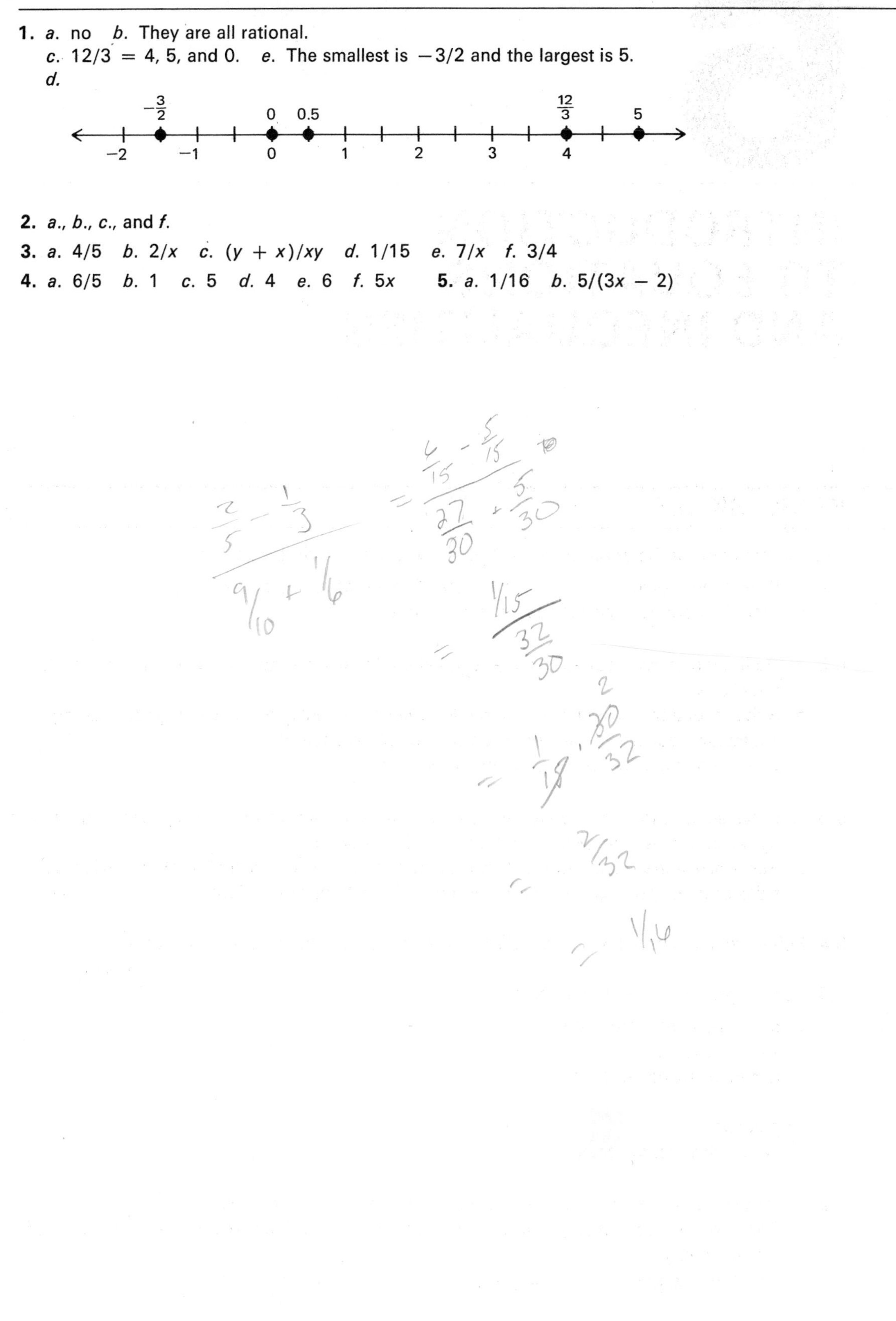

2. *a.*, *b.*, *c.*, and *f.*
3. *a.* 4/5 *b.* $2/x$ *c.* $(y + x)/xy$ *d.* 1/15 *e.* $7/x$ *f.* 3/4
4. *a.* 6/5 *b.* 1 *c.* 5 *d.* 4 *e.* 6 *f.* $5x$ 5. *a.* 1/16 *b.* $5/(3x - 2)$

INTRODUCTION TO EQUATIONS AND INEQUALITIES

BASIC SKILLS

Upon completion of the appropriate section the student should be able to:

6.1 *a.* Determine if given values are in the solution set of an equation.
b. Identify identities and inconsistent equations.

6.2 *a.* Determine if two equations are equivalent by examining the solution sets of the equations.
b. Solve equations in which the variable is raised to only the first power by adding or subtracting equal expressions to both sides of an equation.
c. Use the distributive law to simplify an equation.

6.3 *a.* Solve equations of the above type by multiplying or dividing both sides of an equation by equal expressions that do not contain the variable.
b. Recognize extraneous roots or lost roots that might occur if we multiply or divide both sides of an equation by an expression which contains the variable.

6.4 Solve literal equations with variables to the first power for a specific variable.

6.5 Solve word problems that involve
a. dimensions of a rectangle
b. sum of integers
c. distance, speed, and time
d. rates
e. percent
f. items with various costs.

6.6 *a.* Determine if given values are in the solution set of an inequality.
b. Solve some inequalities by adding or subtracting equal expressions to both sides of the inequality.
c. Graph the solution sets of inequalities on the number line.

6.7 Solve certain inequalities by multiplying or dividing both sides of an inequality by a positive or a negative constant.

PRETEST 6

1. Determine which of the following equations are identities and which are inconsistent.

a. $3(x + 1) = 3x + 1$

b. $x + 6 = x + 5$

c. $x(3x + 2) = 3x^2 + 2x$

2. *a.* What is the solution set of an identity equation?
b. What is the solution set of an inconsistent equation?

3. Find the solution set of each of the following.

a. $x + 3 = 2x - 10$

b. $6(x - 2) = 5(x + 1)$

c. $20x - 4 = 4(5x - 1)$

d. $3(5x + 2) = 4(4x + 3)$

4. Find the solution set of each of the following.

a. $7(x - 2) = 5x + 4$

b. $(3/8)x + 7 = 16$

c. $(x + 2)(x + 1) = (x + 2)2x$

d. $7/(x - 3) = 5$

5. Solve as directed:

a. Solve $P = 25T + 50$ for T.

b. Solve $3LW - 1 = A$ for L.

6. A rectangle has perimeter 52 inches. Its length is 10 inches more than its width. Find the length and width.

7. Solve $2x \geq 5x - 2(2x + 3)$ and shade the solution on the number line.

8. Solve $-8x + 2 < 4x - 7$ and shade the solution on the number line.

Each problem above refers to a section in this chapter as shown in the table.

Problems	*Section*
1 and 2	6.1
3	6.2
4	6.3
5	6.4
6	6.5
7	6.6
8	6.7

PRETEST 6 ANSWERS

1. *a.* inconsistent *b.* inconsistent *c.* identity

2. *a.* the set of all numbers *b.* the empty set

3. *a.* {13} *b.* {17} *c.* all numbers *d.* {−6}

4. *a.* {9} *b.* {24} *c.* {1, −2} *d.* $\{^{22}/_5\}$

5. *a.* $T = \dfrac{P - 50}{25}$ *b.* $L = \dfrac{A + 1}{3W}$

6. length is 18, width is 8

7. $x \geq -6$

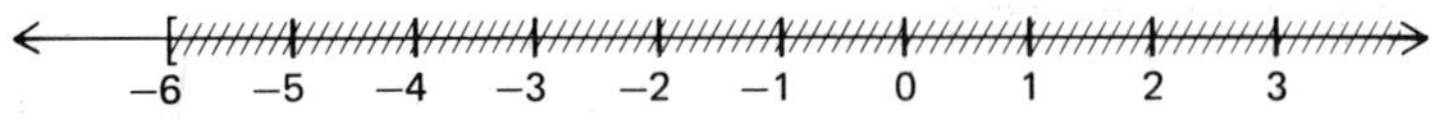

−6 −5 −4 −3 −2 −1 0 1 2 3

8. $x > {}^3/_4$

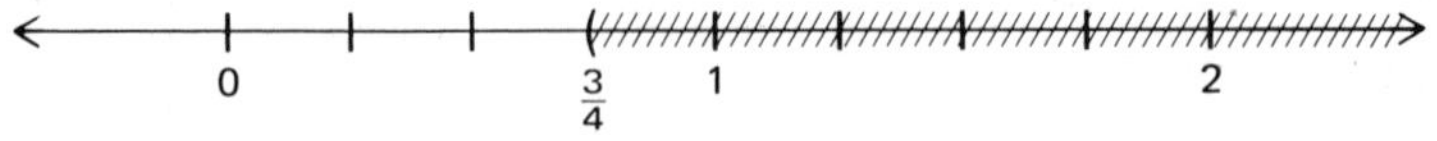

0 $\frac{3}{4}$ 1 2

6.1 EQUATIONS

An equation is a mathematical sentence which claims that two expressions are equal. For example, $x + 4 = 9$ is a claim that the expression $x + 4$ is equal to the number 9 for some value of x. Since $x + 4$ may have many values, the claim may be true or false. If $x = 5$, then the claim that $x + 4 = 9$ is true. But for any value of x other than 5, the claim $x + 4 = 9$ is false.

Equations may take many forms. But they all have the property that one expression is equal to another. An equation is a sentence that may be true or false. Any values of the variable which make the equation true are called *solutions* of that equation. The set of all numbers that make an equation true is called the *solution set* for that equation. We say that a value *satisfies* an equation if it is in the solution set.

Example 1 For some values of x, the equation $5x + 2 = 7$ is false. For instance, if $x = 0$, then $5(0) + 2 = 7$ is false. If $x = 2$, then $5(2) + 2 = 7$ is again false. Therefore, $x = 0$ and $x = 2$ are *not* solutions of the equation $5x + 2 = 7$. The value $x = 1$ makes the equation true, since $5(1) + 2 = 7$ is true. So $x = 1$ is a solution of $5x + 2 = 7$. Furthermore, $x = 1$ is the only solution of $5x + 2 = 7$. Thus the solution set is $\{1\}$.

Example 2 The equation $x^2 - 4 = 0$ is true for only some values of x. If we try $x = 0$, we see that $0^2 - 4 = 0$ is false, so 0 is not a solution. If we put 1 in place of x we get the relation $1^2 - 4 = 0$, which is again false. So 0 and 1 are not solutions of $x^2 - 4 = 0$. However, if $x = 2$, then $x^2 - 4 = 0$ is true, so 2 is a solution. Also, if $x = -2$, we get the true sentence $(-2)^2 - 4 = 0$. So $x = -2$ is also a solution. These values are the only solutions. Thus, the solution set of $x^2 - 4 = 0$ is $\{2, -2\}$.

Example 3 The sentence $x = x + 2$ is an equation. But this equation has no solutions whatsoever. If we add 2 to a number x, it is never the same as x. Therefore, x is never equal to $x + 2$, and $x = x + 2$ has no solutions. We say the solution set for $x = x + 2$ is the *empty set.*

An equation whose solution set is the empty set is called an *inconsistent equation.*

Example 4 The sentence $2x(x + 3) = 2x^2 + 6x$ is an equation. By the distributive law, we know that $2x(x + 3)$ is always equal to $(2x)x + (2x)3$ or $2x^2 + 6x$. Therefore $2x(x + 3) = 2x^2 + 6x$ is always true for any number x. The solution set of this equation is the set of all numbers.

An equation whose solution set is the set of all numbers is called an *identity*.

PROGRAM 6.1

1. The common verbs in mathematics are denoted by the symbols $=$, $<$, and $>$. The symbol $>$ means "is greater than." The symbol $<$ means ____ ____ ____. The symbol $=$ means ____________.

"is less than"

"is equal to" or "equals"

2. The mathematical sentences we will discuss will contain two expressions connected by one of the verbs $=$, $>$, or $<$. Which of the following are mathematical sentences?

a. $4x + 2 = 5$ *b.* $3x - 6$ *c.* $4x > 3$
d. $10x < 2 + 8x$ *e.* $x^2 - 5 = 0$ *f.* $x^2 - 5$

a., c., d., e.

3. Expressions of the form $2x^2 + 8$, $7x^5 + x^4 - 3x^2$, 10, and $x - 1$ (are\are not) mathematical sentences. Since these expressions make no assertions they are neither true nor false for any value of the variable. Is $4x + 8$ true for any value of x? (yes\no). Is $4x + 8$ false for any value of x? (yes\no).

are not

no; no

4. Mathematical sentences can be either true or false. Is $4 = 8$ a mathematical sentence? (yes\no). Is $4 = 8$ true, false, or neither? ________.

yes; false

5. Is $5 + 3 = 8$ a mathematical sentence? (yes\no). Is $5 + 3 = 8$ true, false, or neither? ________.

yes

true

6. A mathematical sentence involving the equal symbol is usually called an equation. We call $2x + 1 = 4x - 3$ an ________.

equation

7. Which of the following are equations?

a. $3x - 2 = 7$ *b.* $2x + 8$ *c.* $4 = 5x^2 + 2x - 3$
d. $8x > 3$ *e.* $5 = 10$

a., c., e.

8. For some values of x, the equation $2x + 1 = 4x - 3$ will be true. For other values of x, it will be false. For $x = 1$, the equation

$$2x + 1 = 4x - 3$$

is $2 \cdot 1 + 1 = 4 \cdot 1 - 3$. When simplified, the equation is $3 =$ ____. This equation is (true\false). Therefore, $x = 1$ makes the equation $2x + 1 = 4x - 3$ (true\false).

1

false

false

9. If $x = 2$, the equation $2x + 1 = 4x - 3$ becomes $2 \cdot 2 + 1 = 4 \cdot 2 - 3$ or ____ $= 5$. Therefore, $x = 2$ makes the equation $2x + 1 = 4x - 3$ (true\false).

5

true

10. In the equation $2x + 1 = 4x - 3$, the variable is ____. Any value of the variable which makes the equation true is called a *solution* of that equation. Since $x = 2$ makes the equation $2x + 1 = 4x - 3$ true, then $x = 2$ is a ________ of the equation ______________.

x

solution;
$2x + 2 = 4x - 3$

11. To determine whether a value is a solution of a particular equation, we substitute that value for the variable and check to see if the resulting equation is true. To determine if 3 is a solution of $2x + 1 = 3x - 2$, we put ____ in place of x.

3

12. When we put 3 in place of x in the equation $2x + 1 = 3x - 2$ we get the new equation $2(\_) + 1 = 3(\_) - 2$. The left side equals ____. The right side equals ____. Therefore, the equation is (true\false) for $x = 3$. So 3 is a ________ of $2x + 1 = 3x - 2$.

3; 3; 7

7; true

solution

13. To determine if 4 is a solution of $2x + 3 = x + 7$, we put ____ in place of x and see if the equation ________ $= 4 + 7$ is true. The left side equals ____. The right side equals ____. So the equation is (true\false) and 4 (is\is not) a solution of $2x + 3 = x + 7$.

4

2(4) + 3

11; 11; true

is

Answers	Frames
	14. To see if -3 is a solution of
	$$2x + 2 = 4 + x$$
	we look at the new equation
$2(-3) + 2$; $4 + (-3)$	________ = ________ (put -3 in place of x)
-4; 1	____ = ____ (simplify both sides)
false; is not	The last equation is (true\false), so -3 (is\is not) a solution.
	15. To see if $x = 2$ is a solution of $2x + 2 = 4 + x$, we look at the equation
$2(2) + 2$; $4 + 2$	________ = ________ (put 2 in place of x)
yes yes	Is this equation true? (yes\no). Is 2 a solution of $2x + 2 = 4 + x$? (yes\no).
Yes; -10 is	**16.** Is -10 a solution of $x + 9 = -1$? (Yes\No), since ____ $+ 9 = -1$ (is\is not) true.
No; 10 false	**17.** Is 10 a solution of $x - 9 = -1$? (Yes\No), because ____ $- 9 = -1$ is (true\false).
{2}	**18.** The set of all solutions of an equation is called the *solution set* of that equation. The equation $2x + 1 = 4x - 3$ has only one solution. It is $x = 2$. Therefore, the solution set for $2x + 1 = 4x - 3$ is ____.
{5, -5}	**19.** The equation $x^2 = 25$ has exactly two solutions. They are 5 and -5. The solution set for the equation $x^2 = 25$ is ________.
y y	**20.** Sometimes it happens that an equation is true for any values of the variables. In the equation $y \cdot y = y^2$ the variable is ____. The equation $y \cdot y = y^2$ is true no matter what value we put in place of ____.
identity	**21.** An equation that is true for all values of the variable is called an *identity*. The equation $y \cdot y = y^2$ is an ________.
is	**22.** The equation $3 = 2 + 1$ has no variables, but the equation (is\is not) always true. We also call the equation $3 = 2 + 1$ an identity.
distributive is identity	**23.** The equation $3(x + 1) = 3x + 3$ is an example of the (commutative\associative\distributive) law. Since this law is always true, the equation $3(x + 1) = 3x + 3$ (is\is not) true for every value of x. Therefore, the equation is an ________.
	24. Which equations are identities?
a., *c.*, *d.*, *f.*	*a.* $4 = 5 - 1$ *b.* $3x = 6$ *c.* $4 + x = x + 4$ *d.* $3(4x) = (3 \cdot 4)x$ *e.* $5 + x = 5 - x$ *f.* $4(1 - x) = 4 - 4x$
is never inconsistent	**25.** It may also happen that an equation is never true. Such an equation is called *inconsistent*. Since $3 = 5 - 4$ (is sometimes\is always\is never) true, it is ________.

26. The equation $x + 3 = x + 5$ is never true no matter what value we put in place of x. If we add x to 3, it is never the same as if we add the same value x to 5. Therefore, the equation $x + 3 = x + 5$ is ________.

inconsistent

27. The solution set of an inconsistent equation is the empty set since there are no solutions. Is $x - 1 = x + 1$ inconsistent? (yes\no). The solution set of $x - 1 = x + 1$ is ________________.

yes

the empty set

28. The equation $x - 1 = 0$ has $x =$ _____ as a solution. Since this is the only solution, the solution set is _____.

1

$\{1\}$

29. The equation $7x^2 3x^4 = 21x^6$ is (an identity\inconsistent\neither). Its solution set is (the set of all numbers\the empty set\some but not all numbers).

an identity

the set of all numbers

30. The equation $x = x + 2$ is (an identity\inconsistent\neither). The reason is that when we add x to 2, we (always\never\sometimes) get x. The solution set of $x = x + 2$ is (all numbers\the empty set\some numbers).

inconsistent

never

the empty set

PRACTICE PROBLEMS

Determine which of the given values of x is a solution for the given equation. Assume that there are no solutions other than those listed. What is the solution set in each case?

1. $2x + 3 = 1$: $x = 1, x = -1, x = 2$
2. $2x = 3x$: $x = 1, x = 0, x = 2$
3. $4x = 3x + 10$: $x = 0, x = -1, x = 10$
4. $x = 5$: $x = 0, x = 5, x = -5$
5. $^3/_2 x = 9$: $x = 0, x = {}^2/_3, x = 6$
6. $^4/_5 x - 1 = {}^1/_2 x$: $x = {}^1/_3, x = {}^2/_5, x = {}^{10}/_3$
7. $2x/(x - 1) = 3$: $x = 1, x = 0, x = 3$
8. $x^2 - 1 = 0$: $x = 0, x = 1, x = -1$
9. $x^2 - 4x + 4 = 0$: $x = 1, x = 5, x = 2$
10. $x^2 + x - 30 = 0$: $x = 1, x = 5, x = -6$
11. $x^3 - 3x^2 + 2x = 0$: $x = 0, x = 1, x = 2$
12. $(x + 5)/(2x^2 + x) = 2/x$: $x = 1, x = 2, x = 5$

Determine which of the following equations are identities or inconsistent equations.

13. $3(x + 1) = 3x + 3$
14. $5 + x = 1 + x$
15. $x^2 - 3 = x^2 + 1$
16. $12x(x - 2) + 1 = 12x^2 - 24x + 1$

SELFTEST

Determine which of the listed values for x are solutions of the given equation. Assuming that there are no solutions other than those listed, what is the solution set in each case?

1. $3x - 1 = 5$: $x = 0, x = 1, x = 2$
2. $(x - 1)x = 0$: $x = 0, x = -1, x = 1$

3. $x^2 - 2x = 3: x = 0, x = -1, x = 3$ **4.** $11x + 4 = 6x - 1: x = -1, x = 2, x = 1$

Which of the following are inconsistent equations? Which are identities?

5. $(x + 1)4 = 4x + 4$ **6.** $x - 6 = x - 8$ **7.** $x - 8 = -8 + x$

SELFTEST ANSWERS

1. $x = 2$; $\{2\}$ **2.** $x = 1$ and $x = 0$; $\{1, 0\}$ **3.** $x = -1$ and $x = 3$; $\{-1, 3\}$
4. $x = -1$; $\{-1\}$ **5.** identity **6.** inconsistent **7.** identity

6.2 EQUIVALENT EQUATIONS I

We do not want to make too great a distinction between equations that have the same solution set. For this reason, if two equations have exactly the same solution set, they are called *equivalent equations*.

The rule below can be used to change a given complicated equation into a simpler equation that is equivalent to the given equation.

Rule of Addition and Subtraction for Equations **When the same algebraic expression is added to or subtracted from both sides of a given equation, the resulting equation is equivalent to the original equation.**

Starting from a given equation the rule may be used repeatedly. Each time the rule is used, the resulting equation is equivalent to the original equation.

We could have stated the rule simply for addition since all subtractions are the same as adding an additive inverse. However, it is sometimes more convenient to use subtraction.

In the following examples we use equivalent equations to find the solution sets.

Example 1 $$x + 5 = 0$$

We wish to find x, but we have $x + 5$. If we subtract 5 from $x + 5$, then we will have just x, for when we subtract 5 we "undo" the operation of "add 5."

$x + 5 - 5 = 0 - 5$ subtract 5 from both sides
$x = -5$ this equation is equivalent to $x + 5 = 0$

The solution set of the equation $x = -5$ is $\{-5\}$ so the solution set of $x + 5 = 0$ is $\{-5\}$. We can verify this result by observing that $-5 + 5 = 0$ is a true statement.

Example 2 $$x - 3 = 7$$

Add 3 to both sides of the equation.

$$x - 3 + 3 = 7 + 3$$
$$x = 10$$

which is equivalent to $x - 3 = 7$. The solution set is $\{10\}$. Again we can check the result by putting 10 in place of x in $x - 3 = 7$. We see that $10 - 3 = 7$ is true.

Example 3 $$x - 2 = 2x$$

Subtract x from both sides and simplify.

$$x - 2 - x = 2x - x$$
$$-2 = x$$

which is equivalent to $x - 2 = 2x$. The solution set of $-2 = x$ is $\{-2\}$, so the solution set of $x - 2 = 2x$ is also $\{-2\}$. When we put -2 in place of x in the original equation, we obtain the true statement $-2 - 2 = 2(-2)$.

Example 4 $$7x - 4 = 6x + 15$$

Subtract $6x$ from both sides and simplify.

$$7x - 4 - 6x = 6x + 15 - 6x$$
$$x - 4 = 15$$

Add 4 to both sides and simplify.

$$x - 4 + 4 = 15 + 4$$
$$x = 19$$

The last equation is equivalent to $7x - 4 = 6x + 15$, so the solution set for the original equation is $\{19\}$. To check, put 19 in place of x. We obtain the true statement $7(19) - 4 = 6(19) + 15$. (Both sides equal 129.)

Example 5

$4(x - 3) = 3(x - 2) + 8$	
$4x - 12 = 3x - 6 + 8$	apply the distributive law
$4x - 12 = 3x + 2$	simplify
$4x - 12 - 3x = 3x + 2 - 3x$	subtract $3x$ from both sides
$x - 12 = 2$	simplify,
$x - 12 + 12 = 2 + 12$	add 12 to both siaes
$x = 14$	simplify

So the solution set is $\{14\}$. Now check:

$$4(14 - 3) \stackrel{?}{=} 3(14 - 2) + 8$$
$$4(11) \stackrel{?}{=} 3(12) + 8$$
$$44 \stackrel{?}{=} 36 + 8$$
$$44 = 44 \qquad \text{true}$$

To solve equations we try to find an equivalent equation with the variable by itself on one side of the equation. If the variable occurs on both sides, we find an equivalent equation with the variable on only one side. It is convenient to collect the terms with a variable on the side that has the variable term with the greatest coefficient. We can add or subtract several terms at a time.

Example 6 Solve $7(x + 5) - 2x = 4(x - 5)$ for x.

$7(x + 5) - 2x = 4(x + 5)$	
$7x + 35 - 2x = 4x - 20$	by the distributive law
$5x + 35 = 4x - 20$	simplify
$5x - 4x + 35 - 35 = 4x - 20 - 4x - 35$	subtract $4x$ and 35 from both sides
$x = -55$	simplify

To check, put -55 in place of x in the original equation.

$$7(-55 + 5) - 2(-55) \stackrel{?}{=} 4(-55 - 5)$$
$$7(-50) - (-110) \stackrel{?}{=} 4(-60)$$
$$-350 + 110 \stackrel{?}{=} -240$$
$$-240 = -240 \qquad \text{true}$$

PROGRAM 6.2

1. The equation $x = 5$ has the number ____ as its only solution. — 5

2. The equation $x + 1 = 6$ has the number ____ as its only solution. — 5

3. The equation $x + 7 = 12$ has the number ____ as its only solution. — 5

{5}

4. The solution set for the equations

$$x = 5, x + 1 = 6 \quad \text{and} \quad x + 7 = 12$$

is ____.

are

5. Equations that have exactly the same solution set are called *equivalent* equations. The equations $x = 5$, $x + 1 = 6$, and $x + 7 = 12$ (are\are not) equivalent equations.

have

equivalent

6. If we are given a list of equations and told that each equation in the list is equivalent to the next equation in the list, then all the equations in the list (have\do not have) the same solution set. Therefore, all the equations in the list are equivalent to each other. In particular, the first and last equations in the list are ________ .

To solve an equation means to compute the solution set of that equation. The next rule will help us solve equations.

Rule for Addition and Subtraction for Equations: When the same algebraic expression is added to or subtracted from both sides of a given equation, the resulting equation is equivalent to the original equation.

9; 9

x; 6

are

{6}; {6}

7. Solve $x + 9 = 15$: We will use the subtraction rule to isolate x on one side of the equation and to isolate the numbers on the other side. Subtract 9 from both sides of the equation. The result is

$$x + 9 - \_\_\_\_ = 15 - \_\_\_\_$$

$$\_\_\_\_ = \_\_\_\_ \qquad \text{simplify each side}$$

By the subtraction rule for equations, the equations $x + 9 = 15$ and $x = 6$ (are\are not) equivalent. The solution set of the last equation is clearly ____. Therefore the solution set for $x + 9 = 15$ is also ____.

2x; 2x

x − 1

1; 1

x

have

are

{10}; {10}

8. Solve $3x - 1 = 2x + 9$: Again we use the addition and subtraction rule for equations to isolate x on one side and all the numbers on the other side.

$$3x - 1 = 2x + 9$$

$$3x - 1 - \_\_\_\_ = 2x + 9 - \_\_\_\_ \qquad \text{subtract } 2x \text{ from both sides}$$

$$\_\_\_\_\_\_\_\_ = 9 \qquad \text{simplify left side}$$

$$x - 1 + \_\_\_\_ = 9 + \_\_\_\_ \qquad \text{add 1 to both sides}$$

$$\_\_\_\_ = 10 \qquad \text{simplify the left side}$$

The equations $3x - 1 = 2x + 9$ and $x = 10$ (have\do not have) the same solution set because they (are\are not) equivalent. The solution set for $x = 10$ is ____. Therefore the solution set for $3x - 1 = 2x + 9$ is ____.

10

10; 10

29; 29

true; is

9. We can check that 10 is a solution of $3x - 1 = 2x + 9$ by putting ____ in place of x. We get the equation

$$3(\_\_\_\_) - 1 = 2(\_\_\_\_) + 9$$

$$\_\_\_\_ = \_\_\_\_ \qquad \text{simplify both sides}$$

The last equation is (true\false), so 10 (is\is not) a solution of the equation $3x - 1 = 2x + 9$.

By repeated use of the rule for addition and subtraction we can start with a given equation and make a list of equations such that each equation in the list is equivalent to the next equation. The solution set of the last equation in the list will always be the same as the solution set of the first equation.

We will organize our work in a two-column format. The left column contains the list of equivalent equations. The right column contains explanations for each step. Supply any missing equations or explanations.

10. Solve $x - 2 = 5$: We will isolate x on one side and all numbers on the other side.

$x - 2 = 5$	given	
$x - 2 + 2 = 5 + 2$	add 2 to both sides	
___ $= 7$	simplify	x

11. The equations $x - 2 = 5$ and $x = 7$ (are\are not) equivalent. — are

12. The solution set for $x - 2 = 5$ and $x = 7$ is ____. — $\{7\}$

13. The solution $x = 7$ (does\does not) check in $x - 2 = 5$ since $7 - 2 = 5$ is (true\false). — does; true

14. Solve $2x + 14 = x + 2$: The first steps isolate all numbers (terms not involving x) on the right side. The next steps isolate all terms involving x on the left side.

$2x + 14 = x + 2$	given	
$2x + 14 - 14 = x + 2 - 14$	subtract 14 from both sides	
$2x =$ ______	simplify	$x - 12$
$2x -$ ____ $= x - 12 - x$	subtract x from both sides	x
____ $= -12$	simplify	x

15. The solution $x = -12$ (does\does not) check since — does

$$2(-12) + 14 = -12 + 2$$

is (true\false). So the solution set for $9x + 14 = 8x + 2$ is ____. — true; $\{-12\}$

16. We place all terms containing the variable on the side that has the variable term with largest coefficient. In $7x - 1 = 3 + 6x$, the coefficient of the x term on the right side is 6. The coefficient of the x term on the left side is ____. The (right\left) side has the largest coefficient of x. So we put all the x terms on the (right\left) side. — 7; left; left

17. Solve $7x - 1 = 3 + 6x$: First isolate all terms involving x on the (right\left) side. Then isolate all terms not involving x on the (right\left) side. — left; right

$7x - 1 = 3 + 6x$	given	
$7x - 1 - 6x = 3 + 6x - 6x$	______________	subtract $6x$ from both sides
$x - 1 =$ ____	simplify	3
$x - 1 + 1 =$ ______	add 1 to both sides	$3 + 1$ or 4
____ $= 4$	simplify	x

Answers	Frames
are	**18.** The equations $7x - 1 = 3 + 6x$ and $x = 4$ (are\are not) equivalent.
true checks	**19.** Since $7(4) - 1 = 3 + 6(4)$ is (true\false), the solution $x = 4$ (checks\does not check).

20. Solve $4 - 2x = 8 - x$: The coefficient of x on the right side is -1. The coefficient of x on the left side is ____. The coefficient ____ is larger. So we isolate the x terms on the (right\left) side and we isolate the terms without x on the (right\left) side.

Answers: -2; -1 / right / left

Answers	Step	Reason
	$4 - 2x = 8 - x$	given
$4 - 2x + 2x$; $8 - x + 2x$	____ = ____	add $2x$ to both sides
$8 + x$	$4 =$ ____	simplify
$4 - 8$; $8 + x - 8$	____ = ____	subtract 8 from both sides
-4	____ $= x$	simplify

21. The solution $x = -4$ checks since, when we put -4 in place of x in $4 - 2x = 8 - x$, we get ____ on both sides.

Answer: 12

22. To solve the equation $2(3x + 5) = 5(x + 1)$, we first simplify both sides as much as possible. Then we proceed as before

Answers	Step	Reason
	$2(3x + 5) = 5(x + 1)$	given
$5x$; $5(1)$	$2(3x) + 2(5) =$ ____ + ____	distributive law

Now we isolate x terms on the (right\left) side and other terms on the (right\left) side.

Answers: left / right

Answers	Step	Reason
$5x + 5 - 5x$	$6x + 10 - 5x =$ ____	subtract $5x$ from both sides
$x + 10$; 5	____ = ____	simplify
$5 - 10$; subtract 10 from both sides	$x + 10 - 10 =$ ____	____
$x = -5$	____ = ____	simplify

So ____ is the solution of $2(3x + 5) = 5(x + 1)$. When we check the solution we see that both sides of the original equation equal ____.

Answers: -5 / -20

23. Solve $x - 1 = 2(x + 1)$.

Answers	Step	Reason
	$x - 1 = 2(x + 1)$	given
$2x + 2$	$x - 1 =$ ____	distributive law

Isolate the x terms on the (right\left) side of the equation and the other terms on the (right\left) side.

Answers: right / left

Answers	Step	Reason
$x + 2$	$-1 =$ ____	subtract x from both sides and simplify
-3	____ $= x$	subtract 2 from both sides and simplify

The solution set is ____.

Answer: $\{-3\}$

24. Solve $4x + 2 = 5x$.

Check your solution.

Answer:
$$4x + 2 = 5x$$
$$4x + 2 - 4x = 5x - 4x$$
$$2 = x$$
$$4(2) + 2 \stackrel{?}{=} 5(2)$$
$$10 = 10 \quad \text{true}$$

25. Solve $4(x + 2) = 5x$.

$$4(x + 2) = 5x$$
$$4x + 8 = 5x$$
$$4x + 8 - 4x = 5x - 4x$$
$$8 = x$$

Check your solution.

$$4(8 + 2) \stackrel{?}{=} 5(8)$$
$$4(10) \stackrel{?}{=} 40$$
$$40 = 40 \quad \text{true}$$

26. Solve $3(x + 5) = 2(x + 2)$.

$$3(x + 5) = 2(x + 2)$$
$$3x + 15 = 2x + 4$$
$$3x + 15 - 2x = 2x + 4 - 2x$$
$$x + 15 = 4$$
$$x + 15 - 15 = 4 - 15$$
$$x = -11$$

Check your solution.

$$3(-11 + 5) \stackrel{?}{=} 2(-11 + 2)$$
$$3(-6) \stackrel{?}{=} 2(-9)$$
$$-18 = -18 \quad \text{true}$$

It is possible to shorten the written work of solving equations by mentally combining several steps. Again we should check the solutions to make sure we have not made a careless mistake.

27. Solve $15x + 12 = 2(8x - 3)$.

$15x + 12 = 2(8x - 3)$	given	
$15x + 12 = 16x - 6$	________ law	distributive
____ $= x$	subtract $15x$ from both sides, add 6 to both sides, and simplify	18

28. The answer $x = 18$ checks since it makes both sides of

$$15x + 12 = 2(8x - 3)$$

equal to ____. 282

29. Solve $9x + 4 = 2(5x + 2)$.

$9x + 4 = 2(5x + 2)$	given	
________ = ________	distributive law and simplify	$9x + 4$; $10x + 4$
____ $= x$	subtract $9x$ and 4 from both sides and simplify	0

30. The solution of $9x + 4 = 2(5x + 2)$ is ____. When we put this value in place of x in the equation we find that both sides equal ____, so the answer does check. 0; 4

31. Solve $4(x - 3) = 5x - 10$.

$$4(x - 3) = 5x - 10$$
$$4x - 12 = 5x - 10$$
$$-2 = x$$

(subtract $4x$ and add 10 to both sides)

Check your solution.

$$4(-2 - 3) \stackrel{?}{=} 5(-2) - 10$$
$$4(-5) \stackrel{?}{=} -10 - 10$$
$$-20 = -20 \quad \text{true}$$

PRACTICE PROBLEMS

Find the solution set for each of the following equations. Check your solutions.

1. $x + 3 = 0$
2. $x - 15 = 40$
3. $x - 8 = 0$
4. $x + 7 = 2$
5. $x + 1 = 6$
6. $-8 = x - 4$
7. $2 = x + 6$
8. $-15 = x + 5$
9. $x - 1 = 1$
10. $x - 4/5 = -1$
11. $x + 2 = 1$
12. $12 = x - 7$
13. $x + 1 = 2x$
14. $x - 8 = 2x$
15. $27x - 5 = 26x$
16. $15x = 16x + 8$
17. $3x - 4 = 2x + 8$
18. $7x + 6 = 8x + 5$
19. $10x + 3 = 9x + 2$
20. $12x - 3 = 11x + 3$
21. $5(x - 2) = 4x$
22. $3(x + 1) = 4x$
23. $4(x - 2) = 3(x + 5)$
24. $-2(x + 5) = -(x + 4)$
25. $-3(x - 3) = -2(x - 1) + 2$
26. $17x - 4 = 5(3x + 1) + x$
27. $20x + 6 = 2(9x - 3) + x$
28. $4(x - 20) = 5(x + 10)$
29. $9x - 6 = 2(4x + 3)$
30. $3(5x + 3) = 4(4x + 4)$
31. $4(2x + 1) = 9(x - 5)$
32. $-5x - 9 = -4(x + 3)$
33. $20x - 5 = 5(4x - 1)$
34. $3x = 2x + 1$
35. $3x + 5x - 7 + (-3x) = 7x + 21 - x$
36. $2x - 3 + x = x + 4 + 3x$
37. $-6x + 9 + 2x = -3x + 2$
38. $2(-3x + 1) + 5x = 8$
39. $-6(3x + 2) + 20x = x + 3$
40. $-2x + 17 + 8x = 3x - 1 + 2x$
41. $16x + 3 = 8x + 9 + 7x$
42. $-3x + 1 - 8x = -4(3x + 2)$
43. $-6x + 5 + 4x = 3(1 - x) + 1$
44. $8x + 31 + (-8x) = -5 + 3(4 + 2x)$
45. $-15x + 6 + 10x = 2(3 - 2x) - 13$

SELFTEST

Find the solution set of each of the following equations. Check your solutions.

1. $x - 7 = 0$
2. $x + 1 = 10$
3. $3x + 1 = 2x$
4. $2x + 3 = x$
5. $12x + 18x = 10 + 29x$
6. $3x - 5 = -1 - x$
7. $5x + 12 = 3(3 + 2x)$
8. $9(x + 1) - 4x = 4(x - 2)$
9. $6(3x - 3) = 4(4x + 2) + x$
10. $17x + 2 = 2(6x + 1) + 4x$

SELFTEST ANSWERS

1. $\{7\}$ 2. $\{9\}$ 3. $\{-1\}$ 4. $\{-3\}$ 5. $\{10\}$
6. $\{1\}$ 7. $\{3\}$ 8. $\{-17\}$ 9. $\{26\}$ 10. $\{0\}$

6.3 EQUIVALENT EQUATIONS II

We can multiply or divide both sides of an equation by a nonzero constant and the new equation is equivalent to the original one. However, when we multiply or divide by an algebraic expression containing the variable, the new equation may not be equivalent to the original one. The situation is described in the next two rules.

Rule of Multiplication for Equations: When the same algebraic expression containing the variable is multiplied on both sides of a given equation, the new equation is such that

1. **any solution of the original equation is still a solution of the new equation, but**
2. **not all solutions of the new equation are necessarily solutions of the original equation.**

Rule of Division for Equations: When the same algebraic expression is divided into both sides of a given equation, the new equation is such that

1. **any solution of the new equation is a solution of the original equation, but**
2. **not all solutions of the original equation are necessarily solutions of the new equation.**

Put briefly, when we multiply, the new equation *might have more solutions* than the original equation, so we might gain extra solutions (see Example 6). But when we divide, the new equation *might have fewer solutions* than the original equation, so we might lose solutions (see Example 3).

If we multiply both sides of an equation by an expression containing a variable, then we must check all solutions in the *original equation*. Solutions that do not work in the original equation are called *extraneous solutions.*

If we divide both sides of an equation by an expression containing a variable, then we must be sure to check the original equation for additional solutions. Any other possible solution will make the divisor 0. In any case, it is always wise to check solutions to make certain an accidental error has not been made.

Example 1 To solve $2x = 6$ for x, divide both sides by the coefficient of the x term and simplify.

$$\frac{2x}{2} = \frac{6}{2}$$

$$x = 3$$

Check: $2(3) = 6$ is true.

Example 2 To solve $5x - 1 = 14$ for x, add 1 to both sides of the equation, simplify, and divide both sides by 5.

$$5x - 1 + 1 = 14 + 1$$

$$5x = 15$$

$$\frac{5x}{5} = \frac{15}{5}$$

$$x = 3$$

Check: $5(3) - 1 = 14$ is true.

Example 3 To solve $2(x - 1)(x + 3) = (x - 1)(4x - 8)$ for x, divide both sides by $(x - 1)$.

$$\frac{2\cancel{(x-1)}(x+3)}{\cancel{(x-1)}} = \frac{\cancel{(x-1)}(4x-8)}{\cancel{(x-1)}}$$

$2(x + 3) = 4x - 8$	simplify
$2x + 6 = 4x - 8$	use the distributive law
$2x + 6 - 2x + 8 = 4x - 8 - 2x + 8$	add 8 to both sides and subtract $2x$ from both sides
$14 = 2x$	simplify
$\frac{14}{2} = \frac{2x}{2}$	divide both sides by 2
$7 = x$	

The solution checks since $2(7 - 1)(7 + 3)$ and $(7 - 1)[(4(7) - 8)]$ both equal 120. Are there any other solutions? We check the values that make the divisor 0. If $x = 1$, then the divisor $(x - 1) = 0$. When 1 is put in place of x in the original equation, both sides are equal to 0. The solution set of the original equation is $\{7, 1\}$.

In this example, we see that the equation $2(x + 3) = 4x - 8$ did not have as many solutions as the original equation. This is because we divided both sides of the original equation by $x - 1$. We would have lost the solution $x = 1$ if we had not checked the value that made the divisor $x - 1$ equal to 0.

Example 4

$$\frac{x-2}{3x} = \frac{x+10}{5x}$$

$15x\left(\frac{x-2}{3x}\right) = 15x\left(\frac{x+10}{5x}\right)$	multiply both sides by $15x$
$5(x - 2) = 3(x + 10)$	simplify
$5x - 10 = 3x + 30$	use the distributive law
$2x - 10 = 30$	subtract $3x$ from both sides
$2x = 40$	add 10 to both sides
$x = 20$	divide both sides by 2

The solution checks since $(20 - 2)/60$ and $(20 + 10)/100$ both equal $3/10$. So the solution set is $\{20\}$.

Example 5

$$\frac{1}{x} + \frac{2}{x-1} = \frac{4}{x-1}$$

$x(x-1)\left(\frac{1}{x} + \frac{2}{x-1}\right) = x(x-1)\left(\frac{4}{x-1}\right)$	multiply both sides by $x(x - 1)$
$\cancel{x}(x-1)\frac{1}{\cancel{x}} + x\cancel{(x-1)}\left(\frac{2}{\cancel{x-1}}\right) = x\cancel{(x-1)}\left(\frac{4}{\cancel{x-1}}\right)$	
$x - 1 + 2x = 4x$	simplify
$3x - 1 = 4x$	
$-1 = x$	subtract $3x$ from both sides

The solution checks. For when we put -1 in place of x in the original equation, we get -2 on both sides. So the solution set is $\{-1\}$.

Example 6

$$\frac{2x-1}{x-1} = \frac{1}{x-1} + 1$$

$(x-1)\left(\frac{2x-1}{x-1}\right) = (x-1)\left(\frac{1}{x-1} + 1\right)$	multiply both sides by $x - 1$
$\cancel{(x-1)}\left(\frac{2x-1}{\cancel{x-1}}\right) = \cancel{(x-1)}\left(\frac{1}{\cancel{x-1}}\right) + (x-1)1$	
$2x - 1 = 1 + x - 1$	
$2x - 1 = x$	simplify
$2x - 1 - x + 1 = x - x + 1$	subtract x and add 1, both sides
$x = 1$	

Does the solution $x = 1$ check in the original equation? No. When we put 1 in place of x in the original equation, we see that the denominator $x - 1$ equals 0. Therefore the original equation has no solution. The solution set is the empty set.

In this example the new equation $2x - 1 = x$ had a solution ($x = 1$) which was not a solution to the original equation. So $x = 1$ is called an *extraneous solution*. It is possible to get extraneous solution whenever we multiply both sides of an equation by an expression containing the variable. In this particular example, we multiplied both sides of the original equation by $x - 1$.

If an equation contains several fractions, we use the multiplication rule to "clear the fractions." We multiply both sides of the equation by a common denominator (preferably the least common denominator) of all the fractions in the equation.

Example 7 Solve ${}^4/_5 x + 2 = {}^3/_4$. The least common denominator of ${}^4/_5$ and ${}^3/_4$ is 20. So we multiply both sides of the equation by 20.

$$\begin{aligned} 20({}^4/_5 x + 2) &= 20({}^3/_4) \\ 20 \cdot {}^4/_5 x + 20 \cdot 2 &= 20 \cdot {}^3/_4 \\ 16x + 40 &= 15 \\ 16x &= -25 \\ x &= {}^{-25}/_{16} \end{aligned}$$

PROGRAM 6.3

1. When both sides of an equation are multiplied by the same nonzero constant, the resulting equation is equivalent to the original equation. The equation $x + 1 = 9$ is equivalent to $2x + 2 = 18$. For we multiply both sides of $x + 1 = 9$ by 2 to get $2x + 2 = 18$. The equation $3x - 5 = 2$ (is\is not) equivalent to the equation $12x - 20 = 8$. For we can multiply both sides of the original equation by ____.	is 4
2. The equation $x + 3 = 2$ is equivalent to ($5x + 3 = 10$\$5x + 15 = 10$) since we can multiply both sides of $x + 3 = 2$ by ____.	$5x + 15 = 10$ 5
3. We can multiply both sides of $-x = -10$ by -1 to obtain the equivalent equation ($x = -10$\$x = 10$\$-x = 10$).	$x = 10$
4. The equation $-x = 7$ is equivalent to ($x = 7$\$-x = -7$\$x = -7$) since we can multiply both sides by ____.	$x = -7$ -1
5. To solve $-3 = -x$, we multiply both sides by ____. The solution is given by the resulting equation ____ = ____.	-1 3; x
6. Solve $-x = 5$. Check your solution.	$(-1)(-x) = (-1)(5)$ $x = -5$ $-(-5) \stackrel{?}{=} 5$ $5 = 5$
7. If we multiply both sides of $x/4 = 5$ by 4 we get ____ = ____. These two equations (are\are not) equivalent so they (have\do not have) exactly the same solutions.	x; 20 are; have
8. To solve equations involving fractions, we multiply both sides of the equation by a common denominator of all the fractions in the equation. The new equation is simpler and it (is\is not) equivalent to the old one. So to solve $x/3 = 2$ we multiply both sides by ____. The new equation is $x =$ ____. So the solution of $x/3 = 2$ is ____.	is 3; 6 6

Answers	Frames
5 $x = 115$; 115	**9.** To solve $x/5 = 23$, we multiply both sides by ____. The new equation is ________. The solution of $x/5 = 23$ is ____.
23	**10.** The solution checks since $115/5 =$ ____.
$8(x/8) = (-10)8$ $x = -80$	**11.** Solve $x/8 = -10$.
$-80/8 = -10$	Check your solution.
	12. Recall that $\frac{a}{b}x = \frac{ax}{b}$, so $\frac{-1}{7}x$ and ${}^{-1}/_{7}x$
$-1x$ or $-x$	are the same as ____/7.
$\frac{-x}{7} = 5$ $7\left(\frac{-x}{7}\right) = 7(5)$ $-x = 35$ $(-1)(-x) = (-1)35$ $x = -35$	**13.** Solve ${}^{-1}/_{7}x = 5$, and check the solution.

14. Solve:

Answers	Step	Reason
	${}^{1}/_{5}x + 10 = 3$	given
5	$5({}^{1}/_{5}x + 10) = 5(3)$	multiply both sides by ____
x; 50	____ + ____ $= 15$	simplify
-35	$x =$ ____	subtract 50 from both sides

15. The solution -35 checks as follows:

Answers	Step	Reason
	${}^{1}/_{5}(-35) + 10 \stackrel{?}{=} 3$	
-7	____ $+ 10 \stackrel{?}{=} 3$	simplify
3	____ $= 3$	simplify

Answers	Frames
is	The last equation (is\is not) true.

16. Solve:

Answers	Step	Reason
	${}^{1}/_{3}x + 4 = 3 + {}^{2}/_{3}x$	given
3; 3	___$({}^{1}/_{3}x + 4) =$ ___$(3 + {}^{2}/_{3}x)$	multiply both sides by 3
x; 12; 9; $2x$	____ + ____ = ____ + ____	simplify

Answers	Frames
3	Now we use the subtraction and addition rule to solve the equation. The solution is ____.
5 does	**17.** When we put 3 in place of x in ${}^{1}/_{3}x + 4 = 3 + {}^{2}/_{3}x$, both sides equal ____, so the solution (does\does not) check.
8; 8	**18.** Solve ${}^{3}/_{8}x - {}^{1}/_{2} = {}^{1}/_{4}x$. The least common denominator of ${}^{3}/_{8}$, ${}^{1}/_{2}$, and ${}^{1}/_{4}$ is ____. So we multiply both sides of the equation by ____.

19. $^{3}/_{8}x - ^{1}/_{2} = ^{1}/_{4}x$ given

___(_______) $= 8(^{1}/_{4}x)$ multiply both sides by 8 — $8(^{3}/_{8}x - ^{1}/_{2})$

_______ = ____ simplify — $3x - 4 = 2x$

Now we use the addition and subtraction rule to find the solution. The solution is $x =$ ____. — 4

20. The solution checks since it makes both sides of the given equation equal to ____. — 1

21. Solve and check $^{3}/_{10}x = ^{1}/_{5}x + 5$. First we multiply both sides by ____ to get an equivalent equation without fractions. The resulting equation is — 10

_______________ — $3x = 2x + 50$

The final solution is _______. Check the solution. — $x = 50$

$$^{3}/_{10}(50) \stackrel{?}{=} ^{1}/_{5}(50) + 5$$
$$15 \stackrel{?}{=} 10 + 5$$
$$15 = 15$$

22. Solve $^{4}/_{7}x - ^{1}/_{2} = ^{9}/_{14}x$.

$$14\left(\frac{4x}{7} - \frac{1}{2}\right) = 14\left(\frac{9x}{14}\right)$$
$$8x - 7 = 9x$$
$$-7 = x$$

Check your solution.

$$^{4}/_{7}(-7) - ^{1}/_{2} \stackrel{?}{=} ^{9}/_{14}(-7)$$
$$-4 - ^{1}/_{2} \stackrel{?}{=} ^{-9}/_{2}$$
$$^{-9}/_{2} = ^{-9}/_{2}$$

23. When both sides of an equation are divided by the same nonzero constant, the resulting equation is equivalent to the original equation. Therefore the two equations have exactly the same _______________. The equation $4x = 12$ is equivalent to $x =$ ____, since we can divide both sides by 4. — solution set; 3

24. The equation $6x = 12$ is equivalent to $x =$ ____ since we can divide both sides by ____. — 2; 6

25. The equations $6x = 12$ and $x = 2$ both have the same solution set {____}. So to solve $6x = 12$ we use (division\multiplication\addition\subtractoin) to remove the coefficient 6 of the x term. — 2; division

26. In the equation $7x = 25$, the coefficient of the x term is ____. To remove the coefficient of the x term, we divide both sides of the equation by ____. The new equation is $x =$ ____. So the solution of $7x = 25$ is ____. — 7; 7; $^{25}/_{7}$; $^{25}/_{7}$

27. In the equation $-8x = 32$ the coefficient of the x term is ____. To solve the equation we divide both sides by ____. The resulting equation is $x =$ ____. So the solution of both equations is ____. — -8; -8; $^{32}/_{-8}$ or -4; -4

28. Solve $-7x = 84$, and check your solution.

$$x = \frac{84}{-7}$$
$$x = -12$$
$$-7(-12) = 84$$

29. Let us solve the equation $-3x + 6 = 15$. We apply the subtraction rule first to isolate the x term on one side of the equation.

	$-3x + 6 = 15$	given
9	$-3x =$ _____	subtract 6 from both sides
-3 (divide by -3)	$x =$ _____	divide both sides by the coefficient of the x term

division or multiplication

30. In general, an efficient way to solve equations is to use the addition and subtraction rules and simplify to isolate the variable term on one side. We use (division\multiplication\addition\subtraction) to remove the coefficient of the variable term.

This is a good place for a study break.

31. Solve:

	$5x + 9 = 4$	given
$5x$; -5	_____ = _____	isolate the x term on the left
-1	$x =$ _____	divide both sides by the coefficient
5		of the x term, which is _____

32. Solve:

	$8x + 4 = 2x - 9$	given
$6x$; -13	_____ = _____	isolate the x terms on the left and simplify
x; $-13/6$	$x =$ _____	divide both sides by the coefficient of the x term

$-40/3$

33. The solution checks since it makes both sides of the original equation equal to _____.

34. Solve $3x - 6 = x + 8$.

$$3x - 6 - x = x + 8 - x$$
$$2x - 6 = 8$$
$$2x - 6 + 6 = 8 + 6$$
$$2x = 14$$
$$x = 7$$

Check your solution.

$$3(7) - 6 \stackrel{?}{=} 7 + 8$$
$$21 - 6 \stackrel{?}{=} 15$$
$$15 = 15$$

addition, subtraction

35. We use (division\multiplication\addition\subtraction) to isolate the variable terms on one side of the equation.

division or multiplication

36. We use (division\multiplication\addition\subtraction) to remove the coefficient of the variable term.

multiplication

37. We use (division\multiplication\addition\subtraction) to change an equation with fractions to an equivalent equation without fractions.

$15x + 4 = 20x - 14$

$18 = 5x$

$18/5 = x$; $18/5$

38. Solve $3/4\,x + 1/5 = x - 7/10$. Multiply both sides by 20 to remove the fraction and simplify. _____________. Isolate the x terms on the right and simplify. _________. Divide both sides by the coefficient of the x term. _________. So the solution is _____.

39. Solve $^2/_3 x - ^1/_9 = ^4/_9 x + 1$. Multiply both sides by _____ to remove the fractions and simplify. _________ = _________. Isolate the x terms on the left side and simplify. _____ = _____. Finish solving the equation. _________.

9

$6x - 1$; $4x + 9$

$2x$; 10; $x = 5$

40. Solve $^1/_6 x - 5 = ^1/_4 x$. Multiply both sides by the least common denominator of $^1/_6$ and $^1/_4$ and simplify. _________ = _________. Complete the solution. _________.

multiply by 12

$2x - 60$; $3x$

$-60 = x$

41. Solve $^5/_4 x + 5 = ^1/_2 x - 4$.

$$4(^5/_4 x + 5) = 4(^1/_2 x - 4)$$
$$5x + 20 = 2x - 16$$
$$3x = -36$$
$$x = -12$$

Check your solution.

$$^5/_4(-12) + 5 \stackrel{?}{=} ^1/_2(-12) - 4$$
$$-15 + 5 \stackrel{?}{=} -6 - 4$$
$$-10 = -10$$

We can multiply or divide an equation by a nonzero constant and the resulting equation is equivalent to the original equation. But when solving equations we must often multiply or divide by an expression containing the variable. In this case, the resulting equation may not be equivalent to the original. For convenience, we repeat the multiplication rule.

Multiplication Rule for Equations: When both sides of an equation are multiplied by the same algebraic expression containing the variable the new equation is such that

1. any solution of the original equation is still a solution of the new equation, and
2. not all solutions of the new equation are necessarily solutions of the original equation.

42. Solutions that do not work in the original equation are not true solutions. They are called *extraneous* solutions. To solve

$$\frac{1}{x-1} = \frac{x}{x-1}$$

we multiply both sides by $x - 1$. The resulting equation is $1 = x$. The solution of the new equation is obviously _____. Is 1 a true solution of the original equation or is it an extraneous solution? When we put 1 in place of x we get

1

$$\frac{1}{1-1} = \frac{1}{1-1}$$

Both sides of the equation are undefined since we are attempting to divide by _____ on both sides. Therefore the solution 1 is (a true\an extraneous) solution.

0; extraneous

43. So when we multiply both sides of an equation by an expression containing the variable, we *must* check the solution in the original equation to be sure that it is a true solution. The equation $10/x = 5$ becomes $10 = 5x$ when we multiply both sides by _____. The solution of $10 = 5x$ is clearly _____. Is this value a solution of $10/x = 5$? (yes\no). It (is\is not) necessary to check the solution in the original equation.

x; 2

yes; is

x **44.** Solve $2/x = 4$. We remove the fraction by multiplying both sides by ____.

	$2/x = 4$	given
x	$(2/x)x = 4(\_\_\_\_)$	multiply both sides by x
2	$\_\_\_\_ = 4x$	simplify
1/2	$\_\_\_\_ = x$	divide both sides by 4

45. We must check the solution in the original equation because we multiplied by an expression which (did\did not) contain the variable. When we put the solution $x = 1/2$ in the original equation $2/x = 4$ we get

did

$$\frac{2}{1/2} = 4$$

true; a true solution

which is (true\false). Therefore $x = 1/2$ is (a true solution\an extraneous solution).

46. Solve $6/(x - 1) = 3$. We remove the fraction by multiplying both sides by ________.

$x - 1$

	$\frac{6}{x-1} = 3$	given
6; $3x - 3$	$\_\_\_\_ = \_\_\_\_\_\_\_\_$	multiply both sides by $x - 1$ and simplify
9; $3x$	$\_\_\_\_ = \_\_\_\_\_\_\_\_$	isolate the x term on the right
$3 = x$	________	divide both sides by the coefficient of the x term

47. When we check the solution in the original equation, we find it makes both sides equal to ____. Therefore the solution is (true\extraneous).

3; true

48. Solve $(3x + 1)/x = 4 + 1/x$. We multiply both sides by ____ to remove the fractions. The resulting equation is

x

$4x + 1$	$3x + 1 = \_\_\_\_\_\_\_\_$	
$1 = x + 1$	$\_\_\_\_\_\_\_\_ = \_\_\_\_\_\_\_\_$	isolate the x terms on the right
$0 = x$	$\_\_\_\_ = \_\_\_\_$	simplify

The solution to the last equation is ____. This (is\is not) a solution of the original equation, because ________________. So the solution is (true\extraneous). Therefore, the solution set for the original equation is ________________.

0; is not

we cannot divide by 0

extraneous

the empty set

For easy reference, let us restate the division rule.

Division Rule for Equations: When both sides of an equation are divided by an algebraic expression containing the variable, the new equation is such that

1. any solution of the new equation is a solution of the original equation, and
2. not all solutions of the original equation are necessarily solutions of the new equation.

49. If we divide both sides of an equation by an expression containing the variable, it (is\is not) possible that the original equation has more solutions than the new equation. — is

50. So we must always check the original equation again to see if any solutions have been lost. The lost solution will always make the expression we used as a divisor equal zero. If we divide both sides of $x(x + 1) = 0$ by $x + 1$, then we lose the solution _____ since that value makes the divisor $x + 1$ equal to _____, and that value also checks in $x(x + 1) = 0$. — -1; 0

51. Solve:

$x(x + 3) = 2x^2$	given	
$\dfrac{x(x + 3)}{x} = \dfrac{2x^2}{x}$	divide both sides by ___ since it is a factor of both sides	x
$x + 3 =$ _____	simplify	$2x$
_____ $= x$	isolate x on the right and simplify	3

52. The solution $x = 3$ (checks\does not check) in the original equation since it makes both sides of $x(x + 3) = 2x^2$ equal to _____. — checks; 18

53. Since we divided both sides of the equation by an expression containing the variable we must check the equation $x(x + 3) = 2x^2$ for another solution. In the previous frame, we divided both sides of the original equation by _____. What value makes the divisor equal to 0? _____. This value (checks\does not check) in $x(x + 3) = 2x^2$, so it (is\is not) another solution. — x; 0; checks; is

54. So the solution set for $x(x + 3) = 2x^2$ is _________. (See the previous three frames.) — $\{3, 0\}$

55. Solve $(x + 2)(x - 1) = 3(x + 2)$. The expression _________ is a factor of both sides. Therefore divide both sides by $x + 2$. — $x + 2$

$$\frac{(x + 2)(x - 1)}{x + 2} = \frac{3(x + 2)}{x + 2}$$

_________ = _____ — $x - 1$; 3

The solution is: _____ — _____ — $x = 4$

Are there any other solutions for the original equation

$(x + 2)(x - 1) = 3(x + 2)$? (yes\no) — yes

If so, what? _________. The solution set for $(x + 2)(x - 1) = 3(x + 2)$ is _________. — $x = -2$ is also a solution; $\{4, -2\}$

56. Solve: $\dfrac{(x - 7)(x + 2)}{x} = 4(x - 7)$

We multiply both sides by _____ to clear the fraction. So we get — x

_________________ $= 4(x - 7)$___ — $(x - 7)(x + 2)$; x

Divide both sides by _________ since it is a factor of both sides. We get _________________. Now we solve the last equation. The result is _____. — $x - 7$; $x + 2 = 4x$; $^2/_3$

	57. Since we multiplied both sides of the original equation by an expression containing the variable, we must make sure that the solution $x = 2/3$ is a true solution. Is $x = 2/3$ a true solution of
yes	$\frac{(x-7)(x+2)}{x} = 4(x-7)$ (yes\no)
	58. Since we also divided both sides of the equation by an expression containing the variable, we must check the original equation for lost solutions.
7	We divided both sides by $x - 7$. So _____ is a possible lost solution. Is this value actually a solution of
yes	$\frac{(x-7)(x+2)}{x} = 4(x-7)$ (yes\no)
$\{2/3, 7\}$	The solution set is therefore _________.

PRACTICE PROBLEMS

Find the solution set of each of the following equations. Check your solutions.

1. $3x = 9$ **2.** $-4x = 16$ **3.** $-6x = 5$

4. $10x = -50$ **5.** $5x + 10 = 0$ **6.** $9x - 3 = 6$

7. $10x - 5 = 2(x + 3)$ **8.** $7(x - 2) = 5x + 4$ **9.** $4x - 6 = 4(2x + 1)$

10. $3(x - 5) = x + 1$ **11.** $x/6 = 2$ **12.** $x/8 = -4$

13. $1/7\,x = 2$ **14.** $1/5\,x = 5$ **15.** $3/2\,x = 6$

16. $-4/9\,x = 36$ **17.** $5/8\,x + 7 = 2$ **18.** $4/5\,x - 6 = 6$

19. $10 - 7/8\,x = 14$ **20.** $3 + 3/8\,x = 9$ **21.** $\frac{x+1}{2x} = 1$

22. $\frac{20x+8}{3x} = 9$ **23.** $\frac{x+5}{x-2} = 8$ **24.** $\frac{x+4}{x+2} = 3$

25. $(x + 2)(x + 1) = (x + 2)(2x)$ **26.** $(x - 1)(x + 3) = (x + 3)x$

27. $\frac{1}{x-3} = 5$ **28.** $\frac{4x+3}{x} = 1$

29. $\frac{6}{x+1} = \frac{10}{x-2}$ **30.** $\frac{2x-7}{4x} = \frac{1+3x}{8x}$

31. $\frac{3x+2}{x-1} = \frac{2x}{x-1} + 2$ **32.** $\frac{15x-12}{x} = \frac{1}{x} + 3$

33. $\frac{8x-3}{x^2} = \frac{1}{x}$ **34.** $\frac{2x-1}{x^2} = \frac{3}{x}$

35. $\frac{x^2-1}{x+1} = x - 1$ **36.** $\frac{2+x}{x} = 1 + \frac{1}{x}$

37. $\frac{(x-1)(x+3)}{x} = x + 2$ **38.** $\frac{(x+1)(x+3)}{x} = x + 1$

39. $(x + 2)(x - 1) = x(x + 2) + (2x + 1)(x + 2)$

40. $x(8x + 3) = (9x - 1)x - x^2$

41. $(x - 5)(3x + 1) = (x - 5)(x + 5)$

42. $(x + 9)2x = (x + 9)(x - 4)$

43. $(2x + 6)(x + 1) = (2x + 6)(2x + 1)$

44. $x^2 = x(1 - x)$

45. $(x + 16)x = x(2x + 8)$

SELFTEST

Find the solution set of each of the following equations. Check your answers.

1. $5x = -40$ **2.** $x/7 = 5$ **3.** $18x - 24 = 6x$ **4.** $5(x + 4) = x$

5. $7(x + 3)(x - 5) = (x + 3)(2x - 1)$ **6.** $10/x = 2$

7. ${}^3/_4 x = 9$ **8.** $(x - 8)/x^2 = 2/x$

9. $(2x - 3)/x = (x + 1)/5x$ **10.** $\dfrac{2}{x - 1} = \dfrac{2}{x - 1}$

SELFTEST ANSWERS

1. $\{-8\}$ **2.** $\{35\}$ **3.** $\{2\}$ **4.** $\{-5\}$ **5.** $\{-3, {}^{34}/_5\}$
6. $\{5\}$ **7.** $\{12\}$ **8.** $\{-8\}$ **9.** $\{{}^{16}/_9\}$ **10.** no solution exists

6.4 LITERAL EQUATIONS

In this section we will study equations which contain several variables. The object will be to solve an equation by writing one of the variables in terms of the others. An equation or formula expressing a relationship among several variables is called a *literal* equation or literal formula. Many laws of science are expressed as literal formulas.

Example 1 $F = ma$. This relationship was discovered by Newton. It says that force F is equal to mass m times acceleration a.

Example 2 $E = mc^2$. This relationship, discovered by Einstein, says that energy E equals mass m times the velocity of light squared c^2.

Example 3 $C = Q/V$. This relationship expresses capacitance in terms of charge and potential. The letter C stands for capacitance measured in Farads; Q stands for electric charge measured in Coulombs; V stands for electrical potential measured in volts.

Example 4 $PV = P'(V')$. This is Boyle's Law for gases. It says that if a gas has pressure P' (read P prime) volume V' (read V prime), and the pressure is changed to P and the volume changed to V while the temperature remains fixed, then the new pressure times the new volume equals the old pressure times the old volume. P and P' are to be considered distinct variables, as are V and V'.

Example 5 Archimedes discovered that the force necessary to lift a stone under water is $F = w - 62.4V$. Here, F is the force in pounds necessary to lift the stone under water; the letter w is the force in pounds necessary to lift the stone on dry land; and V is the volume of the stone in cubic feet.

Example 6 If an object travels at a constant speed v for a time t, the distance s traveled is $s = vt$.

All of the rules of adding, subtracting, multiplying, and dividing for equations also hold for literal equations. Consider the formula $s = vt$ of the last example. This equation defines distance traveled as a product of speed v and time t. But it also defines time as a relationship of distance and speed. Since $s = vt$, we can divide both sides of the equation by v to discover that $t = s/v$. The new equation $t = s/v$ says that the time t required to travel a distance s at a constant speed v is s/v.

Example 7 The area A of a triangle whose base is b units long and whose altitude is h units is $A = {}^1/_2bh$.

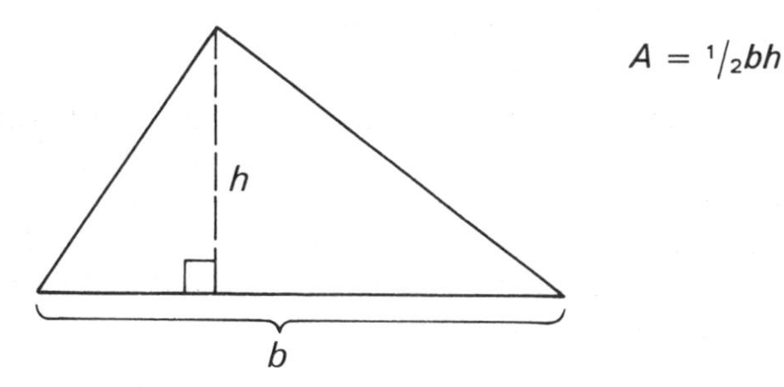

$A = {}^1/_2bh$

Solve $A = {}^1/_2bh$ for b.

${}^1/_2bh = A$	by the symmetric law of equality
$bh = 2A$	multiply both sides by 2
$b = 2A/h$	divide both sides by h

Solve $A = {}^1/_2bh$ for h.

${}^1/_2bh = A$	by the symmetric law of equality
$bh = 2A$	multiply both sides by 2
$h = 2A/b$	divide both sides by b

Example 8 If C represents the temperature measured in degrees Celsius and F represents the temperature measured in degrees Fahrenheit then $F = {}^9/_5C + 32$. When it is 0 degrees Celsius, it is 32 degrees Fahrenheit. A temperature of 100 degrees Celsius corresponds to 212 degrees Fahrenheit. The formula $F = {}^9/_5C + 32$ gives F in terms of C, but it also may be used to find C in terms of F.

$F = {}^9/_5C + 32$	
$F - 32 = {}^9/_5C$	subtract 32 from both sides
$5(F - 32) = 9C$	multiply both sides by 5
$\dfrac{5F - 160}{9} = C$	divide both sides by 9

Example 9 Now we can easily compute the Celsius temperature that corresponds to any Fahrenheit temperature. For example, when the temperature is 0 degrees Fahrenheit, the Celsius reading is

$$C = \frac{5(0) - 160}{9} = \frac{-160}{9} \qquad \text{or about } -17.8$$

Example 10 Is it possible to find the volume of a rock by weighing it? Yes. For we can use the Archimedes formula given in Example 5. The letter F is the weight in pounds of the rock under water. The letter w is the weight in pounds of the rock on dry land and the letter V is the volume in cubic feet of the rock. The formula $F = w - 62.4V$ gives the volume implicitly in terms of F and w. Let us solve this formula for V.

$$F = w - 62.4V$$

$$w - 62.4V = F \quad \text{by the symmetric law of equality}$$

$$-62.4V = F - w \quad \text{subtract } w \text{ from both sides}$$

$$V = \frac{F - w}{-62.4} \quad \text{divide both sides by } -62.4$$

$$V = \frac{w - F}{62.4} \quad \text{simplify}$$

The last formula gives the volume of the rock. For we can weigh the rock on dry land to obtain w. Then we can weigh the rock under water to obtain F. We subtract F from w and divide by 62.4.

An application of Archimedes law was once made for a king. The king of Syracuse ordered a crown made of gold. When the crown was delivered he suspected it was not made of pure gold. The king sent the crown to Archimedes and asked him to test the contents without destroying the workmanship. Archimedes weighed the crown on dry land and then under water. He computed the ratio of these two weights.

$$F = w - 62.4V \qquad \text{So} \qquad \frac{F}{w} = \frac{w - 62.4V}{w}$$

$$= \frac{w}{w} - \frac{62.4V}{w} = 1 - \frac{62.4V}{w}$$

$$= 1 - 62.4\left(\frac{V}{w}\right)$$

The ratio w/V is called the density of a metal and is constant for a given metal no matter what its shape. Since w/V is constant, the multiplicative inverse V/w is also constant. Thus, according to the last equation, F/w is constant for a given metal since $1 - 62.4(V/w)$ is constant for a given metal.

Next, Archimedes weighed a piece of pure gold under water and then on dry land. He computed the ratio F/w for the gold. He concluded that if the ratio F/w of the crown was the same as the ratio F/w of the piece of pure gold, then the crown was made of pure gold. History does not record the decision on the crown. Nevertheless, Archimedes had discovered an important law of science, which he recorded in his book *On Floating Bodies*.

PROGRAM 6.4

1. An equation or formula expressing a relationship among several variables is called a literal equation or literal formula. Many problems are first written as literal equations involving several variables. One variable is solved in terms of the others. Given data may be used to determine the value of this variable.

is	2. The density of an object may be defined as its weight divided by its volume. If D represents density, w represents weight in pounds, and V represents volume in cubic feet, then the definition of density yields the relationship $D = w/V$. This (is\is not) a literal equation.

3. Solve $D = w/V$ for w.

$$D = \frac{w}{V} \quad \text{given}$$

DV

$$\_\_\_\_ = w \quad \text{multiply by } V$$

4. If the density of a rock is 85 pounds per cubic feet and the volume of the rock is 9.5 cubic feet, what is the weight of the rock? From the last frame, the literal formula for weight in terms of density and volume is ________.
Since $D = 85$ and $V = 9.5$, then $w =$ ________ pounds.

$w = DV$

807.5

5. Solve $D = w/V$ for V.

$$D = \frac{w}{V} \quad \text{given}$$

$$DV = w \quad \text{multiply by } V$$

$$V = \_\_\_\_ \quad \text{divide by } D$$

$\frac{w}{D}$

6. If the density of a rock is 85 pounds per cubic foot and the rock weighs 750 pounds, what is the volume of the rock? From the last frame, the literal equation for volume in terms of density and weight is ________. Since $D = 85$ and $w = 750$, then $V =$ ________ cubic feet.

$V = \frac{w}{D}$

8.82

7. An object is falling toward the earth. Let t be the time in seconds since the object was first observed. Let v be the speed of the object when it was first observed. Let u be the speed of the object at time t. Then u, v, and t are related by the literal equation

$$u = 32t + v$$

If $t = 0$, then $u =$ ____. If $t = 2$ and $v = 0$, then $u =$ ____.

v; 64

8. Solve $u = 32t + v$ for v, by ________ ____ to obtain ________.

subtracting; $32t$; $u - 32t = v$

9. Solve $u = 32t + v$ for t. First ________ ____, which gives ________. Then divide by ____, to obtain ________.

subtract; v; $u - v = 32t$

32; $\frac{u - v}{32} = t$

10. The object had a speed v of 50 feet per second when it was first observed and a speed u of 370 feet per second when it hit the earth. How long did it take the object to reach earth after it was first seen? $t =$ ____ seconds. (Use the formula from the previous frame.)

10

11. A rectangle is given. It has length l feet and width b feet. The perimeter of the rectangle is P feet. The variables P, b, and l are related by the literal equation

$$P = 2l + 2b$$

If $l = 2$ and $b = 1$, then $P =$ ____.

6

12. Solve $P = 2l + 2b$ for l.

$P = 2l + 2b$	given	
________	subtract $2b$	$P - 2b = 2l$
________	divide by 2	$\frac{P - 2b}{2} = l$

13. If the perimeter is 10 feet and the width is 4 feet, what is the length?

$l =$ ________ . — 1 foot

14. Solve $P = 2l + 2b$ for b.

$P = 2l + 2b$	given	
$P - 2l = 2b$	subtract $2l$	
________ $= b$	________	$\frac{P - 2l}{2}$; divide by 2

15. When the perimeter is 30 feet and the length is 6 feet, what is the width?

$b =$ ____. — 9 feet

The literal equations in the remaining frames have no particular physical or geometrical meaning. They are given to develop skill in solving literal equations. In each case, we are asked to solve the equation for one of the variables. We isolate this variable on one side of the equation and all terms not involving the variable on the other side.

16. Solve $A + B = 9C/2$ for C.

$A + B = \frac{9C}{2}$	given	
2(________) $= 9C$	multiply by 2	$A + B$
________ $= 9C$	simplify	$2A + 2B$
________ $= C$	divide by 9	$\frac{2A + 2B}{9}$

17. Solve $4R + RS = T$ for R.

$4R + RS = T$	given	
$R(4 + S) = T$	factor out R	
$R =$ ________	divide by $(4 + S)$	$\frac{T}{4 + S}$

18. Solve $PR = 1/S + T$ for S.

$PR = \frac{1}{S} + T$	given	
$SPR =$ ________	multiply by S	$1 + ST$
$SPR - ST = 1$	subtract ____	ST
__$(PR - T) = 1$	factor out S	S
$S =$ ________	divide by $PR - T$	$\frac{1}{PR - T}$

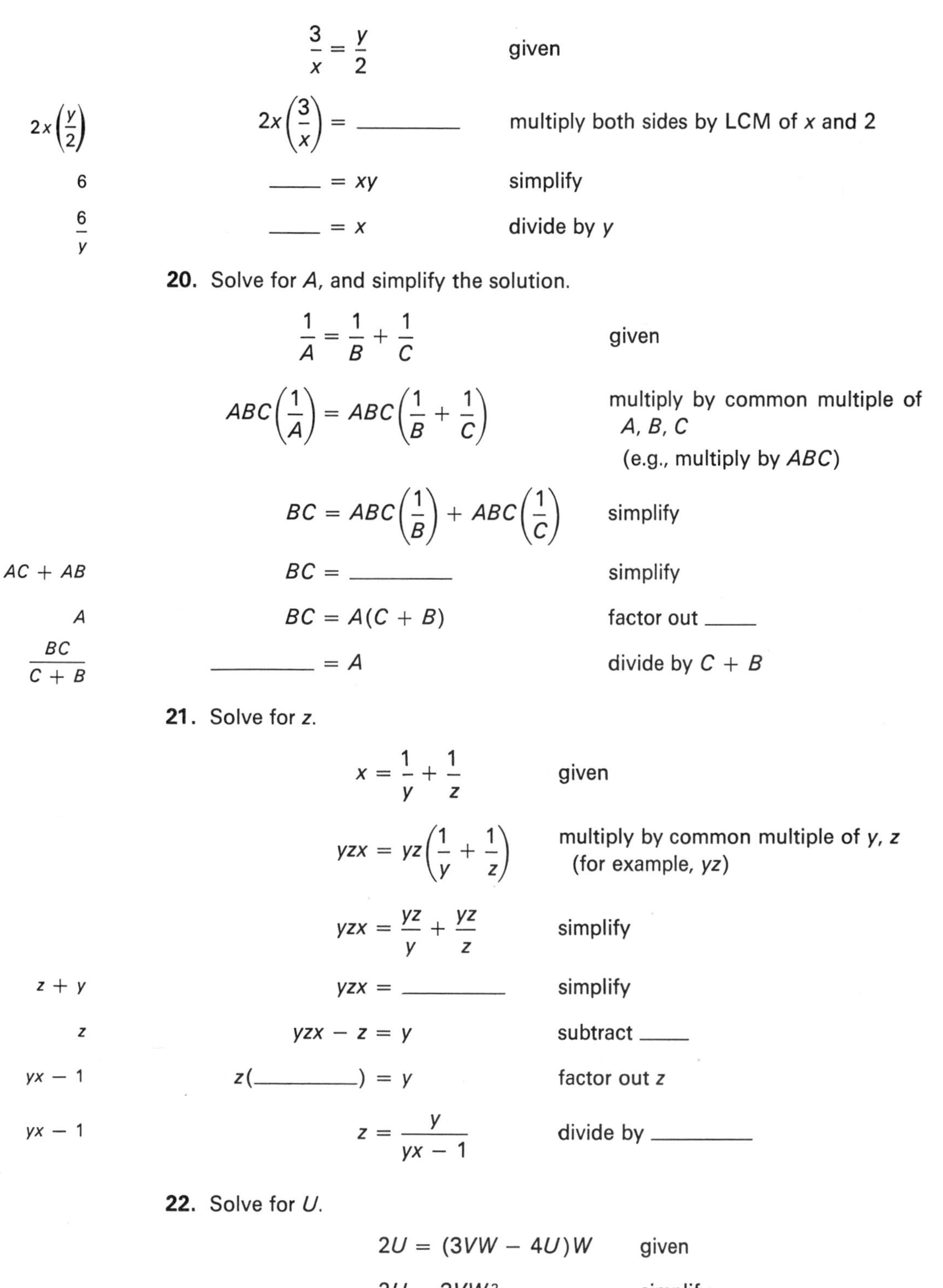

19. Solve for x.

Answer	Step	Reason
	$\dfrac{3}{x} = \dfrac{y}{2}$	given
$2x\left(\dfrac{y}{2}\right)$	$2x\left(\dfrac{3}{x}\right) = $ ______	multiply both sides by LCM of x and 2
6	____ $= xy$	simplify
$\dfrac{6}{y}$	____ $= x$	divide by y

20. Solve for A, and simplify the solution.

Answer	Step	Reason
	$\dfrac{1}{A} = \dfrac{1}{B} + \dfrac{1}{C}$	given
	$ABC\left(\dfrac{1}{A}\right) = ABC\left(\dfrac{1}{B} + \dfrac{1}{C}\right)$	multiply by common multiple of A, B, C (e.g., multiply by ABC)
	$BC = ABC\left(\dfrac{1}{B}\right) + ABC\left(\dfrac{1}{C}\right)$	simplify
$AC + AB$	$BC = $ ______	simplify
A	$BC = A(C + B)$	factor out ____
$\dfrac{BC}{C + B}$	______ $= A$	divide by $C + B$

21. Solve for z.

Answer	Step	Reason
	$x = \dfrac{1}{y} + \dfrac{1}{z}$	given
	$yzx = yz\left(\dfrac{1}{y} + \dfrac{1}{z}\right)$	multiply by common multiple of y, z (for example, yz)
	$yzx = \dfrac{yz}{y} + \dfrac{yz}{z}$	simplify
$z + y$	$yzx = $ ______	simplify
z	$yzx - z = y$	subtract ____
$yx - 1$	$z($______$) = y$	factor out z
$yx - 1$	$z = \dfrac{y}{yx - 1}$	divide by ______

22. Solve for U.

Answer	Step	Reason
	$2U = (3VW - 4U)W$	given
$4UW$	$2U = 3VW^2 - $ ____	simplify
$4UW$	$2U + 4UW = 3VW^2$	add ____
$U(2 + 4W)$	______ $= 3VW^2$	factor out U
$2 + 4W$	$U = \dfrac{3VW^2}{2 + 4W}$	divide by ______

23. Solve for Q and simplify as far as possible.

$\frac{1}{2}V + QV^2 = \frac{1}{3}QV$	given	
$\frac{1}{2}V = \frac{1}{3}QV - QV^2$	subtract QV^2	
$\frac{1}{2}V = Q(\_\_\_\_\_\_\_\_)$	factor out Q	$\frac{1}{3}V - V^2$
$\dfrac{\frac{1}{2}V}{\frac{1}{3}V - V^2} = Q$	divide by $\frac{1}{3}V - V^2$	
$\dfrac{\frac{1}{2}V}{V(\frac{1}{3} - V)} = Q$	factor out V in denominator	
$\dfrac{\frac{1}{2}}{(\frac{1}{3} - V)} = Q$	____________	cancel the V's
$\dfrac{\frac{1}{2}}{\left(\frac{1 - 3V}{3}\right)} = Q$	write denominator as single fraction	
$\_\_\_\_\_\_\_\_ = Q$	rule for division of fractions (invert denominator and multiply)	$\dfrac{3}{2(1 - 3V)}$ or $\dfrac{3}{2 - 6V}$

PRACTICE PROBLEMS

Solve each of the following equations as directed.

1. Solve $7x = y$ for x.
2. Solve $x = yz$ for z; for y.
3. Solve $a + 3 = b$ for a.
4. Solve $q - r = s$ for r; for q.
5. Solve $6A + 9B = C$ for A; for B.
6. Solve $AX + 3B = C$ for X; for B.
7. Solve $PV = P'V'$ for P; for V'.
8. Solve $QR = ST$ for T; for R.
9. Solve $A = \frac{9}{16}B + C$ for B; for C.
10. Solve $X = \frac{5}{9}Y + 32$ for Y.
11. Solve $x = y/z$ for y; for z.
12. Solve $x = (2/a)y$ for a; for y.
13. Solve $1/4 = A + 1/B$ for A; for B.
14. Solve $AB + AC = 2$ for A; for B; for C.
15. Solve $P = 75T + 500$ for T.
16. Solve $(3A - B)/C = 1$ for C; for A; for B.
17. Solve $(2L + 1)W = A$ for L; for W.
18. Solve $V = \frac{1}{2}H(A + B)$ for A; for H; for B.

SELFTEST

1. Solve $I = RV$ for R.
2. Solve $3Y + 2X = Z$ for Y.
3. Solve $P/T = R$ for T.
4. Solve $P = 2L + 2W$ for W.
5. Solve $(A - B)/C = 2$ for A.
6. Solve $X - 4Y = 0$ for X.

SELFTEST ANSWERS

1. $R = I/V$ 2. $Y = (Z - 2X)/3$ 3. $T = P/R$ 4. $W = (P - 2L)/2$
5. $A = 2C + B$ 6. $X = 4Y$

6.5 WORD PROBLEMS

Our work with equivalent equations and literal equations will be useful in solving word problems. Many everyday applications of mathematics are first conceived in the mind as a word problem. The word problem is then put into mathematical symbols and equations. Finally the equations are solved and the solutions are checked in the original problem. The following advice is helpful for solving word problems.

1. Read the problem carefully. Analyze the problem to determine what is given and what must be found.
2. Label the given information and label the quantities to be found. If possible, draw a diagram. Label the various parts of the diagram. Write down all formulas related to the problem, for example, $s = vt$, $A = lb$, etc.
3. Convert the information given in the problem and the diagram to the form of mathematical equations. If you have trouble, go back to step 1.
4. Solve the equations. Check your answers in the equations. Finally, be sure to check your answers in the original word problem. The original problem should make sense with the answers you have provided. If it does not you have made an error in constructing the equations or in solving them. If a mistake occurs, reread the problem and examine the equations.
5. As a last step, be sure you have solved the problem completely. That is, make certain you have found all the unknowns asked for in the problem.

Example 1 A rectangle has perimeter 26 inches. Its length is 5 inches more than its width. Find the width, length, and area of this rectangle.

Solution: If the width is labeled x, then the length must be $x + 5$.

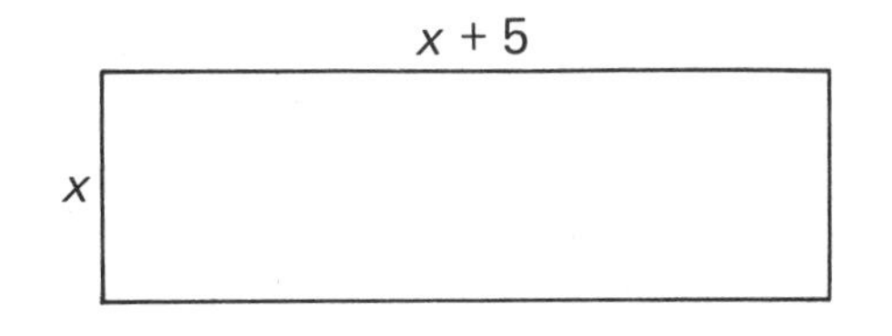

The perimeter is twice the width plus twice the length or $2x + 2(x + 5)$, and is given to be 26 inches. Therefore we have the equation

$$\begin{aligned} 2x + 2(x + 5) &= 26 \\ 2x + 2x + 10 &= 26 \\ 4x + 10 &= 26 \\ 4x &= 16 \\ x &= 4 \end{aligned}$$

So the width x is 4 inches and the length is $x + 5$ or 9 inches. Now that we know the width and the length, we can use the formula A = width times length to find the area.

$$A = 4 \cdot 9 = 36 \text{ square inches}$$

Example 2 An unmanned space capsule has just left earth. It has a velocity of 5,000 miles/hour. It must travel 250,000 miles to reach the moon. How long will it take to reach the moon?

Solution: Let t be the time needed to go 250,000 miles. We use the basic relationship

$$\text{distance} = (\text{speed})(\text{time})$$

We know that the distance is 250,000 miles and the speed is 5,000 miles/hour. So we have the equation

$$250{,}000 \text{ miles} = (5{,}000 \text{ miles/hour})t$$

$$250{,}000 \text{ miles} = \frac{5{,}000 \text{ miles } t}{\text{hour}}$$

$$\frac{250{,}000 \text{ miles}}{\left(\dfrac{5{,}000 \text{ miles}}{\text{hour}}\right)} = t$$

$$50 \text{ hours} = t$$

Notice that the unit "miles" cancels, leaving us with hours.

Example 3 Two joggers start running on a quarter mile oval-shaped track. Jogger A runs 6 miles/hour and jogger B runs 9 miles/hour. How long will it take jogger B to lap jogger A? How far will both have run when this happens?

Solution: Jogger B will lap A when he has run completely around the track and caught up with A. This will happen when B has run exactly $^1/_4$ mile more than A. Let t be the time measured in hours since the joggers started. Let S_A be the distance A has run by time t. Let S_B be the distance B has run by time t. As in the previous example we use the basic relationship

$$\text{distance} = (\text{speed})(\text{time})$$

Therefore

$$S_A = (6 \text{ miles/hour})t \qquad \text{and} \qquad S_B = (9 \text{ miles/hour})t$$

The following equation expresses mathematically the condition "B laps A."

$$S_B = S_A + {}^1/_4$$

By substituting the relations $S_B = 9t$ and $S_A = 6t$ into the previous equation we get

$$9t = 6t + {}^1/_4$$
$$3t = {}^1/_4$$
$$t = {}^1/_{12} \text{ hour}$$

Therefore B laps A after $^1/_{12}$ hour or 5 minutes. How far have they both run in $^1/_{12}$ hour?

$$S_A = 6({}^1/_{12}) = {}^1/_2 \text{ mile}$$
$$S_B = 9({}^1/_{12}) = {}^3/_4 \text{ mile}$$

So A has run $^1/_2$ mile and B has run $^3/_4$ mile.

Example 4 You are in London, England, and it is raining. You dash into a store to buy a rain hat. The hat is marked £1.50 (English pounds). It is on sale at 10 percent off the marked price. You only have United States dollars, but the store agrees to accept them at the exchange rate of \$2.50 U.S. dollars per English pound. How much in U.S. dollars should you pay for the hat?

Solution: First, we need to find out how much the hat costs after the discount. What is 10 percent of £1.50? Remember that if x is any number, then 10 percent of x is $(0.10)x$. Therefore, 10 percent of £1.50 is $(0.10)(1.50) = 0.150$. Since the hat is marked off 10 percent, we deduct 0.150 from 1.50.

$$£1.50 - £0.150 = £1.35$$

The sale price of the hat is £1.35 (English pounds). Now we must find out how much this is in dollars. Let D stand for U.S. dollars and L for English pounds. One pound is worth \$2.50. Therefore,

$$D = (2.50)L$$

L is £1.35 since that is the cost of the hat. So

$$\begin{aligned} D &= (2.50)(1.35) \\ &= 3.375 \end{aligned}$$

The hat costs \$3.38.

Example 5 A man wishes to buy a new car. The price of the new car is \$3,000. The man pays \$500 down. He agrees to pay \$75/month for 3 years. The man estimates that each month the car will depreciate by 1 percent of its new car price. He wants to trade the car in before the money he has paid exceeds the market value of the car. How long can he wait before he must trade in the car?

Solution: Let t represent the time ellapsed in months since the man bought the car. Let P represent the total payments he has made by time t. He puts \$500 down and pays \$75/month. Therefore

$$P = 75t + 500$$

Let M represent the market value of the car at time t. The car has original market value \$3,000 and each month it depreciates (\$3,000)(0.01) = \$30. The new market value is \$3,000 minus the total depreciation. So

$$M = 3{,}000 - 30t$$

The market value M is greater than the money paid P until a certain time t when they are equal. After this time P is greater than M. At what time is $P = M$? Since $P = 75t + 500$ and $M = 300 - 30t$, we have the relation

$$75t + 500 = 3{,}000 - 30t \qquad \text{when} \qquad P = M$$

So

$$\begin{aligned} 105t + 500 &= 3{,}000 \\ 105t &= 2{,}500 \\ t &= 2{,}500/105 \\ t &= 23.8 \qquad \text{(rounded off)} \end{aligned}$$

Therefore the money paid on the car will exceed its market value during the 24th month. He must trade the car in before two years.

PROGRAM 6.5

Answer	Frame
7	**1.** The next group of frames refer to the following word problem: The sum of three consecutive integers is 324. Find the smallest of these integers. Three integers are consecutive if they follow each other in the usual order. The numbers 5, 6, ____ are consecutive integers.
smallest 324	**2.** We must find the ________ of three consecutive integers whose sum is ____. Label this unknown smallest integer x.
$x + 1$ $(x + 1) + 1$ or $x + 2$	**3.** If x is an integer, the next integer after x is ________. The next integer after $x + 1$ is ________.

4. The integers x, $x + 1$, $x + 2$ are ________________ integers. consecutive

5. The problem states that the sum of these consecutive integers is 324. Using our symbols x, $x + 1$, and $x + 2$, this statement becomes the equation

$$x + (x + 1) + (x + 2) = \_\_\_\_$$

324

6. Solve $x + (x + 1) + (x + 2) = 324$.

$3x + 3 = 324$
$3x = 321$
$x = 107$

Since $107 + (107 + 1) + (107 + 2) =$ ____, the solution $x = 107$ (checks\does not check).

324
checks

7. The three consecutive integers whose sum is 324 are 107, ____, ____. The smallest of these integers is ____.

108; 109
107

8. The next group of frames refers to the following word problem: Find the length and width of a rectangle whose perimeter is 20 inches and whose length is 3 inches more than its width. We must find two quantities; they are ________ and ________.

length; width

9. Draw a diagram; label the length l and the width b.

Width = b
Length = l

The perimeter of the rectangle is the distance around the rectangle. From the diagram we see that the distance around the rectangle is $l + b + l +$ ____. When simplified this becomes perimeter = ______________. Since the perimeter is given to be 20, then ____ $= 2l + 2b$.

b
$2l + 2b$
20

10. It is given that the length is 3 inches more than the width. In mathematical symbols this is written $l =$ ________.

$b + 3$

11. In the equation $20 = 2l + 2b$ we replace l by ________. The new equation is $20 = 2($____________$) + 2b$. When this equation is simplified it becomes $20 =$ ________.

$b + 3$
$b + 3$
$4b + 6$

12. Solve $4b + 6 = 20$.

$4b + 6 = 20$
$4b = 14$
$b = 7/2$

13. The width is $b =$ ____. Since the length is 3 more than the width, the length is $l =$ ____. The solution makes sense in the problem because a rectangle of width 3.5 inches and length 6.5 inches has a perimeter of ____ inches.

$7/2$ or 3.5
6.5
20

home; college

14. The next group of frames refer to the following word problem: Sandra normally drives from her home to college in 15 minutes. When she is rushed, she drives 5 miles/hour faster and makes the trip in 12 minutes. How far does Sandra live from the college? We are asked to find the distance between Sandra's ________ and ________. Label this distance s miles.

5 miles/hour

$v + 5$

15. Let v be the average speed Sandra must drive to arrive at her college in 15 minutes. Let u be the average speed she must drive to arrive in 12 minutes. The problem states that to arrive in 12 minutes she must increase her speed by ________________. When stated in mathematical symbols, this becomes the equation $u =$ ________.

hours

$^{15}/_{60}$ or $^{1}/_{4}$; $^{12}/_{60}$ or $^{1}/_{5}$

16. The information given in the problem is summarized in the equation $u = v + 5$. Now we must compute u and v in terms of s (the quantity to be found). The speeds are given in miles/hour, but the times are given in minutes. It is necessary to use the same units. Therefore we must convert minutes to ________. Since there are 60 minutes in each hour, then 15 minutes is _____ hour and 12 minutes is _____ hour.

$4s$

$s/(^{1}/_{5})$

$5s$

17. Remember the relation

$$\text{distance} = (\text{speed})(\text{time})$$

Since s is the distance Sandra travels in $^{1}/_{4}$ hour, she must have an average speed of $v = s/(^{1}/_{4})$. When simplified, the equation is $v =$ _____. When she travels this distance in $^{1}/_{5}$ hour, the average speed is $u =$ _____. When simplified this is $u =$ _____.

$5s = 4s + 5$

18. In the equation $u = v + 5$, we now put $u = 5s$ and $v = 4s$, and get the new equation ________________.

$s = 5$; 5

19. Solve $5s = 4s + 5$ for s: ________. The solution is that Sandra lives _____ miles from college.

20

25; 5

20. This solution makes sense in the original word problem. If Sandra lives 5 miles from college and it takes 15 minutes to go to school her average speed is $5/(^{1}/_{4}) =$ _____ miles/hour. If it takes only 12 minutes, then her average speed is $5/(^{1}/_{5}) =$ _____ miles/hour. Furthermore, 25 is _____ more than 20 as the condition in the problem stated.

Honolulu

San Francisco

the same or equal

21. The next group of frames refer to the following word problem: An airplane is flying from Honolulu to San Francisco. Assume the distance is 3,000 miles and that the plane travels at an air speed of 650 miles/hour. There is a head wind of 30 miles/hour. Find the distance from the starting point to the "point of no return." That is the point beyond which it is faster to fly on to San Francisco than to return to Honolulu. In this problem it is important to know the time required for the airplane to return to ________________ and the time required to fly on to ________________. The airplane is at the "point of no return" when these two times are ________________.

22. Let s be the distance of the airplane from Honolulu. Let H be the time required to return to Honolulu. Let F be the time required to fly on to San Francisco. The time required to travel a certain distance at a constant speed is that distance divided by ________.

the speed

23. If the plane returns to Honolulu it has an air speed of 650 miles/hour plus a tail wind of 30 miles/hour. Therefore, it has a ground speed of ____ miles/hour.

680

24. Since s is the distance to Honolulu and the ground speed is 680, the time required is $H = s/680$.

25. If the plane flies on to San Francisco it has a head wind of 30 miles/hour, so the plane has a ground speed of $650 - 30$ or ____ miles/hour.

620

26. Since the distance between cities is 3,000 miles, and the distance to Honolulu is s, the distance to San Francisco is ________.

$3{,}000 - s$

27. The distance to San Francisco is $3{,}000 - s$ and the ground speed is 620. The time required to go on to San Francisco is $F =$ ________.

$\dfrac{3{,}000 - s}{620}$

28. The "point of no return" is when $F =$ ____.

H

29. Since $F = (3{,}000 - s)/620$ and $H = s/680$, the equation $F = H$ becomes

________ = ________.

$\dfrac{3{,}000 - s}{620} = \dfrac{s}{680}$

30. Solve: $\dfrac{3{,}000 - s}{620} = \dfrac{s}{680}$

$$\frac{3{,}000 - s}{620} = \frac{s}{680}$$
$$2{,}040{,}000 - 680s = 620s$$
$$2{,}040{,}000 = 1{,}300s$$
$$s = 1{,}569$$

The solution $s = 1{,}569$ is rounded off to the nearest mile. To this accuracy the solution (does\does not) check.

does

31. Notice that the "point of no return" is 1,569, which is more than half the distance to San Francisco. This is expected because of the effect of the

________________.

head wind

32. The next group of frames refer to the following word problem: A theater sells tickets at two prices, adult $3.00 and student $1.50. One evening the theater sold 750 tickets and took in $1,425 in cash. How many tickets of each type were sold? We must find the number of ________ tickets and the number of ________ tickets sold.

adult

student

33. Let A be the number of adult tickets sold. Let S be the number of student tickets sold. Since the total number of tickets sold was 750, then

$$A + S = \_\_\_\_$$

750

Therefore, $S =$ ________.

$750 - A$

$3A$

34. Since each adult ticket cost 3 dollars, the cash brought in by all the adult tickets is ________ dollars.

$(1.5)S$

35. Since each student ticket cost 1.5 dollars, the cash brought in by all the student tickets was ________.

$(1.5)S$

36. The total cash taken in by the theater was 1,425 dollars. Therefore

$$3A + \_\_\_\_\_\_\_\_ = 1{,}425$$

$750 - A$

37. Since $S = 750 - A$ and $3A + (1.5)S = 1{,}425$, then

$$3A + 1.5(\_\_\_\_\_\_\_\_) = 1{,}425$$

38. Solve $3A + 1.5(750 - A) = 1{,}425$.

$$1.5A + 1{,}125 = 1{,}425$$
$$1.5A = 300$$
$$A = 200$$

550

39. Since $S = 750 - A$ and $A = 200$, then $S =$ ____.

200, 550, 750, 1,425

40. The number of adult tickets was ____ and the number of student tickets was ____. This makes sense in the original problem since the total number of tickets sold was $200 + 550 =$ ____ and the total cash received was $(200)3 + (550)1.5 =$ ____ dollars.

PRACTICE PROBLEMS

Solve the following problems.

1. The sum of two consecutive integers is 211. Find the integers.
2. The sum of three consecutive odd integers is −39. What are the integers?
3. The sum of four consecutive even integers is 444. Find the smallest of these integers.
4. Together Bob and Jean weigh 340 pounds. Bob weighs 10 pounds more than Jean. How much does each person weigh?
5. Mike has three times as much cash as Marvin. If Mike lends Marvin 5 dollars they will have the same amount of money. How much did Mike have?
6. The length of a certain rectangle is 9 feet more than its width. The perimeter of the rectangle is 24 feet. Find the dimensions of this rectangle.
7. A rectangular lot has a length that is four times its width. The perimeter of the lot is 540 feet. What is the length?
8. Henry has invested \$12,000 at P percent and \$16,000 at $2P$ percent. His annual income from both investments is \$2,200. Find P.
9. A solution of water and antifreeze weighs 200 pounds. It is 12 percent antifreeze by weight. How many pounds of antifreeze must be added to make the solution 18 percent antifreeze by weight?

10. A lake loses water through an irrigation pipe at the rate of 70 gallons/minute. A stream flows into the lake at the rate of 120 gallons/minute. The lake can hold one million gallons of water. If the lake starts out empty, how many hours will it take to fill this lake?

11. Joe deposits $570/month in his account. He spends $465/month from the account. How many months will it take Joe to save an extra $840 in the account?

12. Betty Doe earned $7,000 and put aside $1,200 for taxes on her income. Then she bought a car and used the rest of the money for a vacation. If the vacation cost $1,800 less than the car, how much did she spend on the car?

13. A weather balloon has a maximum capacity of 300 cubic feet. It is $^1/_3$ full of hydrogen when it leaves the ground. Each 1,500 feet, the gas inside the balloon expands 50 cubic feet. At what altitude will the balloon be filled to maximum capacity?

14. A pet shop sells parakeet seed for $1.40/pound. Sunflower seeds sell for $2.00/pound. A mixture of sunflower seeds with parakeet seeds sells for $1.80/pound. If there are 5 pounds of parakeet seed in the mixture, how many pounds of sunflower seeds are there in the mixture?

15. A hardware store makes an assorted mixture of nails. One type of nail costs $0.27/pound. the other type costs $0.14/pound. How many pounds of the first type nail should be mixed with 10 pounds of the second type nail to give a mixture worth $0.18/pound?

16. A hiker walks uphill at 2 miles/hour and downhill at 3 miles/hour. The trail to the top of a mountain is 6 miles. How long does it take him to go round trip without a break? What is his average speed?

17. A cash register contains $40 in nickels, dimes, and quarters. There are twice as many dimes as nickels and $1^1/_2$ times as many quarters as dimes. How many nickels, dimes, and quarters are there?

18. A certain sports arena sells adult tickets for $2.50, children's tickets for $1.00, and military discount tickets for $2.00. One night the arena sold 3 times as many adult tickets as children's tickets and half as many military discount tickets as children's tickets. The total intake from ticket sales was $1,900. Find the number of military discount tickets sold.

19. Georgia and Harriet run a race. Georgia runs at the rate of 3 miles/hour and starts 10 minutes before Harriet. Harriet runs at the rate of 5 miles/hour. How far will Georgia have run when Harriet catches up with her?

*20. In an athletic event, a man rode a bicycle a certain distance and then ran a certain distance. The total distance he covered was 15 miles. It took him 1 hour 45 minutes to do this. He ran 30 minutes longer than he rode the bicycle. He also went 4 times faster on the bicycle than he ran. How far did he run? How far did he ride? How long did he run and how long did he ride?

SELFTEST

1. Find two numbers whose sum is 50 if one number is 12 more than the other.

2. A rectangle is such that its length is 4 feet more than twice its width. The perimeter of the rectangle is 44 feet. Find the dimensions of this rectangle.

* Optional (harder).

SELFTEST ANSWERS

1. Let x be the smaller number. Then $x + 12$ is the other number.

$$\begin{aligned} x + (x + 12) &= 50 \\ 2x &= 38 \\ x &= 19 \text{ is the smaller number} \\ x + 12 &= 31 \text{ is the other number} \end{aligned}$$

2. Perimeter $= P = 2b + 2l = 44$ feet

$$\begin{aligned} 44 &= 2b + 2(2b + 4) \\ 44 &= 6b + 8 \\ 6 &= b \end{aligned}$$

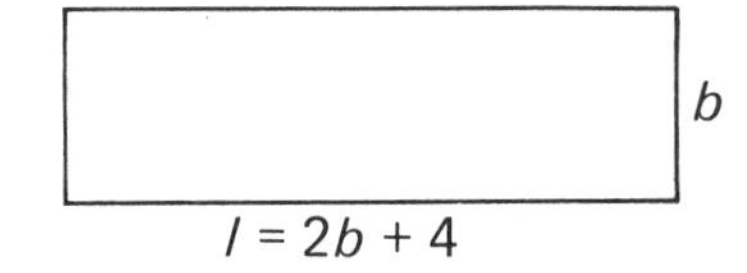

So the width b is 6 feet and the length l is $2b + 4$ or 16 feet.

6.6 INEQUALITIES

In Chapter 1, inequalities were introduced. If the number a is greater than the number b, then we write $a > b$. Of course, if a is greater than b, then b is less than a. We write $b < a$ to mean b is less than a. In this section we will study expressions of the form $3x - 2 > 7$. This is a type of mathematical sentence. It is called an inequality. The inequality symbol, $>$, serves as a verb or action word. The action that takes place in $3x - 2 > 7$ is that $3x - 2$ is made greater than 7.

An inequality such as $3x - 2 > 7$ will sometimes be true and sometimes false depending upon which value of x is used. If $x = 1$ then $3 \cdot 1 - 2 > 7$ is false, since $3 \cdot 1 - 2 = 1$ is not greater than 7. But if $x = 4$, then $3 \cdot 4 - 2 > 7$ is true, since $3 \cdot 4 - 2 = 10$ is greater than 7.

Any values of the variable that make an inequality true are called *solutions* of the inequality. The set of all solutions is called the *solution set* of the inequality.

Inequalities that have the same solution set are called *equivalent inequalities.*

Now we consider the effect of adding or subtracting the same algebraic expression on both sides of an inequality.

Rule of Addition and Subtraction for Inequalities:

1. **If the same algebraic expression is added to both sides of an inequality, the resulting inequality is equivalent to the original inequality. In particular, if $a < b$, and c is any algebraic expression then $a + c < b + c$.**
2. **If the same algebraic expression is subtracted from both sides of an inequality, the resulting inequality is equivalent to the original inequality. In particular, if $a < b$, and c is any algebraic expression, then $a - c < b - c$ is true.**

Example 1 If we add 5 to both sides, $x - 5 > 0$ is equivalent to $x > 5$.

Example 2 If we subtract 6 from both sides, $x + 6 < 2$ is equivalent to $x + 6 - 6 < 2 - 6$ or $x < -4$.

Example 3 Solve:

$$\begin{aligned} 4x &< 3x + 1 && \text{given} \\ 4x - 3x &< 3x + 1 - 3x && \text{subtract } 3x \\ x &< 1 && \text{simplify} \end{aligned}$$

By the rule of subtraction, the inequality $x < 1$ is equivalent to $4x < 3x + 1$. So they have the same solution set. The solution set of $x < 1$ is the set of all numbers less than 1. Therefore, the solution set of $4x < 3x + 1$ is also the set of all numbers less than 1. The solution set is represented on the number line by the portion shaded to the left of 1. We indicate the fact that 1 is *not* included by using a single parenthesis,), as shown in the figure.

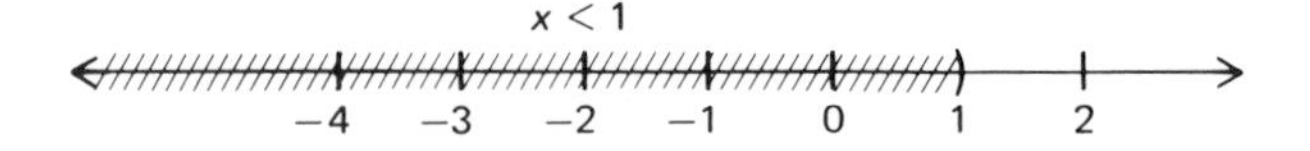

Notice that $x = 1$ is not a solution since $4 \cdot 1 < 3 \cdot 1 + 1$ or $4 < 4$ is false. But all numbers less than 1 are solutions. Let us check $x = -2$. Since

$$4(-2) < 3(-2) + 1$$

(or $-8 < -5$) is true the solution $x = -2$ checks.

Example 4 Solve:

$$\begin{aligned} 9x - 1 &> 8x - 3 && \text{given} \\ x - 1 &> -3 && \text{subtract } 8x \\ x &> -2 && \text{add 1} \end{aligned}$$

The inequalities $x > -2$ and $9x - 1 > 8x - 3$ are equivalent. The solution set of $9x - 1 > 8x - 3$ is the set of all numbers greater than -2. The solution set is represented on this number line by the shaded region to the right of -2.

$-2 < x$ or $x > -2$

−4 −3 −2 −1 0 1 2 3 4

The solution set does not include -2, but it does include all numbers larger than -2. We indicate that -2 is not included by using a single parenthesis, (.

So far we have been studying the relations "greater than" and "less than." Sometimes it is convenient to work with the condition "greater than or equal" or its counterpart "less than or equal." The condition "greater than or equal" combines the "greater than" statement and the "equal" statement with the connective "or." If either statement holds (i.e., either "greater than" or "equal"), then the condition "greater than or equal" is satisfied. The mathematical symbol for "greater than or equal to" is $\geq$. The symbol for the condition "less than or equal to" is $\leq$.

Example 5 If x is greater than or equal to 5 we write $x \geq 5$. If $x = 5$, then $5 \geq 5$ is true. If $x = 21$, then $21 \geq 5$ is also true. We can represent the condition $x \geq 5$ on the number line by shading all numbers to the right of 5 *including* 5. We indicate the fact that we include 5 by use of a single bracket, [.

$5 \leq x$ or $x \geq 5$

−1 0 1 2 3 4 5 6 7 8 9 10

Example 6 Find all numbers x such that $2x + 1$ is less than or equal to $x - 3$. The condition $2x + 1$ is less than or equal to $x - 3$ is written $2x + 1 \leq x - 3$. Therefore we must solve $2x + 1 \leq x - 3$. Since the rule of addition and subtraction holds for equations as well as inequalities it also holds for expressions involving $\geq$ and $\leq$.

$$\begin{aligned} 2x + 1 &\leq x - 3 && \text{given} \\ 2x &\leq x - 4 && \text{subtract 1} \\ x &\leq -4 && \text{subtract } x \end{aligned}$$

The solution set is -4 together with all numbers to the left of -4. The single bracket,], indicates that -4 is *included* in the solution set.

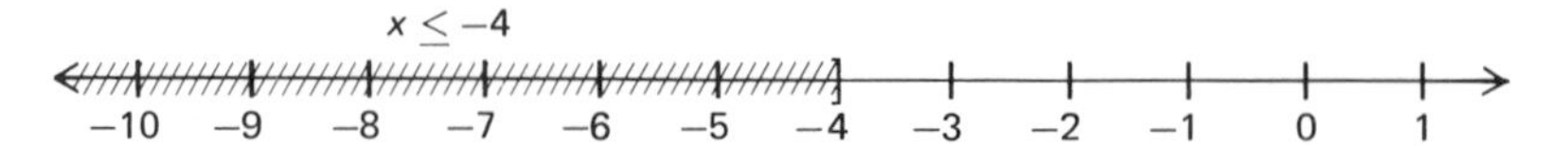

Example 7 Solve:

$$1 - 4x \geq -3x + 2 \quad \text{given}$$
$$1 \geq x + 2 \quad \text{add } 4x$$
$$-1 \geq x \quad \text{subtract } 2$$

The solution set is all numbers less than -1 with -1 included.

$x \leq -1$

−5 −4 −3 −2 −1 0 1 2 3 4 5 6

PROGRAM 6.6

Answer	Frame
b a; right	**1.** The expressions $a > b$ and $b < a$ both mean the same thing. The expression $a > b$ means a is greater than b. The expression $b < a$ means ____ is less than ____. On the number line $a > b$ means a lies to the (left\right) of b.
false; true; true false; false; true	**2.** The expression $a \geq b$ means that either a is greater than b or a equals b. The expression $b \leq a$ means the same thing as $a \geq b$. Label the following as true or false. $6 > 6$ ____ $5 > -1$ ____ $2 \leq 2$ ____ $4 \leq 3$ ____ $1 < 1$ ____ $-3 \leq -3$ ____
positive negative	**3.** A number that is greater than 0 is called a positive number. A number that is less than 0 is called negative. Since $6 > 0$, then 6 is (positive\negative). Since $-2 < 0$, then -2 is (positive\negative).
positive; negative	**4.** If $x > 0$ then x is (positive\negative). If $x < 0$, then x is (positive\negative).
no (it might be zero)	**5.** If $x \geq 0$, then x is called nonnegative. Is a nonnegative number always positive? (yes\no).
greater	**6.** The expression $2x - 1 > x + 5$ is a type of mathematical sentence. It says that $2x - 1$ is made (greater\less) than $x + 5$.
x	**7.** The inequality $2x - 1 > x + 5$ will sometimes be true and sometimes be false depending upon which value of ____ is used.
false	**8.** If $x = 1$, then the inequality $2x - 1 > x + 5$ is (true\false).
true	**9.** If $x = 8$, the inequality $2x - 1 > x + 5$ is (true\false).
true is	**10.** Any values of x which make $2x - 1 > x + 5$ a true statement are called solutions of $2x - 1 > x + 5$. If $x = 10$, then $2x - 1 > x + 5$ is (true\false). Therefore $x = 10$ (is\is not) a solution of $2x - 1 > x + 5$.
set of all positive numbers	**11.** The set of all solutions of an inequality is called the *solution set* for that inequality. The solution set of the inequality $x > 0$ is the ____________ ________.
all numbers greater than 1	**12.** Inequalities that have the same solution set are called *equivalent inequalities*. The inequality $x > 1$ has the solution set (all numbers greater than 1\all

numbers less than 1). The inequality $x - 1 > 0$ has as its solution set all numbers greater than 1. Therefore the inequalities $x > 1$ and $x - 1 > 0$ (are\are not) equivalent. — are

Rule of Addition and Subtraction for Inequalities:

1. If the same algebraic expression is added to both sides of an inequality, the resulting inequality is equivalent to the original inequality. In particular, if c is any algebraic expression and $a < b$ is true, then $a + c < b + c$ is true.
2. If the same algebraic expression is subtracted from both sides of an inequality, the resulting inequality is equivalent to the original inequality. In particular, if c is any algebraic expression, then $a < b$ is true implies $a - c < b - c$ is true.

13. If $a < b$ is true, then $a + 3 < b + 3$ is (true\false). Since $5 > 3$, then $5 - 10 > 3 - 10$ is (true\false). — true; true

14. The rules of addition and subtraction also hold when the greater than symbol, $>$, is replaced with the greater than or equal symbol, $\geq$. The inequality $2x \geq y$ (is\is not) equivalent to $2x + 4 \geq y + 4$. If $14 \leq x$ is true, then $14 - y \leq x - y$ is (true\false). — is; true

15. To solve a given inequality, we will use the rules of addition and subtraction to make a list of inequalities. The first inequality in the list is the given one. Since we use the rules of addition and subtraction, all inequalities in the list (will\will not) be equivalent. Therefore the solution set of the first inequality in the list will be (the same as\different from\sometimes different from) the solution set of the last inequality. — will; the same as

16. Solve:

$2x < x - 1$	given	
$2x - x < x - 1 - x$	subtract x	
$x <$ ____	simplify	-1

The solution sets of $2x < x - 1$ and $x < -1$ (are\are not) the same because the inequalities (are\are not) equivalent. — are; are

17. Referring to Frame 16, the solution set of $x < -1$ is ____________. To represent this set on the number line, we shade that portion of the line which lies to the (right\left) of -1. At the point $x = -1$ we put a single parenthesis,), opening to the left. This indicates that the number $x = -1$ (is\is not) a solution. — all numbers less than -1; left; is not

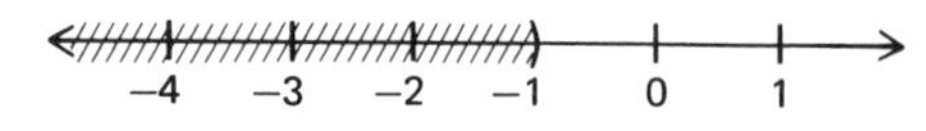

The shaded region represents the condition $x <$ ____. — -1

18. Solve:

$6x - 1 \geq 5x + 2$	given	
$x - 1 \geq$ ____	subtract $5x$	2
$x \geq$ ____	add 1	3

The solution set of $6x - 1 \geq 5x + 2$ is ____________. Does the solution set include the number 3? (yes\no). — all numbers ≥ 3; yes

right
is

19. To represent the condition $x \geq 3$ (from Frame 18) on the number line, we shade that portion to the (right\left) of 3. At the point $x = 3$ we put a single bracket, [. This indicates that the number $x = 3$ (is\is not) included in the solution set for $6x - 1 \geq 5x + 2$.

0 1 2 3 4 5 6

subtract $3x$

−5

all numbers less than −5

20. Solve:

$7 + 4x < 3x + 2$	given
$7 + 4x - 3x < 3x + 2 - 3x$	________
$7 + x < 2$	subtract 7
$x <$ ____	

The solution set of $7 + 4x < 3x + 2$ is ________________.

left; is not

−7 −6 −5 −4

21. To represent this last condition, $x < -5$, on the number line, we shade the region to the (right\left) of $x = -5$. The number $x = -5$ (is\is not) a solution. Represent $x < -5$ on the number line below.

−7 −6 −5 −4 −3 −2 −1 0 1

$$8x - 7x - 1 \geq 3$$
$$x - 1 \geq 3$$
$$x \geq 4$$

4

22. Solve $8x - 1 \geq 7x + 3$.

The solution set is given by the condition $x \geq$ ____.

yes

3 4 5 6

23. Represent the solution set of $8x - 1 \geq 7x + 3$ on the number line. Does the solution set include $x = 4$? (yes\no).

−3 −2 −1 0 1 2 3 4 5

$$12x - 11x + 5 \leq 2$$
$$x + 5 \leq 2$$
$$x \leq -3$$

−3

24. Solve $12x + 5 \leq 11x + 2$.

The solution set is given by the condition $x \leq$ ____.

−5 −4 −3 −2

25. Represent the solution set of $12x + 5 \leq 11x + 2$ on the number line.

−5 −4 −3 −2 −1 0 1 2 3 4 5

PRACTICE PROBLEMS

Answer true or false:

1. $10 > 9$	**2.** $10 > 10$	**3.** $15 < 12$	**4.** $6 \leq 6$	**5.** $6 \leq 9$
6. $4 > -3$	**7.** $0 \leq -1$	**8.** $-7 \leq 7$	**9.** $-12 \leq 1$	**10.** $-5 \geq -10$

Solve each of the inequalities and represent your solution on the number line.

11. $x - 1 < 0$
12. $3x < 2x - 3$
13. $x - 5 > 2x + 1$
14. $-7x + 8x \leq 1$
15. $21x \geq 22x - 5$
16. $1 - 6x < -5x$
17. $3(2x - 1) \leq 5x$
18. $4x - 9 \geq 3(x + 1)$
19. $19x + 3 \leq 6 + 18x$
20. $25x + 5 \geq 2(13x + 1)$
21. $9x < 3(3x + 1)$
22. $9x \geq 3(3x + 1)$
23. $2x \geq 5x - 2(2x - 1)$

SELFTEST

Solve each of the following and represent your solution on the number line.

1. $9 + x < 10$
2. $x + 4 < 0$
3. $5x < 7 + 6x$
4. $2x \geq x + 1$
5. $3x + 5 > 2x - 8$
6. $7x < 2(3x + 1)$
7. $23x + 7 \geq 12(2x - 1)$
8. $8x + 3 < 2(3x + 1) + x$

SELFTEST ANSWERS

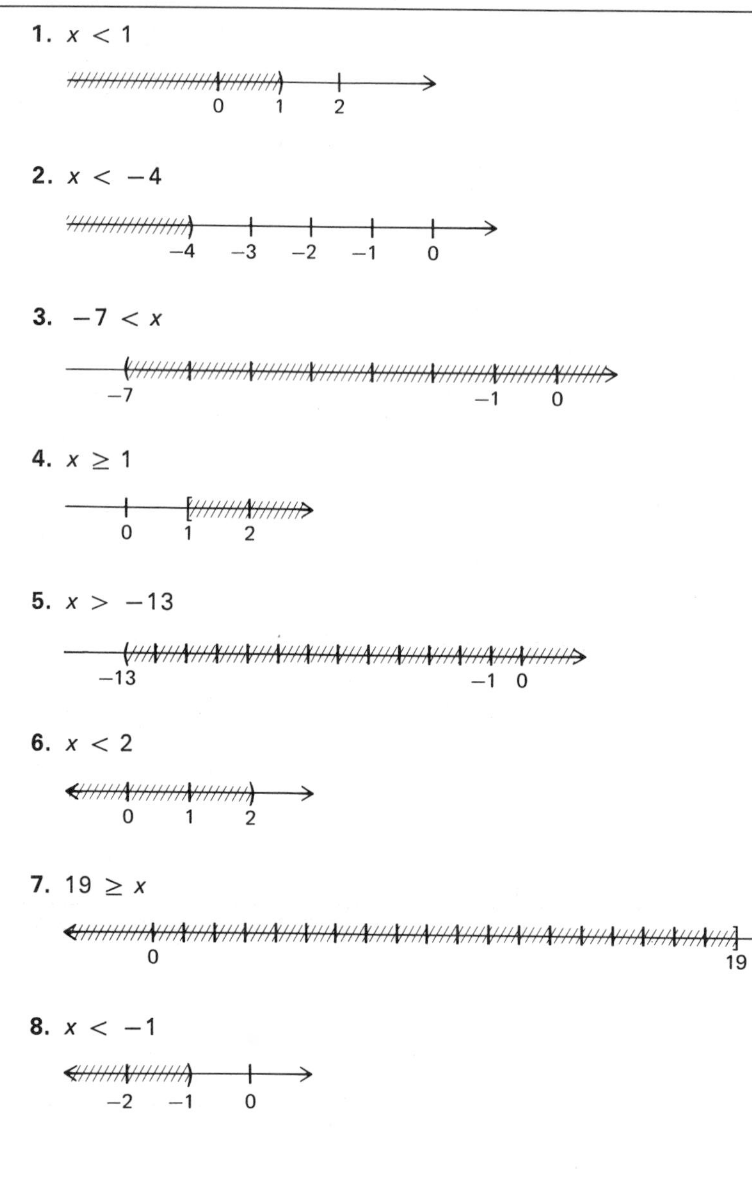

6.7 SOLVING FIRST DEGREE INEQUALITIES IN ONE UNKNOWN

In this section we use the rule for adding and subtracting inequalities together with a new rule for multiplying and dividing.

Rule of Multiplication for Inequalities: **If the same nonzero number c is multiplied on both sides of an inequality, the resulting inequality is equivalent to the original inequality with the additional conditions that**

if c is positive and $a > b$, then $ac > bc$
if c is positive and $a < b$, then $ac < bc$
if c is negative and $a > b$, then $ac < bc$
if c is negative and $a < b$, then $ac > bc$

Notice the rule says that if we multiply an inequality on both sides by a negative number, we reverse the direction of the inequality symbol.

Example 1 Consider the inequality $4 > 1$. Multiply both sides by 3. Since $3 > 0$, the direction of the inequality sign does not change. Thus $3 \cdot 4 > 3 \cdot 1$ or $12 > 3$.

Example 2 Again consider $4 > 1$. Now multiply both sides by -2. Since $-2 < 0$, the direction of the inequality sign must change: $(-2)4 < (-2)1$ or $-8 < 2$.

Example 3 Solve $-3 \leq x/5$ and show the solution on the number line.

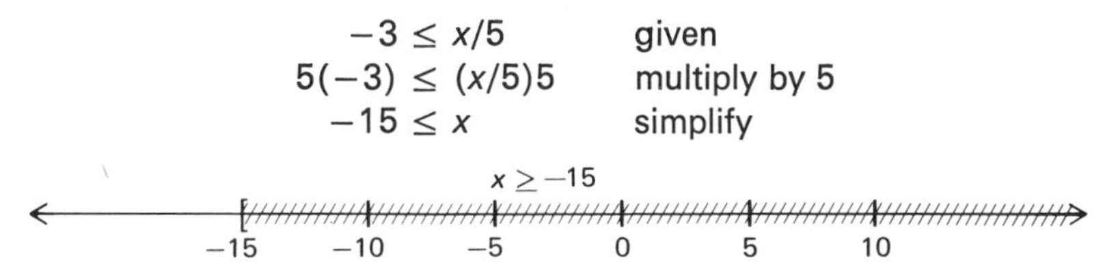

$-3 \leq x/5$	given
$5(-3) \leq (x/5)5$	multiply by 5
$-15 \leq x$	simplify

The rule of division for inequalities is very similar to the rule for multiplication.

Rule of Division for Inequalities: **If the same nonzero number c is divided into both sides of an inequality, the resulting inequality is equivalent to the original one with the additional conditions that**

if c is positive and $a > b$, then $a/c > b/c$
if c is positive and $a < b$, then $a/c < b/c$
if c is negative and $a > b$, then $a/c < b/c$
if c is negative and $a < b$, then $a/c > b/c$

Again notice that if we divide both sides by a negative number, we reverse the direction of the inequality.

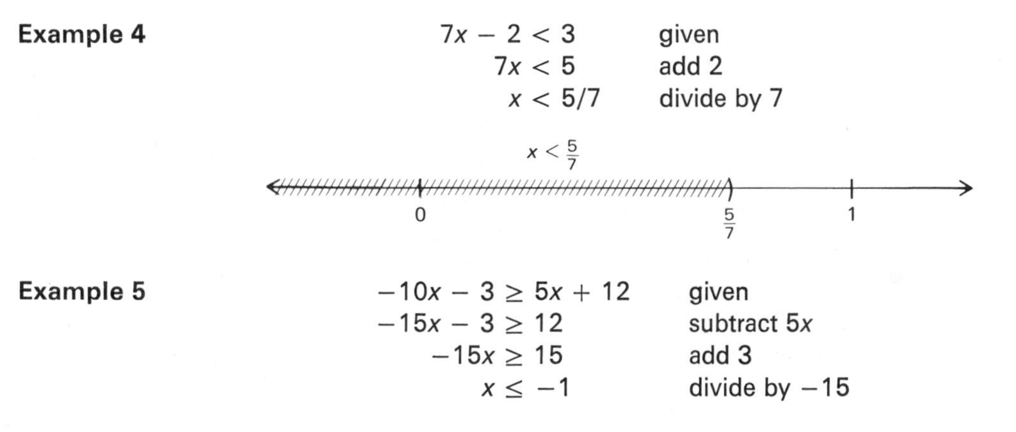

Example 4

$7x - 2 < 3$	given
$7x < 5$	add 2
$x < 5/7$	divide by 7

Example 5

$-10x - 3 \geq 5x + 12$	given
$-15x - 3 \geq 12$	subtract $5x$
$-15x \geq 15$	add 3
$x \leq -1$	divide by -15

In the last step, the direction of the inequality was reversed because the number we divided by (i.e., -15) is negative.

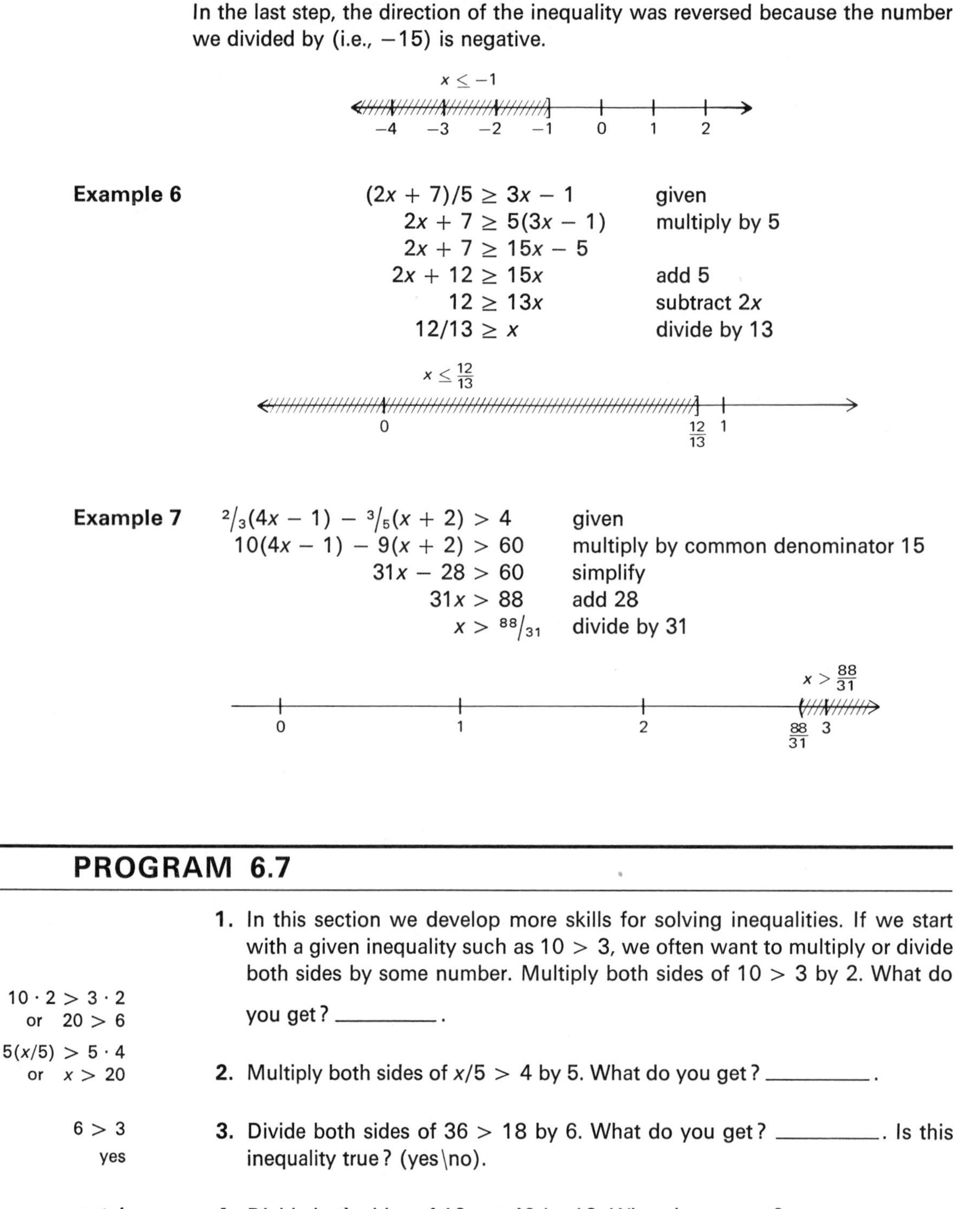

Example 6

$$
\begin{aligned}
(2x + 7)/5 &\geq 3x - 1 && \text{given}\\
2x + 7 &\geq 5(3x - 1) && \text{multiply by 5}\\
2x + 7 &\geq 15x - 5 && \\
2x + 12 &\geq 15x && \text{add 5}\\
12 &\geq 13x && \text{subtract } 2x\\
12/13 &\geq x && \text{divide by 13}
\end{aligned}
$$

$x \leq \frac{12}{13}$

0 $\quad \frac{12}{13}$ 1

Example 7

$$
\begin{aligned}
{}^2/_3(4x - 1) - {}^3/_5(x + 2) &> 4 && \text{given}\\
10(4x - 1) - 9(x + 2) &> 60 && \text{multiply by common denominator 15}\\
31x - 28 &> 60 && \text{simplify}\\
31x &> 88 && \text{add 28}\\
x &> {}^{88}/_{31} && \text{divide by 31}
\end{aligned}
$$

$x > \frac{88}{31}$

0 1 2 $\frac{88}{31}$ 3

PROGRAM 6.7

1. In this section we develop more skills for solving inequalities. If we start with a given inequality such as $10 > 3$, we often want to multiply or divide both sides by some number. Multiply both sides of $10 > 3$ by 2. What do you get? ________.

   Answer: $10 \cdot 2 > 3 \cdot 2$ or $20 > 6$

2. Multiply both sides of $x/5 > 4$ by 5. What do you get? ________.

   Answer: $5(x/5) > 5 \cdot 4$ or $x > 20$

3. Divide both sides of $36 > 18$ by 6. What do you get? ________. Is this inequality true? (yes\no).

   Answer: $6 > 3$; yes

4. Divide both sides of $12x \leq 48$ by 12. What do you get? ________.

   Answer: $x \leq 4$

5. If you multiply or divide both sides of an inequality by the same *positive number*, the direction of the inequality does not change. If you multiply both sides of $8 > -2$ by 3, do you get $24 < -6$ or $24 > -6$? ________.

   Answer: $24 > -6$

   Which inequality is true? ________.

   Answer: $24 > -6$

6. What if you multiplied both sides of an inequality by 4? Does the direction of the inequality change? (yes\no). Why? Because 4 is (positive\negative\zero).

   Answer: no; positive

7. We must be careful when we multiply or divide both sides of an inequality by a negative number. If we multiply or divide both sides of an inequality by the same *negative number* the direction of the inequality sign is *reversed.*

If we multiply both sides of $6 > 4$ by -2, do we get $-12 > -4$ or

$-12 < -4$

$-12 < -4$? ________. There are two reasons why the answer $-12 > -4$ is wrong. First locate -12 and 4 on the number line.

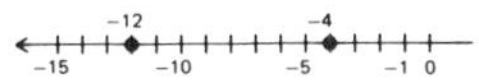

−12
−4
−15
−10
−5
−1 0

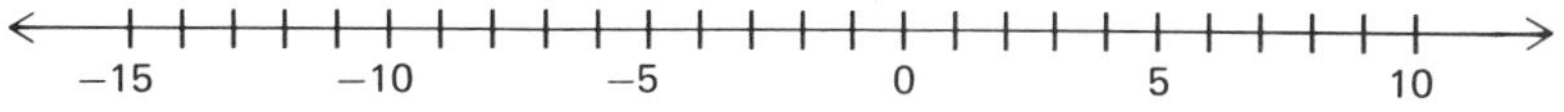

−15
−10
−5
0
5
10

left; smaller

is not

If we locate -12 and -4 on the number line, we see that -12 lies to the (left\right) of -4. Thus -12 is (bigger\smaller) than -4. Therefore $-12 > -4$ (is\is not) true.

negative

must

The second reason that $-12 > -4$ was wrong is that we multiplied $6 > 4$ by -2 which is a (negative\positive) number. When we multiply both sides of an inequality by a negative number we (must\need not) reverse the inequality sign. What do we get when we multiply both sides of

$-30 < -20$

$30 > 20$; positive

$6 > 4$ by -5? ________. When we multiply both sides of $6 > 4$ by 5 we get ($30 > 20$\$30 < 20$) because the multiplier 5 is (negative\positive).

will

8. Divide both sides of $-3x > 9$ by -3. Since -3 is negative the inequality sign (will\will not) be reversed. So we get

$-3x \quad > \quad 9$	given	
$\dfrac{-3x}{-3}\ (>\backslash<)\ \dfrac{9}{-3}$	divide by -3	$<$
$x\ (>\backslash<)\ -3$	simplify	$<$

9. If we carefully observe the above rule on changing the direction of an inequality when we multiply or divide by a negative number, then we can solve inequalities in much the same way we have solved equations. Thus to solve $3x < x + 6$, we have

$3x < x + 6$	given	
$2x <$ ____	subtract x	6
$x <$ ____	divide by 2	3

Represent the solution on the number line below.

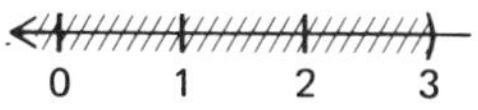

0
1
2
3

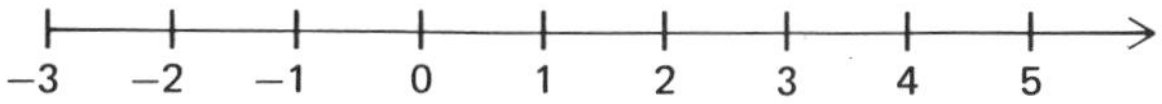

−3
−2
−1
0
1
2
3
4
5

10. Solve:

$-10x + 1 \quad \leq \quad -7$	given	
$-10x \quad \leq$ ____	subtract 1	$-7 - 1$ or -8
$\dfrac{-10x}{-10}\ (\leq\backslash\geq)\ \dfrac{-8}{-10}$	divide by -10	$\geq$

$x \geq \dfrac{+4}{5}$

The solution is ________. Represent the solution on the number line.

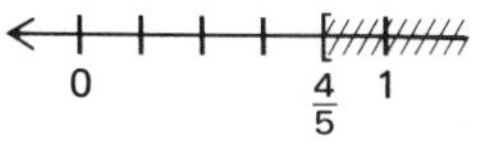

0
4/5
1

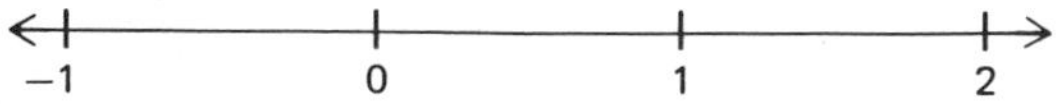

−1
0
1
2

11. Solve:

$\frac{9x}{2} - 2 \geq 4x + 3$	given	
$9x - 4 \geq 8x + 6$	multiply by ____	2
____ $- 4 \geq 6$	subtract $8x$	x
$x \geq$ ____	add 4	10

Represent the solution on the number line.

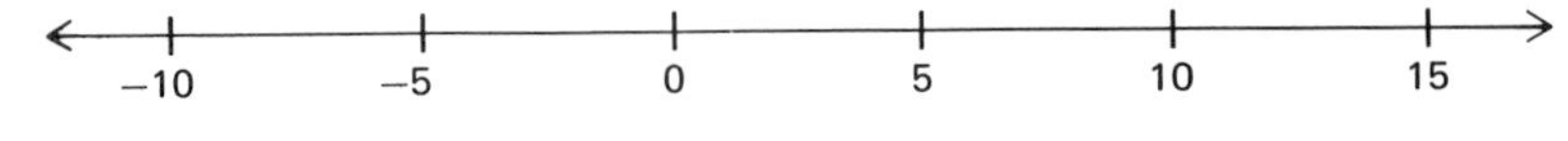

12. Solve:

$3(x-2) \leq \frac{3}{5}(x-1)$	given	
$15(x-2) \leq$ ________	multiply by 5	$3(x-1)$ or $3x-3$
$15x - 30 \leq 3x - 3$	simplify	
$12x - 30 \leq -3$	subtract ____	$3x$
$12x \leq$ ____	add 30	27
$x \leq$ ____	divide by 12	$\frac{27}{12}$ or $\frac{9}{4}$

On the number line below indicate the solution.

13. Solve $-6(x+4) \leq 18x$: If we divide both sides by -6 the direction of the inequality (will\will not) change. — will

$\frac{-6(x+4)}{-6}$ $(\leq \backslash \geq)$ $\frac{18x}{-6}$	divide by -6	$\geq$
$x + 4 \quad \geq$ ________	simplify	$-3x$
$4 \quad \geq$ ________	subtract x	$-3x - x$ or $-4x$

If we divide both sides by -4, will we change the direction of the inequality sign? (yes\no). — yes

$\frac{4}{-4}$ $(\geq \backslash \leq)$ $\frac{-4x}{-4}$	divide by -4	$\leq$
____ $(\geq \backslash \leq)\ x$	simplify	-1; $\leq$

Represent the solution on the number line.

PRACTICE PROBLEMS

Solve each of the following inequalities and represent your solution on the number line.

1. $^4/_3\, x \leq 0$ **2.** $3x \leq 6$ **3.** $5x \leq x$

4. $2x + 4 < 0$ **5.** $16x < 54x$ **6.** $9x < 3x + 1$

7. $5x + 10 \geq 25$
8. $x + 3 < 6x - 8$
9. $2x + 2 \geq x - 4$
10. $-5x + 3 \leq 2x - 1$
11. $x > x + 2(x + 3)$
12. $x < 16 - x$
13. $x + 48 < 9x$
14. $x + 3 > 4x$
15. $-3x \leq -7x + 24$
16. $(3x + 2)/5 \geq x/20 - 1$
17. $-10(x - 2) > 40$
18. $-2(3x + 1) \leq -2x - 14$
19. $16x - 3 > 4x + 9$
20. $(5x - 1)/-2 < x + 1$

SELFTEST

Solve each of the following inequalities and represent your solutions on the number line.

1. $-7x < 21$
2. $2x < 3x$
3. $-2(3x + 2) \geq 0$
4. $30x - 6 < 4$
5. $4x + 6 < 8x$
6. $15x - 3 \geq 6x + 15$
7. $-12x + 7 \leq 5x - 10$
8. $^1/_3(x + 2) < 2x + 4$

SELFTEST ANSWERS

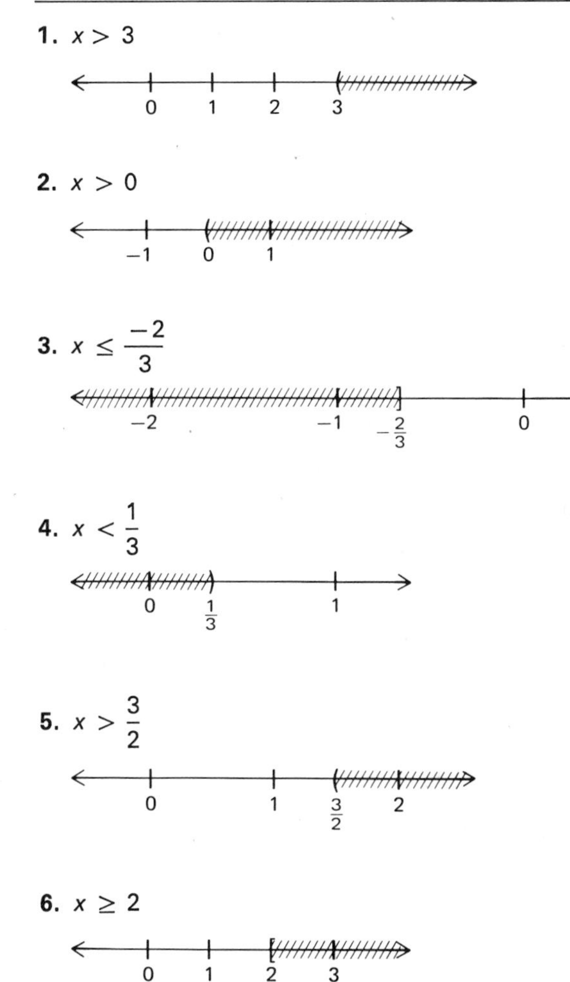

7. $x \geq 1$

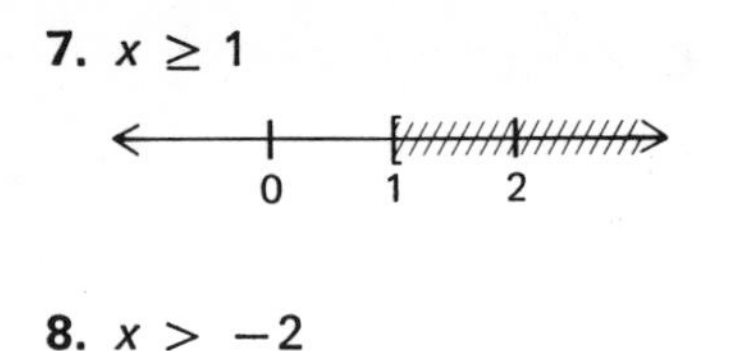

8. $x > -2$

$-3 \quad -2 \quad -1 \quad 0$

POSTTEST 6

1. Determine which of the following equations are identities and which are inconsistent.

a. $2(x - 1) = 2x - 2$
b. $x + 4 = x + 3$
c. $3x + 2 = 1 + 3x + 1$

2. Find the solution set of each of the following.

a. $8x + 2 = 9x - 7$
b. $3(2x - 4) - 5x = 7$
c. $3(2x - 1) = 7x$
d. $2(x + 4) = 3(7 + x)$

3. Find the solution set of each of the following.

a. $2x + 4 + 5x = -2$
b. $(x - 1)/5 = 3$
c. $x(x + 3) = 4(x + 3)$
d. $2/(2x - 1) = 1$

4. Solve as directed:

a. Solve $12x + 3y = 0$ for y.
b. Solve $45S = 15R - 60$ for R.

5. A certain triangle is such that two of its sides have the same length. The remaining side is 15 inches longer than one of the other two sides. The perimeter of the triangle is 75 inches. How long is each side?

6. Solve $8x > 5x + 2(x + 2)$ and shade the solution on the number line.

7. Solve $42x < 2x - 80$ and shade the solution on the number line.

8. Solve $-3x \geq 4$ and shade the solution on the number line.

Each problem above refers to a section in this chapter as shown in the table.

Problems	*Section*
1	6.1
2	6.2
3	6.3
4	6.4
5	6.5
6	6.6
7 and 8	6.7

1. *a*. identity *b*. inconsistent *c*. identity
2. *a*. $\{9\}$ *b*. $\{19\}$ *c*. $\{-3\}$ *d*. $\{-13\}$
3. *a*. $\{-6/7\}$ *b*. $\{16\}$ *c*. $\{-3, 4\}$ *d*. $\{3/2\}$
4. *a*. $y = -4x$ *b*. $R = 3S + 4$
5. sides of 20, 20, and 35 inches
6. $x > 4$

0 1 2 3 4 5

7. $x < -2$

−3 −2 −1 0 1 2 3

8. $x \leq -4/3$

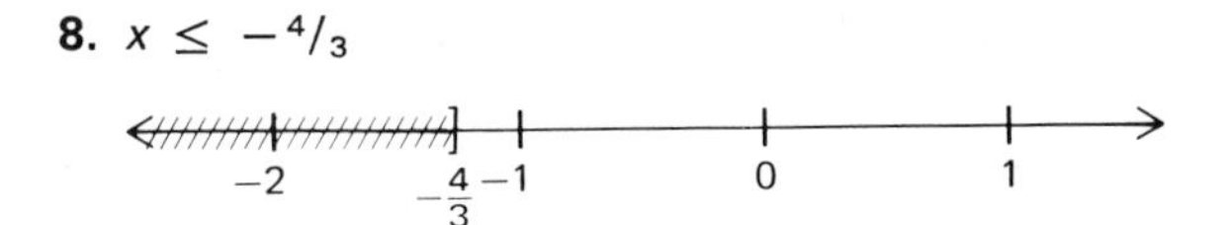

7

EXPONENTS

BASIC SKILLS

Upon completion of the indicated section the student should be able to:

7.1 *a.* Interpret the meaning of x^n for any integer n and $x \neq 0$.
b. Convert an expression with negative exponents to one with positive exponents.
c. Multiply, divide, or simplify expressions involving integer exponents.

7.2 *a.* Convert any expression (greater than, equal to, or less than 1) in decimal notation to one in scientific notation.
b. Convert an expression in scientific notation to one in decimal notation.
c. Multiply or divide some expressions in scientific notation.

7.3 Use the five basic laws of exponents to simplify expressions with integer exponents which involve multiplication, division, a power of a power, a power of a product, or a power of a quotient.

PRETEST 7

1. Simplify and write with positive exponents:

a. x^{-7}
b. x^5x^{-3}
c. $x^8y^{-2}x^{-3}y$
d. $\dfrac{x^{-4}}{x^6}$

2. Convert to scientific notation:

a. 83,100,000
b. 4,381.3
c. 0.0000647

3. Convert to decimal notation:

a. 4.2×10^5
b. 1.9×10^{-3}
c. 6.8×10^{-6}

4. Perform the indicated operations and leave your answer in scientific notation with proper number of significant digits.

a. $\dfrac{6.8 \times 10^8}{2.0 \times 10^{-3}}$

b. $(2.1 \times 10^{-7})(3.0 \times 10^{-3})$

5. Simplify and leave the result with positive exponents:

a. $10x^4y^33y^2x$

b. $\dfrac{16x^7y^4z}{8xy^5z}$

c. $\left(\dfrac{8x^2y^4}{2xy^3}\right)^2$

6. Simplify and leave the result with positive exponents:

a. $(x^2y^{-3})^{-1}$
b. $(3x^2y)^2$
c. $\left(\dfrac{x^5}{y^{-2}}\right)^3$

Each problem above refers to a section in this chapter as shown in the table.

Problems	*Section*
1	7.1
2–4	7.2
5 and 6	7.3

PRETEST 7 ANSWERS

1. *a.* $1/x^7$ *b.* x^2 *c.* x^5/y *d.* $1/x^{10}$

2. *a.* 8.31×10^7 *b.* 4.3813×10^3 *c.* 6.47×10^{-5}

3. *a.* 420,000 *b.* 0.0019 *c.* 0.0000068 **4.** *a.* 3.4×10^{11} *b.* 6.3×10^{-10}

5. *a.* $30x^5y^5$ *b.* $2x^6/y$ *c.* $16x^2y^2$ **6.** *a.* y^3/x^2 *b.* $9x^4y^2$ *c.* $x^{15}y^6$

7.1 NEGATIVE EXPONENTS

In the expression x^n, n is the *exponent* and x the *base*. If n is a positive integer, we know that x^n is the abbreviated notation for the product

$$x^n = \underbrace{x \cdot x \cdot x \cdot \ldots \cdot x}_{\text{use } x \text{ as a factor } n \text{ times}}$$

$$x^5 = x \cdot x \cdot x \cdot x \cdot x$$

If the exponent is 0, then by definition,

$$x^0 = 1 \quad \text{for} \quad x \neq 0 \qquad (0^0 \text{ is undefined})$$

What about negative integer exponents? Certainly, we cannot follow the pattern for positive exponents since using a factor a negative number of times does not make much sense. But we can use previous information to define the meaning of negative exponents. In Chapter 3, we observed two laws for nonnegative exponents.

1. $x^n \cdot x^m = x^{n+m}$
2. $\dfrac{x^n}{x^m} = x^{n-m} \qquad \text{for} \qquad n \geq m$

We placed the restriction $n \geq m$ on the second law so that the exponent $n - m$ would not be negative. Let's see what happens if we ignore that restriction.

Such an application of the second law yields

$$\frac{x^3}{x^5} = x^{3-5} = x^{-2}$$

But by the definition of x^3 and x^5 and the reducing rule, we see that

$$\frac{x^3}{x^5} = \frac{\not{x} \cdot \not{x} \cdot \not{x}}{\not{x} \cdot \not{x} \cdot \not{x} \cdot x \cdot x} = \frac{1}{x \cdot x} = \frac{1}{x^2}$$

Since x^3/x^5 equals x^{-2} and also $1/x^2$, perhaps we should define x^{-2} as

$$x^{-2} = \frac{1}{x^2}$$

What about x^{-3}? By the second law of exponents,

$$\frac{x}{x^4} = \frac{x^1}{x^4} = x^{1-4} = x^{-3}$$

But by the reducing rules and definitions,

$$\frac{x}{x^4} = \frac{\not{x}}{\not{x} \cdot x \cdot x \cdot x} = \frac{1}{x \cdot x \cdot x} = \frac{1}{x^3}$$

The two strings of equations and the transitive law imply that

$$x^{-3} = \frac{1}{x^3}$$

Because of such examples, we make the general definition:

$$x^{-n} = \frac{1}{x^n} \qquad n \text{ a positive integer}$$

So by definition

$$3^{-2} = \frac{1}{3^2} = \frac{1}{9} \qquad x^{-1} = \frac{1}{x} \qquad x^{-10} = \frac{1}{x^{10}}$$

The definition for integer exponents breaks into three cases as follows:

> ***Definition:*** **If n is a positive integer then**
>
> $$x^n = \underbrace{x \cdot x \cdot x \cdot \ldots \cdot x}_{\text{use } x \text{ as a factor } n \text{ times}}$$
>
> $x^o = 1$ **for** $x \neq 0$
>
> $x^{-n} = \dfrac{1}{x^n}$ **for** $x \neq 0$ **(since n is positive, $-n$ is negative)**

We can use our previous laws of exponents with all integer exponents.

Example 1 Simplify and leave the result with positive exponents.

$$
\begin{aligned}
x^2x^{-3} &= x^{-2+(3)} && \text{by law } \mathbf{1} \\
&= x^{-1} \\
&= \frac{1}{x} && \text{by the definition of negative exponent}
\end{aligned}
$$

Example 2 Simplify and leave the result with positive exponents.

$$
\begin{aligned}
\frac{2}{x^{-3}} &= \frac{2}{1/x^3} && \text{by the definition of negative exponent} \\
&= 2 \div \frac{1}{x^3} \\
&= 2 \cdot \frac{x^3}{1} \\
&= 2x^3
\end{aligned}
$$

Example 3 Simplify and leave the result with positive exponents.

$$
\begin{aligned}
\frac{x^{-4}}{x^{-6}} &= x^{-4-(-6)} && \text{by law } \mathbf{2} \\
&= x^{-4+6} \\
&= x^2
\end{aligned}
$$

Example 4 Simplify and leave the result with positive exponents.

$$
\begin{aligned}
\frac{x^4y^5x^{-2}}{x^6y^2} &= \frac{x^{4+(-2)}y^5}{x^6y^2} && \text{by law } \mathbf{1} \\
&= \frac{x^2y^5}{x^6y^2} \\
&= x^{2-6}y^{5-2} && \text{by law } \mathbf{2} \\
&= x^{-4}y^3 \\
&= \frac{y^3}{x^4} && \text{by the definition of negative exponent}
\end{aligned}
$$

PROGRAM 7.1

1. We have studied exponential expressions with positive integer exponents.

x; 6

In x^5, the exponent is 5. The base is ____. In y^6, the exponent is ____.

$y \cdot y \cdot y \cdot y \cdot y \cdot y$

$$y^6 = \_\_ \cdot \_\_ \cdot \_\_ \cdot \_\_ \cdot \_\_ \cdot \_\_$$

6

Use y as a factor ____ times.

4

2. In x^4, the exponent tells us to form a product which has ____ factors of x.

3. We also defined expressions with exponent 0.

$$x^0 = 1 \qquad \text{for all} \qquad x \neq 0$$

1; 1

So $y^0 = 1$, $3^0 = 1$, $z^0 =$ ____, and $6^0 =$ ____.

4. Recall that to multiply expressions with the same base, we add the exponents.

2+3 or 5

$$x^3 \cdot x^2 = x^{\_\_}$$

5. To divide expressions with the same base, we subtract the exponent in the denominator from that in the numerator.

2

$$\frac{x^5}{x^3} = x^{\_\_}$$

$y^{10}x^3$

6. $y^4x^3y^6 =$ ____.

y^3x^2

7. $\dfrac{y^5x^3}{y^2x} =$ ____.

3; 0; y^3

8. $\dfrac{y^4x^5y^3x}{y^4x^6} = y^{\_\_}x^{\_\_} =$ ____.

9. The division rule tells us how to define expressions with *negative integer exponents*. Let's see what x^{-3} means. By the division rule,

x^{-3}

a. $\dfrac{x^2}{x^5} = x^{2-5} =$ ____

But also

$\dfrac{1}{x^3}$

b. $\dfrac{x^2}{x^5} = \dfrac{\not{x} \cdot \not{x}}{\not{x} \cdot \not{x} \cdot x \cdot x \cdot x} =$ ____

equals

By statement *a.*, $x^2/x^5 = x^{-3}$; and by statement *b.*, $x^2/x^5 = 1/x^3$. Therefore x^{-3} (equals\does not equal) $1/x^3$.

10. We use the same type of process to discover the meaning of x^{-4}. By the division rule,

-4

$$\frac{x^3}{x^7} = x^{\_\_}$$

But by the reducing rule,

$\dfrac{1}{x^4}$

$$\frac{x^3}{x^7} = \frac{x \cdot x \cdot x}{x \cdot x \cdot x \cdot x \cdot x \cdot x \cdot x} = \_\_\_\_$$

equals

So x^{-4} (equals\does not equal) $1/x^4$.

11. In general, we use the definition

$$x^{-n} = \frac{1}{x^n}$$

So $x^{-5} =$ ____.

$\frac{1}{x^5}$

12. A negative exponent means we make a fraction with ____ in the numerator and the base to the corresponding positive power in the denominator.

1

$$x^{-8} = \frac{\_\_\_\_}{x^8}$$

1

13. $y^{-3} = 1/$____.

y^3

14. $6^{-2} =$ ____$/6^2 =$ ____$/36$.

1; 1

15. $8^{-2} =$ ____.

1/64

16. We may use either the multiplication rule or the basic definitions to simplify x^3x^{-1}. Thus, $x^3x^{-1} = x^{3+\_\_\_} =$ ____ by the multiplication rule. And by the definition of negative exponents, $x^3 \cdot x^{-1} = x^3 \cdot 1/x =$ ____. Do both techniques give the same result? (yes\no). (The multiplication rule is usually faster to apply.)

(-1); x^2

x^2

yes

17. $x^3x^{-5} = x^{3+\_\_\_} = x^{\_\_\_} =$ ____. (Use only positive exponents in the final result.)

(-5); -2; $1/x^2$

18. $y^5y^{-6} = y^{\_\_\_} =$ ____. (Use only positive exponents in the final result.)

-1; $1/y$

19. $x^4/x^{-2} = x^{4-(-2)} = x^{4+2} =$ ____.

x^6

20. $y^5/y^{-1} = y^{5-\_\_\_} =$ ____.

(-1); y^6

21. Write $1/y^{-1}$ with positive exponent in simplified form.

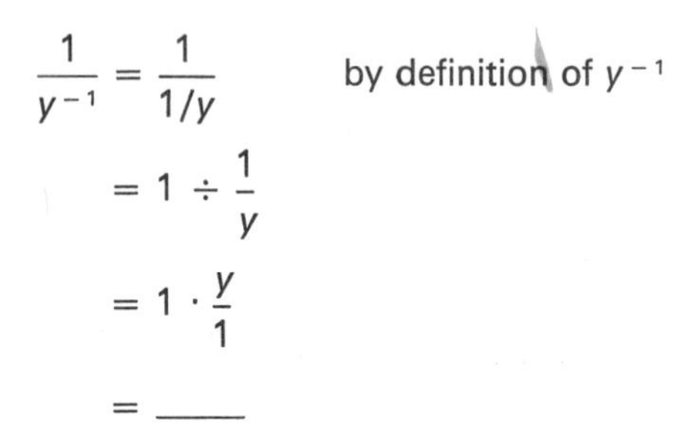

$$\frac{1}{y^{-1}} = \frac{1}{1/y} \quad \text{by definition of } y^{-1}$$

$$= 1 \div \frac{1}{y}$$

$$= 1 \cdot \frac{y}{1}$$

$$= \_\_\_\_$$

y

22. Simplify and write with positive exponent:

$$\frac{2}{x^{-3}} = \frac{2}{1/x^3}$$

$$= \_\_\_\_$$

$2x^3$

23. Simplify and write with positive exponent:

$$\frac{6}{x^{-2}} = \frac{6}{\_\_\_\_} = 6 \cdot \frac{\_\_\_\_}{1} = \_\_\_\_$$

$\frac{1}{x^2}$; x^2; $6x^2$

24. Simplify: $x^3y^4x^{-2}y^{-1} = x^{3+\_\_\_}y^{4+\_\_\_} =$ ____.

(-2); (-1); x^1y^3 or xy^3

$^{-1}$; ^{3}; y^3/x

25. Simplify and write with positive exponents: $x^3y^2x^{-4}y = x^{—}y^{—} =$ ____.

26. Write with positive exponent:

x^2; $\frac{3}{x^2}$

$$3x^{-2} = 3\left(\frac{1}{\_\_\_\_}\right) = \_\_\_\_$$

$\frac{4}{x^3}$

27. Write with positive exponent: $4x^{-3} =$ ____.

PRACTICE PROBLEMS

Simplify and write with positive exponents:

1. x^{-3}
2. x^{-1}
3. x^4x^{-2}
4. x^6x^{-5}
5. $\frac{x^5}{x^7}$
6. $\frac{x^4}{x^5}$
7. $\frac{x^{-3}}{x^4}$
8. $\frac{x^{-7}}{x^2}$
9. $\frac{3}{x^{-5}}$
10. $\frac{2}{x^{-4}}$
11. $x^4y^{-2}x^{-3}y^6$
12. $xy^{-1}y^{-4}x^3$
13. $x^4y^{-3}x^{-4}$
14. $x^2x^{-3}y^{-1}y$
15. $2x^2 3y^4 4x^{-5}$
16. $-9x^3 2x^{-2} 6y^4$
17. $(a + b)^{-1}$
18. $\frac{(x + y)}{(x + y)^{-1}}$
19. $\frac{x(a + b)^2}{x^2(a + b)}$
20. $\frac{x^3yxy}{x^2y^2}$
21. $\frac{18x^2y}{6x^2y^2}$
22. $\frac{12x(x + y)^2}{4x^2(x + y)^{-2}}$
23. $\frac{5x^2y(a + b)^{-1}}{15x(a + b)^{-1}}$
24. $\frac{x^3(x + y)^{-1}xy}{x^4y(x + y)^2}$
25. Is it generally true or false that $x^{-1} + y^{-1} = (x + y)^{-1}$? (*Hint:* Try $x = 1$ and $y = 1$.)

SELFTEST

Simplify and write with positive exponents:

1. x^{-4}
2. $\frac{1}{x^{-3}}$
3. x^4x^{-2}
4. $2x^4 5y^2x^{-3}y^{-1}$
5. $\frac{x^{-5}}{x^6}$
6. $\frac{x}{x^2}$
7. $\frac{x^3y^4}{x^5y^3}$
8. $\frac{y^2}{y^{-3}}$

SELFTEST ANSWERS

1. $1/x^4$ **2.** x^3 **3.** x^2 **4.** $10xy$ **5.** $1/x^{11}$ **6.** $1/x$ **7.** y/x^2 **8.** y^5

7.2 SCIENTIFIC NOTATION

Exponential notation is particularly useful to express very large or small numbers. We use integer powers of 10.

one	$10^0 = 1$	one tenth	$10^{-1} = 0.1$
ten	$10^1 = 10$	one hundredth	$10^{-2} = 0.01$
one hundred	$10^2 = 100$	one thousandth	$10^{-3} = 0.001$
one thousand	$10^3 = 1{,}000$	one millionth	$10^{-6} = 0.000001$
one million	$10^6 = 1{,}000{,}000$		

The nonnegative exponents indicate how many zeros to place after the 1. A U.S. billion is 10^9 or 1,000,000,000. The negative exponents tell the position of the 1 after the decimal. So $10^{-9} = 0.000000001$. In 10^{-9} eight zeros precede the 1. The 1 is thus in the ninth position after the decimal.

It is visually difficult to tell the difference between expressions with many zeros. Scientific notation is much easier to read. To write a number in scientific notation we break the number into two factors. One factor is a number between 1 and 10; the other factor is an appropriate power of 10. For example, in scientific notation, 125 would be 1.25×10^2. For the first factor is a value between 1 and 10 and the second factor is a power of 10. If we write 8,417 in scientific notation, we get 8.417×10^3.

The accuracy of a figure is indicated by the number of digits in the first part. The total payroll of General Motors Detroit in 1971 was $\$8.015071514 \times 10^9$. This figure is accurate to 10 significant figures. The United States national debt reached $\$4.294 \times 10^{11}$ in 1971. This figure is accurate to only four significant digits.

The accuracy of a number depends on both the techniques for measurement and the actual requirements of the situation. When we do calculations with figures in scientific notation, we must be sure that the result does not appear more accurate than the original values.

Example 1 The most remote heavenly body visible to the naked eye is the great galaxy of Andromeda. It is about 2,200,000 light years away or 13,000,000,000,000,000,000 miles away. In scientific notation, this distance is 1.3×10^{19}. Accuracy to two significant figures is indicated in the notation.

Example 2 One large computer can do 3.6×10^7 calculations/second. The exponent 7 tells us to move the decimal 7 places to the right to obtain the decimal value.

$$3.6 \times 10^7 = 36{,}000{,}000$$

This value is quite large. For at the rate of one calculation per second, it would take approximately 28 years to do the same number of calculations. An extremely rapid reader does about 4,000 "calculations" per second.

At the other extreme we find very small values. One of the best balances is so sensitive that it can give measurements accurate to within 2.0×10^{-7} gram. A gram is approximately $^1/_{28}$ ounce. The balance can differentiate between a heavily or lightly inked dot. The notation 2.0×10^{-7} tells us to move the decimal 7 places to the left.

$$2.0 \times 10^{-7} = 0.0000002$$

This figure contains two significant digits in scientific notation.

A nanosecond is 1.0×10^{-9} of a second. This is such a short period of time that light travels less than 12 inches in this period. In decimal notation, $1.0 \times 10^{-9} = 0.000000001$. We move the decimal 9 places to the left.

Scientific notation makes computations easier. For we can use the laws of exponents. When we multiply or divide expressions in scientific notation, we must be sure that the result does not appear more accurate than the original values. The result has only as many significant digits as the least accurate value.

Example 3 Find the product and write it in scientific notation.

$$\begin{aligned}(3.0 \times 10^4)(5.0 \times 10^6) &= (3.0)(5.0) \times 10^4 10^6\\ &= 15 \times 10^{10}\\ &= 1.5 \times 10^{11}\end{aligned}$$

The expression 15×10^{10} is not in scientific notation, which requires that the first number must be between 1 and 10. Clearly 15 is too large. We move the decimal one place to the left and increase the power of 10. The value 1.5×10^{11} is in scientific notation. It has two significant digits since both of the original values have two significant digits.

Example 4 Divide and write the answer in scientific notation.

$$\frac{6.42 \times 10^{-6}}{2.0 \times 10} = \frac{6.42}{2.0} \times \frac{10^{-6}}{10}$$
$$= 3.2 \times 10^{-6-1}$$
$$= 3.2 \times 10^{-7}$$

Here the result has only two significant digits because the denominator had only two. Had the denominator been 2.00 then the result would have been 3.21×10^{-7}. But as the problem was originally stated, we could only use two significant digits in the result.

PROGRAM 7.2

1. An important use of integer exponents occurs in *scientific notation.* In this notation we use the format

$$N \times 10^m$$

where N is a number between 1 and 10 and m is an integer. The following expressions are equal. Which is in scientific notation? ____.

a. 3.0×10^4 *b.* 30,000 *c.* 30×10^3

Answer: *a.*

2. In order to understand scientific notation we need to review powers of 10.

$10^0 = 1$
$10^1 = 10$
$10^2 =$ ____
$10^3 =$ ________

Answers: 100; 1,000

3. $10^6 = 1{,}000{,}000 =$ one million. How many zeros are there after the 1 in one million? ____. In 10^6, the exponent is ____.

Answer: 6; 6

4. In 10^8, the exponent tells how many zeros to place after the 1. So

$10^8 = 100{,}000{,}000$
$10^9 =$ ________________

Answer: 1,000,000,000

5. Some models of the observable universe indicate that the number of atoms in the universe is probably not greater than 10^{85}. To write 10^{85} without exponents we would use ____ zeros after the 1.

Answer: 85

6. $10^{16} =$ ________________ in decimal notation.

Answer: 10,000,000,000,000,000

7. A U.S. billion is 10^9. A British billion is 10^{12}. Write both of these values in decimal notation. ________________________________.

Answer: 1,000,000,000; 1,000,000,000,000

8. $10^{-1} = \frac{1}{10} = 0.1$

$10^{-2} = \frac{1}{10^2} = \frac{1}{100} =$ ____ in decimal notation

Answer: 0.01

$10^{-3} = \frac{1}{10^3} = \frac{1}{\_\_\_\_} =$ ____ in decimal notation

Answer: 1,000; 0.001

9. Negative powers of 10 indicate values (less than\greater than) one. — less than

10. The negative exponent names the position of the "one" after the decimal.

$$10^{-6} = 0.000001$$

The 1 is in the ____ place after the decimal. — 6th

11. 10^{-8} = ________ in decimal notation. — 0.00000001

12. One billionth is 10^{-9}. As a decimal, one billionth is ________. — 0.000000001

13. In scientific notation, 3 million is 3×10^6 or ________ in decimal notation. — 3,000,000

14. To convert 6.7×10^5 to decimal notation, we move the decimal 5 places to the right. 6.7×10^5 = ________. — 670,000

15. Convert 2.56×10^9 to decimal notation. How many places will you move the decimal? ____. In which direction will you move it? ____________. — 9; to the right

$$2.56 \times 10^9 = \_\_\_\_\_\_\_\_\_\_\_\_$$ — 2,560,000,000

16. Convert 4.886×10^7 to decimal notation. ____________. — 48,860,000

17. Convert 3.145×10^4 to decimal notation. ____________. — 31,450

18. To convert 5.0×10^{-3} to decimal notation, we move the decimal 3 places to the (right\left). The result is ________. — left; 0.005

19. Convert 7.3×10^{-8} to decimal notation. ____________. — 0.000000073

20. The number 4.8×10^{-1} is (larger\smaller) than 1. — smaller

21. To convert a number to scientific notation, we write it as a value between 1 and 10 times 10 to some power. Thus, in scientific notation,

$$430{,}000 = 4.3 \times 10^5$$

The power of 10 is 5 since we need to move the decimal in 4.3 (3\4\5) places to the right to obtain 430,000. Write 3,840,000 in scientific notation. — 5

$$3{,}840{,}000 = 3.84 \times 10^{\_\_}$$ — 6

22. A light year is the distance light travels in one year. It is about 5,878,000,000,000 miles. In scientific notation this value is 5.878×10^{__}. — 12

23. The estimate for the world population in 1973 was 3,860,000,000. Write this value in scientific notation. ________. — 3.86×10^9

24. If a positive number is less than one, the corresponding scientific notation will have an exponent that is (positive\negative). — negative

25. The scientific notation for 0.025 is

a. 2.5×10^2 *b.* 2.5×10^{-3} *c.* 2.5×10^{-2} — c.

Answer	Frame
4 -4	**26.** To convert, 0.00098 to scientific notation, we move the decimal ____ places to the right to obtain a number between 1 and 10: $0.00098 = 9.8 \times 10^{—}$.
3.2×10^{-3}	**27.** Convert 0.0032 to scientific notation. ________.
a.	**28.** Which of these values is in scientific notation? *a.* 5.8×10^{-6} *b.* 58×10^{-7} *c.* 0.0000058
two	**29.** In scientific notation of the form $N \times 10^{m}$, the number of digits in N is the number of significant digits. How many significant digits are there in 3.4×10^{8}? ____.
3	**30.** If a 0 is used in a number in scientific notation it is counted as a significant digit. How many significant digits are there in 5.00×10^{4}? ____.
c.	**31.** The accuracy of a value depends on the type of measurements used and the requirements of the situation. If the power of 10 used in two different values is the same, then the figure with more significant digits is the more accurate. Which value is most accurate? *a.* 4.5×10^{3} *b.* 4.50×10^{3} *c.* 4.500×10^{3}
2	**32.** The final result of calculations with figures in scientific notation should have no more significant digits than the original numbers. If we perform the multiplication $(3.8 \times 10^{4})(2.6 \times 10^{3})$, how many significant digits can the result have? ____.
7.2×10^{4}; 2 2	**33.** In the sum $$9.81 \times 10^{4} + 4.66 \times 10^{4} + 7.2 \times 10^{4}$$ the entry with fewest significant digits is ________. It has ____ significant digits. How many significant digits can the actual resulting sum have? ____.
9	**34.** One advantage of scientific notation is that we can use the rules of exponents to make multiplications and divisions easier. $$(3.8 \times 10^{6})(2.0 \times 10^{3}) = (3.8)(2.0) \times (10^{6}10^{3}) = 7.6 \times 10^{—}$$
7.8×10^{8}	**35.** Multiply and put the answer in scientific notation. $$(2.6 \times 10^{3})(3.0 \times 10^{5}) = \_\_\_\_\_\_\_\_$$
11 yes	**36.** Multiply and put the result in scientific notation. $$(4.3 \times 10^{5})(2.0 \times 10^{6}) = 8.6 \times 10^{—}$$ Is 8.6×10^{11} in scientific notation? (yes\no).
left increase 12	**37.** To change 12.6×10^{11} into scientific notation, we need to move the decimal in 12.6 to the (left\right) one place. This change will give us a number between 1 and 10. But then we must (increase\decrease) the exponent of 10 by one. $$12.6 \times 10^{11} = 1.26 \times 10^{—}$$

38. Multiply and put the result in scientific notation. Then round to two significant digits.

$$(8.4 \times 10^6)(3.0 \times 10^5) = \_\_\_\_\_\_$$

2.5×10^{12}

39. We also use the laws of exponents to divide.

$$\frac{8.4 \times 10^6}{2.0 \times 10^4} = 4.2 \times 10^{—}$$

2

40. $\frac{6.9 \times 10^8}{3.0 \times 10^4} =$ ______ in scientific notation.

2.3×10^4

41. $\frac{1.4 \times 10^6}{7.0 \times 10^3} = 0.20 \times 10^{—} = 2.0 \times 10^{—}$.

3; 2

42. $\frac{1.2 \times 10^9}{3.0 \times 10^5} =$ ______ in scientific notation.

4.0×10^3

PRACTICE PROBLEMS

Convert to scientific notation:

1. 10,000,000	**2.** 10,000,000,000	**3.** 6,000,000,000	**4.** 5,000,000
5. 0.00001	**6.** 0.0000001	**7.** 473.5	**8.** 0.0000973
9. 830,000	**10.** 92,100,000	**11.** 168,000	**12.** 0.00035
13. 0.075	**14.** 0.0081	**15.** 190,000	**16.** 0.000189

Convert to decimal notation:

17. 3×10^4	**18.** 5×10^2	**19.** 5.4×10^6	**20.** 7.3×10^9
21. 8.5×10^{-4}	**22.** 9.2×10^{-1}	**23.** 6.8×10^{-7}	**24.** 3.7×10^{-5}
25. 8×10^3	**26.** 13×10^0	**27.** 8.9×10^0	**28.** 15×10^{-1}
29. 1.8×10^{-2}	**30.** 98.4×10^{-4}	**31.** 100×10^2	**32.** 1.89×10^{-3}

33. The longest measure of time is found in Hindu chronology. It is the Kalpa, which is 4,320 million years. Express this value in scientific notation.

34. The American ragweed can generate at least 8,000,000,000 pollen grains in 5 hours. Express this value in scientific notation.

35. In the late 1800s, a swarm of desert locusts covered an estimated 2,000 square miles. There must have been at least 2.5×10^{11} insects. Write this value in decimal notation.

36. The human eye can see an object less than 4×10^{-3} inches across. Write this value in decimal notation.

37. The shortest unit of length is called an attometer. It is 1.0×10^{-16} centimeters. Write this value in decimal notation.

38. The speed of light is 1.86×10^5 miles/second. There are 3.60×10^3 seconds in an hour. How far does light travel in one hour? (Leave the answer in scientific notation and round the answer to three significant digits.)

39. In 1972 the turnover in sales for all branches of the J. C. Penney Company was more than 4,812,000,000 dollars. Write this number in scientific notation.

40. The smallest free-living organism is the Mycoplasma, which is 0.000004 inches long. Write this number in scientific notation.

41. The area of the earth covered by oceans is about 139,670,000 square miles. The area covered by land is about 57,270,000 square miles. Write these numbers in scientific notation.

42. The Amazon River has an average flow of 4.20×10^6 cubic feet/second. What is the average flow for one minute? Write your answer in scientific notation. Use three significant digits.

43. The most massive living thing on earth is a California Redwood tree named General Sherman. The seed of such a tree weighs only about 0.00002 pounds. The growth to maturity represents an increase in weight of 250,000,000,000 fold.

    a. Write the above numbers in scientific notation.
    b. How much do you estimate General Sherman weighs? Leave your answer in scientific notation. Use one significant digit.

44. The mass of the earth is about 6.588×10^{21} tons. A ton is 2.000×10^3 pounds. What is the mass of the earth in pounds? Express the result in scientific notation, and round the answer to four significant digits.

SELFTEST

1. Write 8.73×10^9 in decimal notation.
2. Write 0.00007 in scientific notation.
3. Write 184.3×10^5 in scientific notation.
4. Write 1.61×10^{-3} in decimal notation.
5. Compute $(4.3 \times 10^{16})(2.0 \times 10^{-4})$ and leave the result in scientific notation.
6. Compute $(9.2 \times 10^6)/(2.3 \times 10^{12})$ and leave your result in scientific notation.

SELFTEST ANSWERS

1. 8,730,000,000 **2.** 7×10^{-5} **3.** 1.84×10^7
4. 0.00161 **5.** 8.6×10^{12} **6.** 4.0×10^{-6}

7.3 LAWS OF EXPONENTS

All operations with exponents can be done by using the exponent definitions and the basic multiplication and division laws. But there are other standard laws that enable us to operate more quickly. These laws are valid for all integer exponents.

Basic Laws of Exponents:

1. $x^n \cdot x^m = x^{n+m}$ **Multiplication law**
2. $\dfrac{x^n}{x^m} = x^{n-m}; \quad x \neq 0$ **Division law**
3. $(x^n)^m = x^{n \cdot m}$ **Power-of-a-power law**
4. $(xy)^n = x^n y^n$ **Power-of-product law**
5. $\left(\dfrac{x}{y}\right)^n = \dfrac{x^n}{y^n}; \quad y \neq 0$ **Power-of-quotient law**

We have seen the first two laws applied before. Let's look at some examples to see why the last three laws "work."

Example 1 Simplify $(x^3)^4$. By definition, the exponent 4 has (x^3) as its base. So

$$\begin{aligned}(x^3)^4 &= (x^3)(x^3)(x^3)(x^3)\\ &= (xxx)(xxx)(xxx)(xxx)\\ &= x^{12}\end{aligned}$$

A direct application of the power-of-a-power law **(3)** yields the same result.

$$(x^3)^4 = x^{3 \cdot 4} = x^{12}$$

Example 2 Simplify $(xy)^3$. Here the base is (xy). By definition,

$$\begin{aligned}(xy)^3 &= (xy)(xy)(xy)\\ &= xxxyyy \qquad \text{by the commutative and associative laws}\\ &= x^3y^3\end{aligned}$$

Law **4**, for the power of a product, gives the same result directly.

$$(xy)^3 = x^3y^3$$

Example 3 Simplify $(x/y)^4$. By definition,

$$\begin{aligned}\left(\frac{x}{y}\right)^4 &= \left(\frac{x}{y}\right)\left(\frac{x}{y}\right)\left(\frac{x}{y}\right)\left(\frac{x}{y}\right)\\ &= \frac{x \cdot x \cdot x \cdot x}{y \cdot y \cdot y \cdot y}\\ &= \frac{x^4}{y^4}\end{aligned}$$

By law **5**, we obtain the same result.

$$\left(\frac{x}{y}\right)^4 = \frac{x^4}{y^4}$$

Example 4 Simplify $(x^5)^{-2}$.

$$\begin{aligned}(x^5)^{-2} &= \frac{1}{(x^5)^2} \qquad \text{by the definition of negative exponents}\\ &= \frac{1}{(x^5)(x^5)}\\ &= \frac{1}{x^{10}}\end{aligned}$$

Law **3** gives us the same result.

$$(x^5)^{-2} = x^{5(-2)} = x^{-10} \qquad \text{by law } \mathbf{3}$$

$$= \frac{1}{x^{10}} \qquad \text{by the definition of negative exponents}$$

We often combine laws to simplify more complicated expressions.

Example 5 Simplify and leave the result with positive exponents.

$$(x^4y^{-3})^{-2} = x^{4(-2)}y^{(-3)(-2)} \qquad \text{by laws } \mathbf{3} \text{ and } \mathbf{4}$$

$$= x^{-8}y^6$$

$$= \frac{y^6}{x^8} \qquad \text{by the definition of negative exponents}$$

Example 6 Evaluate:

$$\frac{(3^2)^4}{(3^3)^2} = \frac{3^8}{3^6} \qquad \text{by law } \mathbf{3}$$

$$= 3^2 \qquad \text{by law } \mathbf{2}$$

$$= 9$$

PROGRAM 7.3

1. There are five basic laws for multiplying and dividing exponential expressions. They all follow from definitions and the properties of the real numbers. We have studied the first two.

$n+m$

$x^nx^m = x$—— Multiplication, law **1**

x^{n-m}

$\dfrac{x^n}{x^m} =$ —— Division, law **2**

$(x^n)^m = x^{n \cdot m}$ Power of a power, law **3**

$(xy)^n = x^ny^n$ Power of a product, law **4**

$\left(\dfrac{x}{y}\right)^n = \dfrac{x^n}{y^n}$ Power of a quotient, law **5**

6; 2; 2

By law **3**, $(x^3)^2 = x$——. Law **4** says that $(xy)^2 = x$—y—. By law **5**,

3

3

$\left(\dfrac{x}{y}\right)^3 = \dfrac{x\text{—}}{y\text{—}}$

multiply

2. Law **3** says $(x^n)^m = x^{n \cdot m}$. So if we have an exponential expression raised to a power we (add\multiply) exponents.

add

multiply

3. To multiply exponential expressions we (add\multiply) exponents. But to take a power of a power we (add\multiply) exponents.

4. By basic definitions,

6

$(x^2)^3 = (x^2)(x^2)(x^2) = x$—

3; 6

yes

By law **3**, $(x^2)^3 = x^{2 \cdot}$— $= x$—. Do we obtain the same result whether we use the definition or law **3**? (yes\no).

3; 5; 15

5. $(y^3)^5 = y$—$\cdot$— $= y$—.

6. $(y^6)^2 = y$—. 12

7. $(x^4)^5 =$ ——. x^{20}

8. $(x^3)^6 =$ ——. x^{18}

9. $(x^{-3})^2 = x^{(-3)(2)} = x^{-6} = 1/x^6$. Likewise, $(x^3)^{-2} = x$— $= 1/$——. $^{-6}$; x^6

10. We can evaluate $(2^3)^2$ in two ways. By law **3**: $(2^3)^2 = 2$— $=$ ——. And also, since $2^3 = 8$, as $(2^3)^2 = ($——$)^2 =$ ——. We use the technique which seems easiest. ^{6}; 64 8; 64

11. $(3^2)^2 = 3$— $=$ ——. Also, $(3^2)^2 = ($——$)^2 =$ ——. ^{4}; 81 9; 81

12. Law **4** tells us how to take a product to a power. It says $(xy)^n = x^n y^n$. So $(xy)^5 = x$—y—. ^{5}; ^{5}

13. $(xq)^4 = x$—q—. ^{4}; ^{4}

14. Law **4** follows from the basic definitions and properties of exponents. Thus, by definition,

$$(xz)^2 = (xz)(xz)$$

$=$ —— $\cdot$ —— $\cdot z \cdot z$ by the commutative law x; x

$=$ —— in exponential form x^2z^2

Law **4** gives us the same result faster; that is, $(xz)^2 =$ ——. x^2z^2

15. $(3x)^2 = 3$—x— $=$ ——. ^{2}; ^{2}; $9x^2$

16. $(2y)^3 =$ ————. $2^3y^3 = 8y^3$

17. $(5x)^{-2} = \dfrac{1}{(5x)\text{—}} = \dfrac{1}{\text{——}}$. ^{2}; $25x^2$

18. $(3x)^{-2} =$ ——. (Use only positive exponents.) $\dfrac{1}{9x^2}$

19. We combine laws **3** and **4** to simplify $(5x^2)^2$.

$(5x^2)^2 = 5$—(x^2)— by law **4** ^{2}; ^{2}

$= 5^2x$— by law **3** ^{4}

$=$ ———— $25x^4$

20. $(x^2y^3)^4 = x^{2\cdot}$—$y^{3\cdot}$— by laws **3** and **4** ^{4}, ^{4}

$= x$—y— ^{8}; ^{12}

21. $(x^3y^5)^2 =$ ——. x^6y^{10}

22. $(xy^4)^3 =$ ——. x^3y^{12}

	23. Law **5** is similar to law **4**, only we divide instead of multiply. $$\left(\frac{x}{y}\right)^n = \frac{x^n}{y^n}$$ So
2 ^{2}; x^2	$$\left(\frac{3}{x}\right)^2 = \frac{3^{—}}{x^{—}} = \frac{9}{\_\_\_\_}$$
	24. By law **5** a fraction to a power is the same as the numerator to that power over the denominator to that power
^{3}; $\frac{x^3}{8}$	$$\left(\frac{x}{2}\right)^3 = \frac{x^3}{2^{—}} = \_\_\_\_$$
$x^2/25$	**25.** $(x/5)^2 =$ ____.
no	**26.** When we take the power of an explicit fraction such as $(^5/_4)^2$, there are two approaches we might follow. We can either reduce the fraction first if it does reduce, or we can take the power of the numerator and denominator and reduce that result if it does reduce. Does $^5/_4$ reduce? (yes\no). So
$\frac{25}{16}$	$$\left(\frac{5}{4}\right)^2 = \frac{5^2}{4^2} = \_\_\_\_$$
no	Does the final result reduce? (yes\no).
no	**27.** Evaluate $(^6/_5)^2$: Can you reduce $^6/_5$? (yes\no). So
^{2}; $\frac{36}{25}$	$$\left(\frac{6}{5}\right)^2 = \frac{6^2}{5^{—}} = \_\_\_\_$$
yes	**28.** Evaluate $(^{20}/_{30})^3$: Does $^{20}/_{30}$ reduce? (yes\no).
2	$$\left(\frac{20}{30}\right)^2 = \left(\frac{\_\_\_\_}{3}\right)^2$$
$\frac{4}{9}$	$$= \_\_\_\_$$
	29. Evaluate:
^{3}; $\frac{27}{64}$	$$\left(\frac{27}{36}\right)^3 = \left(\frac{3}{4}\right)^{—} = \_\_\_\_$$
	30. Evaluate $(3^3/3^4)^4$: Again we can reduce the fraction first. For
3	$$3^3/3^4 = 1/\_\_\_\_$$
1/81 (since $3^4 = 81$)	So $(3^3/3^4)^4 = (1/3)^4 =$ ____.
	31. Even if a fraction contains variables as well as specific numbers, it is wise to reduce the fraction if possible before we raise it to a power. In order to simplify $(3xy^2/2x^3)^2$, reduce the fraction.
$\frac{3y^2}{2x^2}$	$$\left(\frac{3xy^2}{2x^3}\right)^2 = \left(\frac{\_\_\_\_}{\_\_\_\_}\right)^2$$ reduce the fraction
$4x^4$	$$= \frac{9y^4}{\_\_\_\_}$$

32. We use the definitions of exponential notation and the laws of exponents to simplify expressions. As usual, we work from the innermost parentheses. Simplify:

$$\left(\frac{(2x^2)^3(2x^5)}{4x^2}\right)^2 = \left(\frac{8x^6 2x^5}{4x^2}\right)^2$$

$$= \left(\frac{16x^{—}}{4x^2}\right)^2$$ 11

$$= (4x^{—})^2$$ 9

$$= \_\_\_\_\_\_$$ $16x^{18}$

33. The five basic laws of exponents are

$x^n \cdot x^m = x^{——}$	Multiplication, law **1**	$n+m$
$\frac{x^n}{x^m} = x^{—}$	Division, law **2**	$n-m$
$(x^n)^m = x^{——}$	Power of a power, law **3**	$n \cdot m$
$(xy)^n = x^{—}y^{—}$	Power of a product, law **4**	n; n
$\left(\frac{x}{y}\right)^n = \frac{x^{—}}{y^{—}}$	Power of a quotient, law **5**	n / n

34. It is important to notice that there (is\is not) a law listed about $x^n + x^m$ or about $y^n - y^m$ — is not

35. Recall that we can add only like terms or terms with exactly the same variable factors to the same powers. $3x^4$ is like

a. $2x^4$ *b.* x^3 *c.* $3x^{-4}$ *d.* x^4 — *a.* and *d.*

36. To add like terms, we add the coefficients but *do not* change the exponent.

$$3x^4 + 5x^4 = 8x^4$$

$$9x^5 + 3x^5 = \_\_\_\_$$ $12x^5$

And $6x^3 + 5y^3$ (can\cannot) be simplified because the terms (are\are not) like. — cannot; are not

37. Simplify:

$$3x^4 + 2x^4 = \_\_\_\_$$ $5x^4$

$$3x^4 \cdot 2x^4 = \_\_\_\_$$ $6x^8$

38. Simplify:

$$2x^3 + 5x = \_\_\_\_$$ already simplified

$$2x^3 \cdot 5x = \_\_\_\_$$ $10x^4$

PRACTICE PROBLEMS

Simplify and leave the results with positive exponents.

1. $3x^4 4x^5$	**2.** $2y^{-3}5y^6$	**3.** $(x^4)^2$	**4.** $(x^3)^4$
5. $(x^{-3})^2$	**6.** $(x^4)^{-2}$	**7.** $(3x)^4$	**8.** $(5y^3)^4$

9. $(2y^3)^4$
10. $(7x^2)^2$
11. $(x^4y^5)^5$
12. $(6x^3y^4)^2$
13. $(x^7y^{-2})^2$
14. $(x^4y^{-5})^2$
15. $(x^3y^{-4})^{-3}$
16. $(x^{-5}y^6)^{-1}$
17. $\left(\frac{4}{x}\right)^2$
18. $\left(\frac{1}{y}\right)^4$
19. $\left(\frac{10}{15}\right)^3$
20. $\left(\frac{16}{8}\right)^2$
21. $\left(\frac{16^7}{16^8}\right)^2$
22. $\left(\frac{5^7}{5^6}\right)^2$
23. $\left(\frac{x^2}{y^3}\right)^4$
24. $\left(\frac{x^4}{y^5}\right)^3$
25. $\left(\frac{3x^2}{x^4}\right)^2$
26. $\left(\frac{2x^4}{x^2}\right)^3$
27. $\left(\frac{x^{-2}}{x}\right)^3$
28. $\left(\frac{x^{-5}}{x^2}\right)^4$
29. $\frac{(3x^4y)^2}{x^5y}$
30. $\frac{(9x^2y^4)^2}{x^5y^4}$
31. $\left(\frac{3x^2y^5}{9xy^4}\right)$
32. $\left(\frac{8x^3y^6}{4xy^7}\right)^3$
33. $\left(\frac{8x}{2x^{-3}}\right)^2$
34. $\left(\frac{5x}{10x^{-2}}\right)^2$
35. $\left(\frac{xyx^{-1}}{y^{-1}}\right)^2$
36. $\left(\frac{18}{6x^{-3}}\right)^{-3}$
37. $\left(\frac{xy^2xy^{-2}}{x^{-2}}\right)^{-1}$
38. $(xyx^{-1}y^{-1})^{-1}$
39. $\left(\frac{xy}{x^{-1}y^{-1}}\right)^{-1}$
40. $\left(\frac{3x^2yx^{-2}}{6y^{-3}}\right)^2$

SELFTEST

Simplify and use positive exponents:

1. $2x^35x^4$
2. $\frac{10x^6}{5x^4}$
3. $\frac{x^3}{x^{-1}}$
4. $\frac{3xy}{9x^2y^{-2}}$
5. $(x^4y^{-2})^3$
6. $\left(\frac{x^3y^5}{x^2y}\right)^2$
7. $\left(\frac{2}{x^5}\right)^3$
8. $(8x^4)^2$

SELFTEST ANSWERS

1. $10x^7$ 2. $2x^2$ 3. x^4 4. $y^3/3x$
5. x^{12}/y^6 6. x^2y^8 7. $8/x^{15}$ 8. $64x^8$

POSTTEST 7

1. Simplify and write with positive exponents:

   a. x^{-4}
   b. $x^8y^2x^{-1}y^3$
   c. $\dfrac{x^{-2}}{x^5}$
   d. $\dfrac{x^{-2}y^3}{xy}$

2. Convert to scientific notation:

   a. 7,430.1
   b. 0.0000958
   c. 44,700,000

3. Convert to decimal notation:

   a. 6.3×10^{-7}
   b. 9.71×10^5
   c. 4.25×10^{-3}

4. Perform the indicated operation and leave your answer in scientific notation with the proper number of significant digits.

   a. $\dfrac{9.6 \times 10^{-13}}{3.0 \times 10^5}$
   b. $(4.2 \times 10^6)(2.0 \times 10^{-4})$

5. Simplify and leave the result with positive exponents:

   a. $3x^4 2x^5y^2$
   b. $\left(\dfrac{15xy^3}{5x^4}\right)^2$
   c. $\dfrac{10x^2y^4}{5x^3y^4}$

6. Simplify and leave the result with positive exponents:

   a. $(x^3y^{-2})^{-1}$
   b. $(2x^3y)^3$
   c. $\left(\dfrac{y^4}{y^2}\right)^{-2}$

Each problem above refers to a section in this chapter as shown in the table.

Problems	*Section*
1	7.1
2–4	7.2
5 and 6	7.3

POSTTEST 7 ANSWERS

1. *a.* $1/x^4$ *b.* x^7y^5 *c.* $1/x^7$ *d.* y^2/x^3

2. *a.* 7.4301×10^3 *b.* 9.58×10^{-5} *c.* 4.47×10^7

3. *a.* 0.00000063 *b.* 971,000 *c.* 0.00425 **4.** *a.* 3.2×10^{-18} *b.* 8.4×10^2

5. *a.* $6x^9y^2$ *b.* $9y^6/x^6$ *c.* $2/x$ **6.** *a.* y^2/x^3 *b.* $8x^9y^3$ *c.* $1/y^4$

RADICALS

BASIC SKILLS

Upon completion of the indicated section the student should be able to:

8.1 *a.* Give an example of an irrational number.
b. Tell which sets of numbers make up the set of real numbers.
c. Locate a multiple of π on the real number line.

8.2 *a.* Indicate a given root of a number by using radical notation.
b. Identify the radicand and index of a radical expression.
c. Use the table to find an approximation of square roots of integers 1–100.
d. Use the table to give approximations to expressions involving addition, subtraction, multiplication, and division of radical expressions.

8.3 *a.* Take the *n*th root of an *n*th power (of a positive value).
b. Take the *n*th power of an *n*th root.
c. Simplify a square root by factoring out perfect squares.

8.4 *a.* Decide if two radical expressions are alike.
b. Add or subtract radical expressions that are alike.

8.5 *a.* Multiply square roots and expressions of the form

$$(a + \sqrt{b})(c + \sqrt{d})$$

b. Divide square roots.
c. Rationalize the denominator of expressions of the form

$$\frac{a}{b\sqrt{c}} \quad \text{or} \quad \frac{a}{b + c\sqrt{d}}$$

8.6 *a.* Factor the sum or difference of radical expressions by factoring out the greatest common factor of the terms.
b. Reduce fractions of the form

$$\frac{a\sqrt{b} + c\sqrt{d}}{e\sqrt{f}}$$

8.7 *a.* Write complex numbers in the standard form $a + bi$.
b. Add or subtract two complex numbers.
c. Multiply two complex numbers.
d. Divide complex numbers (optional).

PRETEST 8

1. *a.* Is every real number a fraction?
   b. Is every fraction a real number?
   c. Give an example of an irrational number.

2. In the notation $\sqrt[4]{50}$ what is the index and what is the radicand?

3. Answer true or false:

   a. $\sqrt{9} = -3$

   b. $\sqrt{16} = 4$

   c. $\sqrt[4]{0} = 0$

   d. $\sqrt[3]{-1} = -1$

   e. $\sqrt{36} = 18$

   f. $\sqrt[3]{8} = -2$

4. Simplify without using tables: (Assume x and y are positive.)

   a. $\sqrt{12^2}$

   b. $\sqrt{4x^2}$

   c. $\sqrt{20}$

   d. $\sqrt{16x^3y^4}$

   e. $\sqrt{2{,}500}$

5. Simplify: (Assume x and y are positive.)

   a. $3\sqrt{6} - \sqrt{6}$

   b. $4\sqrt{y^3} + 3y\sqrt{y}$

   c. $7\sqrt{x^3y^2} - xy\sqrt{x}$

6. Simplify: (Assume x and y are positive.)

   a. $\dfrac{\sqrt{20}}{\sqrt{5}}$

   b. $\dfrac{y}{4\sqrt{3}}$ rationalize denominator

   c. $(3 + \sqrt{x})(3 - \sqrt{x})$

7. Reduce: (Assume x and y are positive.)

   a. $\dfrac{2 + 4\sqrt{3}}{6}$

   b. $\dfrac{x + \sqrt{x^2y}}{xy}$

8. Compute and leave your answer in the form $a + bi$.

a. $(6 + i) + (8 + 3i)$

b. $(8 + \sqrt{-4}) + (2 + \sqrt{-9})$

c. $(2 + i)(5 + 3i)$

Each problem above refers to a section in this chapter as shown in the table.

Problems	*Section*
1	8.1
2 and 3	8.2
4	8.3
5	8.4
6	8.5
7	8.6
8	8.7

PRETEST 8 ANSWERS

1. *a.* no *b.* yes *c.* π is an irrational number

2. the index is 4, the radicand is 50

3. *a.* F *b.* T *c.* T *d.* T *e.* F *f.* F

4. *a.* 12 *b.* $2x$ *c.* $2\sqrt{5}$ *d.* $4xy^2\sqrt{x}$ *e.* 50 **5.** *a.* $2\sqrt{6}$ *b.* $7y\sqrt{y}$ *c.* $6xy\sqrt{x}$

6. *a.* 2 *b.* $(y\sqrt{3})/12$ *c.* $9 - x$ **7.** *a.* $(1 + 2\sqrt{3})/3$ *b.* $(1 + \sqrt{y})/y$

8. *a.* $14 + 4i$ *b.* $10 + 5i$ *c.* $7 + 11i$

8.1 REAL NUMBERS

We have studied various different sets of numbers. In earlier chapters we studied counting numbers, whole numbers, integers, and fractions. Each of these sets of numbers can be represented on the number line. But none of these sets is actually the set of *all* real numbers. If we represent the set of all real numbers on the number line, then every real number must correspond to a point on the number line. Conversely, every point on the number line must correspond to a real number.

Let P be a point on the number line. Since distance can never be negative, the distance from 0 to P is always greater than or equal to zero. We now introduce a new concept of *directed distance*. If P is a point on the number line which lies to the right of 0 we say the directed distance from 0 to P is just the (usual) distance from 0 to P. If P is a point on the number line which lies to the left of 0, we say the directed distance from 0 to P is the negative of the (usual) distance from 0 to P. By convention, the directed distance from 0 to 0 is zero.

With the above discussion in mind, let us make the following definition of the set of all real numbers.

Definition: **The set of all *real numbers* is the set of all directed distances (measured on the number line) from 0 to points on the number line.**

From this definition, it follows that for every point on the number line there is one and only one real number corresponding to that point. For this reason the number line is often called the *real number line*. It is important to notice that every number we have studied so far is a real number. That is, every counting number, whole number, integer, and fraction is a real number.

The set of real numbers obeys all the laws of arithmetic we have previously studied. That is, the real numbers satisfy the

commutative law of addition $a + b = b + a$
commutative law of multiplication $ab = ba$
associative law of addition $a + (b + c) = (a + b) + c$
associative law of multiplication $a(bc) = (ab)c$
distributive law $a(b + c) = ab + ac$
1 is the multiplicative identity $1 \cdot a = a \cdot 1 = a$
0 is the additive identity $a + 0 = 0 + a = a$
every real number R has an additive inverse $-R$
every nonzero real number R has a multiplicative inverse $1/R$.

A question arose in the history of mathematics. "Can all *distances* be expressed as fractions?" The answer to this question was discovered by the Greek mathematician Pythagorus. He found that the set of all real numbers is larger than the set of all fractions. That is, there are some real numbers that are not fractions. This discovery was quite a shock to Pythagorus.

A real number that is not a fraction is called an *irrational number*. We will find many examples of irrational numbers when we study radicals.

In the remainder of this section, we present an example of a real number that is not a fraction. If the student wishes, he may skip this material and proceed directly to the program for this section.

For our construction of an irrational number, we need a theorem from geometry due to Pythagorus. Recall that a right triangle is a triangle that has a 90° angle. The side opposite the 90° angle is called the hypotenuse, the other sides are called legs.

Pythagorean Theorem: **If a and b are the lengths of two legs of a right triangle and c is the length of the hypotenuse of the triangle, then $a^2 + b^2 = c^2$.**

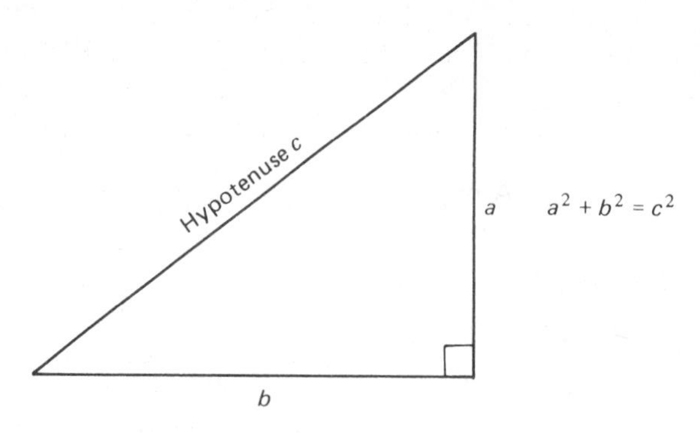

The proof of this theorem will not be presented here. It can be proved by standard techniques of plane geometry.

We now construct a right triangle whose legs have length 1 and whose hypotenuse lies on the real number line. One end of the hypotenuse lies at 0 and the other end at a point labeled N.

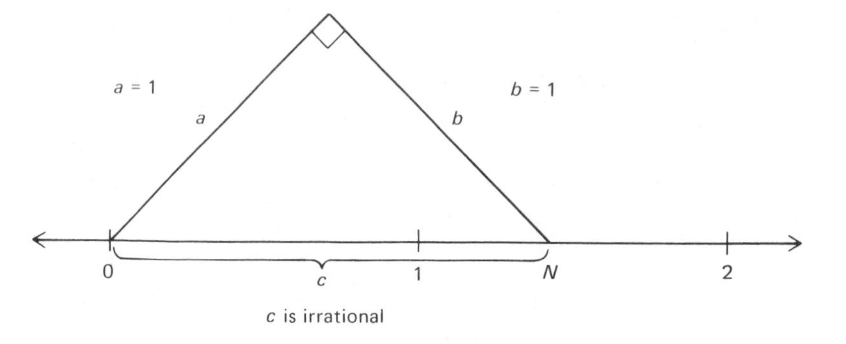

Let c be the real number corresponding to the directed distance from 0 to N. By the Pythagorean theorem it follows that $c^2 = a^2 + b^2$. Since $a = 1$ and $b = 1$ then $c^2 = 1^2 + 1^2 = 2$. So we see that c is the number whose square is 2. However, Pythagorus proved that there is no rational number x such that $x^2 = 2$. But the number c is such that $c^2 = 2$. Therefore c cannot be a rational number. So c is irrational. Pythagorus' method of proof uses no more mathematics than you have already learned. But the proof is very ingenious. You might try this proof on your own, or ask your instructor about it. A discussion of these results can be found in *Euclid's Elements*, book 10.

PROGRAM 8.1

1. We have studied counting numbers, whole numbers, integers, and fractions. Every counting number, whole number, and integer (is\is not) a fraction. Therefore the set of all fractions is the (largest\smallest) set of numbers we have studied in previous chapters.

is

largest

2. For every fraction there is a point on the number line that corresponds to that fraction. Each of the numbers -2, $^{-18}/_4$, and $^5/_2$ is (an integer\a fraction).

a fraction

Represent -2 as a fraction. ________. Represent the numbers -2, $^{-18}/_4$, and $^5/_2$ on the number line.

$-2 = \dfrac{-2}{1}$

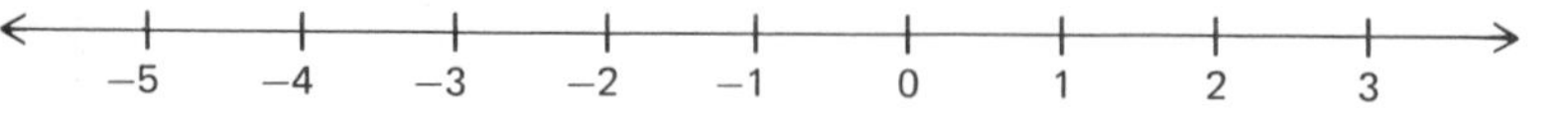

is not

π

irrational; irrational

π is not a fraction

rational

3. Every fraction corresponds to a point on the number line. But there are numbers which are *not* fractions. So the number line (is\is not) completely covered by the set of all fractions. Consider a circle of circumference, C, and diameter, D. The ratio C/D is the same for all circles. It is called the Greek letter pi, which is $(x\backslash\pi\backslash y\backslash a\backslash b)$. To five digits of accuracy, $\pi = 3.1416$. It can be proved that π is *not* a fraction.

A number on the number line that is not a fraction is called *irrational*. A number which is a fraction is called *rational*. It follows that all numbers on the number line are either rational or ________. The number π is (rational\irrational), because ________________. The numbers -2, $^{-18}/_4$, and $^5/_2$ are all (rational\irrational).

4. A circle whose diameter is 1 will have circumference equal to π. Place a circle of diameter 1 on the number line so that it touches the line at the point 0.

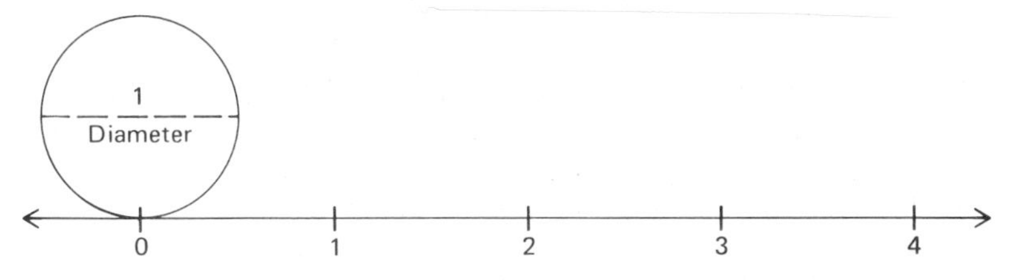

The circle rolled a distance of π on the number line because the circumference of the circle is π.

Roll this circle to the right exactly one full turn. How far did the circle roll on the number line? Explain. *Hint:* Recall that the circumference of a circle is the distance around the circle.

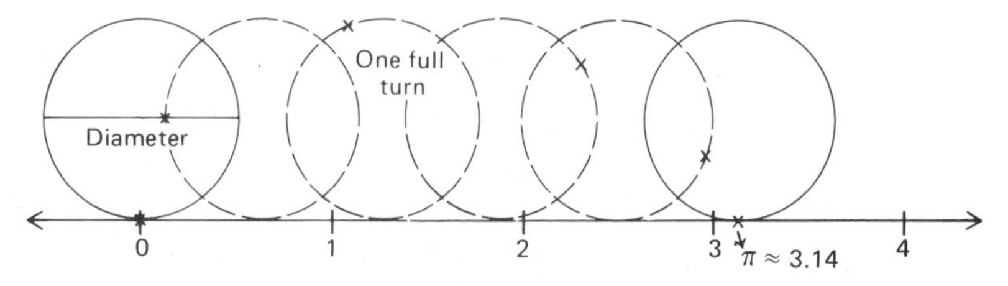

no; The point π is on the number line but it is not a fraction.

5. In the last frame, we constructed the number π on the number line. In the frame before that, we indicated that the number π is not a fraction. Given a point on the number line, is there always a fraction which corresponds to that point? (yes\no). Explain.

irrational

fraction; $-\frac{a}{b}$

6. The number π is irrational. Is π the only irrational number? No, there are many, many irrational numbers. For example, each of the infinitely many numbers π, 2π, 3π, 4π, . . . is irrational. Since π is irrational, what do you think about $-\pi$? It is (rational\irrational). Because if $-\pi$ were rational then it would be a ________. Say, for example that $-\pi = a/b$. Then $\pi =$ ____, so π would be a fraction; and we know this is impossible.

right

left

7. We need a clear definition of the phrase "a number on the number line." To obtain such a definition, we first introduce directed distance. If P is a point on the number line which lies to the right of 0 we say the *directed distance* from 0 to P is just the distance from 0 to P. If P lies to the left of 0 the *directed distance* from 0 to P is the negative of the distance from 0 to P. The directed distance from 0 to 3 is 3, since 3 is to the (left\right) of 0. The directed distance from 0 to -4 is -4, since -4 is to the (left\right) of 0.

8. The directed distance from 0 to 1 is ____. — 1

from 0 to -12 is ____ — -12

from 0 to $^1/_3$ is ____ — $^1/_3$

from 0 to 37.5 is ____ — 37.5

from 0 to -216 is ____ — -216

from 0 to 0 is ____ — 0

from 0 to π is ____ — π

from 0 to 5π is ____ — 5π

from 0 to -2π is ____ — -2π

9. The number line is often called the *real number line*. The numbers corresponding to points on the real number line are called *real numbers*. Or, put in another way, we may say: The set of all real numbers is the set of all directed distances from 0 to points on the number line. Therefore, for every real number there is a corresponding ________ on the real number line. Conversely, to every point on the number line there corresponds (an integer\a real number).

point

a real number

10. Is every real number either rational or irrational? (yes\no). Is every rational and every irrational number a real number? (yes\no). Is every real number irrational? (yes\no). Because, for example, 3 is a real number that is not ________. Is every real number a rational number? (yes\no). Because, for example, π is a ________ number that is not rational. Are there infinitely many rational numbers? (yes\no). Because 1, 2, 3, . . . are ________. Are there infinitely many irrational numbers? (yes\no). Because, for example, π, 2π, 3π, . . . are all ________. Is every integer a real number? (yes\no). Is every whole number a real number? (yes\no). Is every counting number a real number? (yes\no).

yes
yes
no
irrational; no
real
yes; rational
yes
irrational
yes; yes
yes

11. Real numbers obey all the laws of arithmetic we have previously studied. List as many of these laws as you can. The first few are

1. $a + b = b + a$.
2. $a(bc) = (ab)c$.
3. $1 \cdot a = a \cdot 1 = a$ for all a.
4. Each real number R has an additive inverse $-R$ such that $R + (-R) = 0$.

There are 5 more.

5. $ab = ba$
6. $a + (b + c) = (a + b) + c$
7. $0 + a = a + 0 = a$ for all a
8. For all $R \neq 0$ there is a number $1/R$ such that $R(1/R) = 1$
9. $a(b + c) = ab + ac$ Distributive law

12. The sum, product, and difference of any two real numbers is always a real number. Therefore, each of the numbers $^1/_2 + \pi$, $2\pi - 3$, and $(4\pi)(8\pi)$ is again a ____________. The quotient of a real number divided by a nonzero real number is always a real number. Therefore, if A and B are real numbers and $B \neq 0$ then A/B is a ____________. Which of the following are real numbers:

$$^1/_3,\ -6,\ 0,\ 12,\ \pi,\ 7\pi,\ ^9/_2,\ \pi - 1,\ 2\pi + 3,\ 2.1561$$

real number

real number

all of them

PRACTICE PROBLEMS

1. Does every fraction (rational number) correspond to a point on the number line?

2. Does every point on the number line correspond to a rational number (fraction)?

3. Draw a number line so that the unit length is the diameter of a quarter. On this number line the diameter of a quarter is one unit length. Use the quarter to locate π, $-\pi$, and 2π. Are these values rational or irrational?

4. List 10 irrational numbers.

5. Can an irrational number be expressed as a fraction with integer numerator and denominator?

6. What kind of numbers make up the set of real numbers?

7. Does every point on the number line correspond to a real number? Does every real number correspond to a point on the number line?

SELFTEST

1. What is the difference between an irrational number and a rational number?

2. Which of the following are rational, irrational, real?

   a. $^3/_4$ *b.* 2π *c.* 0

3. Answer true or false:

   a. Every real number is rational.
   b. Every rational number is real.
   c. Every real number is irrational.
   d. Every fraction is irrational.

SELFTEST ANSWERS

1. A rational number is the quotient of two integers such as a/b, where a and b are integers ($b \neq 0$). An irrational number is *not* the quotient of two integers.
2. *a.* real and rational *b.* real and irrational *c.* real and rational
3. *a.* F *b.* T *c.* F *d.* F

8.2 INTRODUCTION TO RADICALS

In this chapter, we will learn how to handle algebraic expressions involving radicals. The solution of many equations involve the use of radical expressions, so a thorough mastery of radicals will be of benefit for later work.

First we will define the concept of an nth root of a number. Let n be a positive integer that is greater than or equal to 2. Let a be any real number.

Definition: **An *n*th root of *a* is a number *R* such that $R^n = a$.**

Many numbers have more than one root as may be seen from the definition. For example, a second root of 4 is 2, because $2^2 = 4$. Another second root of 4 is -2, because $(-2)^2 = 4$. A third root of 27 is 3, because $3^3 = 27$. And a third root of -27 is -3 because $(-3)^3 = -27$.

But notice that there is no *real number* x such that $x^2 = -1$ since the square of any real number is not negative. We will learn to deal with this situation when we study a new system of numbers called *complex numbers*.

Let us gather together what we have said so far and at the same time introduce the important notation of radicals.

Another symbol for the *n*th root of a number a is $\sqrt[n]{a}$. The symbol $\sqrt[n]{}$ is called a *radical*; n is called the *index*; and a is called the *radicand*. This definition of radical is subject to the following general principles.

1. If $a \geq 0$ and n is any positive integer, then $\sqrt[n]{a} \geq 0$.
2. If $a < 0$ and n is any positive odd integer, then $\sqrt[n]{a} < 0$.
3. If $a < 0$ and n is any positive even integer, then $\sqrt[n]{a}$ is not defined as a real number. It is a new type of number which is to be studied later.

In cases *1* and *2*, the number $\sqrt[n]{a}$ is commonly called the *n*th principal root of a. In case *3*, there is no principal root, although as we will see later there are *n*th roots in this case.

Traditionally $\sqrt[2]{a}$ is called the "square root of a" and $\sqrt[3]{a}$ is called the "cube root of a." When dealing with the square root, it is permitted and in fact customary to drop the index $n = 2$. So $\sqrt[2]{a}$ and $\sqrt{a}$ really mean the same thing. However, in dealing with the other roots, one must always write out the index explicitly. Thus $\sqrt{a}$ is the square root of a, $\sqrt[3]{a}$ is the cube root, and $\sqrt[4]{a}$ is the fourth root, etc.

Example 1 For any index n, $\sqrt[n]{0} = 0$ since $0^n = 0$.

Example 2 For any index n, $\sqrt[n]{1} = 1$ since $1^n = 1$.

Example 3 Since $4 > 0$ and $4^2 = 16$, $\sqrt{16} = 4$. If we want the negative square root of 16, we would write $-\sqrt{16} = -4$.

Example 4 If you are given a square whose area is a, then the *side* of the square has length $\sqrt{a}$.

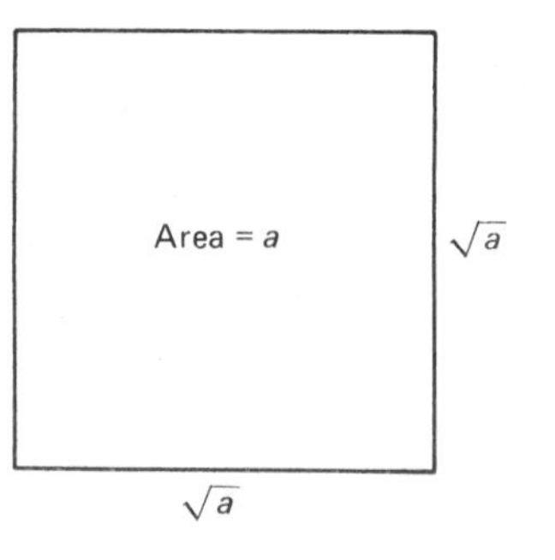

Example 5 $\sqrt[3]{64} = 4$ since $4^3 = 64$

Example 6 $\sqrt[3]{-64} = -4$ since $(-4)^3 = -64$

Example 7 $\sqrt[5]{-1} = -1$ since $(-1)^5 = -1$

Example 8 $\sqrt[4]{81} = 3$ since $3 > 0$ and $3^4 = 81$

If we want the negative fourth root of 81, we would write $-\sqrt[4]{81} = -3$.

Example 9 What is $\sqrt{-2}$? If R is any real number then $R^2 \geq 0$. So there is no real number R such that $R^2 = -2$. Consequently $\sqrt{-2}$ is not a real number. It is another type of number, which we will study later.

Notice that in Examples 3 and 8 there are two possible roots. The square roots of 16 are 4 and -4. But the radical notation $\sqrt[n]{}$ always means for us to take the positive root when there is one. So $\sqrt{16} = 4$. Likewise there are two fourth roots of 81. They are 3 and -3. But $\sqrt[4]{81} = 3$. In the case of $\sqrt[3]{-64}$, there are no positive cube roots. So the radical notation says $\sqrt[3]{-64} = -4$.

How would we find $\sqrt{5}$? By inspection we see that $4 < 5 < 9$. This means that $2 = \sqrt{4} < \sqrt{5} < \sqrt{9} = 3$, so $\sqrt{5}$ is between 2 and 3. To write out $\sqrt{5}$ exactly in decimal form would require infinitely many decimal places. This is because the decimal representation of $\sqrt{5}$ never terminates. The best we can do in decimal form is to approximate $\sqrt{5}$. At the back of this book you will find a square-root table where the square roots of numbers 1 to 100 have been worked out to four digits of accuracy. In a good library you can find tables for cube roots, fourth roots, and higher roots. For the applications we have in mind in this book, a table of square roots will suffice. From our table we see that $\sqrt{5} \approx 2.236$ to four digits of accuracy.

If more accuracy is required, you can consult a table that has more significant digits. When you are solving a problem and come up with a number such as $\sqrt{5}$, what do you do with it? The best answer depends on what you want from the problem. If you are building something, like a bridge, you probably want to know a decimal approximation for $\sqrt{5}$. If four digits accuracy is enough then $\sqrt{5} \approx 2.236$ will do. Of course, $\sqrt{5}$ may just be an intermediate number in a longer calculation. In that case, it may be best to leave $\sqrt{5}$ as it is, and not approximate it with a decimal number. After all, $\sqrt{5}$ has the meaning of being exactly the square root of 5. Unless a decimal approximation is asked for, it is usually best to leave a radical such as $\sqrt{5}$ in its simplest radical form.

Use the square-root table to approximate each of the following examples.

Example 10 $\sqrt{10} \approx 3.162$

Example 11 $2\sqrt{6} \approx 2(2.449) = 4.898$

Example 12 $4 + 2\sqrt{6} \approx 4 + 2(2.449) = 8.898$

Example 13 $2\sqrt{5} - 1 \approx 2(2.236) - 1 = 3.472$

The principal nth root of a positive number is real, but not necessarily rational. It turns out that the principal nth root of any prime number is irrational.

PROGRAM 8.2

1. Let n be an integer which is greater than or equal to 2. Let a be any real number. If R is a number such that

$$R^n = a$$

then we say R is an nth root of a. For example, if $a = 36$ and $n = 2$, then one second root of 36 is 6 since $6^2 =$ ____. Is there another second root of 36? Yes, -6 is another second root of 36 since $(\_\_\_\_)^2 = 36$. So 36 has two second roots, they are ____ and ____.

36

-6

6; -6

2. In general, if a is positive, then there is only one positive nth root of a. What is the positive fourth root of 81? It is 3 since (____)$^4 = 81$. Does 81 have a negative fourth root? (yes\no). If it does what is it? ____. Is there another positive fourth root of 81? (yes\no).

3 | yes; -3 | no

3. What is the third root of 64? It is (2\3\4) since (____)$^3 = 64$. Does 64 have a negative third root? (yes\no). (*Hint:* If you raise a negative value to a third power the result is negative.)

4; 4 | no

4. What is the third root of -64? It is -4 since (____)$^3 = -64$. Does -64 have a positive third root? (yes\no).

-4 | no

5. In general, if a is negative and n is an odd positive integer, then there is only one real nth root of a and it is negative.
   Will the fifth root of -32 be positive or negative? ______. Since $(-2)^5 =$ ____, then the fifth root of -32 is ____.

negative | -32; -2

6. What is the second root of -9? To answer this, we must find a number R such that $R^2 =$ ____. But if R is any real number, then $R^2 \geq 0$. For if $R \geq 0$, then $R^2 \geq 0$; and if $R < 0$, then R^2 ($>$\$<$) 0 as well. So can it ever happen that for some real number R we get $R^2 = -9$? (yes\no).
   Consequently there is no real number which is a second root of -9. In general, if a is negative and n is even, then there is no real number which is an nth root of a. Is there a fourth root of -16? (yes\no). Because $a = -16$ is ______ and $n = 4$ is even.

-9 | $>$ | no | no | negative

7. It gets tiresome to always write out the terms "the nth root of a." We need a shorter notation. Another symbol for the nth root of a number a is

$$\sqrt[n]{a}$$

The symbol $\sqrt[n]{\ }$ is called a *radical*; n is called the *index*; and a is called the *radicand*. Let's rehearse the notation first, and look at restrictions later.
   In $\sqrt[3]{8}$, the index is ____, the radicand is ____. We are to take the ____ root of ____. Use radical notation to express the second root of 26. ______.

3; 8; third | 8 | $\sqrt[2]{26}$ or $\sqrt{26}$

8. When we use the notation $\sqrt[n]{a}$ for "the nth root of a," we attach the following restrictions to the meaning.
   1. If $a \geq 0$ and n is any positive integer, then $\sqrt[n]{a} \geq 0$.
   2. If $a < 0$ and n is any positive *odd* integer, then $\sqrt[n]{a} < 0$.
   3. If $a < 0$ and n is any positive even integer then $\sqrt[n]{a}$ is not defined as a real number.

Thus by case *1*, $\sqrt[2]{16} =$ (4\-4). By case *2*, $\sqrt[3]{-8} =$ (-2\2). By case *3*, $\sqrt[2]{-4}$ (equals 2\equals -2\is not defined as a real number). These three cases are worth remembering.
   They will be referred to later in this program.

4; -2 | is not defined as a real number

9. In cases *1* and *2*, we call $\sqrt[n]{a}$ the "principal nth root of a." In case *3*, there is no principal nth root. Thus, if a number has both a positive and a negative nth root, then the principal nth root is the (positive\negative) root. If a number has only a negative nth root, and no positive nth root, then that negative nth root (is\is not) the principal nth root.

positive | is

2
1; positive
5; 5
−5
5; −5; 5

10. In the case of $\sqrt[2]{25}$, $a = 25$ and $n =$ ____. So we are looking at case (*1**2**3*) in the definition. Therefore $\sqrt[2]{25}$ is (positive\negative). Since $(\_\_\_\_)^2 = 25$ then $\sqrt[2]{25} =$ ____. But there is also a negative root. To indicate the negative root we write $-\sqrt[2]{25} =$ ____. So the two second roots of 25 are ____ and ____. The principal second root is $\sqrt[2]{25}$ or ____.

8; 3
1; positive
2

11. In the case of $\sqrt[3]{8}$, the radicand = ____ and the index $n =$ ____. So we are in case (*1**2**3*). Is $\sqrt[3]{8}$ positive or negative? ________. Since $2^3 = 8$, then $\sqrt[3]{8} =$ ____.

−8; 3
2
negative; positive
−2; −2

12. In the case of $\sqrt[3]{-8}$, the radicand = ____ and the index = ____, so we are in case (*1**2**3*) of the previous definition. Therefore $\sqrt[3]{-8}$ will be (positive\negative) since there is no (positive\negative) root. Since $(\_\_\_\_)^3 = -8$, then $\sqrt[3]{-8} =$ ____.

yes; 10

13. The symbol $\sqrt[2]{a}$ is usually called the "square root of *a*." It is customary to drop the index $n = 2$ for the square root notation. Therefore $\sqrt[2]{a}$ and $\sqrt{a}$ mean the same thing. Is it true that $\sqrt[2]{100}$ and $\sqrt{100}$ are both the same? (yes\no). What is $\sqrt{100}$? ____.

2

14. $\sqrt{64} = 8$ because $(8)^{—} = 64$.

1; 1; 0

15. $\sqrt{1}$ is ____ since $(\_\_\_\_)^2 = 1$. What about $\sqrt{0}$? It is ____.

square
7; cube; 125
5

16. The symbol $\sqrt[3]{a}$ is usually called the "cube root of *a*." In this case, we cannot drop the index $n = 3$. It must be written out explicitly. We can only drop the index when it is 2. Therefore, $\sqrt{49}$ is the (square\cube) root of 49 and $\sqrt{49} =$ ____. But $\sqrt[3]{125}$ is the (square\cube) root of 125. Since $5^3 =$ ____, then $\sqrt[3]{125} =$ ____.

3; 4

17. It is important to think of expressions such as $\sqrt{12}$ as numbers. Let's look a little more closely at $\sqrt{12}$ and figure out about how big it is. Since

$$9 < 12 < 16 \qquad \sqrt{9} < \sqrt{12} < \sqrt{16}$$

We know that $\sqrt{9} =$ ____ and $\sqrt{16} =$ ____. So $\sqrt{12}$ is a number between 3 and 4.

49
7; 6
7

18. Let's find the approximate size of $\sqrt{40}$. The perfect squares between 1 and 100 are

$$1, 4, 9, 16, 25, 36, 49, 64, 81, 100$$

The number 40 is between the perfect squares 36 and ____. So

$$\sqrt{36} < \sqrt{40} < \sqrt{49}$$

Since $\sqrt{36} = 6$ and $\sqrt{49} =$ ____, $\sqrt{40}$ is a number between ____ and ____.

19. Estimate the size of $\sqrt{75}$. Again we look at the perfect squares around 75.

$$64 < 75 < 81$$

Thus $\sqrt{64} < \sqrt{75} <$ ____. We know that $\sqrt{64} =$ ____ and $\sqrt{81} =$ ____. $\sqrt{81}$; 8; 9

Therefore $\sqrt{75}$ is between ____ and ____. 8; 9

20. We have just been giving very rough estimates for the square roots of certain numbers. There is a table in the back of this book that gives the square roots of numbers from 1 to 100 correct to four significant digits. Looking at the table, we find $\sqrt{12} \approx 3.317$. Use the table to approximate the following to 4 digits.

$\sqrt{20} \approx$ ________ $\sqrt{79} \approx$ ________ $\sqrt{96} \approx$ ________ 4.472; 8.888; 9.798

It is important to remember that these are only approximations, but they are good to (1\2\3\4) significant digits. 4

PRACTICE PROBLEMS

1. *a.* In your own words explain the difference in the meaning between the terms "the *n*th root of a number" and "the *n*th power of a number."
   b. What is the second root of 4?
   c. What is the second power of 4?

2. In the notation $\sqrt{5}$, what is the index and what is the radicand?

3. In the notation $\sqrt[3]{20}$, what is the index and what is the radicand?

Answer true or false:

4. $\sqrt{9} = -3$ **5.** $\sqrt{9} = 3$ **6.** $\sqrt[3]{27} = 3$ **7.** $\sqrt[3]{-27} = 3$

8. $\sqrt{-1} = -1$ **9.** $\sqrt[5]{1} = 1$ **10.** $\sqrt{0} = 0$ **11.** $\sqrt[4]{0} = 0$

12. $\sqrt[3]{8} = 2$ **13.** $\sqrt[3]{-8} = -2$ **14.** $\sqrt{16} = -4$ **15.** $\sqrt{25} = 5$

Use the table of square roots to compute the approximate value as a decimal number. (Round to the thousandth place.)

16. $\sqrt{7}$ **17.** $\sqrt{83}$ **18.** $4\sqrt{5}$ **19.** $3\sqrt{2}$

20. $\sqrt{6} + 10$ **21.** $\sqrt{15} - 2$ **22.** $5\sqrt{3} + 1$ **23.** $2\sqrt{6} + 4$

24. $\dfrac{\sqrt{46}}{2}$ **25.** $\dfrac{\sqrt{25}}{5}$

Some roots and powers occur so often that they are worth remembering. Compute the answers to the following problems without use of the tables. Save your answers for later work.

26. Simplify by writing as a single number:

a. 2^2 *b.* 2^3 *c.* 2^4 *d.* 2^5 *e.* 2^6

27. Simplify by writing as a single number:

a. 3^2 *b.* 3^3 *c.* 3^4

28. Simplify by writing as a single number:

a. 4^2 *b.* 4^3 *c.* 4^4

29. Simplify by writing as a single number:

a. 5^2 *b.* 5^3 *c.* 5^4

30. Compute the squares of the numbers 6, 7, 8, 9, 10, 11, 12, 13, 14, 15. Organize your answer in the form of a list.

Use your answers from Problems 26 through 30 to compute each of the following. Do not use tables.

31. $\sqrt{36}$ **32.** $\sqrt{64}$ **33.** $\sqrt[4]{16}$ **34.** $\sqrt[4]{81}$ **35.** $\sqrt[3]{64}$

36. $\sqrt{121}$ **37.** $\sqrt[3]{125}$ **38.** $\sqrt{9}$ **39.** $\sqrt{169}$ **40.** $\sqrt{49}$

41. $\sqrt{100}$ **42.** $\sqrt{196}$ **43.** $\sqrt{144}$ **44.** $\sqrt{81}$ **45.** $\sqrt{225}$

SELFTEST

1. Use radical notation to express the square root of 65.

2. In $\sqrt[3]{64}$, *a.* what is the index? and *b.* what is the radicand?

Use the table of square roots, if necessary, to compute each of the following. (Round to the thousandth position.)

3. $\sqrt{1}$ **4.** $\sqrt[3]{-1}$ **5.** $\sqrt{100}$ **6.** $\sqrt{64}$

7. $\sqrt{20}$ **8.** $3\sqrt{2}$ **9.** $\dfrac{\sqrt{12}}{4}$ **10.** $2\sqrt{3} - 2$

SELFTEST ANSWERS

1. $\sqrt{65}$ **2.** *a.* 3 *b.* 64 **3.** 1 **4.** -1 **5.** 10
6. 8 **7.** 4.472 **8.** 4.242 **9.** 0.866 **10.** 1.464

8.3 SIMPLIFYING RADICAL EXPRESSIONS

We have seen decimal approximations for some square roots. But we often leave radical expressions in radical form rather than converting them to approximate decimal form.

A given radical expression may have several forms that are equal to the original expression. In this section and the next three, we will explore different forms of radical expressions.

Some forms are considered simpler than others. We will always want to put radical expressions in simplest form.

To simplify a radical expression we (1) remove as many factors as possible from the radicand and (2) remove fractions from the radicand.

To remove factors from the radicand means to move them outside the radical symbol. However, we cannot just move a number from inside the radical symbol to outside. We must first take the appropriate root.

There are four basic laws that help us simplify radicals. We will state these for *positive radicands* only. Consequently, in this section and the next three, we will assume that all radicands are positive. Some of the laws can be extended to the case of negative radicands, but a little more caution is necessary. We will leave that work to a later course.

Laws of Radicals:

Assume $a > 0$ and $b > 0$

1. $\sqrt[n]{a^n} = a$

2. $(\sqrt[n]{a})^n = a$

3. $\sqrt[n]{ab} = \sqrt[n]{a} \cdot \sqrt[n]{b}$

4. $\sqrt[n]{\frac{a}{b}} = \frac{\sqrt[n]{a}}{\sqrt[n]{b}}$

The first two laws say that either the nth root of an nth power or the nth power of the nth root give us the original value.

Example 1 Simplify $\sqrt{6^2}$: There are two ways to complete the simplification. The notation indicates that we are to first square 6 and then take the square root.

$$\sqrt{6^2} = \sqrt{36} = 6$$

But if we observe that we are to take the square root of a squared quantity, we can use law **1** to obtain the result directly.

$$\sqrt{6^2} = 6 \qquad \text{by law } \mathbf{1} \qquad \sqrt[n]{a^n} = a \qquad \text{for} \qquad a > 0$$

Example 2 Simplify $(\sqrt[3]{64})^3$: Here we are to take the cube root of 64 first and then raise that value to the third power.

$$(\sqrt[3]{64})^3 = 4^3 = 64$$

But since we are to cube a cube root, law **2** enables us to obtain the result immediately.

$$(\sqrt[3]{64})^3 = 64 \qquad \text{by law } \mathbf{2} \qquad (\sqrt[n]{a})^n = a$$

Instead of taking the root of an entire quantity, we may take the root of each factor according to law **3** ($\sqrt[n]{ab} = \sqrt[n]{a}\,\sqrt[n]{b}$).

Example 3 Simplify $\sqrt{500}$: We want an expression exactly equal to $\sqrt{500}$. But the square-root table gives only an approximation. We must use another technique. This time we factor 500 and use law **3** to factor out any perfect squares.

$$\begin{aligned} \sqrt{500} = \sqrt{10 \cdot 10 \cdot 5} &= \sqrt{10 \cdot 10}\,\sqrt{5} && \text{by law } \mathbf{3} \\ &= \sqrt{10^2}\,\sqrt{5} \\ &= 10\sqrt{5} && \text{by law } \mathbf{1} \end{aligned}$$

Example 4 Simplify $3\sqrt{a^2b^3}$: We wish to factor the radicand into factors that are perfect squares.

$$\begin{aligned} 3\sqrt{a^2b^3} &= 3\sqrt{a^2}\,\sqrt{b^2}\,\sqrt{b} && \text{by law } \mathbf{3} \\ &= 3ab\sqrt{b} && \text{by law } \mathbf{1} \end{aligned}$$

Example 5 Simplify:

$$\begin{aligned} 4\sqrt{18y^5} &= 4\sqrt{9 \cdot 2 \cdot y^2 \cdot y^2 \cdot y} \\ &= 4\sqrt{9} \cdot \sqrt{y^2} \cdot \sqrt{y^2} \cdot \sqrt{2y} && \text{by law } \mathbf{3} \\ &= 4 \cdot 3 \cdot y \cdot y \cdot \sqrt{2y} && \text{by law } \mathbf{1} \\ &= 12y^2(\sqrt{2y}) \end{aligned}$$

We remove fractions from the radicand by the use of law **4**.

$$\sqrt[n]{\frac{a}{b}} = \frac{\sqrt[n]{a}}{\sqrt[n]{b}}$$

Example 6 Simplify:

$$\sqrt{\frac{7}{16}} = \frac{\sqrt{7}}{\sqrt{16}} \qquad \text{by law } \mathbf{4}$$

$$= \frac{\sqrt{7}}{4}$$

PROGRAM 8.3

1. There are four basic laws of radicals. We make the restriction that the radicands be positive.

Laws of Radicals	*Examples*
1. $\sqrt[n]{a^n} = a$	$\sqrt{25} = 5$
2. $(\sqrt[n]{a})^n = a$	$(\sqrt[3]{2})^{—} = 2$
3. $\sqrt[n]{ab} = \sqrt[n]{a}\sqrt[n]{b}$	$\sqrt{5 \cdot 2} = \sqrt{5}\sqrt{\_\_\_}$
4. $\sqrt[n]{\frac{a}{b}} = \frac{\sqrt[n]{a}}{\sqrt[n]{b}}$	$\sqrt{\frac{3}{5}} = \frac{\sqrt{3}}{\sqrt{\_\_\_}}$

3

2

5

2. We use these laws to simplify radicals. That is to (1) make the value under the radical symbol as small as possible, and (2) remove fractions from under the radical symbol.

Each of the following pairs consist of two equal values. Which member of each pair is in simplified form?

a. $\sqrt{4}$; 2   *b.* $2\sqrt{2}$; $\sqrt{8}$ *c.* $\sqrt{5/4}$; $\sqrt{5}/2$

a. 2; *b.* $2\sqrt{2}$; *c.* $\frac{\sqrt{5}}{2}$

3. The first law says

$$\sqrt[n]{a^n} = a \qquad \text{for } a > 0$$

So if we take the nth root of an nth power, we get the original value.

$$\sqrt{x^2} = x; \quad \sqrt[4]{5^4} = \_\_\_\_; \quad \sqrt[3]{7^{—}} = 7$$

5; 3

4. The first law makes it easy to take the fifth root of a quantity to the ________ power.

fifth

5. To simplify $\sqrt[4]{6^4}$ we do not need to evaluate 6^4 first and then take the fourth root of that quantity. We just observe that in $\sqrt[4]{6^4}$ we are to take the ________ root of 6 to the ________ power. By the first law of radicals, $\sqrt[4]{6^4} = \_\_\_\_$.

fourth; fourth

6

6. $\sqrt{x^2} = x$; likewise $\sqrt{y^2} = \_\_\_\_$.

y

7. The second law is similar to the first law. But it says that if we take the nth power of an nth root, we get the original value.

$$(\sqrt[n]{a})^n = a \qquad \text{for } a > 0$$

$$(\sqrt[3]{14})^3 = 14 \qquad \text{by law } \mathbf{2}$$

Similarly $(\sqrt{6})^2 =$ ____. 6

8. As in the first law, the power and the root (must be\need not be) the same before we can apply law **2**. must be

9. By law **2**, $(\sqrt{x})^2 = x$. Likewise,

$$(\sqrt{y})^2 = \_\_\_\_ \qquad \text{and} \qquad (\sqrt[3]{z})^3 = \_\_\_\_$$

y; z

10. Law **3** says that we can either take a root of an entire radicand, or we can take roots of the factors of the radicand.

$$\sqrt[n]{ab} = \sqrt[n]{a}\sqrt[n]{b};\ a > 0,\ b > 0$$

$$\sqrt{24} = \sqrt{4 \cdot 6} = \sqrt{4}\sqrt{\_\_\_\_} = \_\_\_\_\sqrt{6}$$

6; 2

11. We use the third law to reduce the values under the radical symbol. Suppose we are dealing with square roots. If we can find factors of the radicand which are perfect squares (such as 4, 9, 16, 25, 36, 49, ____, ____, ____, etc.), we can eventually take the roots of these factors. Let's see how we do this. 64, 81, 100

$$\sqrt{20} = \sqrt{4 \cdot 5}$$

the number ____ is a perfect square. So 4

$$\sqrt{20} = \sqrt{4 \cdot 5} = \sqrt{4}\sqrt{\_\_\_\_} \qquad \text{by law } \mathbf{3}$$

5

$$= 2\sqrt{5} \qquad \text{since} \qquad \sqrt{4} = \_\_\_\_$$

2

Thus $\sqrt{20} = 2\sqrt{5}$. The expression $2\sqrt{5}$ is considered to be in simpler form.

For the number under the radical symbol in $\sqrt{20}$ is ____ while the number under the radical symbol in $2\sqrt{5}$ is ____. Which is the smaller number under the radical symbol? ____. 20 5 5

12. Simplify:

$$\sqrt{300} = \sqrt{100 \cdot 3}$$

Which value is a perfect square? ____. Therefore 100

$$\sqrt{300} = \sqrt{100}\sqrt{\_\_\_\_} \qquad \text{by law } \mathbf{3}$$

3

$$= \_\_\_\_\sqrt{3}$$

10

13. Now we will simplify the square root of 90.

$$\sqrt{90} = \sqrt{9 \cdot 10}$$

Which value is a perfect square? ____. Therefore 9

$$\sqrt{90} = \sqrt{\_\_\_\_}\sqrt{\_\_\_\_} \qquad \text{by law } \mathbf{3}$$

9; 10

$$= \_\_\_\_\sqrt{\_\_\_\_} \qquad \text{in simplified form}$$

3; 10

	14. Let's see if we can simplify $\sqrt{10}$.
	$$\sqrt{10} = \sqrt{5 \cdot 2}$$
no no cannot	Are either factors perfect squares? (yes\no). Can we take the square root of either factor without using a decimal approximation? (yes\no). Therefore we (can\cannot) simplify $\sqrt{10}$ any more.
	15. To simplify $9x$, we take the following steps.
x; 3; x	$$\sqrt{9x} = \sqrt{9 \cdot x} = \sqrt{9}\sqrt{\_\_\_\_} = \_\_\_\_\sqrt{\_\_\_\_}$$
	16. Simplify:
a	$$\sqrt{16a^3} = \sqrt{16 \cdot a^2 \cdot \_\_\_\_}$$
a	$$= \sqrt{16}\sqrt{a^2}\sqrt{\_\_\_\_}$$
a; 4	$= 4a\sqrt{\_\_\_\_}$ since $\sqrt{16} = \_\_\_\_$ and $\sqrt{a^2} = a$
	17. Simplify:
	$$\sqrt{8b^4} = \sqrt{4 \cdot 2 \cdot b^2 \cdot b^2}$$
	$$= \sqrt{4}\sqrt{b^2}\sqrt{b^2}\sqrt{2}$$
2; b; b	$$= \_\_\_\_ \cdot \_\_\_\_ \cdot \_\_\_\_\sqrt{2}$$
perfect squares	**18.** Whenever we have a square root of an expression, we factor the expression into (perfect squares\perfect cubes\prime factors). Then we take the square roots of all the perfect squares.
y	$$\sqrt{x^4y^5} = \sqrt{x^2}\sqrt{x^2}\sqrt{y^2}\sqrt{y^2}\sqrt{\_\_\_\_}$$
x; x; y; y; y	$$= \_\_ \cdot \_\_ \cdot \_\_ \cdot \_\_\sqrt{\_\_}$$
^{2}; ^{2}; y	$$= x^{\_\_}y^{\_\_}\sqrt{\_\_}$$
	19. To simplify $\sqrt{75b^3}$, we factor the radicand as follows.
3	$$\sqrt{75b^3} = \sqrt{25 \cdot b^2 \cdot \_\_\_\_ \cdot b}$$
$3b$	$$= \sqrt{25}\sqrt{b^2}\sqrt{\_\_\_\_}$$
5; $3b$	$$= \_\_\_\_b\sqrt{\_\_\_\_}$$
	20. The fourth law of radicals helps us to remove fractions from the radicand.
	$\sqrt[n]{\dfrac{a}{b}} = \dfrac{\sqrt[n]{a}}{\sqrt[n]{b}}$ $\quad a > 0$ and $b > 0$
does	Essentially, law **4** (does\does not) let us take the roots of the numerator and denominator separately.
3	**21.** $\sqrt{\dfrac{9}{4}} = \dfrac{\sqrt{9}}{\sqrt{4}} = \dfrac{\_\_\_\_}{2}$
	22. By law **4**,
x^2; $\dfrac{x}{4}$	$$\sqrt{\frac{x^2}{16}} = \frac{\sqrt{\_\_\_\_}}{\sqrt{16}} = \_\_\_\_$$
x^3; x; x 25	**23.** $\sqrt{\dfrac{x^3}{25}} = \dfrac{\sqrt{\_\_\_\_}}{\sqrt{\_\_\_\_}} = \dfrac{\_\_\_\_\sqrt{\_\_\_\_}}{5}$

PRACTICE PROBLEMS

You may assume that x, y, and a are positive.

Simplify:

1. $\sqrt{11^2}$ 2. $\sqrt{29^2}$ 3. $\sqrt{x^2}$ 4. $\sqrt{y^4}$ 5. $\sqrt{3^4}$

6. $\sqrt[3]{26^3}$ 7. $(\sqrt[3]{10})^3$ 8. $(\sqrt[3]{x})^3$ 9. $(\sqrt{x})^2$ 10. $(\sqrt{2y})^2$

11. $\sqrt{3a^2}$ 12. $\sqrt{5x^2}$ 13. $\sqrt{4a}$ 14. $\sqrt{9x}$ 15. $\sqrt{8}$

16. $\sqrt{12}$ 17. $\sqrt{18}$ 18. $\sqrt{50}$ 19. $\sqrt{20}$ 20. $\sqrt{3,200}$

21. $\sqrt{1,600}$ 22. $\sqrt{900}$ 23. $\sqrt{63x}$ 24. $\sqrt{8y}$ 25. $\sqrt{4x^2y}$

26. $\sqrt{16x^3y^4}$ 27. $\sqrt{x^3}$ 28. $\sqrt{y^5}$ 29. $\sqrt{28x^3y^5}$ 30. $\sqrt{50x^2y^4}$

SELFTEST

Simplify:

1. $\sqrt{61^2}$ 2. $(\sqrt{19x})^2$ 3. $\sqrt{48}$ 4. $\sqrt{20x^2y^5}$ 5. $\sqrt{90x^3y^4}$ 6. $\sqrt{2,500}$

SELFTEST ANSWERS

1. 61 2. $19x$ 3. $4\sqrt{3}$ 4. $2xy^2\sqrt{5y}$ 5. $3xy^2\sqrt{10x}$ 6. 50

8.4 ADDITION AND SUBTRACTION OF RADICALS

To carry out addition or subtraction of polynomial expressions we had to have "like" terms. That is, the exponential part of the terms had to have the same bases and exponents. Likewise, radical expressions must be "alike" before we add.

Two *radical expressions are alike* if the radicands and the indices are the same. We can indicate the addition of unlike radicals by using a plus symbol between the expressions. But we must have like radicals before we can simplify an addition or subtraction.

Example 1 The expressions $\sqrt{8}$ and $3\sqrt{8}$ are alike because both involve the square root of 8.

Example 2 The expressions $6\sqrt{3}$ and $6\sqrt{2}$ are not alike because the radicands 3 and 2 differ.

Example 3 The expressions $\sqrt[3]{7}$ and $\sqrt{7}$ are not alike because the indices differ. The first expression is a cube root, while the second one is a square root.

We add like radicals by adding the coefficients of the radicals. The radicand or value under the radical symbol remains unchanged.

Example 4 Simplify $\sqrt{5} + 2\sqrt{5}$. The coefficient of $\sqrt{5}$ is an understood 1. The coefficient of $2\sqrt{5}$ is 2.

$$\sqrt{5} + 2\sqrt{5} = 3\sqrt{5}$$

Sometimes it is not immediately apparent that two radical expressions are like. We may have to simplify one or the other or both before they appear in like form.

Example 5 Simplify $5\sqrt{2} + 3\sqrt{8}$. First we simplify $3\sqrt{8}$.

$$3\sqrt{8} = 3\sqrt{4 \cdot 2} = 3\sqrt{4}\sqrt{2} = 6\sqrt{2}$$

$$5\sqrt{2} + 3\sqrt{8} = 5\sqrt{2} + 6\sqrt{2}$$

$$= 11\sqrt{2}$$

Example 6 Simplify $6\sqrt{12} - 2\sqrt{27}$. Both terms simplify.

$$6\sqrt{12} = 6\sqrt{4}\sqrt{3} = 12\sqrt{3} \qquad 2\sqrt{27} = 2\sqrt{9}\sqrt{3} = 6\sqrt{3}$$

Therefore $6\sqrt{12} - 2\sqrt{27} = 12\sqrt{3} - 6\sqrt{3} = 6\sqrt{3}$.

Example 7 Simplify $3a^2\sqrt{2b^3a} - b\sqrt{18ba^5}$. Again we can simplify both terms.

$$3a^2\sqrt{2b^3a} = 3a^2\sqrt{b^2}\sqrt{2ab} = 3a^2b\sqrt{2ab}$$

$$b\sqrt{18ba^5} = b\sqrt{9}\sqrt{a^2}\sqrt{a^2}\sqrt{2ab} = 3ba^2\sqrt{2ab}$$

The two previous results imply that

$$3a^2\sqrt{2b^3a} - b\sqrt{18ba^5} = 3a^2b\sqrt{2ab} - 3a^2b\sqrt{2ab} = 0$$

PROGRAM 8.4

$7x$; not possible

1. In our work with polynomials we could simplify an addition involving like terms. Simplify each of the following if possible.

$3x + 4x =$ ________ $\qquad 7x + 3y =$ ________

$3\sqrt{5}$

2. Radicals are alike if they both have the same index and the same radicand. Therefore $4\sqrt{5}$ is like ($\sqrt[3]{5}$ \ $3\sqrt{5}$ \ $\sqrt{4}$).

no; The indexes differ

3. Are $\sqrt{x}$ and $\sqrt[3]{x}$ like? (yes\no). Why? ________________.

is

4. $3\sqrt{x}$ (is\is not) like $5\sqrt{x}$.

$8\sqrt{10}$

5. We can simplify radical addition or subtraction if the expressions involved have like radicals.

$$2\sqrt{10} + 3\sqrt{10} = 5\sqrt{10}$$

$$6\sqrt{10} + 2\sqrt{10} = \_\_\_\_$$

$8\sqrt{7}$

6. When we add or subtract like radicals, the radicand remains unchanged. We just add the values outside the radical symbol.

$$3\sqrt{7} + 5\sqrt{7} = \_\_\_\_$$

$2\sqrt{x}$

7. Simplify: $4\sqrt{x} - 2\sqrt{x} =$ ________.

No

alike

8. Can we simplify $3\sqrt{5} + 4\sqrt{7}$? (Yes\No), because the radicals are not ________.

9. Sometimes terms do not appear to be alike. But if we simplify them we see that they are.

$$\sqrt{2} + \sqrt{8} = \sqrt{2} + \sqrt{4}\sqrt{2}$$
$$= \sqrt{2} + \_\_\_\_\sqrt{2}$$
$$= \_\_\_\_\sqrt{2}$$

2

3

10. To simplify $4\sqrt{5} + \sqrt{45}$, we first simplify $\sqrt{45}$.

$$\sqrt{45} = \sqrt{9}\sqrt{\_\_\_\_} = \_\_\_\_\sqrt{5}$$
$$4\sqrt{5} + \sqrt{45} = 4\sqrt{5} + \_\_\_\_\sqrt{5}$$
$$= \_\_\_\_$$

5; 3

3

$7\sqrt{5}$

11. Simplify:

$$4\sqrt{6} + \sqrt{54} = 4\sqrt{6} + \_\_\_\_\sqrt{6}$$
$$= \_\_\_\_\sqrt{6}$$

3

7

12. Simplify:

$$5\sqrt{3} + 2\sqrt{12} = 5\sqrt{3} + 2 \cdot \_\_\_\_\sqrt{3}$$
$$= \_\_\_\_$$

2

$9\sqrt{3}$

13. Simplify: $2x\sqrt{5} + 3x\sqrt{5} = \_\_\_\_x\sqrt{5}$.

5

14. Simplify:

$$2x\sqrt{x} + 7\sqrt{x^3} = 2x\sqrt{x} + \_\_\_\_\sqrt{x}$$
$$= \_\_\_\_\sqrt{x}$$

$7x$

$9x$

15. Simplify:

$$6a\sqrt{3a^2b} - 2\sqrt{3a^4b} = 6a^2\sqrt{\_\_\_\_} - 2a^2\sqrt{\_\_\_\_}$$
$$= \_\_\_\_\sqrt{\_\_\_\_}$$

$3b$; $3b$

$4a^2$; $3b$

16. Simplify: $3\sqrt{x^2} - 3x = \_\_\_\_ - 3x = \_\_\_\_$.

$3x$; 0

17. Simplify:

$$5\sqrt{25a^2b^5} + 3ab\sqrt{b^3} = \_\_\_\_\_\_\sqrt{b} + \_\_\_\_\_\_\sqrt{b} = \_\_\_\_\_\_\sqrt{b}$$

$25ab^2$; $3ab^2$; $28ab^2$

18. Simplify as much as possible:

$$4\sqrt{x} - 5\sqrt{x^2} = 4\sqrt{x} - \_\_\_\_$$

$5x$

Does the last expression simplify any more? (Yes\No), because the terms (are\are not) alike.

No

are not

19. Simplify as much as possible:

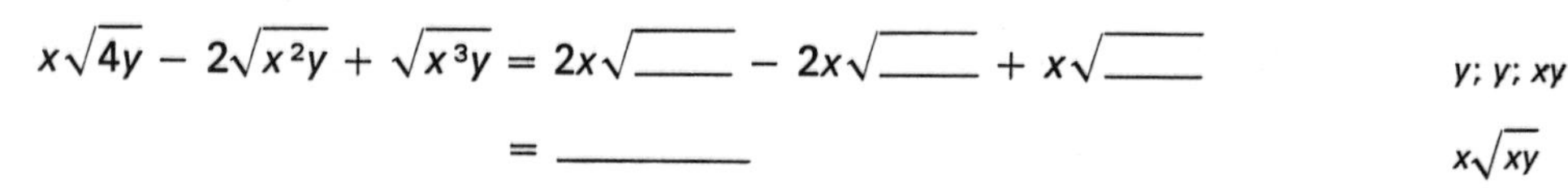

$$x\sqrt{4y} - 2\sqrt{x^2y} + \sqrt{x^3y} = 2x\sqrt{\_\_\_\_} - 2x\sqrt{\_\_\_\_} + x\sqrt{\_\_\_\_}$$
$$= \_\_\_\_\_\_$$

y; y; xy

$x\sqrt{xy}$

PRACTICE PROBLEMS

Add or subtract and simplify:

1. $\sqrt{3} + 5\sqrt{3}$
2. $2\sqrt{7} + \sqrt{7}$
3. $4\sqrt{x} + 5\sqrt{x}$
4. $\sqrt{y} - 2\sqrt{y}$
5. $\sqrt{8} - \sqrt{2}$
6. $\sqrt{2} + \sqrt{18}$
7. $4\sqrt{72} + 5\sqrt{32}$
8. $\sqrt{250} - \sqrt{40}$
9. $2\sqrt{20} - 3\sqrt{45}$
10. $4\sqrt{12} + 5\sqrt{27}$
11. $6a\sqrt{4ab} + 3\sqrt{a^3b}$
12. $7x\sqrt{5} - 3\sqrt{20x^2}$
13. $b\sqrt{4b} + 3\sqrt{b^3}$
14. $7\sqrt{x^3} - x\sqrt{9x}$
15. $2b\sqrt{a^2b} - a\sqrt{b^3}$
16. $5xy\sqrt{3} + \sqrt{12x^2y^2}$
17. $4x^2\sqrt{xy^3} + 3y\sqrt{x^5y}$
18. $7x\sqrt{y^3x^2} - 5y\sqrt{yx^4}$
19. $\sqrt{16} + \sqrt{3} - \sqrt{48}$
20. $3\sqrt{5} - \sqrt{20} + \sqrt{45}$
21. $3\sqrt{2a} + 4a\sqrt{2} - \sqrt{8a}$
22. $\sqrt{20x} - 2\sqrt{5x} - 2\sqrt{x}$

SELFTEST

Simplify:

1. $\sqrt{5} + 3\sqrt{5}$
2. $7\sqrt{50} - \sqrt{98}$
3. $\sqrt{8} - 3\sqrt{2} + \sqrt{4}$
4. $6x\sqrt{xy^3} + 3y\sqrt{x^3y}$
5. $4xy^2\sqrt{3} - \sqrt{3x^2y^4}$
6. $5x^2\sqrt{3y} + \sqrt{12x^4y}$

SELFTEST ANSWERS

1. $4\sqrt{5}$ 2. $28\sqrt{2}$ 3. $2 - \sqrt{2}$ 4. $9xy\sqrt{xy}$ 5. $3xy^2\sqrt{3}$ 6. $7x^2\sqrt{3y}$

8.5 MULTIPLYING AND DIVIDING RADICAL EXPRESSIONS

Multiplying or dividing radical expressions with the same index is fairly simple. For we just multiply or divide the radicands. But then we must simplify the results.

Multiplication Law:

$$\sqrt[n]{a}\,\sqrt[n]{b} = \sqrt[n]{ab} \qquad (a > 0, \quad b > 0)$$

Division Law:

$$\frac{\sqrt[n]{a}}{\sqrt[n]{b}} = \sqrt[n]{\frac{a}{b}} \qquad (a > 0, b > 0)$$

Example 1 Simplify:

$$\begin{aligned} \sqrt{3}\,\sqrt{15} &= \sqrt{3 \cdot 15} && \text{by the multiplication law} \\ &= \sqrt{45} \\ &= \sqrt{9}\,\sqrt{5} = 3\sqrt{5} \end{aligned}$$

Example 2 Simplify:

$$
\begin{aligned}
2\sqrt{8}\,\sqrt{20} &= 2\sqrt{8 \cdot 20} && \text{by the multiplication law}\\
&= 2\sqrt{4 \cdot 2 \cdot 4 \cdot 5}\\
&= 2\sqrt{16}\,\sqrt{10}\\
&= 2 \cdot 4 \cdot \sqrt{10}\\
&= 8\sqrt{10}
\end{aligned}
$$

Example 3 Simplify $\sqrt{6}(4 + \sqrt{3})$. Since we cannot simplify $4 + \sqrt{3}$ we use the distributive law to carry out the multiplication.

$$
\begin{aligned}
\sqrt{6}(4 + \sqrt{3}) &= 4\sqrt{6} + \sqrt{6}\,\sqrt{3} && \text{by the distributive law}\\
&= 4\sqrt{6} + \sqrt{18} && \text{by the multiplication law}\\
&= 4\sqrt{6} + \sqrt{9}\,\sqrt{2}\\
&= 4\sqrt{6} + 3\sqrt{2}
\end{aligned}
$$

Example 4 Simplify $(2 + \sqrt{x})(4 + \sqrt{y})$. Neither factor simplifies. We multiply these factors just as we multiply binomials.

$$
\begin{aligned}
(2 + \sqrt{x})(4 + \sqrt{y}) &= (2 + \sqrt{x})4 + (2 + \sqrt{x})\sqrt{y}\\
&= 8 + 4\sqrt{x} + 2\sqrt{y} + \sqrt{x}\,\sqrt{y}\\
&= 8 + 4\sqrt{x} + 2\sqrt{y} + \sqrt{xy}
\end{aligned}
$$

The final sum does not simplify further.

Example 5 Simplify $(3 + \sqrt{5})(3 - \sqrt{5})$. This example is similar to the previous one. But the final result simplifies quite a bit.

$$
\begin{aligned}
(3 + \sqrt{5})(3 - \sqrt{5}) &= (3 + \sqrt{5})3 - (3 + \sqrt{5})\sqrt{5}\\
&= 9 + 3\sqrt{5} - 3\sqrt{5} - \sqrt{5}\,\sqrt{5}\\
&= 9 - \sqrt{25}\\
&= 9 - 5\\
&= 4
\end{aligned}
$$

Example 6 Simplify:

$$
\begin{aligned}
\frac{\sqrt{10}}{\sqrt{5}} &= \sqrt{\frac{10}{5}} && \text{by the division law}\\
&= \sqrt{2}
\end{aligned}
$$

Example 7 Simplify:

$$
\begin{aligned}
\frac{\sqrt{18x^2y}}{2\sqrt{2x}} &= \frac{1}{2}\sqrt{\frac{18x^2y}{2x}}\\
&= \frac{1}{2}\sqrt{9xy}\\
&= \frac{3}{2}\sqrt{xy}
\end{aligned}
$$

When we divide radicals we often find either a radical in a denominator or a fraction under the radical symbol. Simplified form calls for

1. no radical in a denominator, and
2. no fraction as a radicand

If we wish to use decimal approximations, the simplified form is easier to work with. We do not want a radical in the denominator because it is rather tedious to divide by a four-, five-, or six-place decimal approximation. Most square-root tables have square roots of integer values rather than of fractions. Thus we would not want a fraction under the radical.

Let's simplify $\sqrt{3}/\sqrt{5}$. Although $\sqrt{3}/\sqrt{5} = \sqrt{3/5}$, we still do not have a simplified form, for in $\sqrt{3}/\sqrt{5}$ there is a radical in the denominator. In $\sqrt{3/5}$ there is a fraction under the radical symbol.

A process called *rationalizing the denominator* allows us to simplify $\sqrt{3}/\sqrt{5}$. This process involves multiplying both the numerator and denominator by a value that will make the denominator rational (i.e., one that will enable us to remove the radical symbol.)

$$\frac{\sqrt{3}}{\sqrt{5}} = \frac{\sqrt{3} \cdot \sqrt{5}}{\sqrt{5} \cdot \sqrt{5}} = \frac{\sqrt{15}}{\sqrt{25}} = \frac{\sqrt{15}}{5}$$

The last expression is in simplified form.

In general, if a denominator is of the form $a\sqrt{b}$, we multiply numerator and denominator by the radical part $\sqrt{b}$. This process will rationalize the denominator.

Example 8 Simplify $\sqrt{6x}/(2\sqrt{7y})$. The denominator is $2\sqrt{7y}$. It is of the form $a\sqrt{b}$. Thus we multiply numerator and denominator by $\sqrt{7y}$.

$$\frac{\sqrt{6x}}{2\sqrt{7y}} = \frac{\sqrt{6x}\,\sqrt{7y}}{2\sqrt{7y}\,\sqrt{7y}} = \frac{\sqrt{42xy}}{2 \cdot 7y} = \frac{\sqrt{42xy}}{14y}$$

In the final result the denominator has no radical. The numerator will not simplify further.

How do we rationalize a fraction with a denominator of the form $a + \sqrt{b}$? If we multiply numerator and denominator by $\sqrt{b}$ the denominator will still contain a radical.

$$\frac{x}{4 + \sqrt{3}} = \frac{x\sqrt{3}}{(4 + \sqrt{3})\sqrt{3}} = \frac{x\sqrt{3}}{4\sqrt{3} + 3}$$

The last expression is not in simplified form because there is still a radical in the denominator. But if we multiply numerator and denominator by an expression of the form $a - \sqrt{b}$ we free the denominator of radicals.

$$\frac{x}{4 + \sqrt{3}} = \frac{x(4 - \sqrt{3})}{(4 + \sqrt{3})(4 - \sqrt{3})} = \frac{4x - x\sqrt{3}}{16 + 4\sqrt{3} - 4\sqrt{3} - \sqrt{3}\,\sqrt{3}}$$

$$= \frac{4x - x\sqrt{3}}{13}$$

The expressions $a + \sqrt{b}$ and $a - \sqrt{b}$ are *conjugates* of each other. When we multiply conjugates, the radical part disappears.

Example 9 Simplify $6/(9 + \sqrt{2})$: The conjugate of $9 + \sqrt{2}$ is $9 - \sqrt{2}$. We multiply numerator and denominator by $9 - \sqrt{2}$.

$$\frac{6}{9 + \sqrt{2}} = \frac{6(9 - \sqrt{2})}{(9 + \sqrt{2})(9 - \sqrt{2})}$$

$$= \frac{54 - 6\sqrt{2}}{81 + 9\sqrt{2} - 9\sqrt{2} - \sqrt{4}}$$

$$= \frac{54 - 6\sqrt{2}}{79}$$

The final fraction does not reduce.

PROGRAM 8.5

1. The third law of radicals says

$$\sqrt[n]{a \cdot b} = \sqrt[n]{a}\sqrt[n]{b} \qquad (a > 0, \quad b > 0)$$

We used this law to simplify radicals. Now simplify $\sqrt{18}$.

$$\sqrt{18} = \sqrt{9 \cdot 2} = \sqrt{9} \cdot \sqrt{2} = \_\_\sqrt{2}$$

3

2. If we read the third law from right to left, we get the multiplication law:

$$\sqrt[n]{a}\sqrt[n]{b} = \sqrt[n]{ab}$$

Are the indexes the same in all the radicals in the multiplication law? (yes\no). Multiply $\sqrt{2} \cdot \sqrt{5} = \sqrt{\_\_\_}$.

yes; 10

3. To multiply radicals with the same index, we multiply the radicands and use the same index.

$$\sqrt[3]{4}\sqrt[3]{5} = \sqrt[3]{\_\_\_}$$

20

4. A product of two radicals often can be simplified.

$$\sqrt{15}\sqrt{3} = \sqrt{45}$$
$$= \sqrt{9}\sqrt{\_\_\_}$$

5

$$= \_\_\_\_$$

$3\sqrt{5}$

5. Multiply and simplify:

$\sqrt{5}\sqrt{20} = \sqrt{\_\_\_\_}$ by the multiplication law — $\sqrt{100}$

$= \_\_\_\_$ in simplest form — 10

6. Multiply and simplify:

$\sqrt{a}\sqrt{ab} = \sqrt{\_\_\_}$ by the multiplication law — a^2b

$= \_\_\_\sqrt{b}$ in simplest form — a

7. Multiply: $2\sqrt{3}\sqrt{5} = 2\sqrt{\_\_\_}$.

15

8. We multiply values under the radical symbol with other values under the radical symbol. Also we multiply values outside the radical symbol with values outside the radical symbol.

$$3\sqrt{5} \cdot 4\sqrt{2} = 3 \cdot 4\sqrt{5}\sqrt{2}$$
$$= \_\_\_\sqrt{\_\_\_}$$

12; 10

9. Multiply: $4\sqrt{a} \cdot 6\sqrt{b} = \_\_\_\_$.

$24\sqrt{ab}$

10. Multiply and simplify:

$2\sqrt{3} \cdot 4\sqrt{6} = 8\sqrt{\_\_\_}$ by the multiplication law — 18

$= \_\_\_\sqrt{2}$ in simplest form — 24

11. We use the distributive law to multiply.

$$\sqrt{3}(2 + \sqrt{6}) = 2\sqrt{3} + \sqrt{3}\sqrt{6}$$

18 $\quad = 2\sqrt{3} + \sqrt{\_\_\_\_}$

3 $\quad = 2\sqrt{3} + \_\_\sqrt{2}$

12. Multiply and simplify:

$3\sqrt{2}$ $\quad \sqrt{2}(3 + \sqrt{6}) = \_\_\_\_ + \sqrt{2}\sqrt{6}$

$3\sqrt{2}$; 2 $\quad = \_\_\_\_ + \sqrt{2}\sqrt{\_\_\_\_}\sqrt{3}$

$3\sqrt{2}$; $2\sqrt{3}$ $\quad = \_\_\_\_ + \_\_\_\_$ in simplest form

13. Let's review binomial multiplication.

$$(3 + x)(4 + y) = (3 + x)4 + (3 + x)y$$

12; $4x$; $3y$; xy $\quad = \_\_\_\_ + \_\_\_\_ + \_\_\_\_ + \_\_\_\_$

14. Multiply:

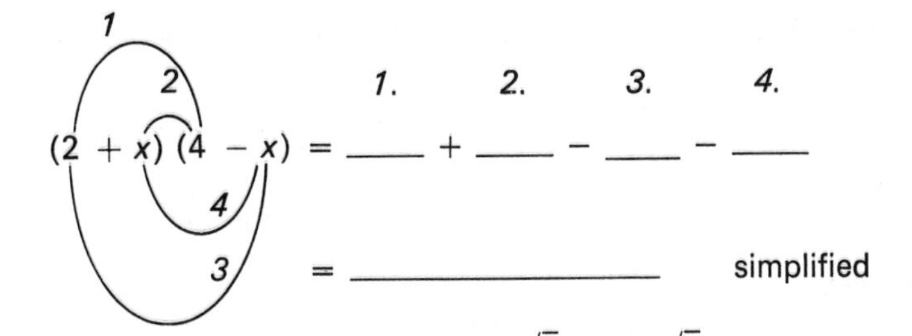

8; $4x$; $2x$; x^2

$8 + 2x - x^2$

15. We multiply the radical expression $(2 + \sqrt{x})(3 + \sqrt{y})$ in the same way.

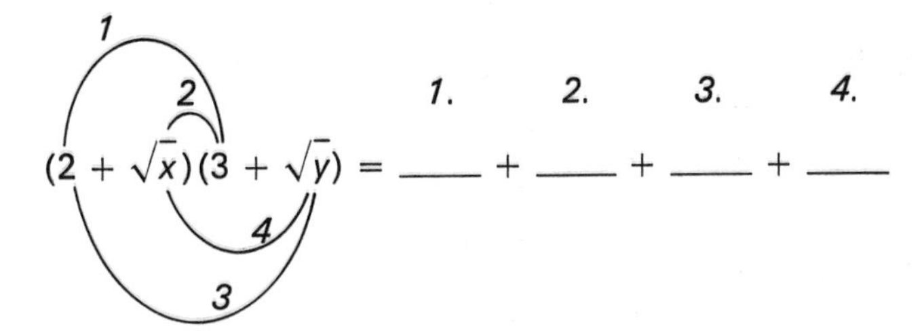

6; $3\sqrt{x}$; $2\sqrt{y}$; $\sqrt{xy}$

16. Multiply:

1. 2. 3. 4.

$$(4 + \sqrt{a})(3 + \sqrt{y}) = \_\_\_\_ + \_\_\_\_ + \_\_\_\_ + \_\_\_\_$$

12; $3\sqrt{a}$; $4\sqrt{y}$; $\sqrt{ay}$

17. Multiply and simplify:

$$(2 + \sqrt{x})(3 + \sqrt{x}) = 6 + 3\sqrt{x} + 2\sqrt{x} + \sqrt{x^2}$$

6; $5\sqrt{x}$; x $\quad = \_\_\_\_ + \_\_\_\_ + \_\_\_\_$

18. Multiply and simplify: $(2 - \sqrt{5})(2 + \sqrt{5}) = \_\_\_\_\_\_\_\_\_\_\_\_$.

$4 + 2\sqrt{5} - 2\sqrt{5} - 5$
or -1

19. To divide two radicals with the same index, we divide the radicands. By the division law,

$$\frac{\sqrt[n]{a}}{\sqrt[n]{b}} = \sqrt[n]{\frac{a}{b}} \quad \text{and} \quad \frac{\sqrt{15}}{\sqrt{3}} = \sqrt{\frac{\ \ }{\ \ }} = \underline{\quad}$$

$\frac{15}{3}$; $\sqrt{5}$

20. Simplify:

$$\frac{\sqrt{12}}{\sqrt{3}} = \sqrt{\frac{\ \ }{\ \ }} = \underline{\qquad}$$

$\frac{12}{3}$; $\sqrt{4}$ or 2

21. Simplify:

$$\frac{\sqrt{6x^2y}}{\sqrt{3x}} = \sqrt{\frac{\ \ }{\ \ }} = \underline{\qquad}$$

$\frac{6x^2y}{3x}$; $\sqrt{2xy}$

22. A value under a radical symbol can be divided by any other value under a radical symbol, but not by a value outside a radical symbol. The expression $\sqrt{10}/2$ is in simplest form because 2 (is\is not) under a radical symbol (that is, 2 is not part of a radicand).

is not

23. Is $\sqrt{10}/5$ simplified? (Yes\No), because 5 is outside the ____________.

Yes; radical sign

24. We can simplify $\sqrt{10}/\sqrt{5}$, since 10 and 5 are both under radical symbols.

$$\frac{\sqrt{10}}{\sqrt{5}} = \sqrt{\frac{\ \ }{\ \ }} \quad \text{by the division law}$$

$\frac{10}{5}$

$$= \underline{\quad} \quad \text{in simplest form}$$

$\sqrt{2}$

25. Simplify:

$$\frac{\sqrt{10}}{2\sqrt{2}} = \frac{1}{2} \cdot \frac{\sqrt{10}}{\sqrt{2}} = \frac{1}{2}\sqrt{\underline{\quad}} \quad \text{or} \quad \frac{\sqrt{\underline{\quad}}}{2}$$

5; 5

26. Simplify:

$$\frac{\sqrt{15x^3}}{x\sqrt{3x}} = \frac{1}{x} \cdot \frac{\sqrt{15x^3}}{\sqrt{3x}}$$

$$= \frac{1}{x} \cdot \sqrt{\frac{\ \ }{\ \ }}$$

$\frac{15x^3}{3x}$ or $5x^2$

$$= \underline{\qquad\qquad} \quad \text{in simplest form}$$

$\frac{1}{x} \cdot \sqrt{5x^2} = \frac{x\sqrt{5}}{x} = \sqrt{5}$

Reduce the final fraction.

27. Simplified form calls for

1. no fraction under a radical symbol, and
2. no radical in a denominator.

Which of the following equal expressions is in simplest form?

a. $\frac{\sqrt{2}}{\sqrt{7}}$ *b.* $\sqrt{\frac{2}{7}}$ *c.* $\frac{\sqrt{14}}{7}$

c.

28. We need a technique other than the division process to simplify some fractions with radicals.

$$\frac{\sqrt{2}}{\sqrt{5}} = \sqrt{\frac{2}{5}} \quad \text{by the division law}$$

there is a radical in the denominator
there is a fraction under the radical

But $\sqrt{2}/\sqrt{5}$ is not in simplified form because ________________. Nor is $\sqrt{2/5}$ in simplified form because ________________.

5

29. If we multiply both numerator and denominator of $\sqrt{2}/\sqrt{5}$ by a carefully chosen factor, we can remove the radical from the denominator.

$$\frac{\sqrt{2}}{\sqrt{5}} = \frac{\sqrt{2}\cdot\sqrt{5}}{\sqrt{5}\cdot\sqrt{5}} = \frac{\sqrt{10}}{\sqrt{5^2}} = \frac{\sqrt{10}}{\_\_\_\_}$$

$\dfrac{\sqrt{35}}{7}$

30. To *rationalize a denominator*, we multiply both numerator and denominator by a factor that will make the denominator rational. Rationalize the denominator:

$$\frac{\sqrt{5}}{\sqrt{7}} = \frac{\sqrt{5}\cdot\sqrt{7}}{\sqrt{7}\cdot\sqrt{7}} = \_\_\_\_\_\_$$

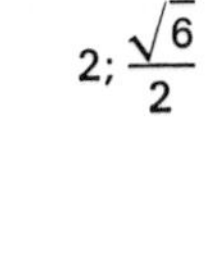

31. Rationalize the denominator:

$$\frac{\sqrt{3}}{\sqrt{2}} = \frac{\sqrt{3}\cdot\sqrt{2}}{\sqrt{2}\cdot\sqrt{\_\_\_}} = \_\_\_\_\_\_$$

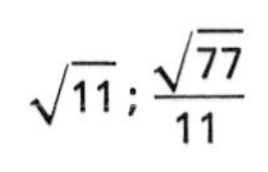

32. Rationalize the denominator:

$$\frac{\sqrt{7}}{\sqrt{11}} = \frac{\sqrt{7}\cdot\_\_\_\_}{\sqrt{11}\cdot\sqrt{11}} = \_\_\_\_\_\_$$

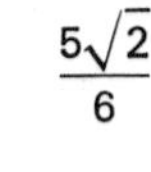

33. To rationalize the denominator of $5/(3\sqrt{2})$, it suffices to multiply numerator and denominator by $\sqrt{2}$ instead of $3\sqrt{2}$.

$$\frac{5}{3\sqrt{2}} = \frac{5\cdot\sqrt{2}}{3\sqrt{2}\cdot\sqrt{2}} = \_\_\_\_\_\_$$

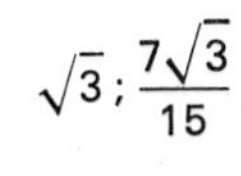

34. Rationalize the denominator:

$$\frac{7}{5\sqrt{3}} = \frac{7\cdot\sqrt{3}}{5\sqrt{3}\cdot\_\_\_\_} = \_\_\_\_\_\_$$

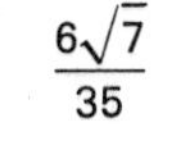

35. Rationalize the denominator: $6/(5\sqrt{7}) =$ ______.

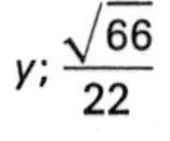

36. Rationalize the denominator and simplify:

$$\frac{\sqrt{6y}}{2\sqrt{11y}} = \frac{\sqrt{6y}\sqrt{11y}}{2\sqrt{11y}\sqrt{11y}} = \frac{\_\_\sqrt{66}}{2\cdot 11y} = \_\_\_\_$$

$4-\sqrt{2}$

37. The expressions $a+\sqrt{b}$ and $a-\sqrt{b}$ are *conjugates* of each other. What is the conjugate of $4+\sqrt{2}$? ______.

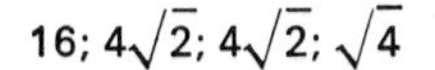

38. When we multiply conjugates, the radical part disappears.

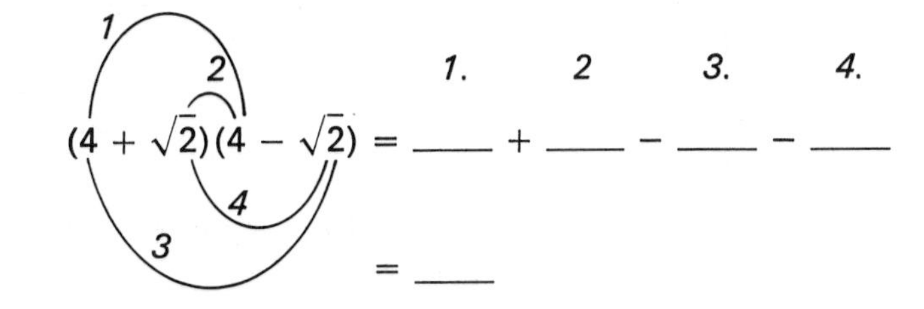

$$(4+\sqrt{2})(4-\sqrt{2}) = \_\_\_\_ + \_\_\_\_ - \_\_\_\_ - \_\_\_\_$$
$$= \_\_\_\_$$

14

39. What is the conjugate of $1 - \sqrt{3}$? ________.

$(1 - \sqrt{3})(1 + \sqrt{3}) =$ ____

$1 + \sqrt{3}$

-2

40. To simplify $4/(1 - \sqrt{3})$, we multiply numerator and denominator by the conjugate of $1 - \sqrt{3}$.

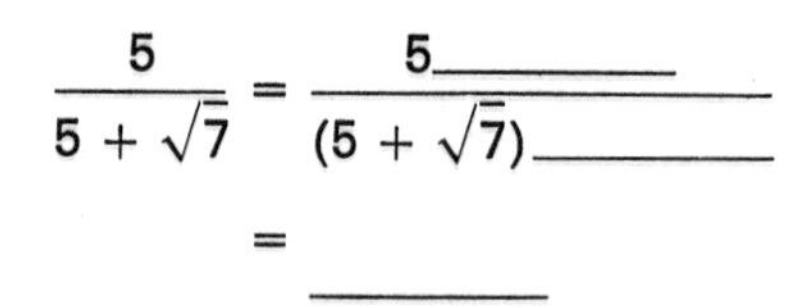

$$\frac{4}{1-\sqrt{3}} = \frac{4(1+\sqrt{3})}{(1-\sqrt{3})(1+\sqrt{3})}$$

$=$ ________

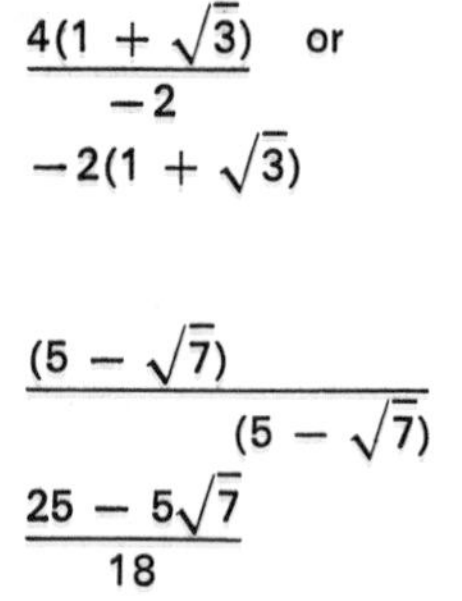

$\frac{4(1+\sqrt{3})}{-2}$ or

$-2(1 + \sqrt{3})$

41. Rationalize the denominator:

$$\frac{5}{5+\sqrt{7}} = \frac{5\_\_\_\_}{(5+\sqrt{7})\_\_\_\_}$$

$=$ ________

$\frac{(5 - \sqrt{7})}{(5 - \sqrt{7})}$

$\frac{25 - 5\sqrt{7}}{18}$

PRACTICE PROBLEMS

Multiply and simplify:

1. $\sqrt{2}\sqrt{7}$ 2. $\sqrt{5}\sqrt{3}$ 3. $\sqrt{5}\sqrt{10}$ 4. $\sqrt{2a}\sqrt{6a}$

5. $(3 + \sqrt{x})(2 + \sqrt{x})$ 6. $(5 + \sqrt{y})(3 + \sqrt{y})$

7. $(2 + \sqrt{6})(3 - \sqrt{4})$ 8. $(1 - \sqrt{5})(1 + \sqrt{5})$

Divide and simplify:

9. $\frac{\sqrt{10}}{\sqrt{2}}$ 10. $\frac{\sqrt{30}}{\sqrt{5}}$ 11. $\frac{\sqrt{40a^3}}{\sqrt{10a}}$ 12. $\frac{\sqrt{50}}{\sqrt{2}}$

13. $\frac{3\sqrt{6}}{2\sqrt{3}}$ 14. $\frac{5\sqrt{15a}}{2\sqrt{5a}}$ 15. $\frac{9\sqrt{20}}{6\sqrt{5}}$ 16. $\frac{3\sqrt{50a^2}}{5a\sqrt{2}}$

Rationalize the denominator:

17. $\frac{1}{\sqrt{3}}$ 18. $\frac{2}{\sqrt{x}}$ 19. $\frac{\sqrt{3}}{\sqrt{y}}$ 20. $\frac{\sqrt{a}}{\sqrt{2}}$

21. $\frac{5}{4\sqrt{3}}$ 22. $\frac{6}{5\sqrt{7}}$ 23. $\frac{2}{3+\sqrt{2}}$ 24. $\frac{5}{4-\sqrt{3}}$

25. $\frac{1}{1-\sqrt{x}}$ 26. $\frac{3}{6+\sqrt{x}}$ 27. $\frac{4}{\sqrt{2}}$ 28. $\frac{3}{\sqrt{2x}}$

29. $\frac{\sqrt{5}}{\sqrt{7}}$ 30. $\frac{8}{3\sqrt{6}}$ 31. $\frac{-2}{1+\sqrt{x}}$ 32. $\frac{1}{3-\sqrt{2}}$

33. $\frac{4}{2+\sqrt{2x}}$ 34. $\frac{-1}{1-\sqrt{5}}$ 35. $\frac{1}{\sqrt{x}-\sqrt{y}}$ 36. $\frac{1}{\sqrt{x}+\sqrt{y}}$

SELFTEST

Simplify:

1. $3\sqrt{2x} \cdot 5\sqrt{6x}$ 2. $\frac{8\sqrt{10}}{4\sqrt{2}}$ 3. $\frac{\sqrt{7}}{\sqrt{2}}$ 4. $\frac{2}{3\sqrt{x}}$

5. $\frac{1}{2+\sqrt{x}}$ 6. $(3 + \sqrt{2})(3 - \sqrt{2})$ 7. $(4 + \sqrt{x})(1 + \sqrt{x})$

SELFTEST ANSWERS

1. $30x\sqrt{3}$ 2. $2\sqrt{5}$ 3. $\dfrac{\sqrt{14}}{2}$ 4. $\dfrac{2\sqrt{x}}{3x}$ 5. $\dfrac{2-\sqrt{x}}{4-x}$ 6. 7 7. $x + 5\sqrt{x} + 4$

8.6 FACTORING AND REDUCING RADICAL EXPRESSIONS

To reduce a fraction we normally factor numerator and denominator and then look for common factors. The distributive law allows us to factor radical expressions. The process is similar to that of factoring polynomials. We try to factor out the greatest common "monomial" factor. But often we must change the form of the radical before the greatest factor becomes apparent.

Example 1 Factor $4\sqrt{5} + \sqrt{35}$.

$$\begin{aligned} 4\sqrt{5} + \sqrt{35} &= 4\sqrt{5} + \sqrt{7}\sqrt{5} \\ &= \sqrt{5}(4 + \sqrt{7}) \end{aligned}$$

Example 2 Factor $3\sqrt{5} + \sqrt{54}$.

$$\begin{aligned} 3\sqrt{5} + \sqrt{54} &= 3\sqrt{5} + \sqrt{9}\sqrt{6} \\ &= 3\sqrt{5} + 3\sqrt{6} \\ &= 3(\sqrt{5} + \sqrt{6}) \end{aligned}$$

Now let us see how we use this type of factoring to reduce fractions involving radicals.

Example 3 Reduce $\dfrac{4+\sqrt{8}}{6}$.

$$\begin{aligned} \frac{4+\sqrt{8}}{6} &= \frac{4+2\sqrt{2}}{6} \\ &= \frac{\not{2}(2+\sqrt{2})}{\not{2}\cdot 3} \\ &= \frac{2+\sqrt{2}}{3} \end{aligned}$$

Example 4 Reduce $(3\sqrt{5} - \sqrt{70})/\sqrt{5}$. In this type of problem we could rationalize the denominator first. But we will reduce the expression directly.

$$\begin{aligned} \frac{3\sqrt{5}-\sqrt{70}}{\sqrt{5}} &= \frac{3\sqrt{5}-\sqrt{5}\sqrt{14}}{\sqrt{5}} \\ &= \frac{\sqrt{\not{5}}(3-\sqrt{14})}{\sqrt{\not{5}}} \\ &= 3 - \sqrt{14} \end{aligned}$$

Example 5 Reduce:

$$\begin{aligned} \frac{4x^2 - \sqrt{24x^2y}}{8x} &= \frac{4x^2 - \sqrt{4}\sqrt{x^2}\sqrt{6y}}{8x} && \text{simplify numerator} \\ &= \frac{4x^2 - 2x\sqrt{6y}}{8x} \\ &= \frac{2x(2x - \sqrt{6y})}{2\cdot 4x} && \text{factor} \\ &= \frac{2x - \sqrt{6y}}{4} && \text{reduce} \end{aligned}$$

PROGRAM 8.6

1. We factor radical expressions much the same way as we factor polynomials.

$$3x + 12 = 3x + 3 \cdot 4 = 3(x + \_\_\_\_)$$

in factored form. — 4

2. We use the distributive law to factor $4 + 2\sqrt{5} = 2(2 + \_\_\_\_)$. — $\sqrt{5}$

3. Factor: $12 + 6\sqrt{3} = \_\_(\_\_\_\_\_\_\_\_)$. — 6; $2 + \sqrt{3}$

4. Sometimes we must change the form of the radical before we can factor.

$$8 + \sqrt{20} = 8 + \sqrt{4}\sqrt{5} = 8 + \_\_\_\_\sqrt{5}$$ — 2

$$= 2(\_\_\_\_\_\_\_\_) \quad \text{in factored form}$$ — $4 + \sqrt{5}$

5. $\sqrt{2} + \sqrt{6} = \sqrt{2} + \sqrt{2}\sqrt{\_\_\_\_}$ — 3

$= \sqrt{2}(\_\_\_\_\_\_\_\_)$ in factored form — $1 + \sqrt{3}$

6. $6\sqrt{5} + \sqrt{15} = 6\sqrt{5} + \sqrt{3}\sqrt{\_\_\_\_}$ — 5

$= \_\_\_\_(6 + \sqrt{3})$ in factored form — $\sqrt{5}$

7. Factor:

$$15 + \sqrt{50} = 15 + \sqrt{2}\sqrt{\_\_\_\_}$$ — 25

$$= 15 + \_\_\_\_\sqrt{2}$$ — 5

$$= \_\_\_\_\_\_\_\_ \quad \text{in factored form}$$ — $5(3 + \sqrt{2})$

8. Factor $\sqrt{3} + \sqrt{21}$:

$\sqrt{3} + \sqrt{7}\sqrt{\_\_\_\_} = \_\_\_\_\_\_\_\_$ in factored form — 3; $\sqrt{3}\,(1 + \sqrt{7})$

9. To reduce fractions, we factor numerator and denominator and divide out the common factors.

$$\frac{3 + 6\sqrt{2}}{6} = \frac{3(\_\_\_\_\_\_\_\_)}{3 \cdot 2}$$ — $1 + 2\sqrt{2}$

$$= \_\_\_\_\_\_\_\_ \quad \text{in reduced form}$$ — $\frac{1 + 2\sqrt{2}}{2}$

10. Reduce:

$$\frac{5 + 10\sqrt{3}}{20} = \frac{\_\_(\_\_\_\_\_\_\_\_)}{20}$$ — 5; $1 + 2\sqrt{3}$

$$= \_\_\_\_\_\_\_\_$$ — $\frac{1 + 2\sqrt{3}}{4}$

11. Reduce:

$$\frac{4 + \sqrt{12}}{8} = \frac{4 + 2\sqrt{\_\_\_\_}}{8}$$ — 3

$$= \frac{2(\_\_\_\_\_\_\_\_)}{8}$$ — $2 + \sqrt{3}$

$$= \_\_\_\_\_\_\_\_$$ — $\frac{2 + \sqrt{3}}{4}$

2 or $\sqrt{4}$

$\frac{3+\sqrt{5}}{2}$

$\frac{2+\sqrt{3}}{3}$

$3+\sqrt{5}$

$3+\sqrt{5}$

$\sqrt{4}+\sqrt{5}$

$2+\sqrt{5}$

12. Reduce:

$$\frac{6+\sqrt{20}}{4} = \frac{6+\_\_\_\_\sqrt{5}}{4}$$

$$= \_\_\_\_\_\_\_\_$$

13. Reduce: $(6+\sqrt{27})/9 =$ ________.

14. Reduce:

$$\frac{3\sqrt{2}+\sqrt{10}}{\sqrt{2}} = \frac{\sqrt{2}(\_\_\_\_\_\_\_\_)}{\sqrt{2}}$$

$$= \_\_\_\_\_\_\_\_$$

15. Reduce:

$$\frac{\sqrt{12}+\sqrt{15}}{\sqrt{3}} = \_\_\_\_\_\_\_\_$$

$$= \_\_\_\_\_\_\_\_ \quad \text{in simplest form}$$

PRACTICE PROBLEMS

Factor completely:

1. $8+10\sqrt{6}$ **2.** $16-12\sqrt{2a}$ **3.** $4-\sqrt{8}$

4. $6+\sqrt{9}$ **5.** $3\sqrt{5a}+\sqrt{10a}$ **6.** $\sqrt{15}+5\sqrt{3}$

7. $\sqrt{12}-\sqrt{15}$ **8.** $\sqrt{18}+\sqrt{10}$ **9.** $\sqrt{4a^3}+a\sqrt{ab}$

10. $\sqrt{40}-4\sqrt{5}$ **11.** $6\sqrt{10}-2\sqrt{18}$ **12.** $3\sqrt{a^2b^2}-6a\sqrt{b^3}$

Reduce:

13. $\frac{2-4\sqrt{3}}{6}$ **14.** $\frac{6+9\sqrt{2}}{12}$ **15.** $\frac{4x+\sqrt{8x^2}}{2x}$ **16.** $\frac{9-\sqrt{18}}{3}$

17. $\frac{\sqrt{2}+\sqrt{10}}{3\sqrt{2}}$ **18.** $\frac{\sqrt{a}-\sqrt{ab}}{\sqrt{a}}$ **19.** $\frac{10+3\sqrt{50}}{6}$ **20.** $\frac{12+3\sqrt{48}}{8}$

21. $\frac{a+b\sqrt{a^2b}}{ab}$ **22.** $\frac{3x-\sqrt{18x^2}}{6x}$

SELFTEST

1. Factor $5-10\sqrt{5}$ **2.** Factor $10+\sqrt{75}$ **3.** Reduce $\frac{3+9\sqrt{2}}{3}$

4. Reduce $\frac{8+2\sqrt{12}}{20}$ **5.** Reduce $\frac{2x+\sqrt{2x^2y}}{6x}$

SELFTEST ANSWERS

1. $5(1-2\sqrt{5})$ **2.** $5(2+\sqrt{3})$ **3.** $1+3\sqrt{2}$

4. $\frac{2+\sqrt{3}}{5}$ **5.** $\frac{2+\sqrt{2y}}{6}$

8.7 COMPLEX NUMBERS

So far we have studied quite a few different number systems. We started with counting numbers and whole numbers. Then we saw a need for negative counting numbers and so developed the integers. We later worked with fractions and finally with real numbers, such as $\sqrt{2}$, that were not fractions.

Each of these number systems can be thought of as coming about from a specific need. To use a well-worn expression, necessity is the mother of invention. The work of solving equations has occupied a great deal of our attention. Let us look at the number systems needed to solve some equations.

Example 1 What type of number is needed to solve $x - 3 = 1$? We know the solution is 4. It is a counting number.

Example 2 Solve $x + 5 = 0$, and describe the type of number needed. We see that -5 is the solution. This number is an integer.

Example 3 Solve $3x + 5 = 0$: To solve an equation of this type, it is necessary to use a new number system called fractions or rational numbers, for the solution is $x = -5/3$.

Example 4 Solve $x^2 - 2 = 0$: This equation is equivalent to $x^2 = 2$. The solutions are $x = \sqrt{2}$ and $x = -\sqrt{2}$. Let's check these solutions.

$$x^2 - 2 = 0 \qquad x^2 - 2 = 0$$
$$(\sqrt{2})^2 - 2 = 0 \qquad (-\sqrt{2})^2 - 2 = 0$$
$$2 - 2 = 0 \quad \text{so } \sqrt{2} \text{ checks} \qquad 2 - 2 = 0 \quad \text{so } -\sqrt{2} \text{ checks}$$

As we pointed out in the first section of this chapter, numbers such as $\sqrt{2}$ are not fractions. So it was necessary to invent a larger number system which included such numbers. This system is called the *real number system*.

Example 5 Solve $x^2 + 1 = 0$: This equation is equivalent to $x^2 = -1$. But if x is any real number then $x^2 \geq 0$. So there can be no real number x which satisfies $x^2 = -1$. What do we do?

The necessity of solving equations of the type $x^2 + 1 = 0$ brings us to the invention of a new number system called the *complex number system*. It was proved by a famous mathematician named Gauss (1777–1855) that the complex number system is large enough to solve all equations of polynomial type with complex coefficients. In fact, this result of Gauss is called the *Fundamental Theorem of Algebra*.

To solve $x^2 + 1 = 0$, we need a number x such that $x^2 = -1$. Let us invent such a number and call it the imaginary unit i. So in symbols, i is a new number defined so that $i^2 = -1$. We may also write $i = \sqrt{-1}$. Then $x = i$ solves the equation $x^2 + 1 = 0$ since

$$i^2 + 1 = -1 + 1 = 0$$

Example 6 Let us look at the number $i = \sqrt{-1}$ a little more.

$i = \sqrt{-1}$

$i^2 = -1$ by definition

$i^3 = i^2 i = (-1)i = -i$

$i^4 = i^2 i^2 = (-1)(-1) = 1$

$i^5 = i^4 i = (1)i = i$

$i^6 = i^5 i = ii = -1$

$i^7 = i^6 i = (-1)i = -i$

$i^8 = i^7 i = (-i)i = -i^2 = -(-1) = 1$

Notice that as we take higher powers of i we get a cycle which repeats itself. The first four powers of i yield i, -1, $-i$, 1. The higher powers of i just repeat in this established cycle. The number i is the imaginary unit. Numbers of the type bi, where b is any nonzero real number, are called *pure imaginary numbers.* For example, each of the following are pure imaginary numbers: i, $2i$, $3i$, $-5i$, $\frac{7}{4}i$, $\sqrt{2}i$, and $(\sqrt{5}/2)i$.

A *complex number in standard form* is any number of the type $a + bi$ where a and b are any real numbers. Each of the following are complex numbers: $0 + 2i$, $5 - i$, $\sqrt{2} + \sqrt{10}i$, $8 + 0i$, $-\sqrt[3]{12} + 4i$.

There is some overlapping of definitions here. Any pure imaginary number can be written $0 + bi$. So pure imaginary numbers are complex numbers. Also any real number a can be written as $a + 0i$ (since $0i = 0$). We see that every real number is thus a complex number. In fact, the complex number system includes all the number systems studied in this book. It is the largest number system we shall study.

The following diagram may help you understand how the number systems are related. As we read down, the sets get smaller; as we read up, they get bigger.

- Complex numbers
  - Real numbers
    - Integers
      - Whole numbers
        - Counting numbers
    - Irrational numbers
  - Pure imaginary numbers

In the complex number $a + bi$, the number a is called the *real part* and bi is called the *imaginary part.*

Two complex numbers are equal if and only if their real parts are equal and their imaginary parts are equal. That is

$$a + bi = c + di \quad \text{if and only if} \quad a = c \quad \text{and} \quad b = d$$

Example 7 If $5 + 3i = a + bi$, what are a and b? Since the real parts must be equal, $5 = a$. The imaginary parts must also be equal, so $3 = b$.

Example 8 If $a + bi = 0 + 0i$, what are a and b? It follows that $a = 0$ and $b = 0$.

In all problems involving complex numbers, each complex number should first be expressed in the form $a + bi$. Then we can proceed according to the usual rules of algebra and substitute $i^2 = -1$ when appropriate.

Example 9 Express $\sqrt{-3}$ in standard form: Since $\sqrt{-3} = \sqrt{3(-1)} = \sqrt{3}\sqrt{-1} = \sqrt{3}i$, the standard form is $0 + \sqrt{3}i$.

Example 10 Express $4 - \sqrt{-16}$ in standard form:

$$\sqrt{-16} = \sqrt{16(-1)} = \sqrt{16}\sqrt{-1} = 4i$$

Thus $4 - \sqrt{-16} = 4 - 4i$ in standard form.

Addition and Subtraction: **To add (or subtract) two complex numbers such as $a + bi$ and $c + di$, we just add (or subtract) the real and imaginary parts separately.**

$$(a + bi) + (c + di) = (a + c) + (b + d)i$$
$$(a + bi) - (c + di) = (a - c) + (b - d)i$$

Example 11 Add $6 + \sqrt{-9}$ and $7 + 2i$. First we change $6 + \sqrt{-9}$ to the form $a + bi$. Since

$$\sqrt{-9} = \sqrt{9(-1)} = \sqrt{9}\sqrt{-1} = 3i$$

the imaginary part of $6 + \sqrt{-9}$ is $3i$. The real part of $6 + \sqrt{-9}$ is 6, so we can write $6 + \sqrt{-9}$ as $6 + 3i$. The number $7 + 2i$ is already in complex form. Now we add the real and imaginary parts separately.

$$\begin{aligned}(6 + 3i) + (7 + 2i) &= (6 + 7) + (3 + 2)i \\ &= 13 + 5i\end{aligned}$$

Example 12 Subtract $\sqrt{-25}$ from $4 + 11i$. The first step is to write $\sqrt{-25}$ in complex form.

$$\sqrt{-25} = \sqrt{25(-1)} = \sqrt{25}\sqrt{-1} = 5i = 0 + 5i$$

So the real part of $\sqrt{-25} = 0 + 5i$ is 0 and the imaginary part is $5i$.

$$\begin{aligned}(4 + 11i) - (0 + 5i) &= (4 - 0) + (11 - 5)i \\ &= 4 + 6i\end{aligned}$$

Multiplication: **To form the product of two complex numbers such as $a + bi$ and $c + di$, multiply them according to the usual laws of algebra and substitute $i^2 = -1$.**

$$\begin{aligned}(a + bi)(c + di) &= ac + adi + bci + bdi^2 \\ &= ac + (ad + bc)i - bd \qquad \text{since} \qquad i^2 = -1 \\ &= (ac - bd) + (ad + bc)i\end{aligned}$$

Example 13 Multiply $2 + i$ and $3 - 4i$:

$$\begin{aligned}(2 + i)(3 - 4i) &= 6 - 8i + 3i - 4i^2 \\ &= 6 - 5i - 4i^2 \\ &= 6 - 5i + 4 \qquad \text{since} \qquad -4i^2 = -4(-1) = +4 \\ &= 10 - 5i\end{aligned}$$

For our purposes the rule for division of two complex numbers is an optional topic. If you wish to omit the rule for division you may. However, if you are interested, you will find the general formula below. Your instructor can easily provide the proof of this formula.

Division: **Let $a + bi$ and $c + di$ be complex numbers such that $c + di \neq 0 + 0i$. Then**

$$\frac{a + bi}{c + di} = \left(\frac{ac + bd}{c^2 + d^2}\right) + \left(\frac{bc - ad}{c^2 + d^2}\right)i$$

Example 14 Divide $3 - 2i$ by $1 + 3i$. In this example, $a + bi = 3 - 2i$ so $a = 3$ and $b = -2$. Also, $c + di = 1 + 3i$ so $c = 1$ and $d = 3$. By use of the division formula we obtain

$$\begin{aligned}\frac{3 - 2i}{1 + 3i} &= \left(\frac{3 - 6}{1 + 9}\right) + \left(\frac{-2 - 9}{1 + 9}\right)i \\ &= \frac{-3}{10} - \frac{11}{10}i\end{aligned}$$

PROGRAM 8.7

1. We define a new number i called the *imaginary unit*. The number i has the special property that $i^2 = -1$. Since $i^2 = -1$, we also sometimes write $i = \sqrt{-1}$. Because $i^2 = -1$ then

$i^3 = (i^2)i = (\_\_\_\_)i = \_\_\_\_$ — -1; $-i$

$i^4 = (i^2)(i^2) = (-1)(\_\_\_\_) = \_\_\_\_$ — -1; 1

$i^5 = (i^4)i = (\_\_\_\_)i = \_\_\_\_$ — 1; i

$i^6 = (i^5)i = (\_\_\_\_)i = \_\_\_\_$ — i; -1

In this way we can compute higher powers of i. But any power of i will always be one of the four numbers i, -1, $-i$, or 1.

b., e., f.

2. Let b be any nonzero real number. Then a number of the form bi is called a *pure imaginary number*. So the numbers $5i$, $\sqrt{2}\,i$, $-8i$, and $-\sqrt{10}\,i$ are all pure imaginary numbers. Which of the following are pure imaginary numbers?

a. $0i$ *b.* $12i$ *c.* -8 *d.* -1 *e.* $\frac{i}{2}$ or $\frac{1}{2}i$ *f.* $\sqrt{6}\,i$

complex

all of them

3. A *complex number in standard form* is any number of the type $a + bi$ where a and b are real numbers. Each of the numbers $8 - 3i$, $0 + 4i$, $2 + 0i$ is a ________ number. Which of the following are complex numbers?

a. $18 - 2i$ *b.* $4 - \sqrt{2}\,i$ *c.* $6 + 0i$
d. $0 + 0i$ *e.* $0 - 2i$ *f.* $-4 + 6i$

0

can

0

-12

$\frac{3}{2}$; 0

yes

4. Since $0i =$ _____, every real number a can be written

$$a = a + 0 = a + 0i$$

In this way, every real number (can\cannot) be thought of as a complex number. Write 5 in the form of a complex number.

$$5 = 5 + \_\_i$$

Write -12 in the form of a complex number

$$-12 = \_\_\_\_ + 0i$$

Now write $^3/_2$ in the form of a complex number.

$$\frac{3}{2} = \_\_\_\_ + \_\_i$$

Can every real number be written in the form of a complex number? (yes\no).

can

6*i*

$0; \sqrt{2}$

5. Let us see if every pure imaginary number is also a complex number. Suppose bi is a pure imaginary number. Then $bi = 0 + bi$. So bi (can\cannot) be written as a complex number. Write $6i$ in the form of a complex number.

$$6i = 0 + \_\_\_\_$$

Write $\sqrt{2}\,i$ in the form of a complex number.

$$\sqrt{2}\,i = \_\_\_\_ + \_\_\_\_\,i$$

i

0; 2

6. Let's see what type of number $\sqrt{-4}$ is. Since we are to take the square root of a negative number, we know we will not get a real number. In fact $\sqrt{-4}$ is a pure imaginary number. For

$$\sqrt{-4} = \sqrt{4(-1)} = \sqrt{4}\,\sqrt{-1} = 2i \qquad \text{since} \qquad \sqrt{-1} = \_\_\_\_$$

Since $\sqrt{-4} = 2i$, we see that $\sqrt{-4}$ is also a complex number. Let's write it in the form of a complex number.

$$\sqrt{-4} = 2i = \_\_\_\_ + \_\_\_\_i$$

-1; *i*

7. Write $\sqrt{-6}$ in the form of a pure imaginary number.

$$\sqrt{-6} = \sqrt{6(-1)} = \sqrt{6}\,\sqrt{\_\_\_\_} = \sqrt{6}\,\_\_\_\_$$

In complex form $\sqrt{-6}$ looks like

$$0 + \_\_\_\_ i$$

$\sqrt{6}$

8. In the complex number $a + bi$ the number a is called the *real part* and bi is called the *imaginary part*. The real part of $3 + 6i$ is ____. What is the imaginary part of $8 - 43i$? It is ______. What is the real part of $2i$? Since $2i = 0 + 2i$, the real part is ____.

3

$-43i$

0

9. Two complex numbers are equal if and only if their real parts are equal and their imaginary parts are equal. That is, $a + bi = c + di$ if and only if $a = c$ and $b = d$.

If $a + bi = 8 + 7i$, then $a = 8$ and $b =$ ____. If $x + yi = 3 - 10i$, then $x =$ ____ and $y =$ ____. If $a + 3i = -4 + bi$, then $a =$ ____ and $b =$ ____.

7

3; -10; -4

3

10. When doing arithmetic on complex numbers, we must first express each complex number in the standard form $a + bi$. Then we can proceed according to the usual rules of algebra and substitute $i^2 = -1$ when appropriate. Let us first practice putting complex numbers in standard form.

11. Express $\sqrt{-9}$ in the standard form $a + bi$.

$$\sqrt{-9} = \sqrt{9(-1)} = \sqrt{9}\sqrt{-1} = 3\_\_\_\_$$

Therefore $\sqrt{-9} = 0 + \_\_\_\_ i$ in standard form.

i

3

12. Express $83 + \sqrt{-25}$ in the standard form. Since

$$\sqrt{-25} + \sqrt{25(\_\_\_\_)} = \sqrt{25}\sqrt{\_\_\_\_} = 5\_\_\_\_$$

then $83 + \sqrt{-25} = 83 + \_\_\_\_$ in standard form.

-1; -1; i

$5i$

13. Addition and subtraction:

To add (or subtract) two complex numbers we just add (or subtract) the real and imaginary parts separately.

add real parts

$$(8 + 3i) + (7 + 2i) = (8 + \_\_\_\_) + (3 + \_\_\_\_)i$$

add imaginary parts

$$= 15 + \_\_\_\_ i$$

7; 2

5

14. Compute:

$$(12 + 6i) + (5 - 2i) = (12 + \_\_\_\_) + (6 + \_\_\_\_)i$$

$$= \_\_\_\_ + \_\_\_\_ i$$

5; (-2)

17; 4

15. Compute:

$$(2 + 9i) - (1 + 6i) = (2 - \_\_\_\_) + (\_\_\_\_ - 6)i$$

$$= \_\_\_\_ + \_\_\_\_ i$$

1; 9

1; 3

In this example we (added\subtracted) the real parts and (added\subtracted) the imaginary parts.

subtracted; subtracted

	16. Add $8 + \sqrt{-36}$ to $-5 + 2i$. First we must convert $8 + \sqrt{-36}$ to standard form.
$-1; -1; i$	$\sqrt{-36} = \sqrt{36(\_\_\_\_)} = \sqrt{36}\sqrt{\_\_\_\_} = 6\_\_\_\_$
$6i$	So $8 + \sqrt{-36} = 8 + \_\_\_\_$, and
2	$(8 + 6i) + (-5 + 2i) = (8 - 5) + (6 + \_\_\_\_)i$
$3; 8$	$= \_\_\_\_ + \_\_\_\_i$
	17. Subtract $12 + 2i$ from $\sqrt{-81}$. Again we must first write $\sqrt{-81}$ in standard form.
$-1; 9$	$\sqrt{-81} = \sqrt{81}\sqrt{\_\_\_\_} = \_\_\_\_i$
$0; 9i$	So in standard form, $a + bi$, we have $\sqrt{-81} = \_\_\_\_ + \_\_\_\_$. Our subtraction problem is thus
$12; 2$	$(0 + 9i) - (12 + 2i) = (0 - \_\_\_\_) + (9 - \_\_\_\_)i$
$-12 + 7i$	$= \_\_\_\_\_\_\_\_$
	18. Multiplication: To form the product of two complex numbers we multiply them according to the usual rules of algebra and then substitute $i^2 = -1$.
	$(2 + i)(3 + 5i) = 6 + 3i + 10i + 5i^2$
(-5)	$= 6 + 13i + \_\_\_\_$ (use $i^2 = -1$)
$1; 13$	$= \_\_\_\_ + \_\_\_\_i$
	19. Multiply $4 - i$ and $1 + 3i$.
	$(4 - i)(1 + 3i) = 4 + 12i - i - 3i^2$
11	$= 4 + \_\_\_\_i - 3i^2$
-3	$= 4 + 11i - (\_\_\_\_)$
3	$= 4 + 11i + \_\_\_\_$
$7 + 11i$	$= \_\_\_\_\_\_\_\_$
	20. Multiply $3 + \sqrt{-49}$ and $1 - 2i$. First we must convert $3 + \sqrt{-49}$ to the standard form $a + bi$. Since
$-1; 7; 3 + 7i$	$\sqrt{-49} = \sqrt{49}\sqrt{\_\_\_\_} = \_\_\_\_i$, then $3 + \sqrt{-49} = \_\_\_\_\_\_\_\_$
	Now we can do the multiplication.
14	$(3 + 7i)(1 - 2i) = 3 - 6i + 7i - \_\_\_\_i^2$
$1; 14$	$= 3 + \_\_\_\_i - \_\_\_\_i^2$
$1; 14$	$= 3 + \_\_\_\_i + \_\_\_\_$
-1	and since $i^2 = \_\_\_\_$, therefore
$17 + i$	$(3 + \sqrt{-49})(1 - 2i) = \_\_\_\_\_\_\_\_$

PRACTICE PROBLEMS

Simplify and write in the form bi (b is a real number; $i^2 = -1$).

1. $\sqrt{-1}$ **2.** $\sqrt{-36}$ **3.** $\sqrt{-49}$

4. $\sqrt{-7}$ **5.** $\sqrt{-81}$ **6.** $\sqrt{-20}$

7. $\sqrt{-x^2}$ $(x > 0)$ **8.** $\sqrt{-a^2b}$ $(a > 0, b > 0)$ **9.** $\sqrt{-ab^2}$ $(a > 0, b > 0)$

Simplify and write in the standard form of a complex number.

10. 3 11. $4i$ 12. i^2

13. i^3 14. $2 + \sqrt{-4}$ 15. $3 - \sqrt{-9}$

16. $(2 + 3i) + (4 + 5i)$ 17. $(4 + 2i) + (3 + 7i)$

18. $(2 - 4i) + (9 + 5i)$ 19. $(6 + 3i) + (7 - 8i)$

20. $(4 + 8i) - (3 + 2i)$ 21. $(7 + 5i) - (3 + i)$

22. $(2 + 7i) - (4 + 3i)$ 23. $(6 + 3i) - (4 + 8i)$

24. $(9 - 5i) + (8 + 3i)$ 25. $(7 - 3i) - (2 + 4i)$

26. $(2 + 3i)(1 + 4i)$ 27. $(6 + 4i)(3 + 2i)$

28. $(4 - 2i)(3 + 2i)$ 29. $(5 - 4i)(2 - i)$

30. $(4 + \sqrt{-4}) + (3 + 2i)$ 31. $(5 + \sqrt{-9}) - (4 + \sqrt{-4})$

32. $(3 + \sqrt{-16})(2 + \sqrt{-25})$ 33. $(7 + \sqrt{-4})(3 + \sqrt{-1})$

34. Is every real number also a complex number?

35. Is every pure imaginary number also a complex number?

36. Is every complex number a real number?

Simplify and write in the standard form of a complex number.

37. $(5 - i) \div (1 - i)$ 38. $2 \div (1 + i)$

39. $(7 - 11i) \div (4 + i)$ 40. $(4 + 17i) \div (6 - 5i)$

SELFTEST

1. Which of the following are real, pure imaginary, or complex numbers in standard complex form?

   a. i *b.* $8i$ *c.* $3 + i$ *d.* $\sqrt{9}$

   e. $3 + 2i$ *f.* $\frac{1}{2} + \frac{2}{3}i$ *g.* $\sqrt{25}$ *h.* $4 - \sqrt{16}$

Simplify and write in standard form for a complex number.

2. $\sqrt{-36}$ 3. 7 4. $5 + \sqrt{-16}$

5. $(4 + 2i) + (6 - 5i)$ 6. $(8 + 7i) - (4 + i)$

7. $(5 + 2i)(3 - 6i)$ 8. $(2 + \sqrt{-9})(3 + i)$

SELFTEST ANSWERS

1. *a.* pure imaginary *b.* pure imaginary *c.* complex in standard form *d.* real *e.* complex in standard form *f.* complex in standard form *g.* real *h.* real

2. $6i$ or $0 + 6i$ in standard form 3. $7 + 0i$ 4. $5 + 4i$ 5. $10 - 3i$

6. $4 + 6i$ 7. $27 - 24i$ 8. $3 + 11i$

POSTTEST 8

1. *a.* For every point on the number line is there a corresponding real number?
   b. Are all these numbers integers or fractions?
   c. Are there real numbers besides integers and fractions?
   d. If so, give an example.
   e. What are they called?

2. In the notation $\sqrt[3]{85}$ what is the index and what is the radicand?

3. Answer true or false:

   a. $\sqrt{0} = 0$

   b. $\sqrt[3]{27} = 3$

   c. $\sqrt{-1} = 1$

   d. $\sqrt[3]{-8} = 2$

   e. $\sqrt{25} = 5$

   f. $\sqrt{1{,}600} = 1.6$

4. Simplify without using tables: (assume that x and y are positive)

   a. $\sqrt{x^2}$

   b. $\sqrt{16y^4}$

   c. $\sqrt{50}$

   d. $\sqrt{x^3y^5}$

   e. $\sqrt{400}$

5. Simplify: (assume x and y are positive)

   a. $6\sqrt{5} - 2\sqrt{5}$

   b. $2\sqrt{x^3} + 3x\sqrt{x}$

   c. $8\sqrt{x^3y} - 2x\sqrt{xy}$

6. Simplify: (assume x and y are positive)

   a. $\dfrac{\sqrt{8}}{\sqrt{2}}$

   b. $\dfrac{x}{6\sqrt{5}}$ rationalize the denominator

   c. $(1 + 2\sqrt{y})(1 - 2\sqrt{y})$

7. Reduce:

   a. $\dfrac{3 + 6\sqrt{2}}{9}$

   b. $\dfrac{4 + \sqrt{8}}{10}$

8. Compute and leave your answer in the form $a + bi$:

a. $(10 - i) + (3 + 2i)$

b. $(4 + \sqrt{-1}) + (2 + \sqrt{-9})$

c. $(1 + i)(3 - 4i)$

Each problem above refers to a section in this chapter as shown in the table.

Problems	*Section*
1	8.1
2 and 3	8.2
4	8.3
5	8.4
6	8.5
7	8.6
8	8.7

POSTTEST 8 ANSWERS

1. *a.* yes *b.* no *c.* yes *d.* π *e.* irrational **2.** index is 3, radicand is 85

3. *a.* T *b.* T *c.* F *d.* F *e.* T *f.* F **4.** *a.* x *b.* $4y^2$ *c.* $5\sqrt{2}$ *d.* $xy^2\sqrt{xy}$ *e.* 20

5. *a.* $4\sqrt{5}$ *b.* $5x\sqrt{x}$ *c.* $6x\sqrt{xy}$ **6.** *a.* 2 *b.* $(x\sqrt{5})/30$ *c.* $1 - 4y$

7. *a.* $(1 + 2\sqrt{2})/3$ *b.* $(2 + \sqrt{2})/5$ **8.** *a.* $13 + i$ *b.* $6 + 4i$ *c.* $7 - i$

FUNCTIONS, RECTANGULAR COORDINATES, AND GRAPHING

BASIC SKILLS

Upon completion of the indicated section the student should be able to:

9.1 *a.* Distinguish between a set of two elements and an ordered pair.
b. Decide if two ordered pairs are equal.
c. Fill in the missing element of an ordered pair $(x, \_\_)$ or $(\_\_, y)$ so that the pair satisfies a given equation in x and y.

9.2 *a.* Draw an xy-coordinate system for real numbers x and y, number the unit lengths, and label the origin.
b. Locate an ordered pair of real numbers on the xy plane.
c. Give the coordinates of any point indicated on the xy plane.
d. Find the regions or quadrants described by any combination of the following inequalities

$$x < 0, \quad x > 0, \quad y < 0, \quad y > 0$$

9.3 *a.* Give an example of a relation and write it as a set of ordered pairs.
b. Identify the domain and range of a relation.
c. Distinguish between a relation and a function.
d. Give an example of a function and write it as a set of ordered pairs.
e. Identify the domain and range of a function.

9.4 *a.* Graph equations that can be put in the form $y = mx + b$.
b. Graph equations of the type $y = b$ or $x = b$ for some real number b.

9.5 *a.* Find the slope of a line if two points on the line are given or if the equation is given.
b. Find the y intercept of a line either from the graph or the equation of the line.
c. Answer the question: Do all equations have graphs that are straight lines?

9.6 *a.* Graph regions of the plane corresponding to inequalities of the type

$$y < mx + b, \quad y > mx + b, \quad y \leq mx + b, \quad y \geq mx + b$$

b. Graph regions of the plane described by combinations of inequalities such as

$$x < a, \quad x \geq b, \quad y \leq c, \quad y > d, \quad \text{etc.}$$

for real numbers a, b, c, d.

PRETEST 9

1. If $(x, 3)$ is to satisfy $3x + 2y = -3$, find x.

2. Shade the portion of the xy plane described by the conditions $x > 0$ and $y < 0$.

3. Consider the set

$$\{(1, 5), (2, 8), (3, 4), (1, 2)\}$$

   a. Is this set a relation?
   b. Is it a function? Why?
   c. What are the domain and range?

4. Graph $y = 2x + 3$.

5. Graph $y = 3$.

6. Find the slope of the line through (2, 3) and (8, 5).

7. Graph $x + 2y = 8$.
   a. What is the slope?
   b. What is the y-intercept?

8. Graph the region $y \leq 2.5x - 4$.

9. Graph the region in which $x \geq 2$ and $y < -4$.

Each problem above refers to a section in this chapter as shown in the table.

Problems	*Section*
1	9.1
2	9.2
3	9.3
4 and 5	9.4
6 and 7	9.5
8 and 9	9.6

PRETEST 9 ANSWERS

1. -3

2.

3. *a*. yes *b*. No, (1, 5) and (1, 2) are both in the set.
   c. domain $= \{1, 2, 3\}$,
   range $= \{2, 4, 5, 8\}$

4.

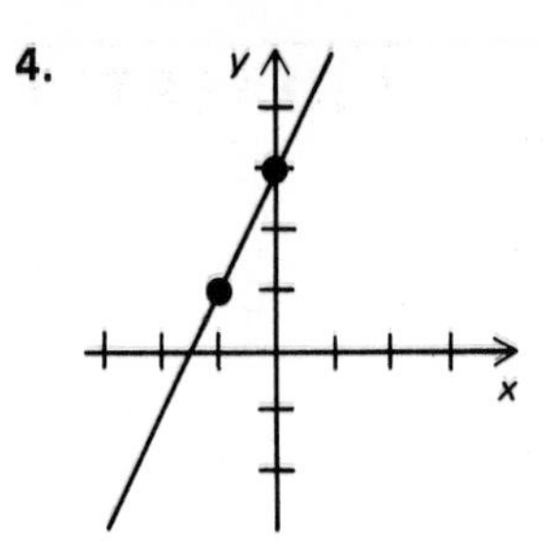

5.

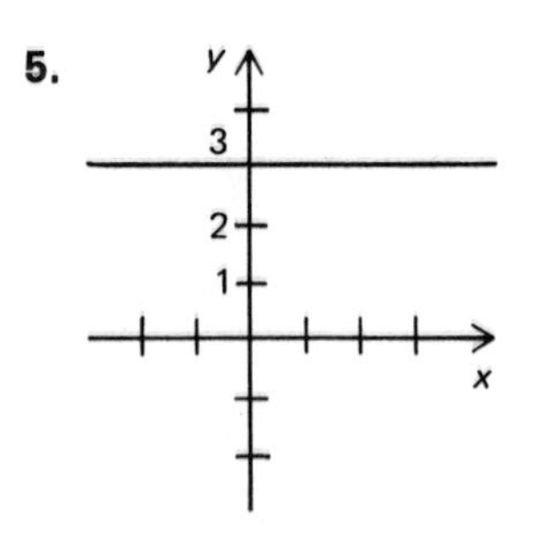

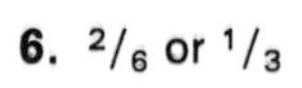

6. $^2/_6$ or $^1/_3$

7. *a.* $^-{}^1/_2$
b. (0, 4)

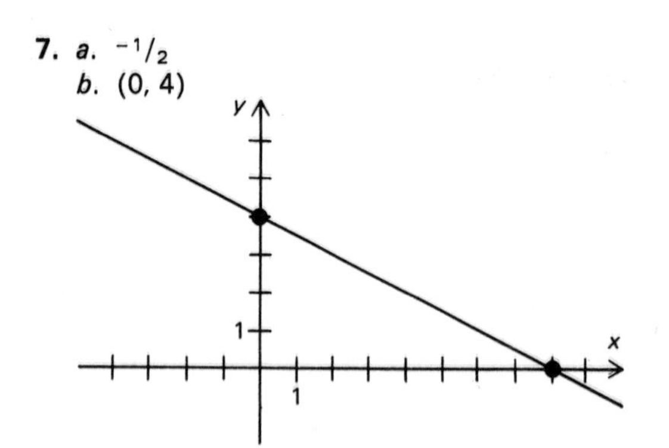

8.

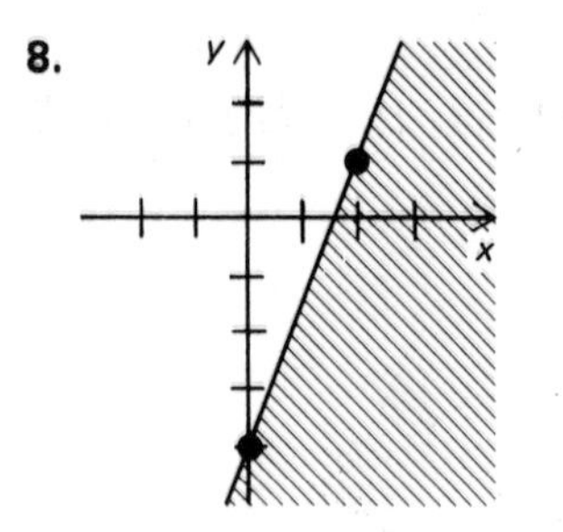

9.

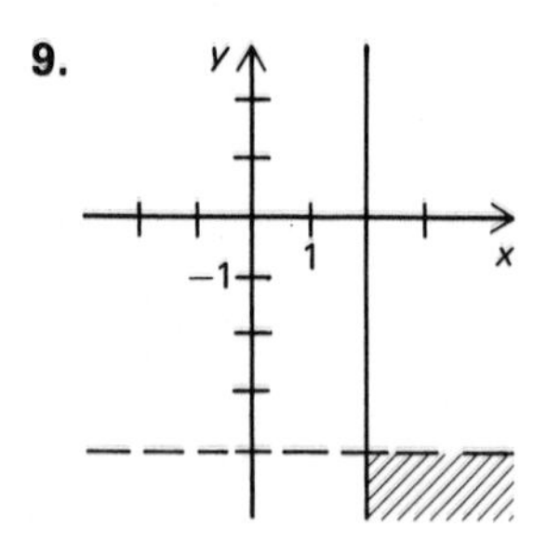

9.1 ORDERED PAIRS

In this section we will introduce a new concept called *ordered pairs*. Ordered pairs will be useful in later sections when we study the relationship between algebra and plane geometry.

When we list the members of a set, the order in which the members are listed does not matter. Therefore

$$\{x, y\} = \{y, x\} \qquad \text{and} \qquad \{a, b, c\} = \{a, c, b\} = \{b, c, a\} = \{c, b, a\}$$

Since ordered pairs are not the same as sets, we will not use the set notation for ordered pairs. Instead of using set braces { }, we use parentheses ( ) to denote ordered pairs. This convention is commonly used by most mathematicians.

Definition: **We call (a, b) the *ordered pair* whose first member is a and whose second member is b.**

The ordered pair whose first member is 10 and whose second member is 3 is (10, 3). The ordered pair whose first member is 3 and whose second member is 10 is (3, 10). It is especially important to note that the pairs (10, 3) and (3, 10) are not the same ordered pair. The reason is of course that the ordering of the members in each pair is different. Therefore (10, 3) is not the same as (3, 10). However, the set {3, 10} equals the set {10, 3} since in a set the ordering of the members does not matter.

The next definition tells us when two ordered pairs are equal.

Definition: **Two ordered pairs (a, b) and (x, y) are equal if and only if $a = x$ and $b = y$. That is, if and only if they have equal first members and equal second members.**

Example 1 $(7, 11) = (7, 11)$ but $(7, 11) \neq (11, 7)$.

Example 2 $(4, 3) \neq (4, 4)$ since the second members are not the same.

Example 3 $(10, x) = (10, 15)$ is true if and only if $x = 15$.

Example 4 $(a, 3) = (5, 3)$ is true if and only if $a = 5$.

Example 5 $(x, y) = (12, 19)$ is true if and only if $x = 12$ and $y = 19$.

Example 6

$$(2 + 2, 7 - 2) = (4, 5)$$

Example 7 $(6, -3) \neq \{6, -3\}$ since ordered pairs and sets are different.

Example 8 Given that $(x, y) = (\_\_, 8)$, find the missing first member of the ordered pair $(\_\_, 8)$ which will satisfy the equation $2x + 5y = 60$. The condition $(x, y) = (\_\_, 8)$ says that $y = 8$ and x is to be found. Since $y = 8$ and $2x + 5y = 60$, then $2x + 5 \cdot 8 = 60$. Therefore $2x + 40 = 60$. It follows that $x = 10$. Therefore $(x, y) = (10, 8)$. We say the ordered pair (10, 8) *satisfies* the equation $2x + 5y = 60$.

Example 9 Given $(x, y) = (1, \_\_)$, find the missing second member of the ordered pair $(1, \_\_)$ which will satisfy the equation $3x - y = 5$. The condition $(x, y) = (1, \_\_)$ says that $x = 1$ and y is to be found. Since $x = 1$ and $3x - y = 5$, then $3 \cdot 1 - y = 5$ and $y = -2$. Therefore $(x, y) = (1, -2)$. The ordered pair $(1, -2)$ satisfies the equation $3x - y = 5$.

PROGRAM 9.1

1. We call (a, b) the *ordered pair* whose first member is a and whose second member is b. So (5, 3) is an ordered pair. Its first member is 5 and its second member is 3. Likewise (8, 12) is an ________ ________. The first member of (8, 12) is ____ and the second member is ____.

ordered pair

8; 12

2. In ordered pairs, the order in which the members are written down makes a difference. The ordered pairs (3, 4) and (4, 3) are such that their members are written in reverse order. Therefore (3, 4) and (4, 3) (are\are not) equal.

are not

Because their members are written in ________ order.

reverse

3. Ordered pairs are different from sets. The sets {6, 15} and {15, 6} are the same since in a set the order of occurrence of the members does not matter. But the order of occurrence does matter for ordered pairs. Because sets and ordered pairs are different, we use a different notation for each of them. Parentheses denote ordered pairs, and braces denote sets. Thus {5, 17} is (a set\an ordered pair) and {5, 17} = {17, 5} is (true\false). But (5, 17) is (a set\an ordered pair) and (5, 17) = (17, 5) is (true\false).

a set; true

an ordered pair; false

4. The ordered pairs (1, 3) and (1, 2) are not equal since their (first\second) members are different. Also the ordered pairs (4, 6) and (0, 6) are not equal since their (first\second) members are different.

second

first

5. The ordered pairs (7, 2) and (7, 2) (are\are not) equal because both first members are ________ and both second members (are\are not) the same.

are

the same; are

6. The ordered pairs (5, 1) and (6, 1) (are\are not) equal because the first members (are\are not) the same.

are not

are not

7. In general $(x, y) = (a, b)$ if and only if $x = a$ and $y = b$. That is if and only if they have equal first members and equal second members. So $(x, y) = (9, 5)$ if and only if $x = 9$ and $y = 5$.

$(x, 3) = (12, 3)$ if and only if $x =$ ____

12

$(-7, y) = (-7, 0)$ if and only if $y =$ ____

0

$(x, y) = (10, 19)$ if and only if $x =$ ____ and $y =$ ____

10; 19

8. $(x, y) = (4, 4)$ if and only if both x and y are equal to ____. The statement $(x, y) = (y, x)$ is true (never/always/only when x and y are the same).

4

only when x and y are the same

9. $4x + y = 7$ is an equation in x and y. Let $(x, y) = (1, 3)$. Then $x =$ ____ and $y =$ ____. Putting these values of x and y back into the equation gives

1

3

$$4(1) + 3 = 7$$

So substituting the ordered pair $(x, y) = (1, 3)$ into the equation $4x + y = 7$ makes the equation (true\false).

true

10. Suppose we are given an equation involving x and y. Let $(x, y) = (a, b)$ so $x = a$ and $y = b$. If the values $x = a$ and $y = b$ make the equation true, we say that the ordered pair (a, b) *satisfies* the equation.

1; 2

To discover if the ordered pair (1, 2) satisfies the equation $x^2 + y^2 = 5$, let $(x, y) = (1, 2)$. Then $x =$ ____ and $y =$ ____. With these values of x and y substituted into the equation, we get

1; 2

$$x^2 + y^2 = 5 \quad \text{or} \quad (\_\_\_)^2 + (\_\_\_)^2 = 5$$

does

Therefore the ordered pair (does\does not) satisfy the equation $x^2 + y^2 = 5$.

2; 3

11. To find whether the ordered pair (2, 3) satisfies the equation $x^2 + y^2 = 5$, we set $(x, y) = (2, 3)$ to get $x =$ ____ and $y =$ ____. Substituting these values in $x^2 + y^2 = 5$ we get

2; 3

$$(\_\_\_)^2 + (\_\_\_)^2 = 5 \qquad \text{or} \qquad 13 = 5$$

no
false

Does the ordered pair (2, 3) satisfy the equation $x^2 + y^2 = 5$? (yes\no). Substituting $x = 2$ and $y = 3$ into the equation gives the (true/false) statement $13 = 5$.

−2

12. Given $(x, y) = (-2, \_\_\_)$, find the missing second member of the ordered pair $(-2, \_\_\_)$ which will satisfy the equation $4x + 7y = 6$. Since $(x, y) = (-2, \_\_\_)$, then $x =$ ____.

It is our job to find y. Substitute the above value of x into $4x + 7y = 6$. We get

$$4(-2) + 7y = 6$$

14

$$7y = \_\_\_$$

2

$$y = \_\_\_$$

2

Therefore the ordered pair $(-2, \_\_\_)$ satisfies the equation $4x + 7y = 6$.

13. Determine x so that the ordered pair $(x, 12)$ satisfies the equation

$$3x - y = 0$$

12

Since $(x, y) = (x, 12)$, then $y =$ ____. If $y = 12$, the equation $3x - y = 0$ becomes

12

$$3x - \_\_\_ = 0$$

12

$$3x = \_\_\_$$

4

$$x = \_\_\_$$

4

So the ordered pair $(x, y) = (\_\_\_, 12)$ satisfies the equation $3x - y = 0$.

PRACTICE PROBLEMS

1. Answer as true or false:

a. $\{1, 3\} = \{3, 1\}$ *b.* $(1, 3) = (3, 1)$ *c.* $(1, 3) = (1, 3)$ *d.* $(1, 3) = \{1, 3\}$
e. $(x, y) = (2, 5)$ only if $x = 2$ and $y = 5$ *f.* $(x, y) = (y, x)$ only if $x = y$

2. Answer as true or false:

a. (1, 0) satisfies $x^2 + y^2 = 1$
b. (3, 2) satisfies $y = 2x$
c. (−7, 10) satisfies $2y - 3x + 1 = 1$
d. (0, 0) satisfies $2x^2 - 3y + 1 = 1$
e. (0, 4) satisfies $y = 3x^2 - 2x + 4$
f. (1, 1) satisfies $x^3 + 2x^2 - 1 = y$

3. Given that $(x, 3)$ satisfies $2x + y = 5$, find x.

4. Given that $(x, 0)$ satisfies $y^2 + 2x + 1 = 9$, find x.

5. Given that $(10, y)$ satisfies $3x - y = 5$, find y.

6. Given that $(-8, y)$ satisfies $4x + 2y = 0$, find y.

SELFTEST

1. Answer as true or false:

a. $(3, 4) = (4, 3)$ *b.* $(5, 7) = \{5, 7\}$
c. $(8, 12) = (6 + 2, 12)$ *d.* $\{7, 2\} = \{2, 7\}$

2. Does $(1, 0)$ satisfy $x^2 + y = 1$?

3. If $(x, 3)$ satisfies $2x + 4y = 2$, find x.

SELFTEST ANSWERS

1. *a.* F *b.* F *c.* T *d.* T **2.** yes **3.** $x = -5$

9.2 RECTANGULAR COORDINATES

In this section we will show how ordered pairs of real numbers can be used to represent points in the plane.

Construct two perpendicular lines, one horizontal and the other vertical. The point of intersection of these lines is called the *origin*. Choose a convenient unit of length and mark this length on both lines. The positive units of length lie to the right of the origin on the horizontal line and above the origin on the vertical line. We complete the numbering of unit lengths on both lines and thereby obtain two perpendicular real-number lines.

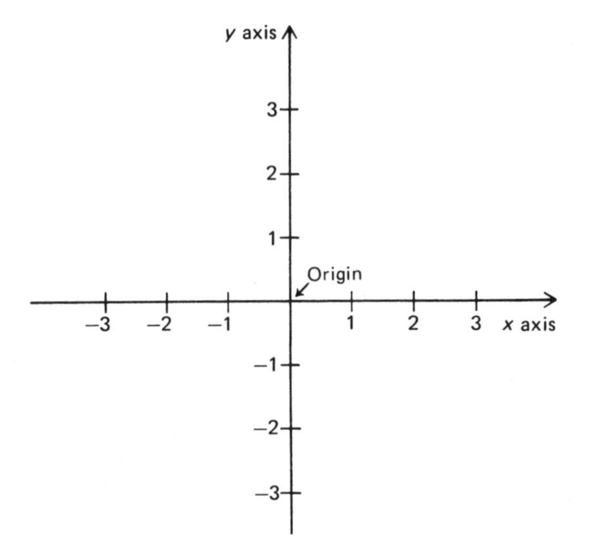

The horizontal line is called the *horizontal axis* or the *x axis*. The vertical line is called the *vertical axis* or the *y axis*. The plane which contains the x axis and the y axis is called the *xy plane*.

Consider an ordered pair (x, y) of real numbers. The first member, x, of the ordered pair (x, y) is called the *abscissa* or x coordinate. The second member, y, of the ordered pair (x, y) is called the *ordinate* or y coordinate.

Next we find the point in the xy plane which corresponds to the ordered pair (x, y). First the abscissa x is located on the x axis. Then a line L is drawn perpendicular to this axis at point x. The ordinate y is located on the y axis at point y. A line M is drawn perpendicular to this axis at point y. The intersection of the two lines L and M is the point on the plane which corresponds to the ordered pair (x, y).

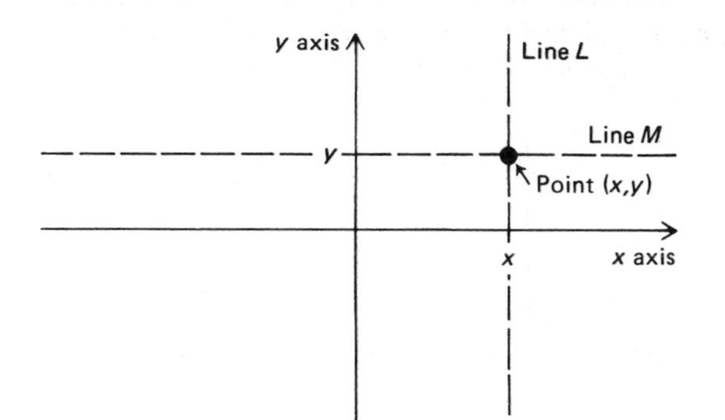

Thus we see that for every ordered pair (x, y) of real numbers there is a unique point on the xy plane which corresponds to (x, y). Conversely, corresponding to every point on the plane there is a uniquely determined ordered pair (x, y) of real numbers.

Example 1 Locate the point (1, 3) in the xy plane. First we locate the number 1 on the x axis. Then we draw a line L through this point on the x axis and perpendicular to the x axis. Next we locate 3 on the y axis. Draw a line M through this point and perpendicular to the y axis. The point (1, 3) is located at the intersection of L and M.

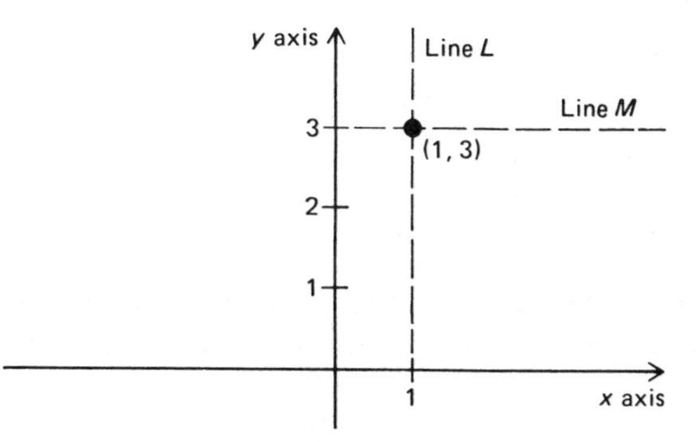

Example 2 Locate each of the following points on the xy plane.

$$(-2, 4),\ (-6, -2),\ (4, -3),\ (0, {}^1/_2),\ (3, 3),\ (-4, 0),\ (0, 0)$$

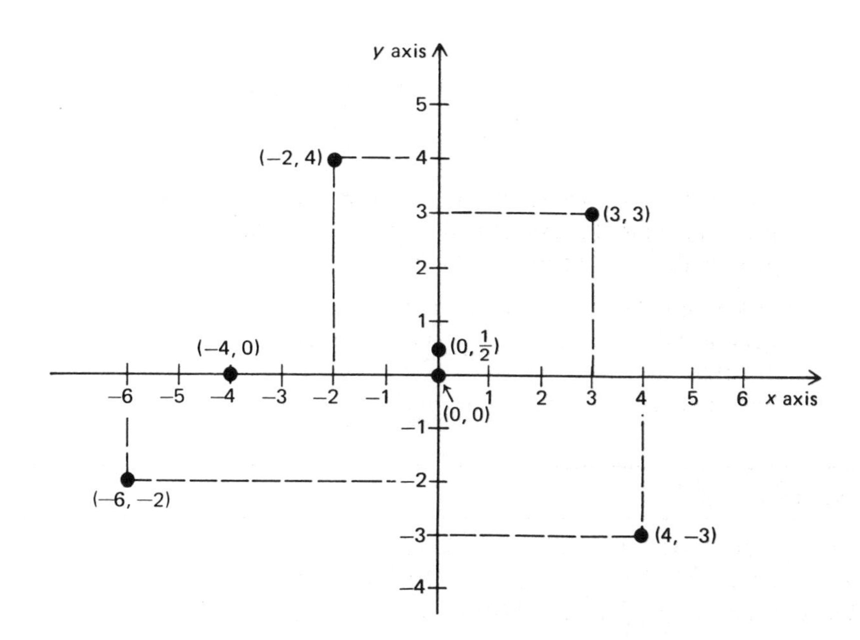

Example 3 Describe that portion of the xy plane for which $y > 0$. The condition $y > 0$ is satisfied by all points (x, y) for which the y coordinate is positive. This region is shaded in the figure.

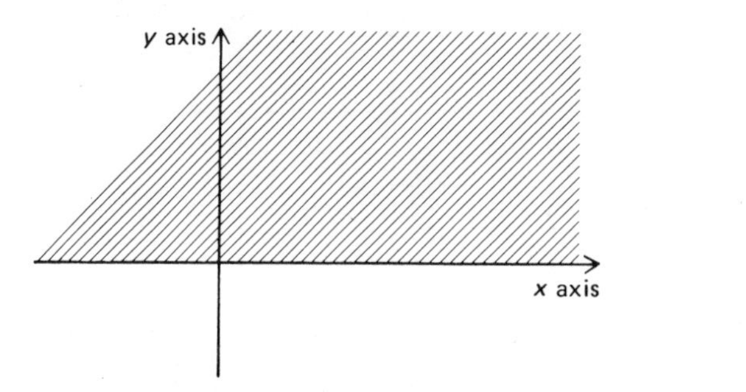

Example 4 Describe that portion of the xy plane for which $x > 0$ and $y > 0$. The condition $x > 0$ and $y > 0$ is satisfied by all points (x, y) for which both the x and y coordinates are positive. The region is shaded in the figure.

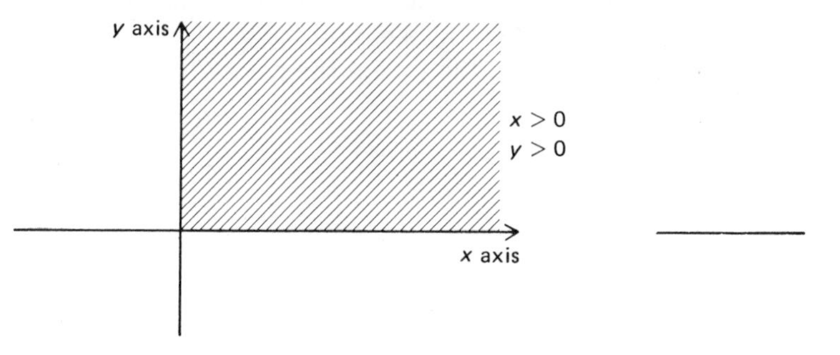

Example 5 The xy plane is divided into four regions by the x axis and y axis. These regions are called quadrants I to IV and are labeled as shown here.

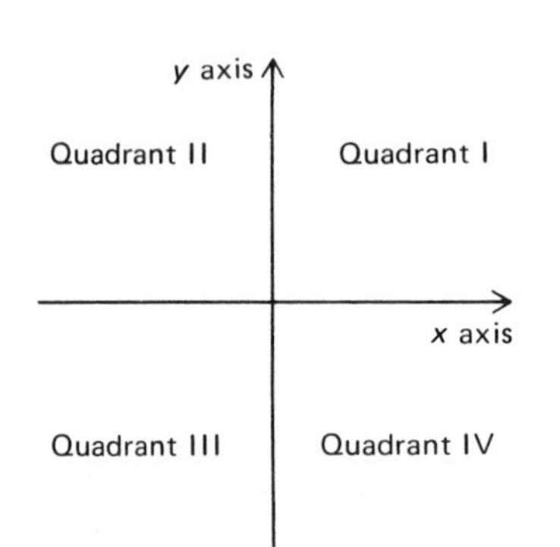

For each of the quadrants find a pair of inequalities that can be used to describe that quadrant. In Example 4, the first quadrant can be represented by the inequalities $x > 0$ and $y > 0$. Thus

quadrant I	$x > 0$	and	$y > 0$
quadrant II	$x < 0$	and	$y > 0$
quadrant III	$x < 0$	and	$y < 0$
quadrant IV	$x > 0$	and	$y < 0$

PROGRAM 9.2

1. A plane is a flat surface which extends indefinitely. Consider the plane that contains the piece of paper on which we are working. Draw two perpendicular lines on this plane. Make one line horizontal (level) and the other vertical

(straight up and down). Let us call the point of intersection of these lines the *origin*. Label the origin in this diagram.

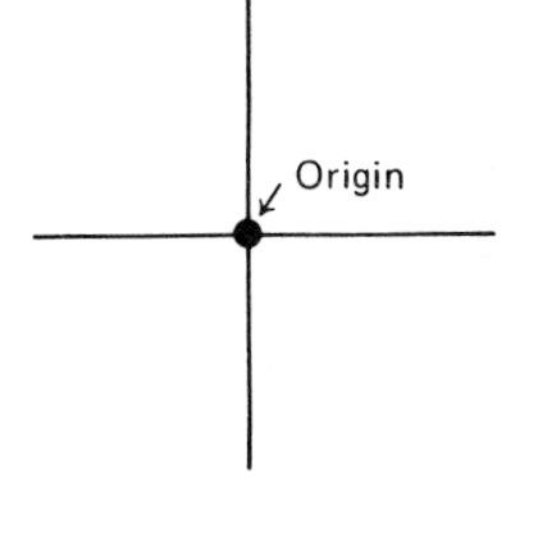

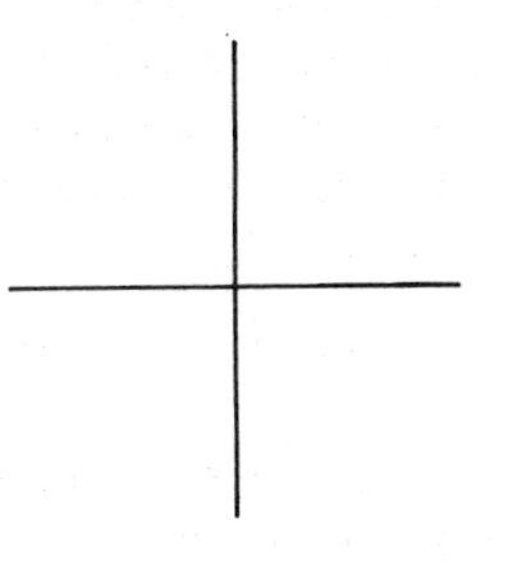

2. In the above diagram, one line is horizontal. This horizontal line is called the x axis. The vertical line is called the y axis. Label the x axis and the y axis.

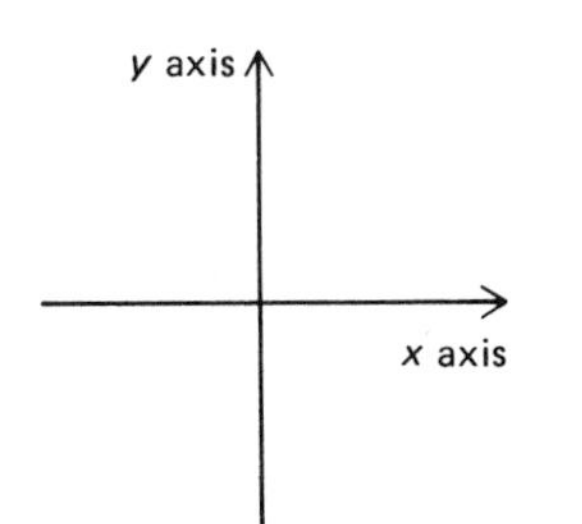

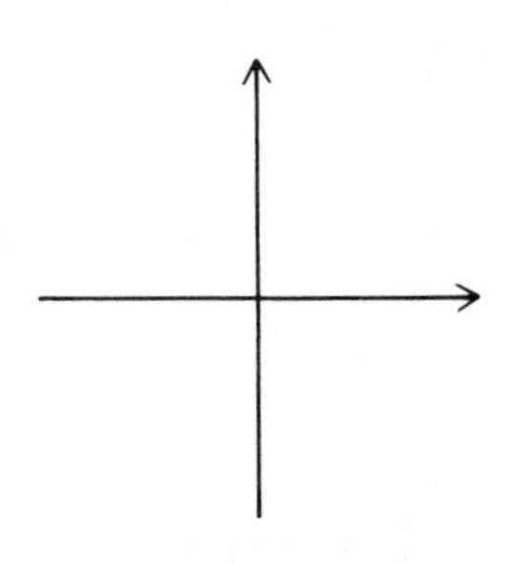

3. We choose a convenient unit length and construct a number line on the x axis. Positive numbers are located to the right of the origin. Negative numbers are to the left of the origin. The origin is the point which corresponds to 0. Locate the numbers -3, -2, $-1/2$, 0, 2, $5/2$, 3, and 4 on the x axis below. The number 1 is already marked.

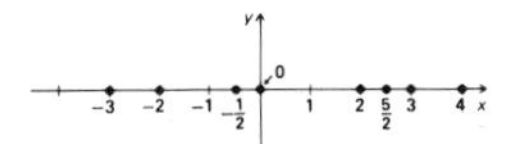

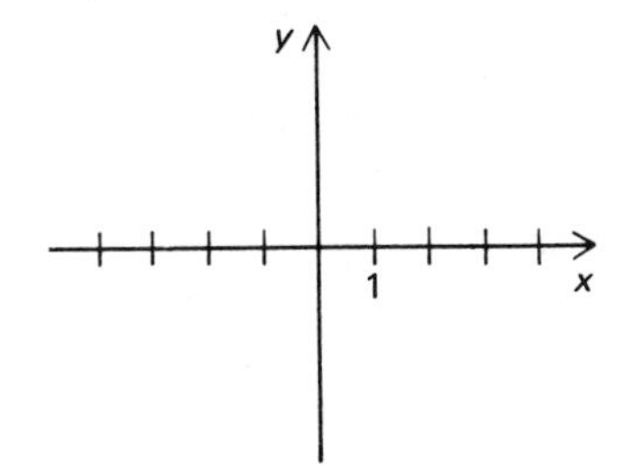

4. Now with the same unit length we used for the x axis we construct a number line on the y axis. Positive numbers are located above the origin; negative numbers are located below the origin; and the origin corresponds to 0. Locate -4, -2, -1, 0, 2, and 2.5 on the y axis.

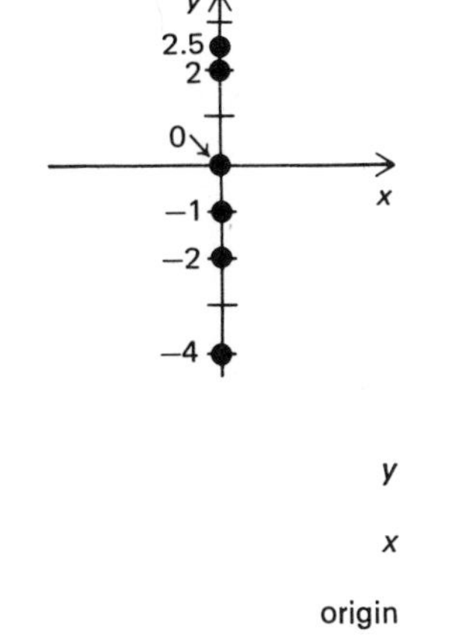

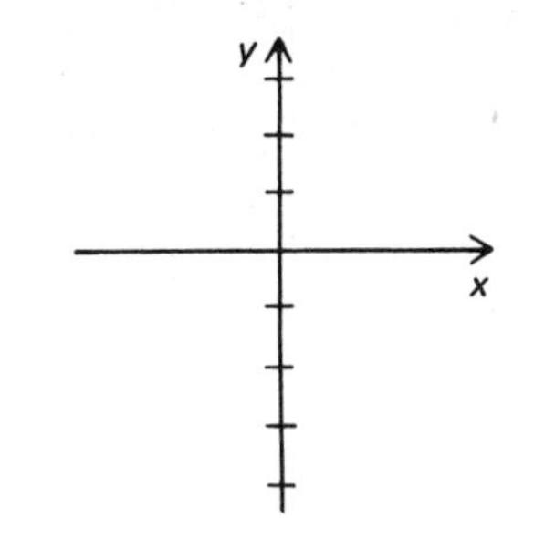

y

x

origin

5. We have constructed two number lines. The vertical line is called the _____ axis. The horizontal line is the _____ axis. The point at which these two lines meet is called the _________.

6. Whenever two distinct lines in space intersect, these lines determine a unique plane. It is the plane which contains the two lines. The x axis and y axis determine a plane which we call the *xy-coordinate plane* or more briefly the xy plane. Now suppose you are already given a plane, for example the plane of a chalk board. Is it possible to make this plane into an xy coordinate plane? (yes it is/not it isn't). We can draw perpendicular lines in the plane. We call one the x axis and the other the ________.

yes it is

y axis

7. Consider the points P and Q in the diagram below. We can describe the location of each with respect to the origin. If we begin at the origin and go a horizontal distance of 7 units and then a vertical distance of 5 units we reach P. To reach Q from the origin, we go a horizontal distance of ____ units and then a vertical distance of ____ units.

4

3

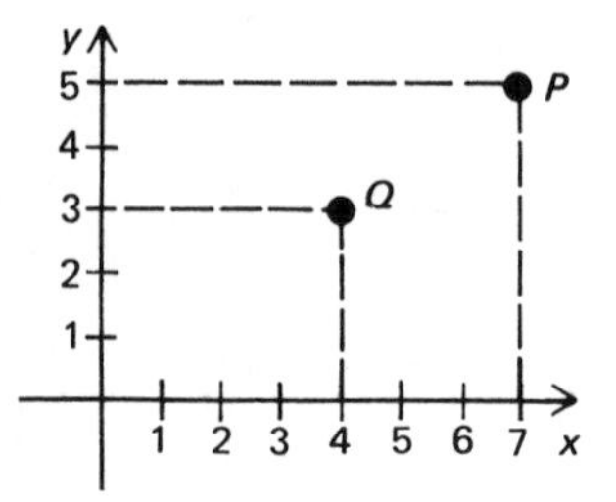

8. The location of the point P on the xy plane may be concisely written by use of the ordered-pair notation (a, b). The first member is the x coordinate or *abscissa*. It tells how to move horizontally from the origin. The second member is the y coordinate or *ordinate*. It tells us how to move vertically.

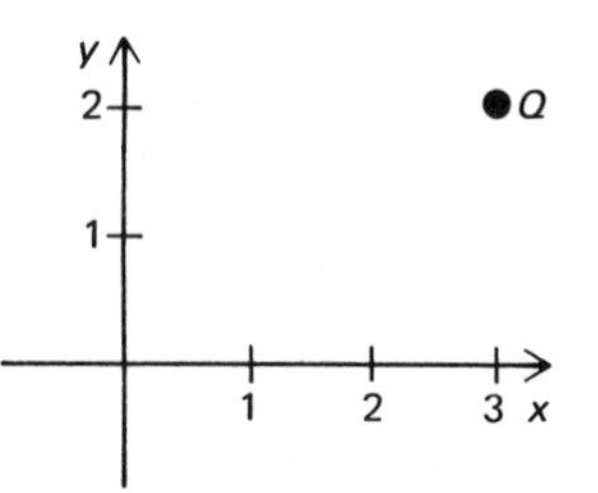

The point Q may be described by the ordered pair (3, 2) because from the origin we move ____ units horizontally and ____ units vertically to reach Q. The ordinate is ____ and the abscissa is ____.

3; 2

2; 3

9. The point R below is described by the ordered pair (____, ____). The x coordinate is ____ and the y coordinate is ____.

5; 3

5; 3

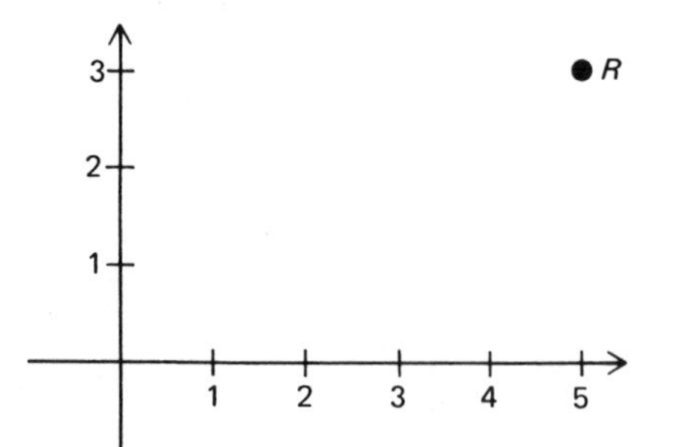

2

5

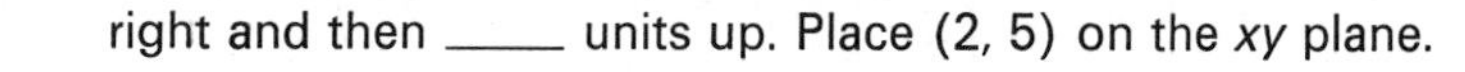

10. To find the point (2, 5), we begin at the origin and move ____ units to the right and then ____ units up. Place (2, 5) on the *xy* plane.

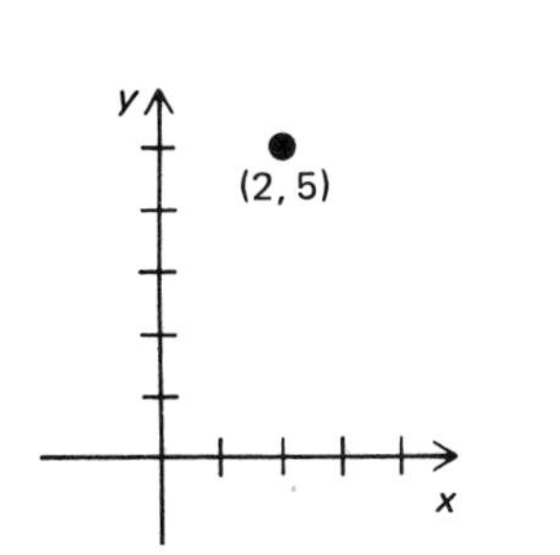

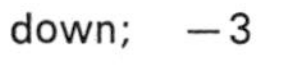

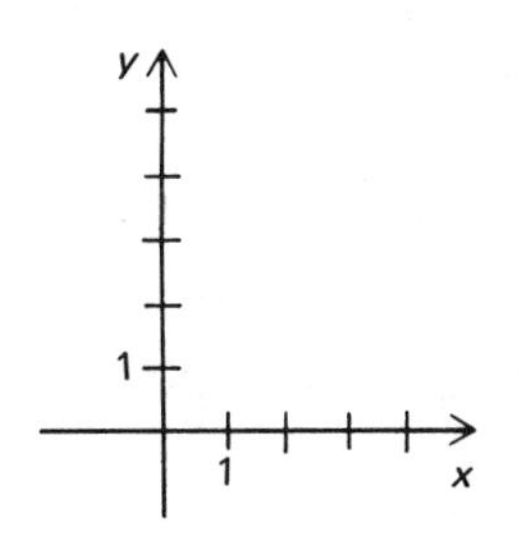

down; -3

11. Let's locate the point $(-2, -3)$ on the *xy* plane. The *x* coordinate -2 tells us to move 2 units horizontally. The negative sign of -2 tells us to move left. Then we move (up\down) 3 units because the *y* coordinate is ____. Now plot the point $(-2, -3)$.

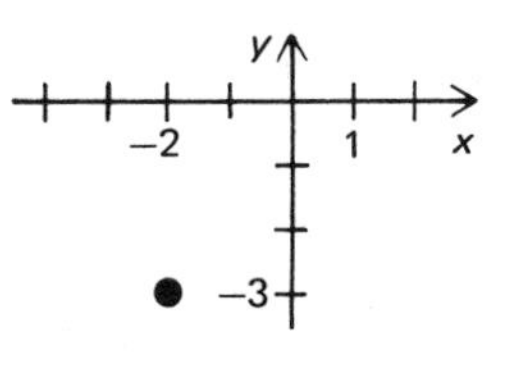

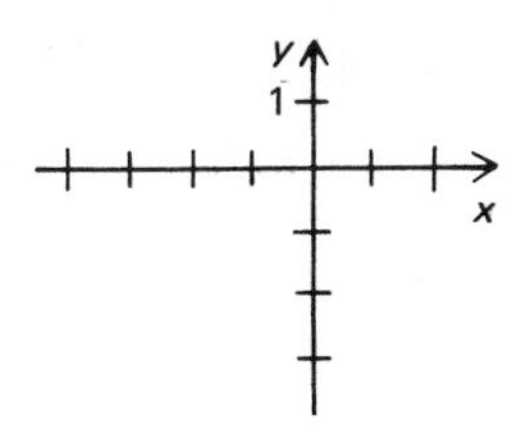

12. Locate these points on the *xy* plane above.

a. (4, 2) *b.* $(-3, -5)$ *c.* $(1, -2)$ *d.* $(-6, 3)$

(*b.* -2, 1)
c. $(-1, -4)$
d. $(5, -2)$

13. Give the coordinates (abscissa first and then the ordinate) of the points in this figure:

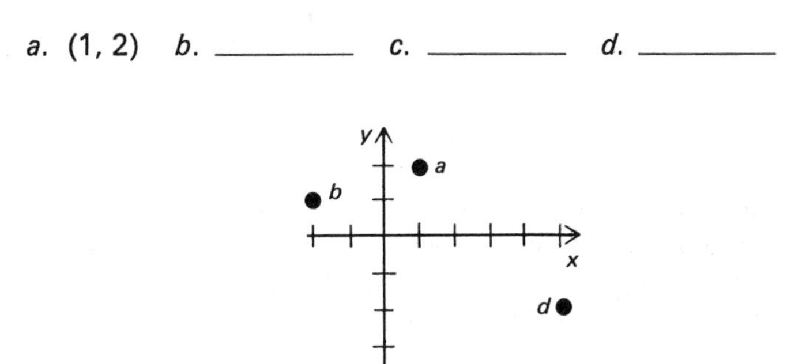

a. (1, 2) *b.* ________ *c.* ________ *d.* ________

0

14. The origin corresponds to the pair (0, 0). To locate (0, 0), we begin at the origin and move 0 units horizontally and ____ units vertically. Since we really didn't move at all the point (0, 0) remains at the origin.

15. Locate (0, 4) on the xy plane. To find (0, 4), we make no (horizontal\vertical) move. Then we move 4 units (vertically\horizontally).

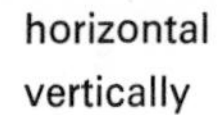
horizontal
vertically

16. Locate $(0, -3)$ and (5, 0) on the xy plane.

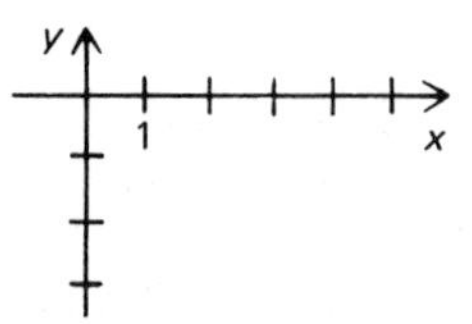

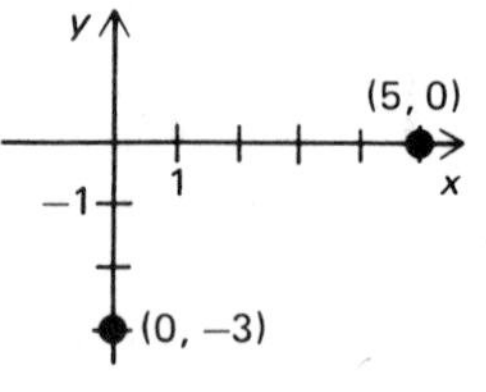

17. Not only can we plot points on the xy plane, we can also indicate entire regions in which all the points satisfy a specified condition. Let's consider that region of the xy plane for which $x > 0$. The condition $x > 0$ is satisfied by all points (x, y) for which x is (positive\negative\zero). The points with positive x coordinate lie to the (left\right) of the y axis. Shade the region for which $x > 0$ is true.

positive
right

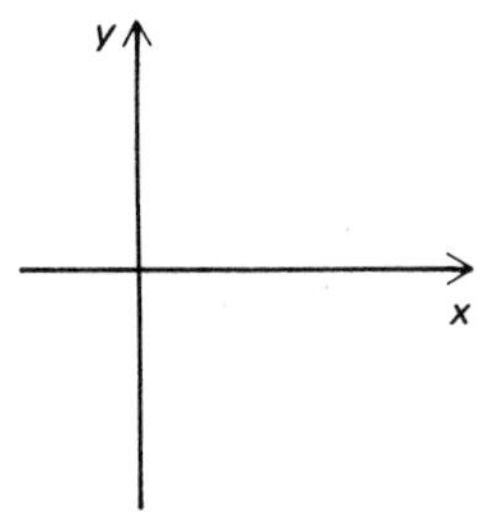

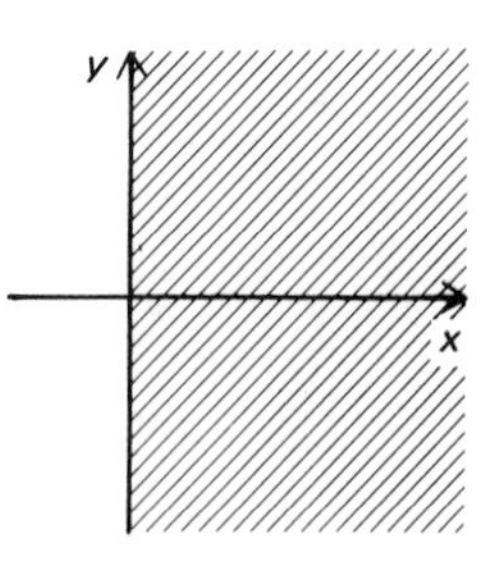

18. Consider that region for which $y < 0$ is true. A point (x, y) satisfies $y < 0$ if the y coordinate is (positive\negative\zero). The points with negative y coordinate lie (below\above\on) the x axis. Shade the region where $y < 0$.

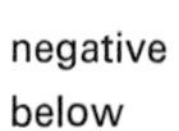
negative
below

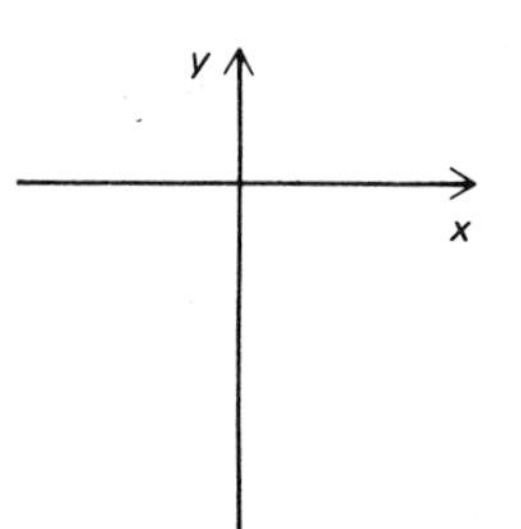

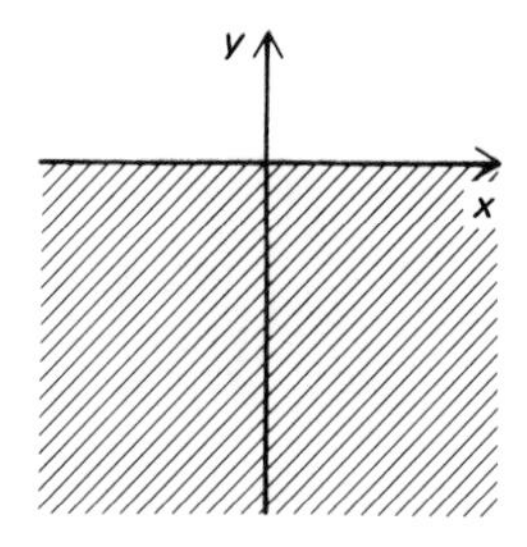

19. Consider the region in which both $x > 0$ and $y < 0$ are true. The region $x > 0$ and $y < 0$ were treated separately above. What is the region in which both $x > 0$ and $y < 0$ is true? It must be to the (left\right) of the y axis and (above\below) the x axis. Shade this region.

right
below

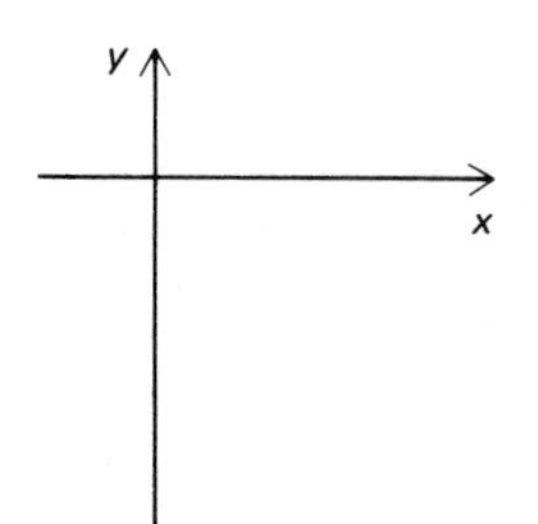

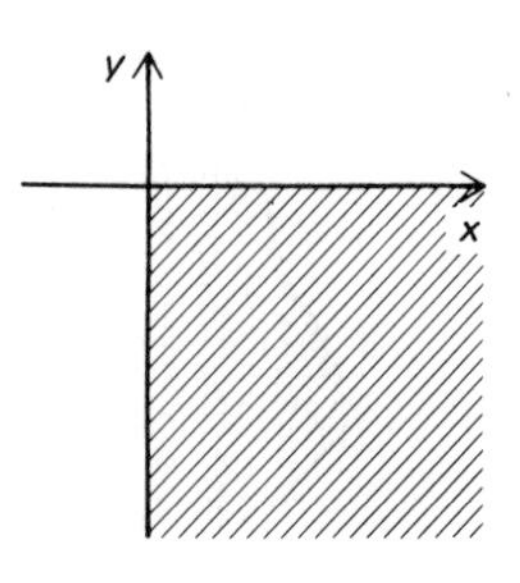

PRACTICE PROBLEMS

1. Suppose you are given an xy-axis system and an ordered pair (x, y) of real numbers. In your own words explain how you would locate the point corresponding to (x, y). Draw a diagram.

2. Suppose you are given an xy-axis system and a point P in the xy plane. In your own words explain how you would find an ordered pair (x, y) which corresponds to that point.

3. Answer as true or false:

a. The x and y axis intersect at the ordinate.
b. In the ordered pair (x, y), the abscissa is x.
c. The point $(0, 0)$ corresponds to the origin.
d. The point $(3, 0)$ lies on the x axis.
e. The point $(0, -2)$ lies on the y axis.
f. The point $(1, 1)$ lies in the first quadrant.
g. The point $(-1, -2)$ lies in the second quadrant.
h. The point $(1, -5)$ lies in the fourth quadrant.

4. Locate each of the following points on the xy plane. If the point lies on one of the axes, say which axis it lies on. Otherwise say which quadrant the point is in.

a. $(1, 3)$ *b.* $(7, -9)$ *c.* $(5.1, 4)$ *d.* $(-3.2, 6.9)$ *e.* $(-12, -15)$
f. $(0, 5)$ *g.* $(-1, 0)$ *h.* $(3, 3)$ *i.* $(4, 0)$ *j.* $(0, -2)$

5. Give the coordinates of each point indicated.

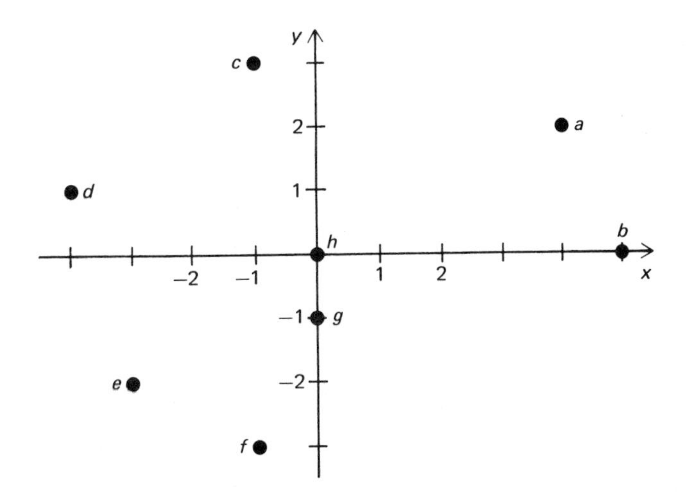

6. Shade that portion of the xy plane for which $y > 0$.

7. Shade that portion of the xy plane for which $x < 0$.

8. Shade that portion of the xy plane for which both $x < 0$ and $y > 0$.

SELFTEST

1. Locate these points on the xy plane.

a. $(3, 4)$ *b.* $(-2, 1)$ *c.* $(-1, -4)$ *d.* $(2, -3)$ *e.* origin

2. Give the coordinates of the points.

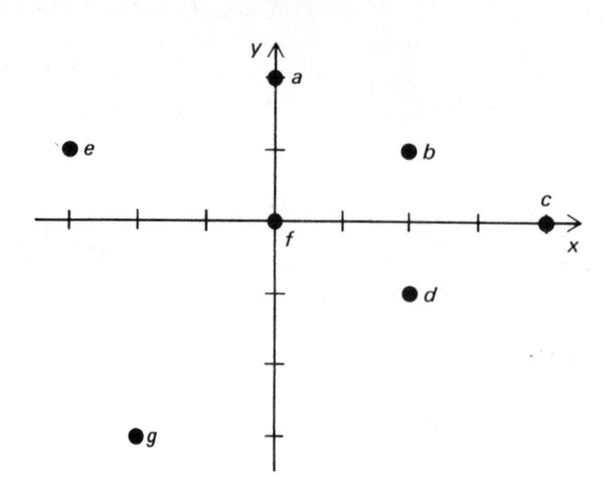

3. *a.* Shade the portion of the plane for which $x < 0$ and $y > 0$.
   b. Which quadrant is this?

SELFTEST ANSWERS

1.

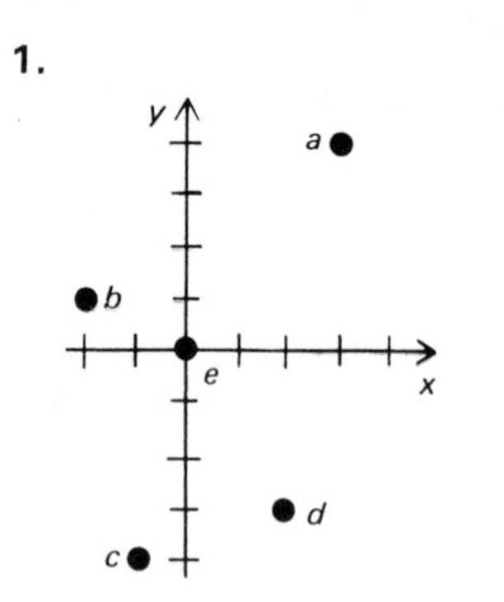

2. *a.* (0, 2) *b.* (2, 1) *c.* (4, 0) *d.* (2, −1) *e.* (−3, 1) *f.* (0, 0) *g.* (−2, −3)

3. *a.*

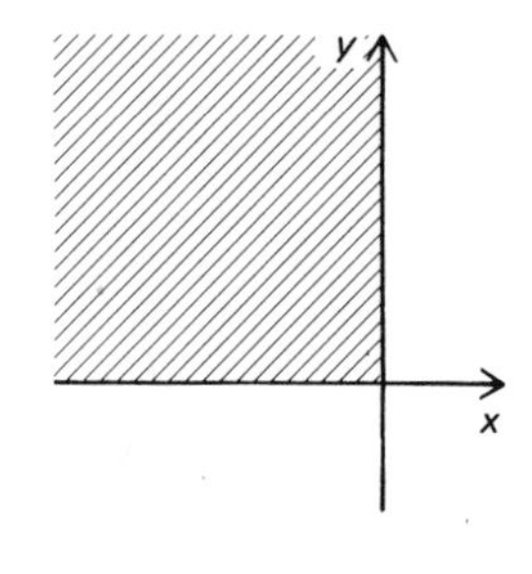

b. Quadrant II

9.3 RELATIONS AND FUNCTIONS

Consider the two sets A and B.

$$A = \{1, 2\} \qquad B = \{5, 6\}$$

There are several ways to pair the elements from set A with those from set B. We can use a set of ordered pairs (x, y) where $x \in A$ and $y \in B$ to describe the different pairings. Recall that the notation $x \in A$ means x is a member of A. We read $x \in A$ as "x belongs to A."

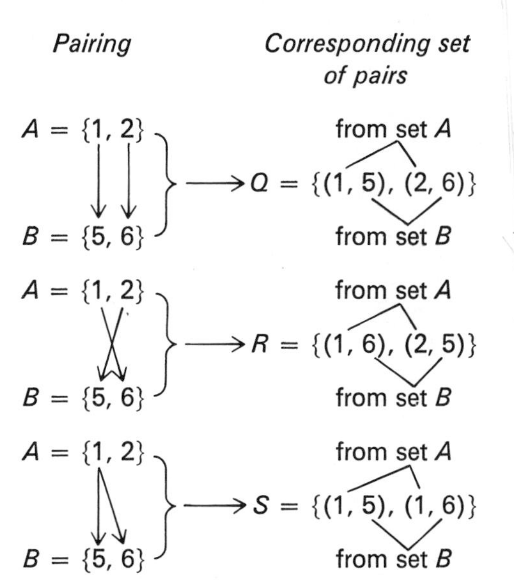

The sets Q, R, and S are all different, for the ordered pairs in the sets are different. But the sets Q, R, and S all have elements of the form (x, y), where x is a member of A and y is a member of B. The sets Q, R, and S are all examples of relations on A and B.

Definition: **Let A and B be two sets. Any set of ordered pairs (x, y) such that x is a member of the set A and y is a member of the set B is called a *relation on A and B*.**

Whenever we have two nonempty sets, we can form a relation. The relation will tell us exactly how to pair the elements of the first set with the elements of the second set. Associated with every relation are two special sets. The *domain* of a relation is the set of all first members of the ordered pairs of the relation. The *range* is the set of all second members of the ordered pairs of the relation.

Notice that if a relation is defined from a set A to a set B, the domain is always a subset of A and the range is always a subset of B. The domain is the part of A that is *actually used* as first elements in the relation. Likewise, the range is that part of B that is *actually used* as second elements in the relation.

Example 1 Let $A = \{1, 2, 3\}$ and $B = \{2, 4, 6\}$. Let's form a relation R on A and B such that a member x from A is "related to" or "paired with" a member y from B if and only if $y = 2x$.

$$R = \{(x, y) \mid x \in A, y \in B, \text{ and } y = 2x\}$$

Another way to show R is to actually list the pairs in the relation.

elements from A
form the *domain*

$$R = \{(1, 2), (2, 4), (3, 6)\}$$

elements from B
form the *range*

Since we used all the elements from A as first elements in the ordered pairs of R, the domain is A. The range is the set B, since we actually used all the elements of B as second elements.

Example 2 Let $A = \{1, 2, 3\}$ and $B = \{1, 4, 9, 16\}$. This time form a relation Q on A and B such that an element x from A is related to an element y of B if and only if $y = x^2$.

$$S = \{(x, y) \mid x \in A, y \in B, \text{ and } y = x^2\}$$

$$\text{domain} = A = \{1, 2, 3\}$$
$$S = \{(1, 1), (2, 4), (3, 9)\}$$
$$\text{range} = \{1, 4, 9\}$$

This time the range is not B. For 16 is an element of B and it was not used as a second element in any pair in S.

Example 3 Let A and B both be the set of counting numbers.

$$A = B = \{1, 2, 3, 4, 5, \ldots\}$$

The relation Q will consist of pairs (x, y) where $x \in A$ and $y \in B$ and $y = x + 1$.

$$Q = \{(1, 2), (2, 3), (3, 4), (4, 5), \ldots, (x, x + 1), \ldots\}$$

Notice that since $y = x + 1$, then $(x, y) = (x, x + 1)$.

The pair (2, 5) is not in Q since $5 \neq 2 + 1$. But the pair (100, 101) is in S. For $101 = 100 + 1$. Our domain is the set of counting numbers. The range of Q is the set of all counting numbers from 2 on. Why is 1 not included in the range?

Example 4 Let A be the set of all real numbers and let B be the set of all real numbers. Let us say that a number x of A is related to a number y of B if and only if $x > y$. This relation on A and B may also be written in ordered-pair form. Some of the (infinitely many) ordered pairs would be

$$\{(1.5, 0), (0, -3), (5, 1), (10, 0), (4, 3), (-3, -6), \ldots\}$$

The domain and range are both the set of real numbers.

Example 5 Let $A = \{1, 2, 3, 4, 5, 6, 7\}$ and $B = \{0, 2, 3, 4, \ldots\}$. Let us think of A as a listing of the days in a given week (Sunday corresponds to 1, Monday to 2, etc.). Let us think of the members of the set B as values representing the possible number of babies born on a certain day in a given week. The county clerk of Maui (in the state of Hawaii) is interested in the number of babies born in his county during a given week. Let us write this relationship as a set of ordered pairs (x, y). The first member x of the ordered pair indicates the day of the week. So $1 \leq x \leq 7$. The second member y indicates the number of babies born on day x. For a typical week the listing might look like

$$\{(1, 3), (2, 9), (3, 6), (4, 2), (5, 0), (6, 15), (7, 4)\}$$

The domain is the set of first elements in the pairs or

$$\{1, 2, 3, 4, 5, 6, 7\}$$

The range consists of the numbers used as second elements in the pairs of the relation.

$$\text{range} = \{3, 9, 6, 2, 0, 15, 4\}$$

In Examples 1–4, the relation was well expressed by a formula. For instance, the pairs (x, y) were in the relation of Example 1 if and only if $y = 2x$. In Example 2, pairs (x, y) were in the relation if and only if $y = x^2$.

Most relations discussed in this book will come from an underlying formula. In our treatment we will follow the usual custom and often treat the formula and the relation (remember a relation is a set of ordered pairs) as one and the same thing. The formula just describes how to pick the second element y for a given element x.

However, it is important to realize that not every relation can be expressed by a simple formula (or any formula at all). For example, what formula would you use for the "number of babies" relation of Example 5? What formula is there that connects time of day and temperature? These are examples of relations for which there is no concrete formula. In these cases, the best we can say is that the relationship is simply defined by the ordered pairs.

Next we will investigate an important relation of a special type. It is called a function.

Definition: **A *function* is a relation with the following property. For no member x of the domain are there two different members y and z of the range such that both (x, y) and (x, z) are part of the relation.**

This definition says that in a function, if the first members of two ordered pairs are the same, then the second members of the ordered pairs must also be the same. That is, if (x, y) and (x, z) are part of a function, then $y = z$.

Example 6 Which of the following relations represent functions?

a. $$\{(1, 1), (3, 0), (5, 2)\} = F$$

The relation F tells us how to draw arrows from domain elements to range elements. This represents a function because each domain element has only one arrow leading to a range element.

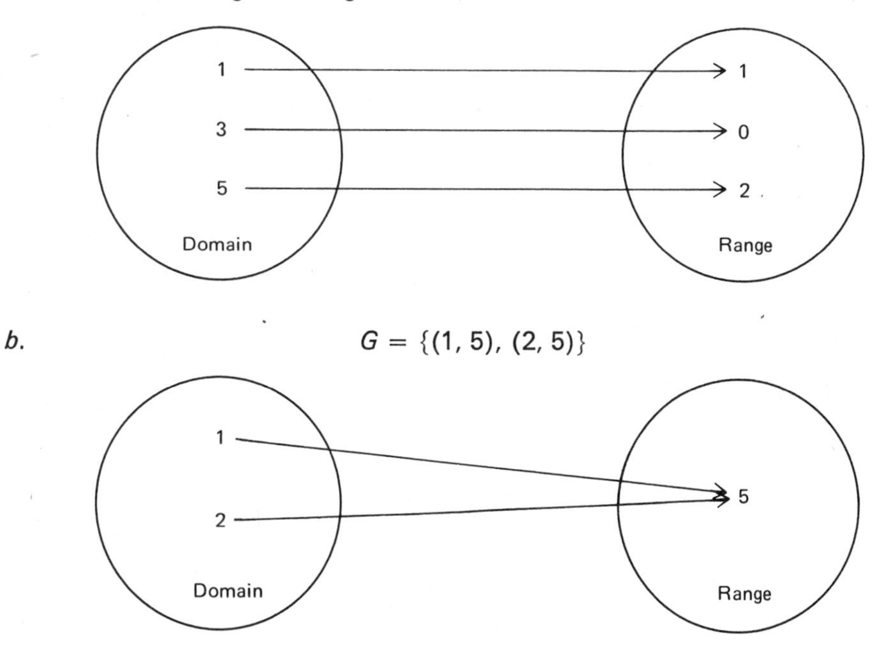

b. $$G = \{(1, 5), (2, 5)\}$$

The relation G is also a function because there is only one arrow from each domain element.

c. $$H = \{(1, 5), (1, 2), (3, 4)\}$$

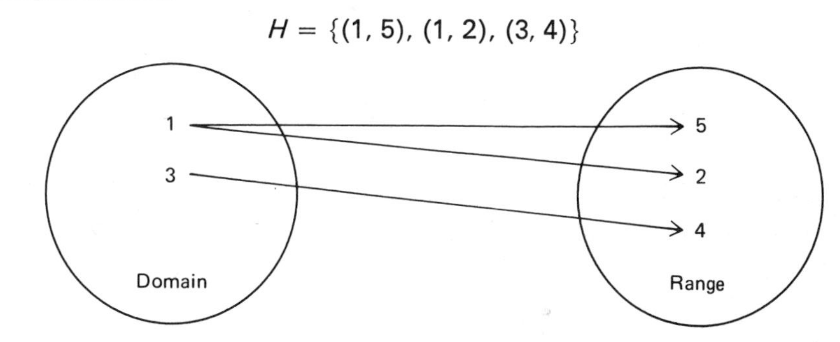

The relation H is *not* a function since the domain element 1 has two arrows which go to different range elements (the elements 5 and 2).

Let the ordered pair (x, y) be part of a relation. Then we sometimes say that y is a value of the relation at x. If the relation is not a function, there might be several y values corresponding to a given x in the domain. However, if the relation is in fact a function then for each x in the domain there is one and only one y value corresponding to x. So for a function we may say that y is *the* (unique) value of the function at x. For example, if our function contains the pair $(2, 3)$, we say that 3 is the value of the function at 2.

We bring this section to a close by describing the commonly used "functional notation."

We may think of a function as a special type of correspondence between the members of its domain and range. It is the ordered pairs that tell which members of the range correspond to specific members of the domain. If (x, y) is a typical ordered pair belonging to a function, then we say that y is *the value of the function at* x.

Since x is a typical member of the domain, we call x the *independent variable*. The y is now determined by the x we have selected. Consequently, we call y the *dependent variable*. To emphasize the fact that y depends on x and on the function, we use a special notation. Let's call our function f and let x represent a typical element of the domain. Then our range values y will be written as

$$y = f(x)$$

In this context the symbol $f(x)$ does not mean f times x. Instead the symbol $f(x)$ is used to represent the range value y that goes with x in the function f. The symbol $f(x)$ is read "f of x" or "the value of the function f at x." Thus we can think of our function as the set

$$f = \{(x, f(x)) \mid x \in \text{domain and } f(x) \in \text{range}\}$$

Example 7 Let $f = \{(1, 2), (3, 4), (5, 6)\}$, and compute $f(1)$, $f(3)$, and $f(5)$. We know that $f(1)$ means "f of 1," or the range value that goes with 1. Here, the range value 2 goes with 1. Thus $f(1) = 2$.

$f(3) = 4$ since the pair $(3, 4)$ is in f

$f(5) = 6$ since the pair $(5, 6)$ is in f

Sometimes, instead of listing the ordered pairs in a function, we just give the formula to compute the second values and the domain.

Example 8 Let h be the function with domain $= \{1, 2, 3, 4, 5\}$ and $h(x) = x + 3$. List the pairs in h.

Domain element x	*Corresponding range element* $h(x) = x + 3$
1	$h(1) = 1 + 3 = 4$
2	$h(2) = 2 + 3 = 5$
3	$h(3) = 3 + 3 = 6$
4	$h(4) = 4 + 3 = 7$
5	$h(5) = 5 + 3 = 8$

$$h = \{(1, 4), (2, 5), (3, 6), (4, 7), (5, 8)\}$$

The range is $\{4, 5, 6, 7, 8\}$.

PROGRAM 9.3

1. Let $A = \{3, 4\}$ and $B = \{8, 9\}$. There are several ways to pair elements from A to those in B. Below is one way.

$$A = \{3, 4\}$$

$$B = \{8, 9\}$$

8; 9	The number 3 is paired with ____ and 4 is paired with ____. We can use a set of ordered pairs instead of arrows to show the pairing.

from set A

8; 9 $Q = \{(3, \_\_), (4, \_\_)\}$

B from set ____

2. Here is another way to pair elements from A to those of B.

$$A = \{3, 4\}$$
$$B = \{8, 9\}$$

The set R of ordered pairs that describes this pairing is

A from set ____

9; 8 $R = \{(3, \_\_), (4, \_\_)\}$

B from set ____

3. Whenever we have any two nonempty sets, we can pair elements of one to elements of the other.

$$C = \{6, 7\}$$
$$D = \{8\}$$

Show the set Q of ordered pairs that describes the above relation or pairing.

8; 8 $Q = \{(6, \_\_), (7, \_\_)\}$

4. Let A and B be any two nonempty sets. Then any set of the form

$$\{(x, y) \mid x \text{ is from } A \text{ and } y \text{ is from } B\}$$

is called a *relation on A and B*. To form a relation on sets A and B, we use

a. ordered pairs (x, y) such that

A; B *b.* x is a member of set ____ and y is a member of set ____

5. Let $A = \{3, 7\}$ and $B = \{1, 5\}$. Which of the following are relations on A and B?

a. $\{(3, 1), (7, 5)\}$ *b.* $\{(3, 7), (1, 5)\}$

Only *a.* and *c.* *c.* $\{(3, 5), (7, 1)\}$ *d.* $\{3, 1, 7, 5\}$

sets In *b.*, the elements are from the wrong ____; and in *d.*, we do not have a

pairs set of ____.

6. Let $A = \{1, 2, 3\}$ and $B = \{6, 7, 8\}$. Give the relation consisting of pairs (x, y) such that $y = x + 5$. The pairs will be $(x, x + 5)$.

5; 5 $T = \{(1, 1 + 5), (2, 2 + \_\_), (3, 3 + \_\_)\}$

7; 8 $T = \{(1, 6), (2, \_\_), (3, \_\_)\}$

yes Is T the requested relation on A and B? (yes\no).

7. Let $A = \{2, 4, 6\}$ and $B = \{1, 3, 5\}$. Which of the following is the relation on A and B that has pairs (x, y) such that $y = x - 1$?

a. $\{(2, 3), (4, 1), (6, 5)\}$

b. *b.* $\{(2, 1), (4, 3), (6, 5)\}$

c. $\{(2, 5), (4, 1), (6, 3)\}$

8. Let $A = \{2, 3, 4\}$ and $B = \{1, 4, 9, 16\}$. Form the relation S that has pairs (x, y) such that $y = x^2$.

$$S = \{(2, 2^2), (3, (\_\_)^2), (4, (\_\_)^2)\}$$
$$\downarrow \downarrow \quad \downarrow \downarrow \quad \downarrow \downarrow$$
$$S = \{(2, 4), (3, \_\_), (4, \_\_)\}$$

3; 4

9; 16

Did we use all the elements of B in this relation? (yes\no). If not, which one was not used? ____.

no

1

9. There are two important sets associated with every relation.

Domain = the set of first elements of the pairs in the relation
Range = the set of second elements of the pairs in the relation

Let R be the relation $\{(0, 0), (1, 3), (2, 8), (4, 10)\}$. The domain of R is $\{0, 1, \_\_, \_\_\}$. The range of R is $\{0, 3, \_\_, \_\_\}$.

2, 4; 8, 10

10. For the relation $S = \{(4, 3), (2, 3), (8, 3)\}$, domain $= \{\_\_, \_\_, \_\_\}$ and range $= \{\_\_\}$.

4, 2, 6

3

11. We can specify a relation by

a. giving a formula that tells us how to compute the second element of each pair, and
b. giving the domain of the relation.

Let Q be the relation consisting

a. of pairs (x, y) where $y = 2x$
b. with domain $= \{4, 5, 6\}$

then

$$Q = \{(4, 2 \cdot 4), (5, 2 \cdot \_\_), (6, 2 \cdot \_\_)\}$$

5; 6

$$Q = \{(\_\_ \_\_), (\_\_ \_\_), (\_\_ \_\_)\}$$

4, 8; 5, 10; 6, 12

The range of Q is $\{\_\_ \_\_ \_\_\}$.

8, 10, 12

12. Show the relation T of pairs (x, y) if $y = x + 3$ and the domain of T is $\{0, 1, 2, 3\}$.

$$T = \{(0, \_\_), (1, \_\_), (2, \_\_), (3, \_\_)\}$$

3; 4; 5; 6

13. Next we turn our attention to an important type of relation.

Definition: A *function* is a relation in which no two pairs with the same first elements have different second elements.

Can the pairs (1, 7) and (1, 8) both be in a function? (yes\no). Can the pairs (2, 4) and (2, 1 + 3) both be in a function? (yes\no). Because $4\ (= \backslash \neq)\ 1 + 3$.

no

yes

=

14. Let's use arrows to show how to pair domain and range elements. The relation F is a function since each domain element goes to (only one\several) range element(s).

only one

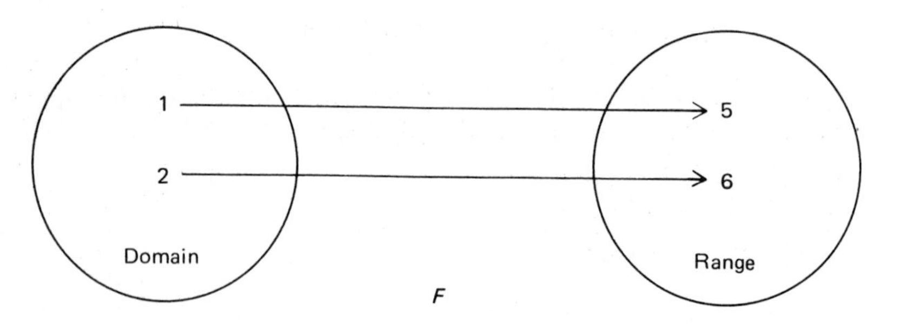

is not
two different
elements

3; 9

15. Let us consider the relation H indicated below. The relation H (is\is not) a function since some domain elements go to (only one element\two different elements) in the range. In particular the element 1 in the domain is paired with both ____ and ____.

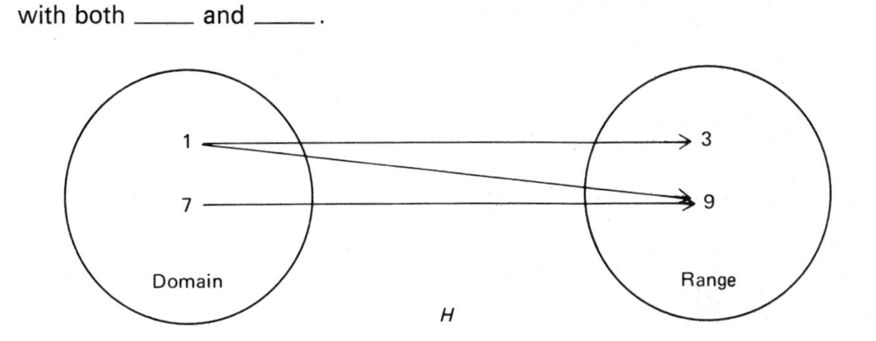

is
only one
is

16. Consider the relation G indicated below. The relation G (is\is not) a function. For each domain element goes to (only one\several) range element(s). In a function it (is\is not) possible to use the same range element for several different domain elements.

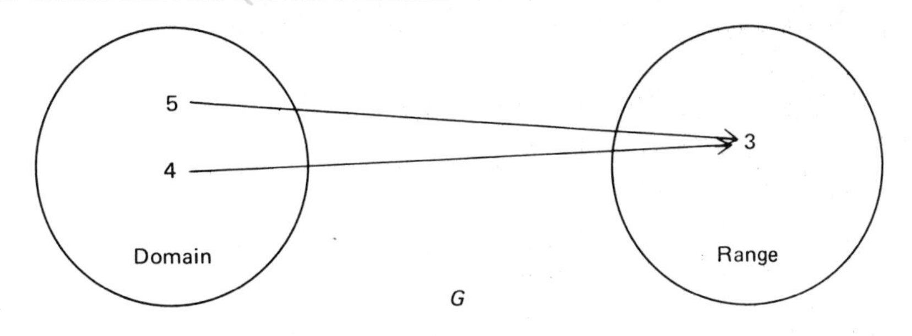

a., c., and d.

17. Which of the following are functions?

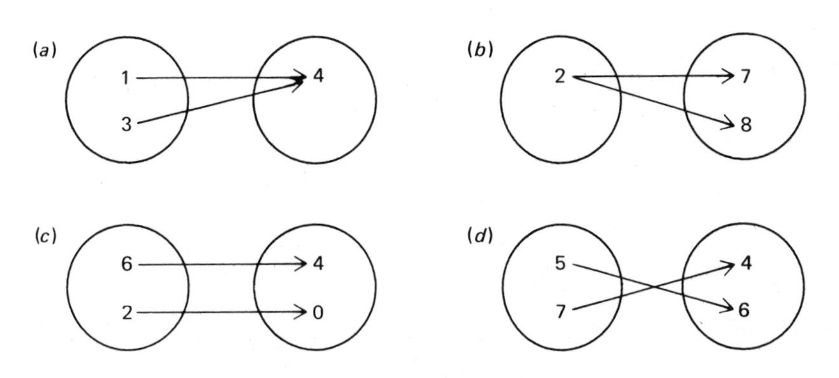

no

1, 2; yes
range elements

18. Look at the functions $f = \{(1, 4), (2, 8)\}$ and $g = \{(1, 7), (2, 10)\}$. Are the functions f and g the same? (yes\no). The domain of f is $\{1, 2\}$. The domain of g is {__, __}. Are the domains the same? (yes\no). These functions are different because the (domain elements\range elements) are different.

19. The second element y in a pair (x, y) of a function depends on two things: the function itself, and the first element. There is a special notation for the second element that shows both the function name and the first element. If (x, y) is a pair in the function f, then we write

$$y = f(x)$$

$f(x)$ is read "f of x", and refers to the range value y paired with the domain element x in the function f.

Thus the pairs in f are $(x, f(x))$. Let

$$f = \{(1, 3), (2, 5)\} = \{(1, f(\_\_)), (2, f(\_\_))\}$$ 1; 2

Then $f(1)$ reads "f of 1" and

$f(1) = 3$ since 3 is paired with 1 in the function f

$f(2) =$ ____ since 2 is paired with ____ in f 5; 5

$f(2)$ means ("f of 2"\"f times 2"). "f of 2"

20. Let g be the function $\{(1, 2), (3, 4), (5, 6)\}$. Then $g(1)$ reads "____ of 1" and g

$g(1) = 2$ since 2 is paired with ____ in the function g 1

$g(3) =$ ____ 4

$g(5) =$ ____ 6

21. To describe the function h we can give the domain and the formula that shows us how to compute $h(x)$ for any x. Suppose h has domain $\{3, 4, 5\}$ and $h(x) = x + 2$. Write the ordered pairs in h. First we must compute the range element that goes with each domain element.

Domain element	*Corresponding range element*	
x	$h(x) = x + 2$	
3	$h(3) = 3 + 2 =$ ____	5
4	$h(4) = 4 + 2 =$ ____	6
5	$h(5) =$ ____ $+ 2 =$ ____	5; 7

$$h = \{(3, \_\_), (4, \_\_), (5, \_\_)\}$$ 5; 6; 7

22. Let g have domain $\{1, 2, 4\}$ and $g(x) = 3x$. To compute $g(1)$ we put 1 wherever we have x in $g(x) = 3x$. Thus

$g(1) = 3(1) =$ ____ 3

$g(2) = 3(\_\_) =$ ____ 2; 6

$g(4) = 3(\_\_) =$ ____ 4; 12

$g = \{(1, \_\_), (2, \_\_), (4, \_\_)\}$ 3; 6; 12

23. Suppose $f(x) = 2x + 1$, and the domain is the set of all whole numbers.

$f(0) = 2(0) + 1 =$ ____ 1

$f(1) = 2(\_\_) + 1 =$ ____ 1; 3

$f(2) = 2(\_\_) + 1 =$ ____ 2; 5

$f(3) = 2(3) +$ __ $=$ ____ 1; 7

Some of the pairs in f are (0, __), (1, __), (2, __), (3, __). 1; 3; 5; 7

PRACTICE PROBLEMS

1. Relations and functions occur often in everyday life. For example, the time-of-day–temperature function, and the mother–child relation are familiar to us. List several other common relations. What is the domain and range for each relation? Which relations are functions?

2. The following set of ordered pairs is a relation: $\{(1, 2), (2, 3), (3, 4), (4, 5), (5, 6)\}$.
   a. What is the domain of this relation?
   b. What is the range of this relation?
   c. Is this relation a function? Explain.

3. Given, the set of ordered pairs $\{(10, 9), (8, 7), (6, 5), (4, 3), (2, 1)\}$.
   a. What is the domain?
   b. What is the range?
   c. Is this relation a function? Explain.

4. Given, the set $\{(1, 0), (2, 0), (3, 0), (4, 0), (5, 0)\}$.
   a. What is the range?
   b. What is the domain?
   c. Is this a function? Explain.

5. Given, the set $\{(0, 2), (1, 3), (2, 4), (3, 5), (4, 6), (5, 7)\}$.
   a. What is the domain?
   b. What is the range?
   c. Is this relation a function? Explain.

6. Given, the set $\{(0, 0), (1, 3), (2, 6), (3, 9), (4, 12)\}$. Let (x, y) represent a typical ordered pair from this set.
   a. If $x = 0$, what is y?
   b. If $x = 1$, what is y?
   c. If $x = 2$, what is y?
   d. If $x = 3$, what is y?
   e. If $x = 4$, what is y?
   f. In each of the above is it true that $y = 3x$?
   g. Does each ordered pair in the given set satisfy the formula $y = 3x$?

7. Given, the set $\{(-2, 3), (-1, 2), (0, 1), (1, 0), (2, -1)\}$. Let (x, y) represent a typical ordered pair from this set.
   a. If $x = -2$, what is y?
   b. If $y = 1$, what is x?
   c. If $y = 0$, what is x?
   d. If $x = 2$, what is y?
   e. For each of the above is it true that $y = -x + 1$?
   f. Does each ordered pair in the given set satisfy the formula $y = -x + 1$?
   g. Is the set a function?

8. Given, the set $\{(0, 0), (1, 4), (2, 8), (3, 12), (4, 16)\}$. Let (x, y) represent a typical element. Is it true that $y = 4x$?

9. Given, the function h with domain $\{0, 2, 3\}$ and $h(x) = x + 5$.
   a. Compute $h(0)$, $h(1)$, and $h(2)$.
   b. List the three ordered pairs in h.

10. Given, the function g with domain $\{4, 5\}$ and $g(x) = 2x$.
   a. Compute $g(4)$ and $g(5)$.
   b. List the two ordered pairs in g.

11. Given, the function f with domain $\{0, 2, 3\}$ and $f(x) = x^2$.
   a. Compute $f(0)$, $f(2)$, $f(3)$.
   b. List the three ordered pairs in f.

12. List the pairs in the function h if $h(x) = 3x - 1$ if the domain of h is $\{3, 4, 5, 6\}$.

SELFTEST

1. Consider the relation $\{(2, 0), (3, 5), (4, 2), (5, 1), (6, 7)\}$.
   a. Find the domain and range.
   b. Is the relation a function? Why?
   c. If $(5, 2)$ were included in the relation, would it still be a function? Why?

2. Given, the formula $y = 2x - 1$.

a. Fill in the missing parts of the ordered pairs (x, y):

$$\{(0, \_\_), (\_\_, 1), (3, \_\_), (4, \_\_)\}$$

b. Is this set a function? *c.* What is the domain? *d.* What is the range?

3. Is H a function?

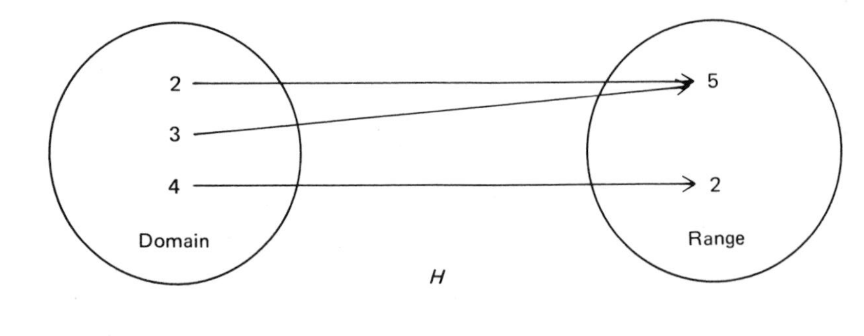

4. If $f(x) = 3x$, then

a. $f(1) = \_\_\_$ *b.* $f(2) = \_\_\_$ *c.* $f(-3) = \_\_\_$

SELFTEST ANSWERS

1. *a.* domain $= \{2, 3, 4, 5, 6\}$; range $= \{0, 1, 2, 5, 7\}$.
b. yes, because each domain element is paired with exactly one range element.
c. no, because both pairs (5, 1) and (5, 2) would be in the set.

2. *a.* $\{(0, -1), (1, 1), (3, 5), (4, 7)\}$. *b.* yes
c. domain $= \{0, 1, 3, 4\}$ *d.* range $= \{-1, 1, 5, 7\}$.

3. yes **4.** *a.* 3 *b.* 6 *c.* -9

9.4 GRAPHING LINES

Let us review some of the things we have done in this chapter. First we introduced ordered pairs and saw how they can be located on the *xy* plane. Later we saw that every relation and function can be thought of as a collection of ordered pairs. In this section we will bring these concepts together to study a fundamental relationship between algebra and plane geometry. From now on, unless otherwise said, all our relations and functions will be such that their domain and range are sets of real numbers.

Definition: **The *graph* of a relation is the set of all points in the *xy* plane that correspond to some ordered pair of the relation.**

In other words, given a relation involving x and y, the graph of that relation is the set of points $\{(x, y) \mid x \text{ and } y \text{ together make the relation true}\}$.

We recall that a function is just a special kind of relation, so the above definition also gives us the definition of the graph of a function. It is just the set of all points in the *xy* plane which correspond to ordered pairs of the function.

So a graph is a set of points. Therefore a graph is, in a sense, a geometrical display of the information contained in a relation or function. A simple but important geometrical figure is a straight line. In this section we will study lines in the *xy* plane.

Example 1 Find the graph corresponding to the equation

$$y = 3x + 1$$

The equation $y = 3x + 1$ expresses y as a function of x. To graph this equation we must express it as a set of ordered pairs and then graph each of the ordered pairs on the xy plane. The ordered pairs will all be solution pairs for the equations. For the values of each pair will make the equation true.

A table will help us organize our work. We are free to pick any value we like for x. But once we choose an x value, we must use it in the equation to *solve* for the corresponding y value.

Pick any value for x x	*Solve the resulting equation for y* $y = 3x + 1$	*Solution pairs* (x, y)
-3	$y = 3(-3) + 1 = -8$	$(-3, -8)$
-2	$y = 3(-2) + 1 = -5$	$(-2, -5)$
-1	$y = 3(-1) + 1 = -2$	$(-1, -2)$
0	$y = 3(0) + 1 = 1$	$(0, 1)$
1	$y = 3(1) + 1 = 4$	$(1, 4)$
2	$y = 3(2) + 1 = 7$	$(2, 7)$
3	$y = 3(3) + 1 = 10$	$(3, 10)$

Of course we could use more values of x and y if we wanted to do so. But the above seven values of x and y seem to be enough to start with. Next we locate the ordered pairs on the xy-plane and then connect the points with a smooth curve. In our case the smooth curve actually turns out to be a straight line.

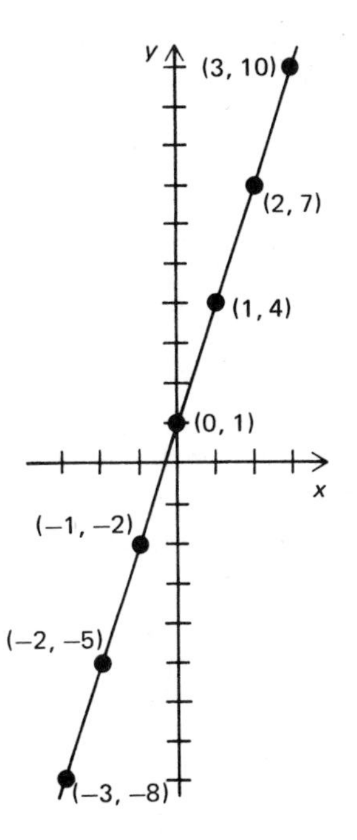

Example 2 Graph $x - y = -4$: Again we must find pairs (x, y) that make the equation true. We plot these points and connect them with a smooth curve. Our job will be easier if we solve the equation for one of the variables. Let's solve for y.

$$x - y = -4$$
$$x + 4 = y$$

Here x is the independent variable and y is the dependent variable. In other words, we can pick any value we like for x and solve the resulting equation for the corresponding y value.

Pick any value for x x	*Solve the resulting equation for y* $y = x + 4$	*Solution pairs* (x, y)
4	$y = 4 + 4 = 8$	(4, 8)
3	$y = 3 + 4 = 7$	(3, 7)
2	$y = 2 + 4 = 6$	(2, 6)
1	$y = 1 + 4 = 5$	(1, 5)
0	$y = 0 + 4 = 4$	(0, 4)
−2	$y = -2 + 4 = 2$	(−2, 2)
−4	$y = -4 + 4 = 0$	(−4, 0)

We plot the solution pairs on the *xy*-plane and then connect them. Again we have a straight line.

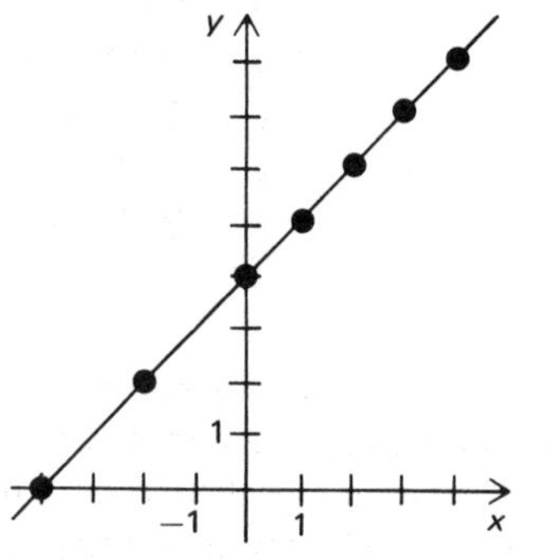

All solution pairs (x, y) of $x - y = -4$ must be on this line. Since the pair $(-1, 3)$ is a solution of $x - y = -4$, it should be on the line. If you look at the graph you will see it on the line.

Likewise, if a pair (x, y) is not a solution of $x - 4 = -4$, it will not be on the line. Since $(1, 1)$ does not satisfy $x - y = -4$, it should not be on the line. Again check the graph. You will see that the line does not go through $(1, 1)$.

Example 3 Graph $x = -y - 3$: This equation is already solved for x. So this time let's pick y values and solve for corresponding x values.

Solve for x $x = -y - 3$	*Pick y* y	*Solution pairs* (x, y)
$x = -3 - 3 = -6$	3	(−6, 3)
$x = -1 - 3 = -4$	1	(−4, 1)
$x = -0 - 3 = -3$	0	(−3, 0)
$x = -(-2) - 3 = -1$	−2	(−1, −2)
$x = -(-4) - 3 = 1$	−4	(1, −4)

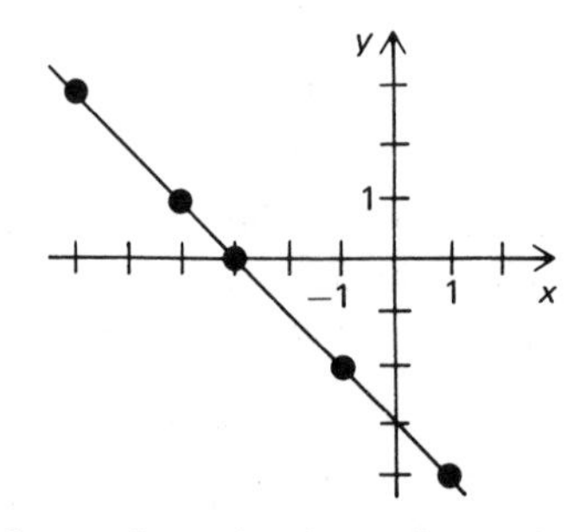

In each of our examples we have plotted at least five solution pairs before we drew the line. But all the examples of this section have graphs that are straight lines. We have been doing extra work! For we need only *two points to graph a line.* From now on we will use only two points to find the line and then use a third solution pair to check our work.

Example 4 Graph $4x - 2y = 6$: Let's first solve for y.

$$4x - 2y = 6$$

$$4x - 6 = 2y$$

$$\frac{4x}{2} - \frac{6}{2} = \frac{2y}{2}$$

$$2x - 3 = y$$

x	y	
2	1	since $2(2) - 3 = 1$
0	-3	since $2(0) - 3 = -3$

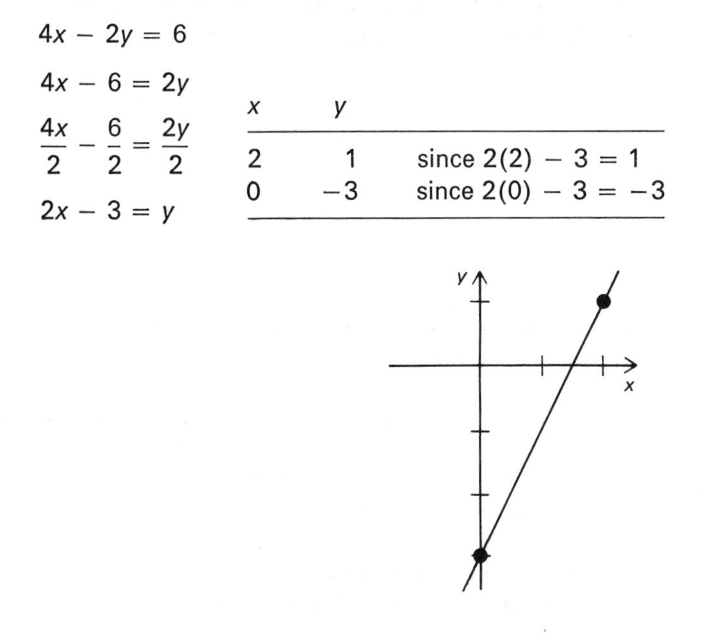

We use a third solution pair as a check. The pair $(1, -1)$ satisfies $4x - 2y = 6$. It is on the line.

There are two special cases which bear mentioning. In the next two examples we look at vertical and horizontal lines.

Example 5 Find the graph corresponding to the equation $x = 5$. In this case the x value is given to be 5. The y value is not specified, so it is intended to be an arbitrary number. The ordered pairs for the equation $x = 5$ are $(5, y)$, where y is any arbitrary number. For instance, (5, 0), (5, 1), (5, 2), (5, 3), (5, −2), (5, −3) are some ordered pairs that correspond to the equation $x = 5$. If we plot these ordered pairs in the xy plane, we get a vertical (straight up and down) line. The line is parallel to the y axis and is 5 units to the right of the y axis.

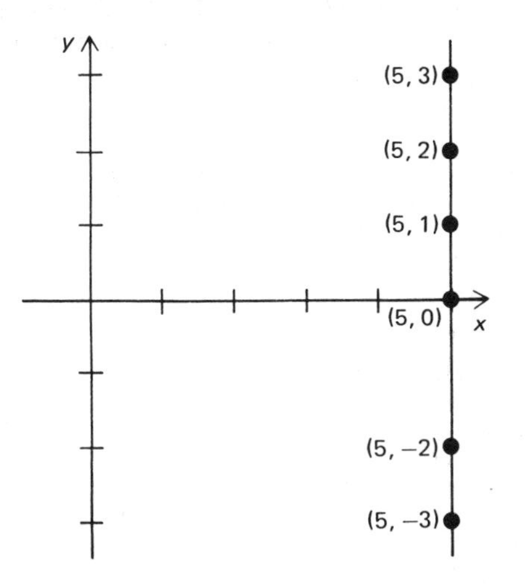

Example 6 Find the graph corresponding to the equation $y = -2$. The y value is given to be $y = -2$ and the x values are intended to be arbitrary. The ordered pairs for the equation $y = -2$ are of the form $(x, -2)$, where x is any arbitrary number. It follows that (−3, −2), (−1, −2), (2, −2), (3, −2), and (0, −2) are some ordered pairs corresponding to the equation $y = -2$.

If we plot these ordered pairs and connect them with a smooth curve, we get a horizontal (that is flat or level) line. The line is parallel to the x axis and 2 units below the x axis.

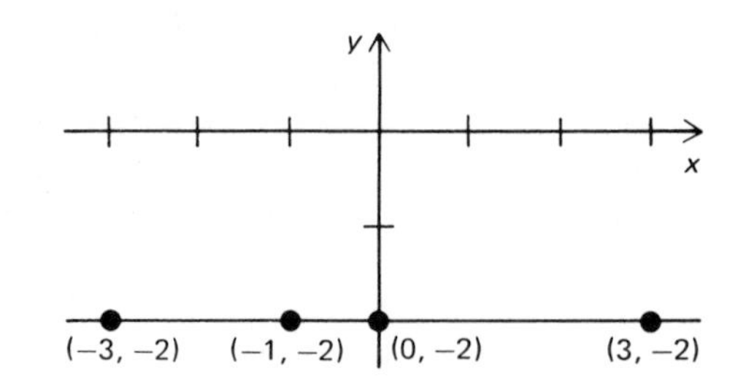

PROGRAM 9.4

1. In previous sections we have discussed ordered pairs and functions. We found that functions were sets of special, ordered pairs. If we let (x, y) represent the pairs of a given function whose domain and range are sets of real numbers, we can easily graph the function on the xy plane. For we just plot all the pairs of the function on the plane.

Graph the function S on the xy plane

$$S = \{(-1, -2), (0, 0), (1, 2), (2, 4)\}$$

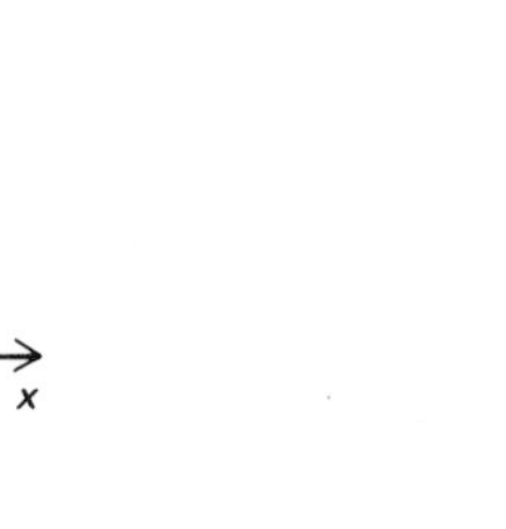

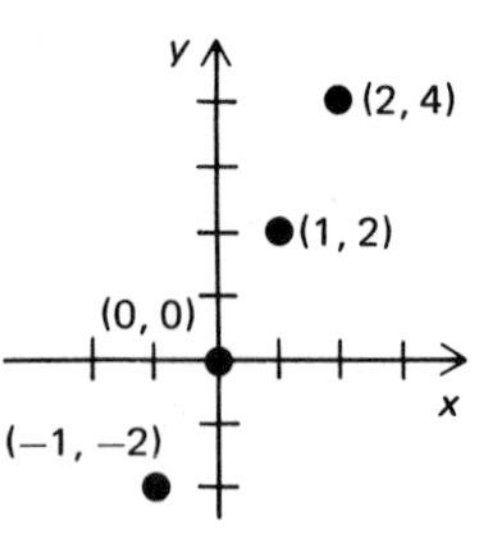

2. Let us graph the function given by the equation $y = 2x + 1$. The domain is the set of *all real numbers*. Thus x can be (any real number\only an integer\only a fraction). We make a table where the x values are assigned by us. Then we compute the corresponding y values.

any real number

Pick x	*Compute* $y = 2x + 1$	*Corresponding ordered pairs*
−3	$y = 2(-3) + 1 = -5$	(−3, −5)
−2	$y = 2(-2) + 1 =$ ___	(−2, ___)
−1	$y = 2($___$) + 1 =$ ___	(___, ___)
0	$y = 2($___$) + 1 =$ ___	(___, ___)
1	______	(___, ___)
2	______	(___, ___)
3	______	(___, ___)

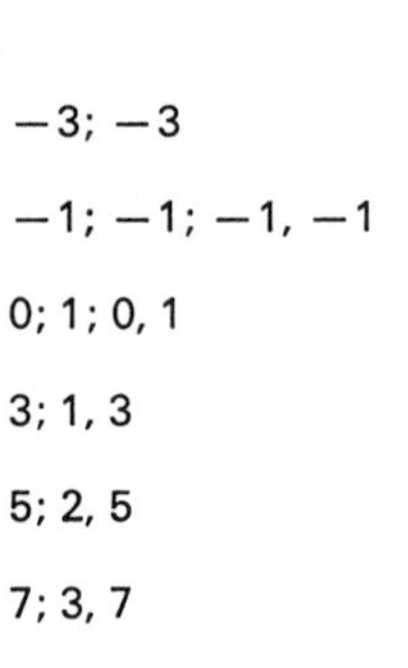
−3; −3
−1; −1; −1, −1
0; 1; 0, 1
3; 1, 3
5; 2, 5
7; 3, 7

From the tables we have constructed we see that each of the corresponding ordered pairs (does\does not) satisfy the equation $y = 2x + 1$. Of course these are only a few of the infinitely many possible ordered pairs. But in this case there will be enough.

does

Next we locate each of the above ordered pairs on the xy plane. The points $(-3, -5)$ and $(-2, -3)$ have already been plotted. Plot the other points.

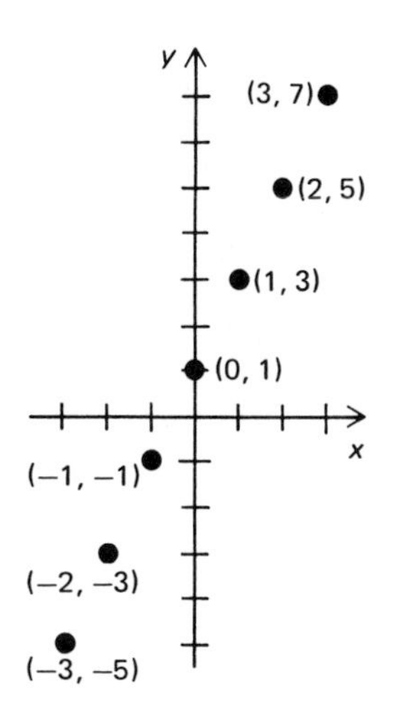

To complete the graph, we need to indicate the location of *all* the other pairs that satisfy $y = 2x + 1$ when x is a real number. Do this below by drawing a smooth curve through the points we have already located. The curve has the shape of a (circle\line\squiggle) that keeps going in both directions.

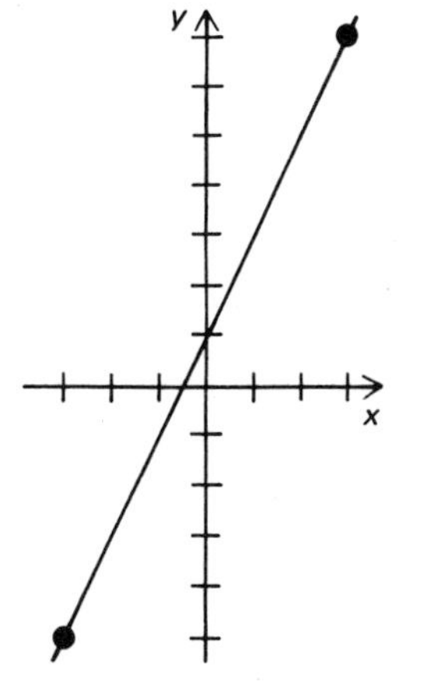

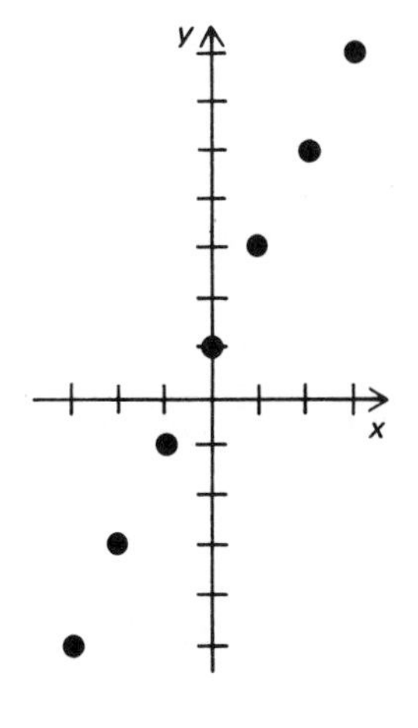

3. From now on, the domain and range of our functions will be the set of real numbers. Now let's graph the equation

$$y = 3x - 2$$

There are two major steps. First, find some ordered pairs which satisfy $y = 3x - 2$. Second, plot these ordered pairs on the xy plane and connect the points with a smooth curve.

Step one: To find ordered pairs that satisfy the equation, let x be the independent variable. Choose values for x, then use the equation to compute the corresponding values for y.

-8; -8
-1; -5; $(-1, -5)$
-2; $(0, -2)$
1; $(1, 1)$

x	$y = 3x - 2$	*Resulting ordered pairs*
-2	$y = 3(-2) - 2 =$ __	$(-2,$ __$)$
-1	$y = 3($__$) - 2 =$ __	______
0	$y =$ ____	______
1	____	______

Step two: Now plot the ordered pairs and draw a smooth curve connecting them. Should each point (x, y) on the line satisfy the equation $y = 3x - 2$? (yes\no).

yes

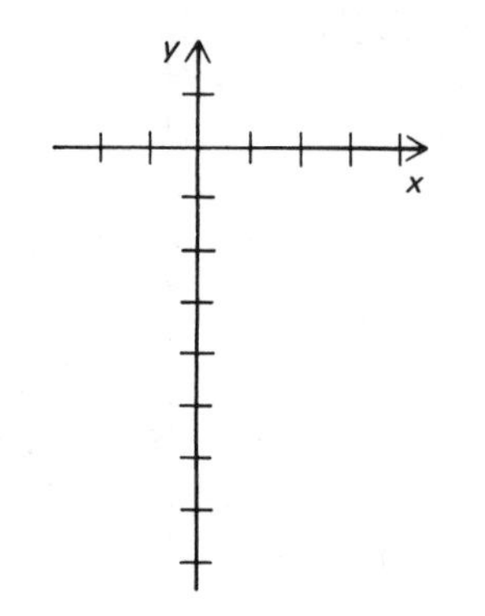

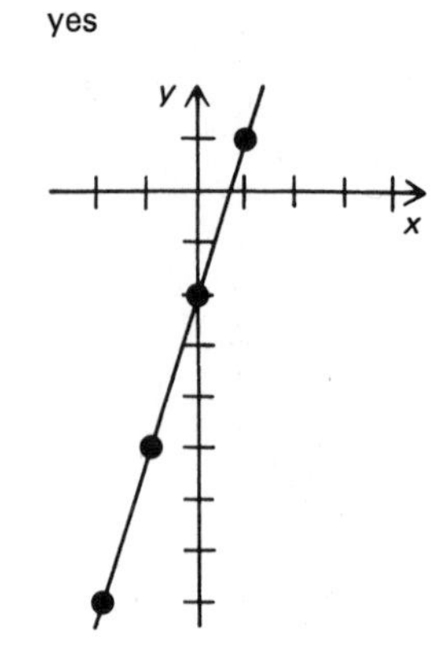

4. Graph $x + y = 4$: First find solution pairs (x, y) for this equation. This will be easier if we solve the equation for y.

$$x + y = 4$$
$$y = 4 - \_\_\_\_$$

x

Pick values x	*Solve for y* $y = 4 - x$	*Solution pairs* (x, y)
3	1 since $y = 4 - 3$	(3, 1)
0	__ since $y = 4 - 0$	(0, __)
1	__ since $y = 4 - 1$	(__, __)
−2	__ since $y = 4 - (-2)$	________

4; 4

3; 1, 3

6; (−2, 6)

Plot your solution pairs on the xy plane. Then connect the points with a line. Is the point (0, 0) on your line? (yes\no). Does the point (0, 0) satisfy $x + y = 4$? (yes\no).

no

no

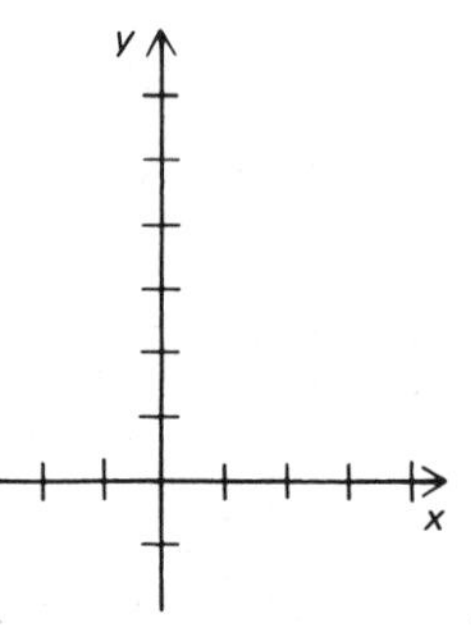

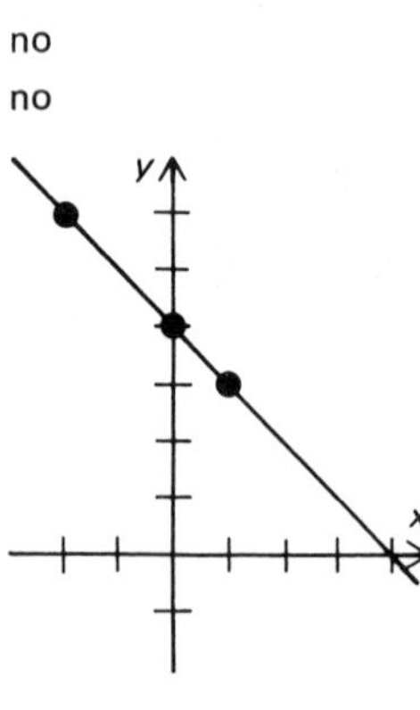

5. This graph is the graph of the equation $2y + 3x = 4$. Is (0, 1) a point on the graph? (yes\no). Is (0, 1) a solution of $2y + 3x = 4$? (yes\no). Is (2, −1) on the line? (yes\no). Is it a solution of the equation? (yes\no).

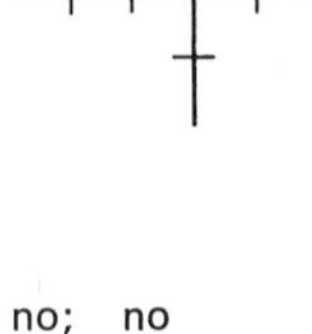

no; no

yes; yes

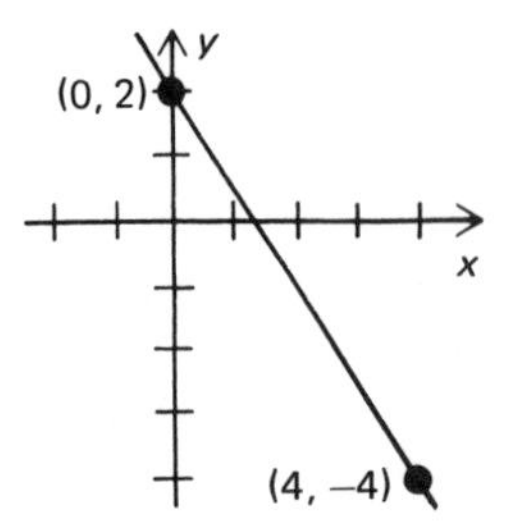

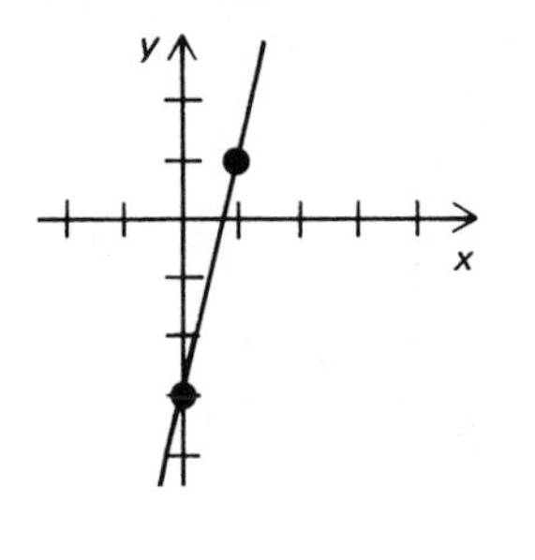

6. The graphs we have just drawn have all been straight lines. *The rest of the graphs in this section will also be straight lines.* We need only two points to determine the location of a line. So, in order to graph a line we can get by with only two solution pairs.

We can graph the line $y = 4x - 3$ by using the solution pairs $(0, -3)$ and $(1, 1)$. Do so here.

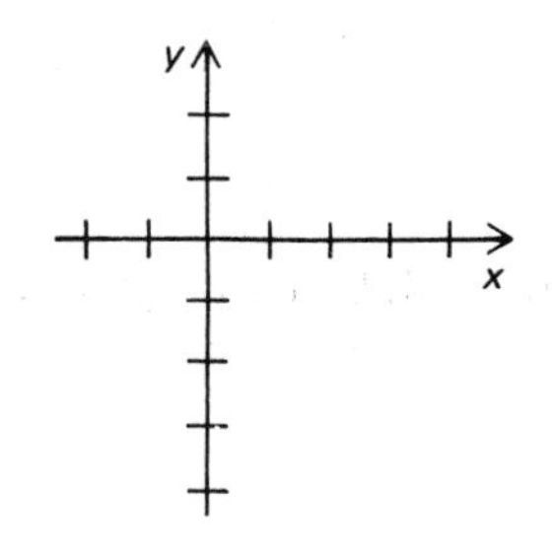

The pairs $(2, 5)$ and $(-1, -7)$ are also solutions of $y = 4x - 3$. Plot these points also and draw a line through them. Is it the same line as the one just drawn? (yes\no).

yes

7. Graph the line determined by the equation $6x + 3y = 9$. First solve for y.

$$6x + 3y = 9$$

$$3y = 9 - \_\_\_\_$$

6x

$$y = \frac{9}{3} - \frac{6x}{3}$$

$$y = \_\_\_\_ - \_\_\_\_$$

3; 2x

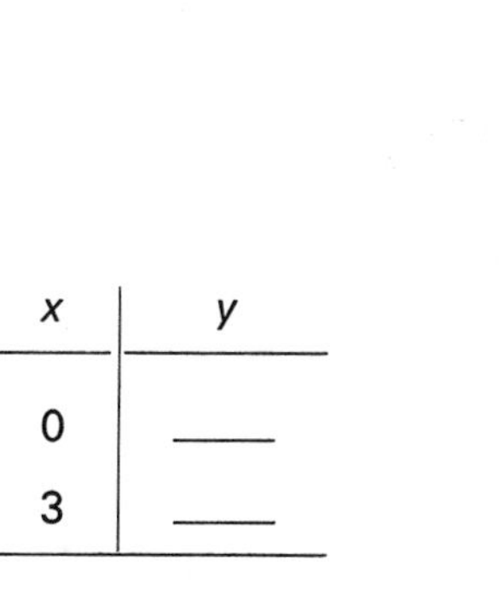

x	y
0	____
3	____

3

−3

Now draw the graph. As a check, we could use a third solution pair $(1, 1)$. Is this point on your graph? (yes\no).

yes

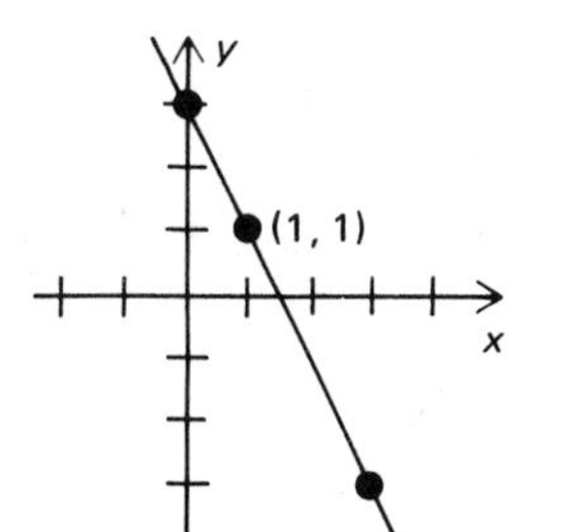

8. Graph the line determined by the equation $x = 2y - 1$. This equation is solved for x. So this time pick values for y and solve the resulting equation for x.

Solve for x	*Pick y*
$x = 2y - 1$	y
−1	0
____	____

Pick any value for y and solve for x.

Now draw the graph. Is the point (7, 4) on the graph? (yes\no). Is it a solution of $x = 2y - 1$? (yes\no).

yes
yes

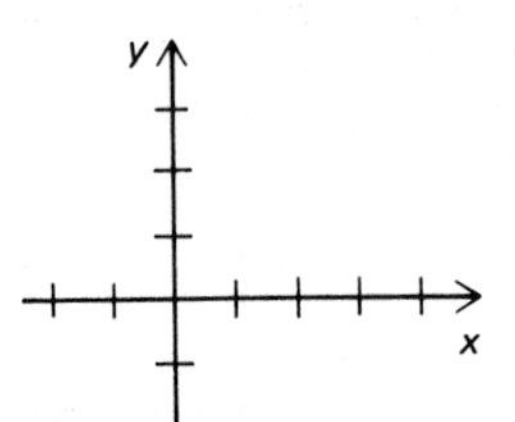

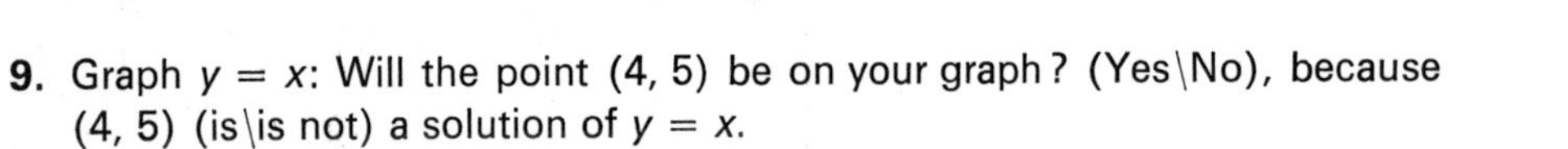

9. Graph $y = x$: Will the point (4, 5) be on your graph? (Yes\No), because (4, 5) (is\is not) a solution of $y = x$.

no
is not

x	y
2	2
−4	___

−4

Now draw the graph. Is the origin (0, 0) a point on this line? (yes\no).

yes

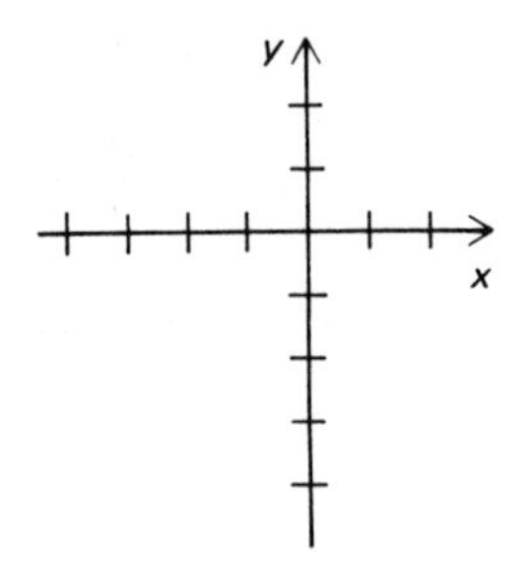

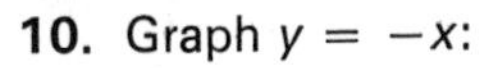

10. Graph $y = -x$:

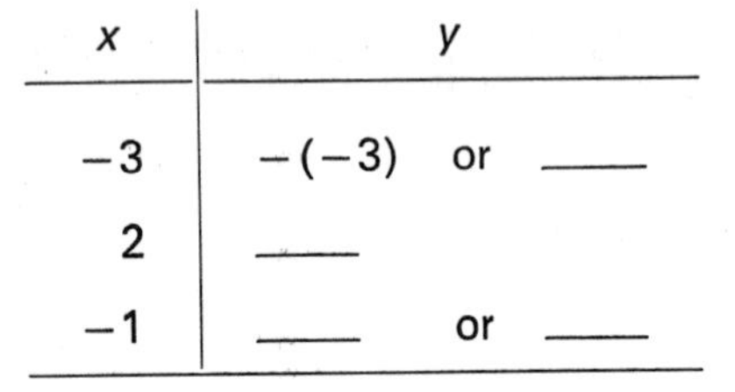

x	y		
−3	−(−3)	or	___
2	___		
−1	___	or	___

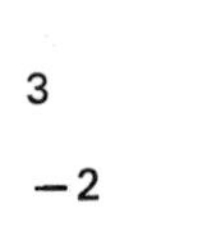

3
−2
−(−1); 1

Does the graph go through the origin? (yes\no).

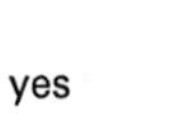

yes

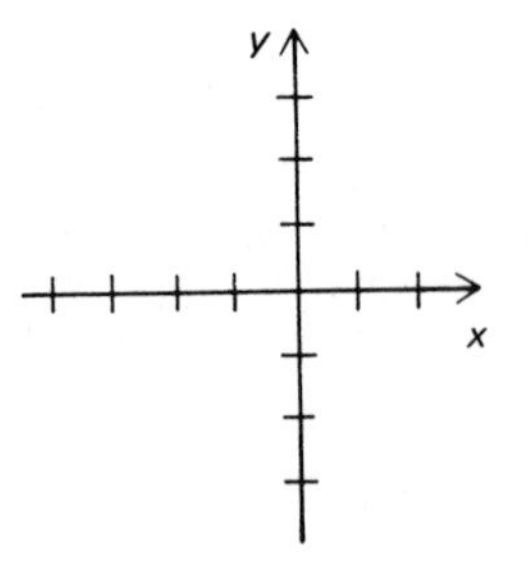

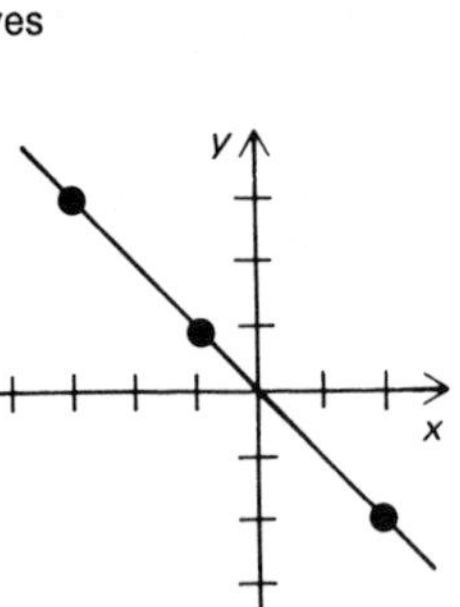

11. The line in the next graph is (vertical\horizontal). Give the coordinates of points A: ________, B: ________, and C: ________. The value of the x coordinate for each point is ____.

vertical

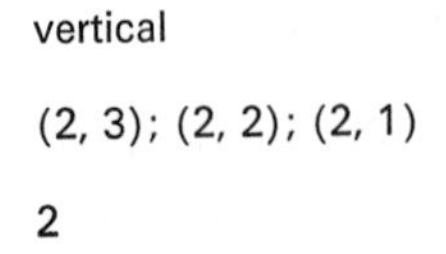

(2, 3); (2, 2); (2, 1)
2

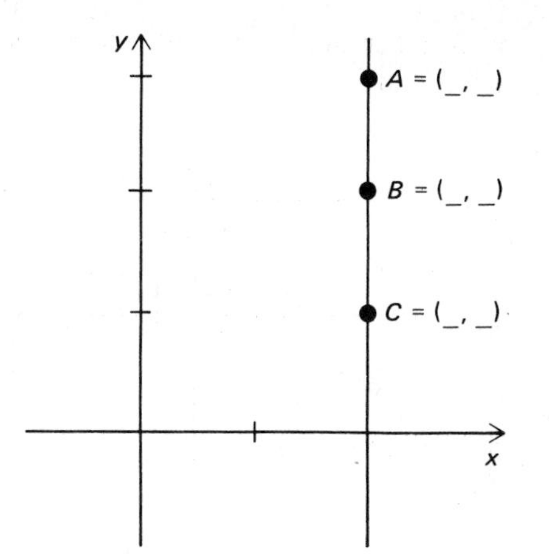

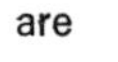

are

x axis

12. If a line is vertical, then all of the x coordinates (are\are not) the same. The equation of a vertical line just tells us the value of x. The equation $x = 3$ graphs as a vertical line through 3 on the (x axis\y axis). Graph $x = 3$.

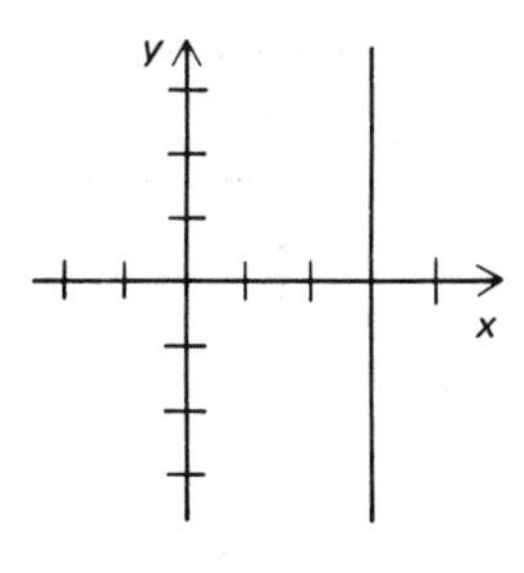

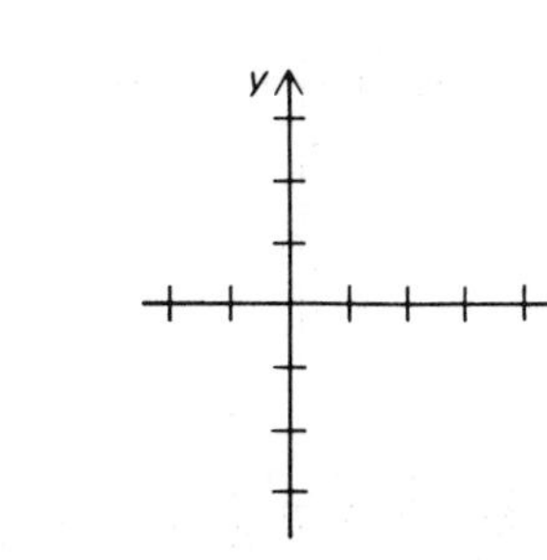

−2

13. In the equation $x = -2$ no y values are specified. This means that y can take on any value. But for every solution pair (x, y), the value for x is (-2\unknown\also any value). Graph $x = -2$.

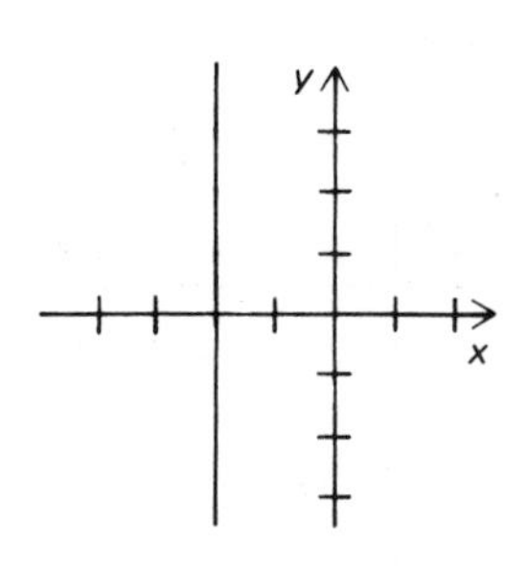

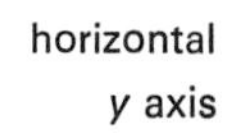

horizontal

y axis

2

14. The line shown on the graph is (vertical\horizontal). Which axis does it go through? The (x axis\y axis.) What is the y value for each point on the line? _____.

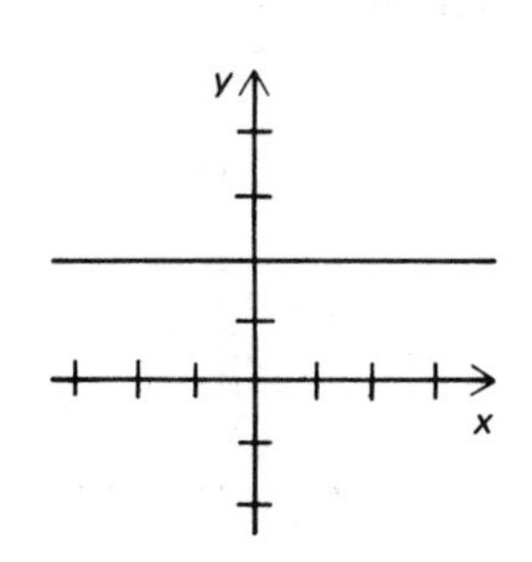

15. The equation $y = 3$ describes a horizontal line through the point (0, 3). Which line is the graph of $y = 3$? _____.

B

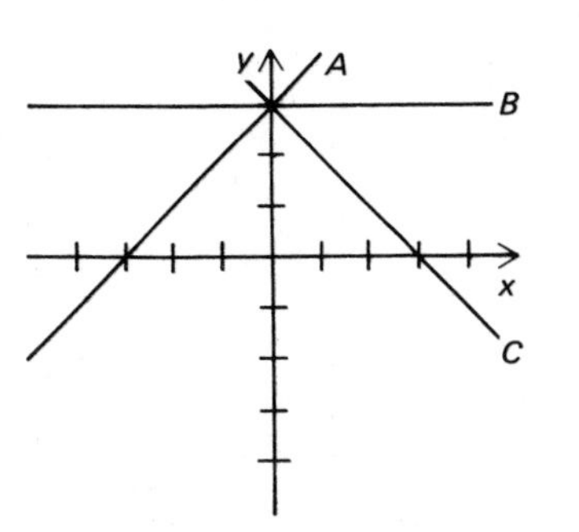

16. The graph of $y = -2$ is a (vertical\horizontal) line. Graph this line.

horizontal

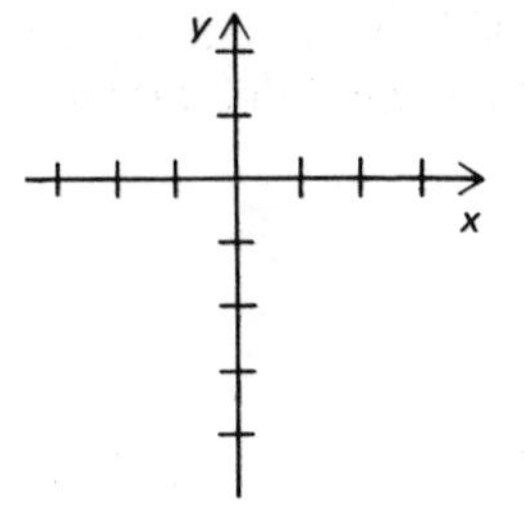

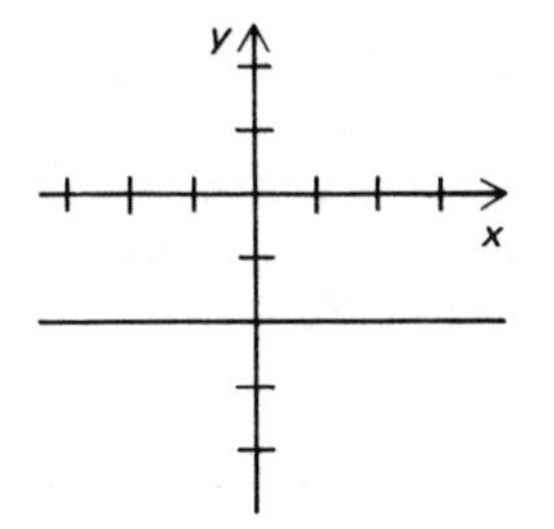

17. If an equation is in both x and y, such as $y = x - 5$, we make a table of solution pairs and then graph the line. Complete the table and graph $y = x - 5$.

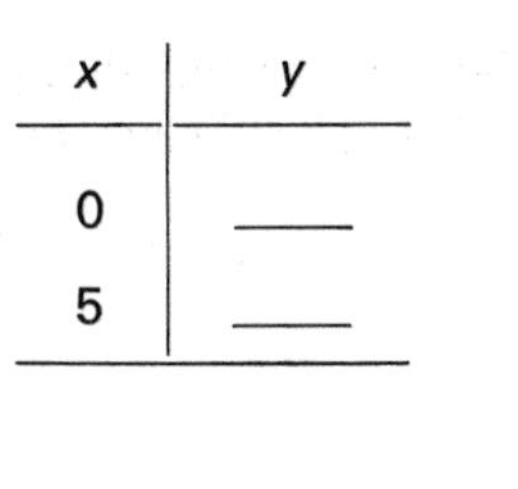

x	y
0	____
5	____

-5

0

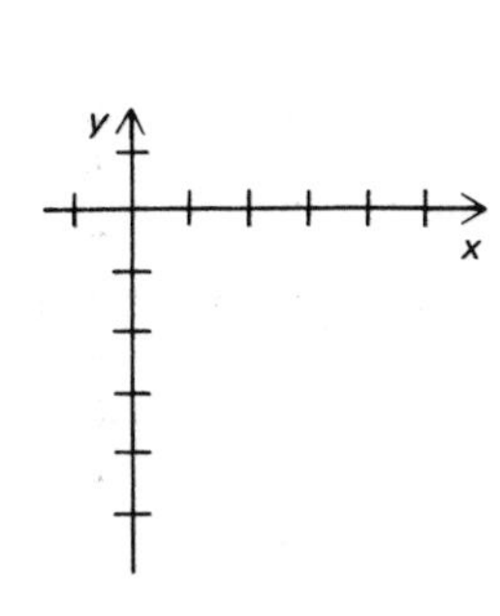

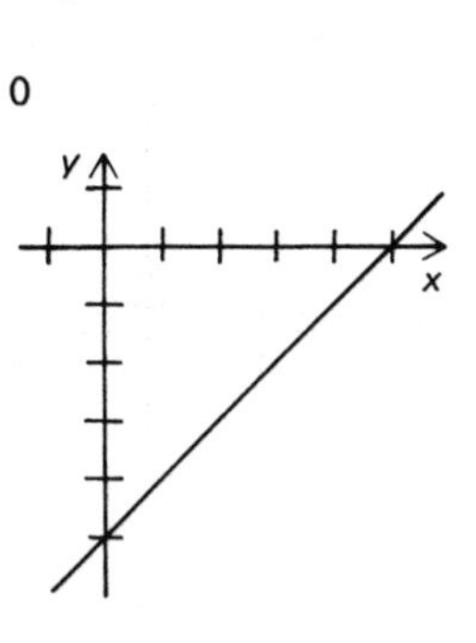

18. The equation $y = 4$ has (one\two) variables. The graph is a (vertical\horizontal) line through points $(x, 4)$. Graph $y = 4$:

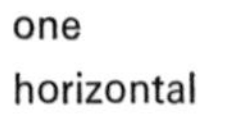

one

horizontal

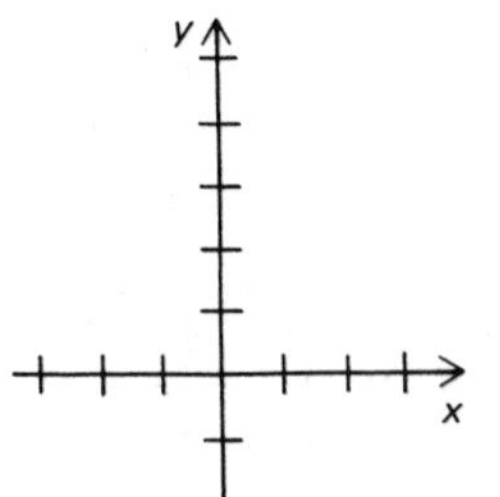

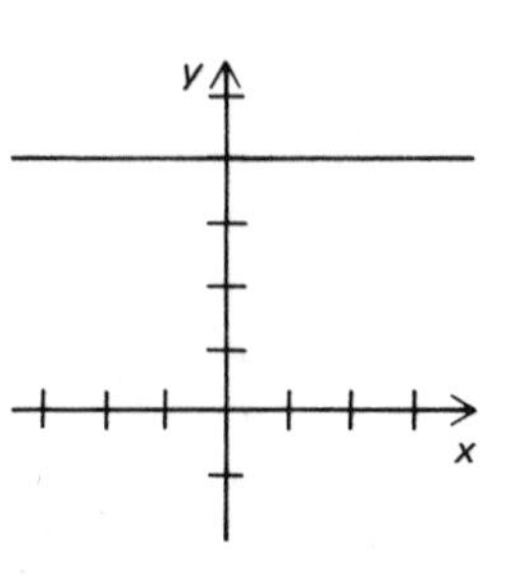

(5, y)

19. The equation $x = 5$ graphs as a vertical line through points ((x, 5) \ (5, y)). Graph $x = 5$:

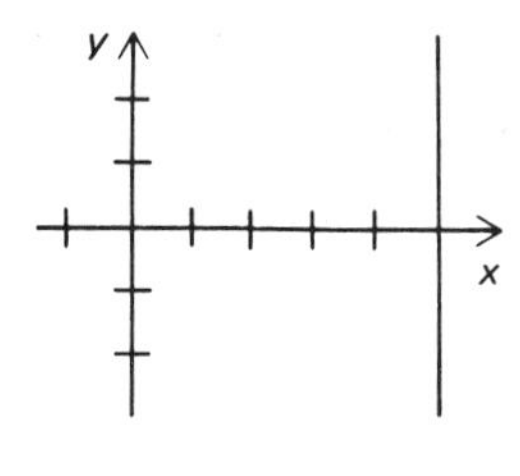

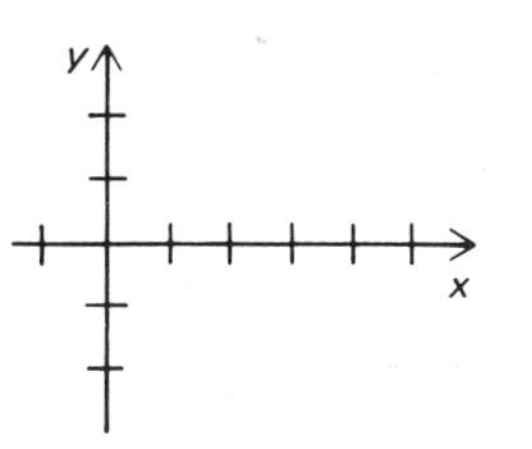

PRACTICE PROBLEMS

Graph each of the following equations.

1. $y = 3x$	**2.** $y = 2x$	**3.** $y = -x$
4. $y = -2x$	**5.** $y = x + 1$	**6.** $y = x - 1$
7. $y = 2x + 1$	**8.** $y = 3x - 2$	**9.** $y = 6$
10. $y = -3$	**11.** $x = -4$	**12.** $x = 3$
13. $x + y = 3$	**14.** $x + y = 4$	**15.** $2x + y = 1$
16. $x + 2y = 1$	**17.** $3x + 2y = 6$	**18.** $3y + 2x = 6$
19. $3y + 6 = x$	**20.** $2y - 4 = x$	**21.** $5y + 25 = -10x$
22. $4y + 20 = 4x$	**23.** $12y + 16x = 9$	**24.** $2x - 3y = 6$

25. Which of the following points correspond to solution pairs of the equation graphed below?

a. (0, 1) *b.* (−2, −3) *c.* (4, 2) *d.* (−4, 0)

26. Which of the following points correspond to solution pairs of the equation graphed below?

a. (4, 0) *b.* (4, 2) *c.* (3, 4) *d.* (2, 4)

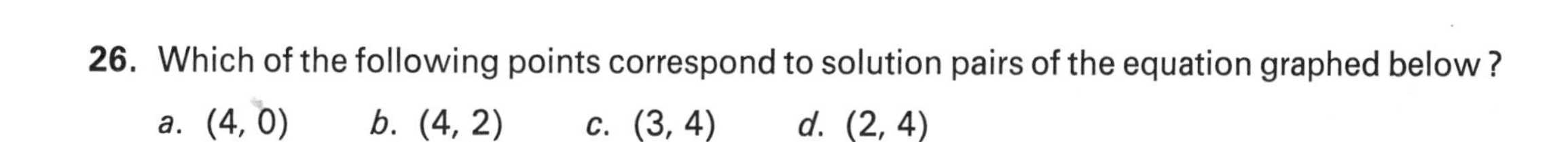

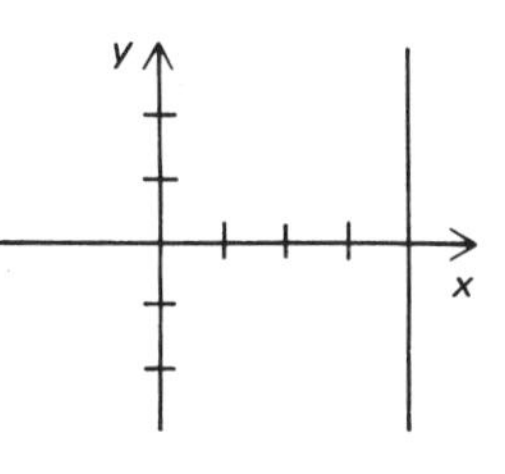

SELFTEST

Graph on the *xy* plane:

1. $y = x + 1$ **2.** $y = 3x$ **3.** $x + y = 4$ **4.** $2x + y = 3$ **5.** $x = -1$

SELFTEST ANSWERS

1.

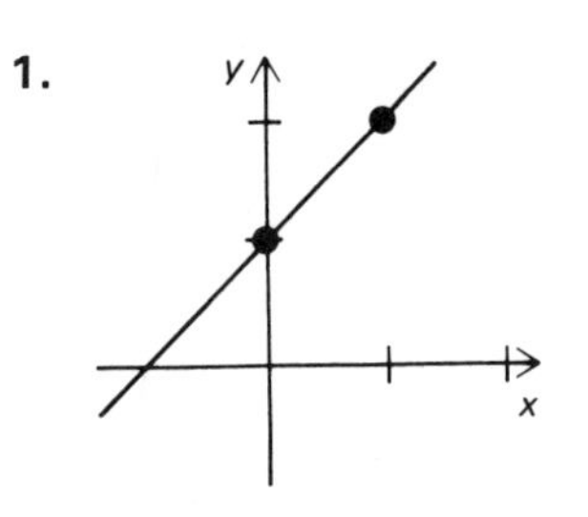

2.

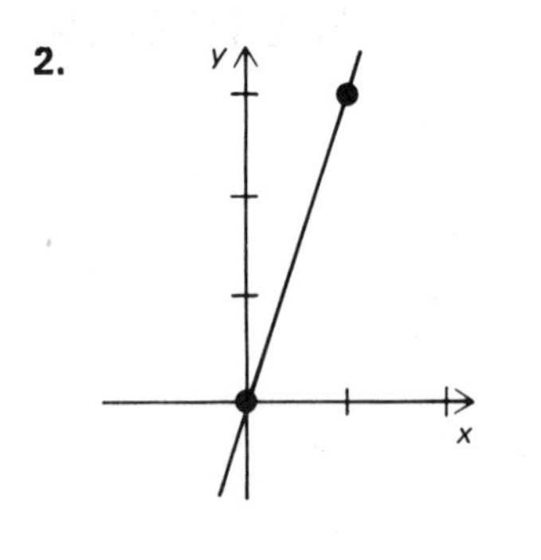

3.

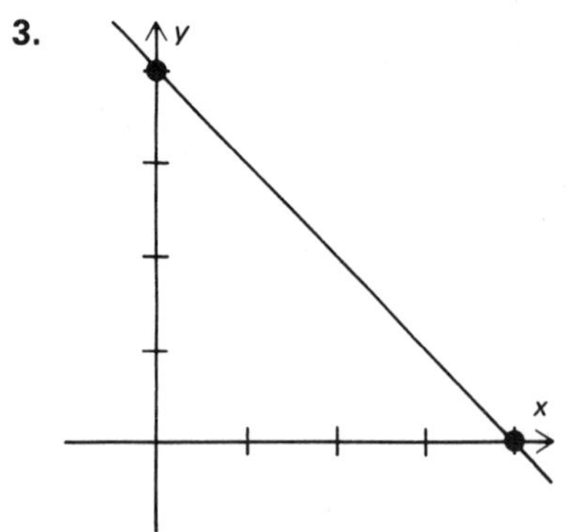

4.

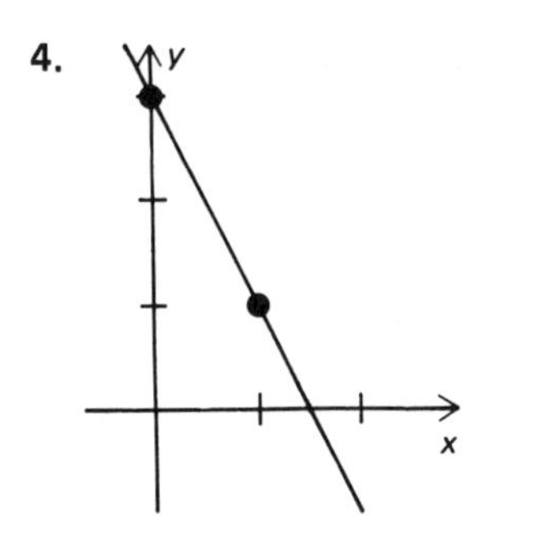

5.

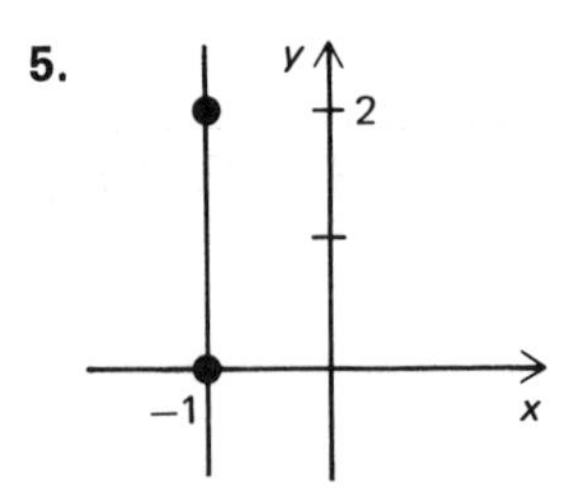

9.5 SLOPE-INTERCEPT FORM OF A LINE

What property of a line makes it different from all other curves? It has the property that it does not bend. But how do we express this property mathematically? In this section we will see how to do this, using a new concept called the "slope" of a line.

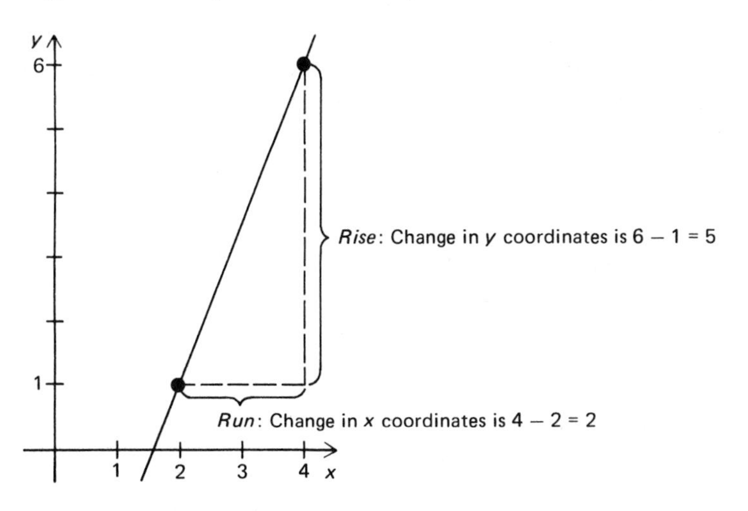

Let's look at the line through the points (2, 1) and (4, 6). In going from (2, 1) to (4, 6), how much do the x and y coordinates change? We call the change in the y coordinates the *rise* and the change in the x coordinates the *run*.

Definition: **The *slope* of a line through the points (x_1, y_1) and (x_2, y_2) is the rise over the run. (*Note:* We read x_1 as "x sub-one." In general x_1 and x_2 represent different points.)**

$$\text{slope} = \frac{\text{rise}}{\text{run}} = \frac{\text{change in } y \text{ coordinates}}{\text{change in corresponding } x \text{ coordinates}}$$

$$= \frac{y_1 - y_2}{x_1 - x_2}$$

Example 1 Let's determine the slope of the line through the points (4, 6) and (2, 1). We need to call one of the points the "first" point. It does not matter which point we call the first. But once we decide on a first point, we must be consistent. Here, we will call (4, 6) the first point. So in our formula we use (4, 6) in place of (x_1, y_1), and (2, 1) in place of (x_2, y_2).

$$\text{slope} = \frac{\text{rise}}{\text{run}} = \frac{y_1 - y_2}{x_1 - x_2} = \frac{6 - 1}{4 - 2} = \frac{5}{2}$$

The slope $^5/_2$ means that every time we move 2 units to the right on the x axis, we move up 5 units on the y axis. Our landing point will again be on the line.

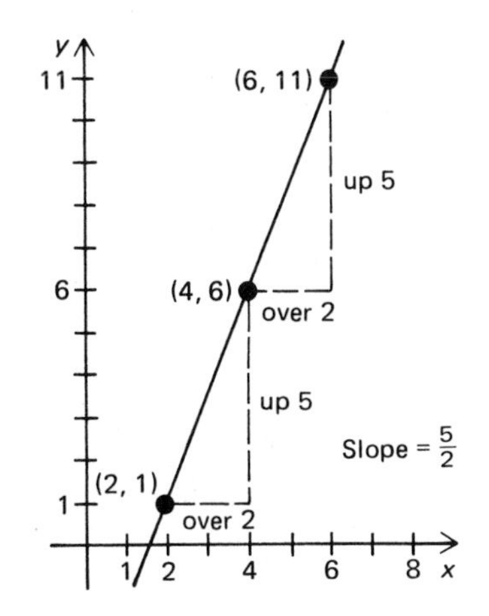

A line has the property that it does not bend. Algebraically, this is the same as saying that *the slope of a line does not change.* That is, the slope of a line will be the same no matter what two points on the line we use to compute it.

We use this property of slopes and lines to find the equation of a line through two points. Let's find the equation of our line through (2, 1) and (4, 6).

Example 2 Find the equation of the line through (2, 1) and (4, 6). Let (x, y) be any other point on the line. The slope of the line will be the same no matter what points we use. So let's use the general point (x, y) as our first point on the line and (4, 6) as the second point.

$$\text{slope} = \frac{\text{rise}}{\text{run}} = \frac{y - 6}{x - 4}$$

But in Example 1 we computed the slope of this line to be $^5/_2$. Since the slope of the line does not change, we get the equation

$$\frac{5}{2} = \frac{y - 6}{x - 4}$$

Now solve the equation for y.

$$^5/_2(x - 4) = y - 6$$
$$^5/_2x - 10 = y - 6$$
$$^5/_2x - 4 = y$$

The last equation is the equation that corresponds to the line through (2, 1) and (4, 6).

Besides the slope, there is another important feature of the graph of a line. It is called the y intercept of the line. The *y intercept* of a line is that point where the line crosses the y axis. Points on the y axis are those that correspond to ordered pairs of the type $(x, y) = (0, y)$. This means that the y value of the y intercept is that value of y for which $x = 0$. If we let $x = 0$ in our equation $y = {}^5/_2x - 4$, we get $y = -4$. Therefore, the y intercept of our line is located at $x = 0$ and $y = -4$ or the point $(0, -4)$.

Example 3 Graph the line $y = 2x + 1$. Find the y intercept and the slope.

x	$y = 2x + 1$
0	1
2	5

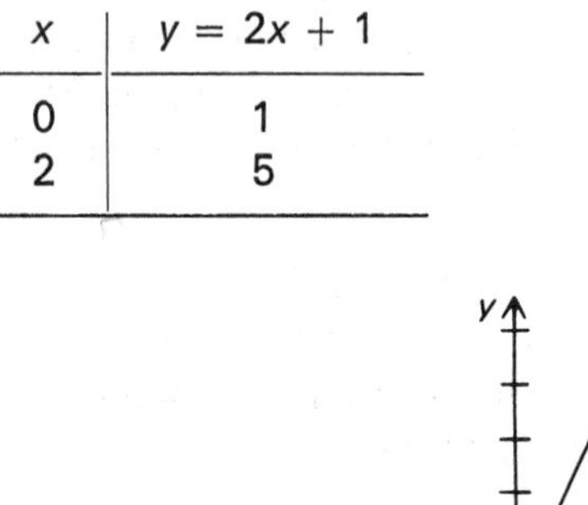

The line crosses the y axis at (0, 1). This is the y intercept.

To find the slope we can use any two points on the line. Let's use (0, 1) and (2, 5). We will let (0, 1) be the first point. So (0, 1) is (x_1, y_1), and (2, 5) is (x_2, y_2).

$$\text{slope} = \frac{y_1 - y_2}{x_1 - x_2} = \frac{1 - 5}{0 - 2} = \frac{-4}{-2} = \frac{2}{1} = 2$$

It is interesting to note that in our equation $y = 2x + 1$ the coefficient of x (that is, 2) is the slope of the line, and the constant term (that is, 1) is the y value of the y intercept.

The previous two examples have a lot in common. Let's abstract their common feature into a general statement about lines and their graphs.

> Every line corresponds to an equation of the form $y = mx + b$ where the coefficient of x (that is, m) is the slope of the line and the constant term (that is, b) is the y value of the y intercept.

Conversely:

> The graph of an equation $y = mx + b$ is a line. The slope of the line is m and the y value of the y intercept is b.

Example 4 Graph $y = -2x + 1$: Whenever we have an equation of the form $y = mx + b$, we know that the graph is a line. Two points determine a line. Therefore we need find only two ordered pairs (x, y) that satisfy the relation $y = -2x + 1$. As before, we are free to pick any value for x; once we choose a value for x, we use the equation $y = -2x + 1$ to determine the corresponding value for y. Thus, for the choice $x = 0$, we have $y = -2(0) + 1 = 1$. And for $x = 2$ we have $y = -2(2) + 1 = -3$.

x	$y = -2x + 1$
0	1
2	−3

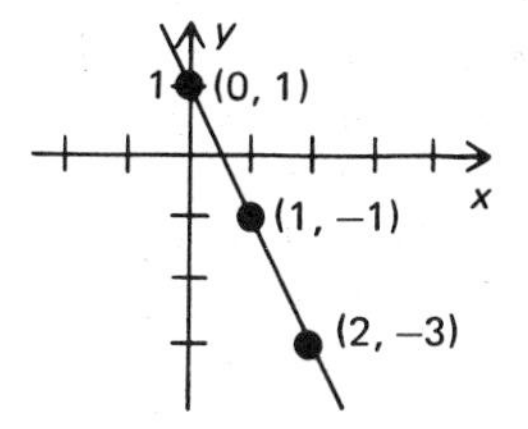

Notice that the line goes through the point $(1, -1)$, and that this ordered pair also satisfies the relation $y = -2x + 1$. In fact any point on the line satisfies the relation. Since the equation is of the form $y = mx + b$, we see that the slope m is -2 and the y value of the y intercept is 1 (that is, $b = 1$).

Example 5 Graph $2y - 3x = 6$: First we solve the equation for y. If it fits the form $y = mx + b$, we know the graph is a line.

$$2y - 3x = 6$$

$$2y = 3x + 6$$

$$y = \frac{3x}{2} + \frac{6}{2}$$

$$y = \frac{3}{2}x + 3$$

The equation does fit the form with $m = 3/2$ and $b = 3$. Now we find two solution pairs (x, y) of $y = 3/2x + 3$; locate the points on the xy plane, and connect them with a line.

x	$y = 3/2x + 3$
0	3 since $3/2(0) + 3 = 3$
-4	-3 since $3/2(-4) + 3 = -3$

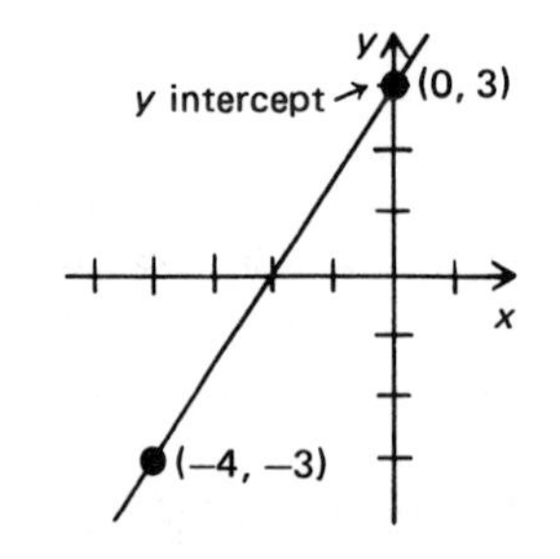

The slope is $3/2$ and the y value of the y intercept is 3.

Any equation in two variables that does not fit the form $y = mx + b$ is not a line. Consequently, we will need to plot more than two solution pairs to graph such an equation.

Example 6 Graph $y = x^2 + 1$: This equation is not a line because x is to the second power. The equation does not fit the form $y = mx + b$. We will need quite a few points to graph this equation.

x	$y = x^2 + 1$
0	$y = 0^2 + 1 = 1$
1	$y = 1^2 + 1 = 2$
-1	$y = (-1)^2 + 1 = 2$
2	$y = 2^2 + 1 = 5$
-2	$y = (-2)^2 + 1 = 5$
3	$y = 3^2 + 1 = 10$
-3	$y = (-3)^2 + 1 = 10$

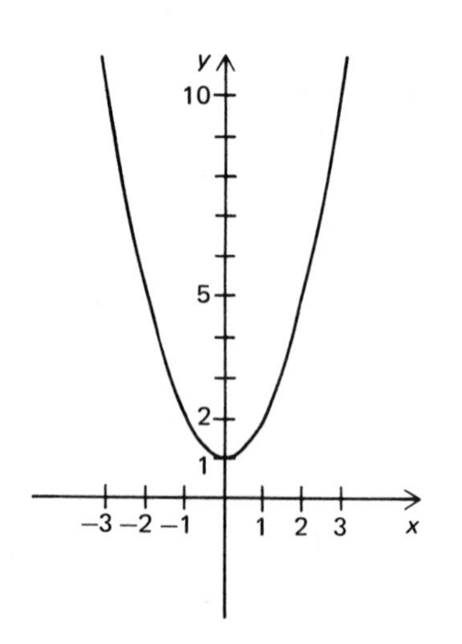

A curve of this shape is called a *parabola*.

PROGRAM 9.5

1. We have been graphing lines. The property that distinguishes a line from other graphs is that it does not bend anywhere. Which of the following are graphs of a line?

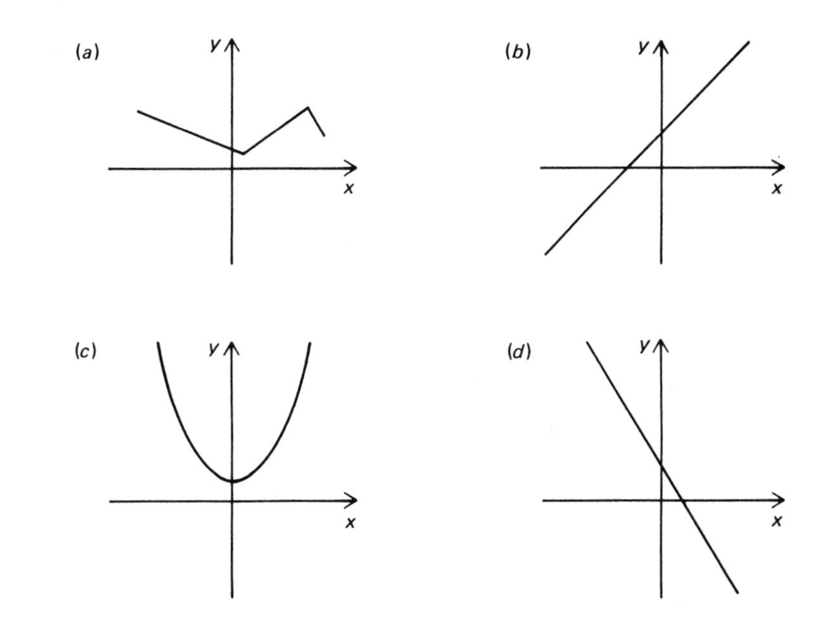

(b); (d)

2. Consider two points on a line, (4, 6) and (1, 2). The *rise* corresponds to the difference of the y coordinates. The *run* is the difference of the x coordinates.

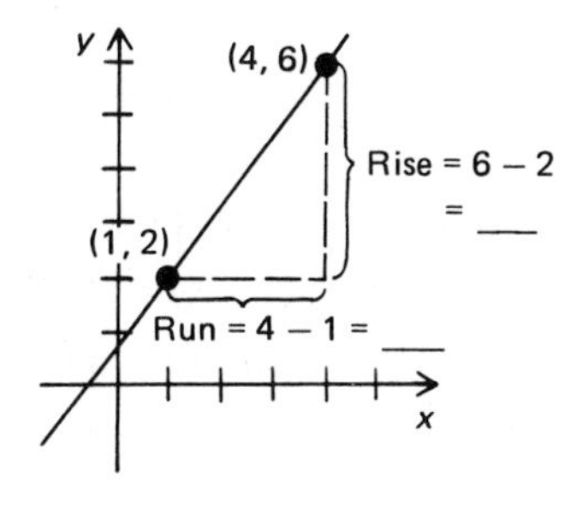

4 (rise)
3 (run)

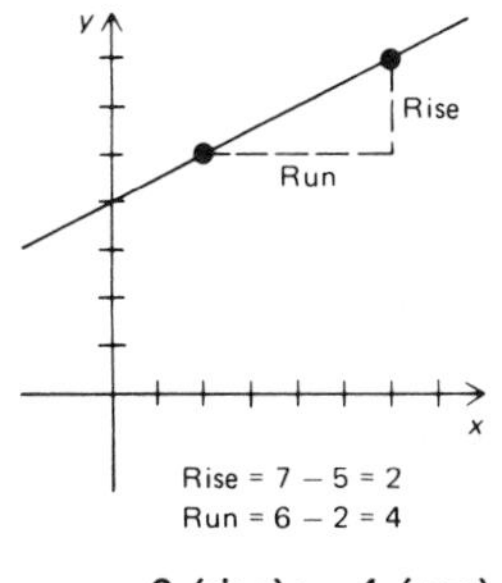

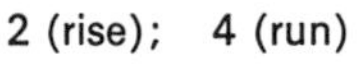

2 (rise); 4 (run)

3. Compute the rise and run between the points (2, 5) and (6, 7).

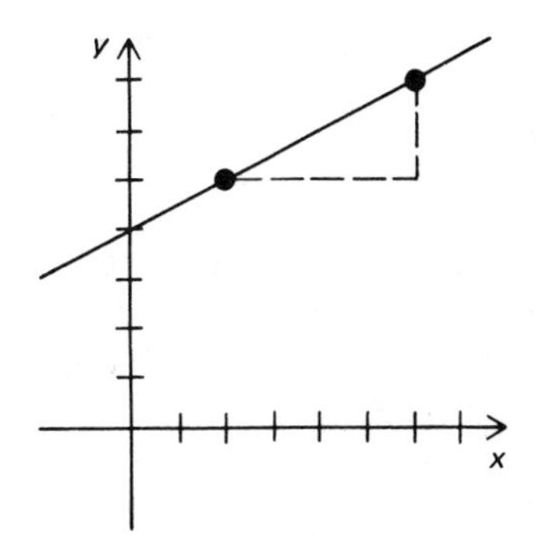

rise

4. The (rise\run) is a vertical measure or the difference of the y coordinates.

run
horizontal

5. The difference of the x coordinates gives the (rise\run). The run is a (horizontal\vertical) measure.

6. The ratio or fraction of the rise over the run is a very important concept in the study of lines, for no matter what two points on the line we select, this fraction will be the same.

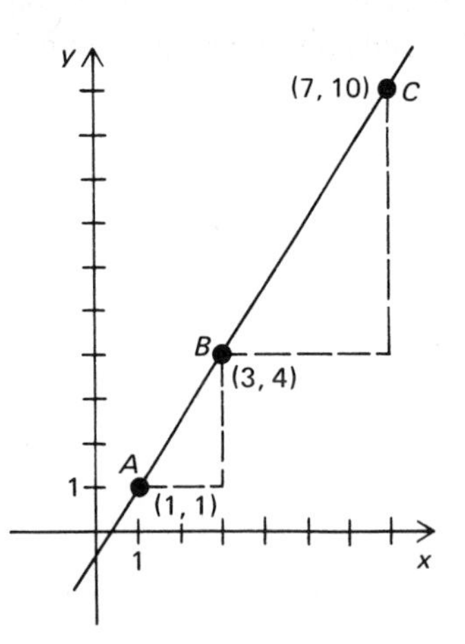

Here, we see that using the points A and B,

$$\frac{\text{rise}}{\text{run}} = \frac{\rule{2cm}{0.4pt}}{2}$$

4 − 1 or 3

If we use B and C, we get

$$\frac{\text{rise}}{\text{run}} = \frac{6}{\rule{2cm}{0.4pt}} = \rule{2cm}{0.4pt}$$

7 − 3 or 4; $^3/_2$

Are the two fractions equal? (yes\no).

yes

7. The ratio of the rise to the run is called the *slope of a line*

$$\text{slope} = \frac{\rule{2cm}{0.4pt}}{\text{run}}$$

rise

8. To compute the slope of a line, we need two points. We designate the first point as (x_1, y_1). We read x_1 as "x sub one" and y_1 as "y sub one." We call the second point (x_2, y_2). The notation x_2 reads "x sub ________" while y_2 reads "y (sub\lower) 2."

two

sub

$$\text{slope} = \frac{\text{rise}}{\text{run}} = \frac{y_1 - y_2}{x_1 - x_2}$$

Compute the slope of the line through (4, 8) and (2, 6). Let (4, 8) be the first point or (x_1, y_1). Then (2, 6) is the second point or (________).

x_2, y_2

$$\text{slope} = \frac{y_1 - y_2}{x_1 - \rule{1cm}{0.4pt}} = \frac{8 - 6}{4 - \rule{1cm}{0.4pt}} = \rule{1cm}{0.4pt}$$

2; 2; 1

9. It does not matter which point we designate as the first point or (x_1, y_1). But once we call a point the first point, we must stick to the designation.

Let's compute the slope of the line through (1, 5) and (2, 8). First let (1, 5) be (x_1, y_1), and (2, 8) be (x_2, y_2).

$$\text{slope} = \frac{5 - \rule{1cm}{0.4pt}}{1 - 2} = \rule{1cm}{0.4pt}$$

8; $\frac{-3}{-1}$ or 3

Now let (2, 8) be the first point or (x_1, y_1), and (1, 5) be (x_2, y_2).

$$\text{slope} = \frac{8 - \rule{1cm}{0.4pt}}{2 - 1} = \rule{1cm}{0.4pt}$$

5; $\frac{3}{1}$ or 3

yes

no

Are the slopes the same? (yes\no). Does it matter which point is the "first" point? (yes\no).

10. The slope of a line does not change no matter what two points on the line we use. This is another way of saying the line does not bend. Compute the slope of this line by using the designated points.

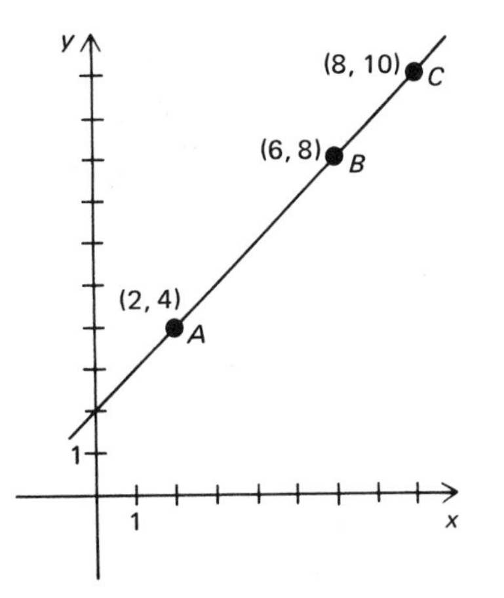

For points A and B, with B as the first point,

6; 2; $^4/_4 = 1$

$$\text{slope} = \frac{\text{rise}}{\text{run}} = \frac{8 - 4}{\_\_ - \_\_} = \_\_\_\_\_\_$$

For points B and C, with B as the first point,

$6 - 8; \dfrac{-2}{-2} = 1$

$$\text{slope} = \frac{\text{rise}}{\text{run}} = \frac{8 - 10}{\_\_\_\_\_\_} = \_\_\_\_\_\_$$

yes

Are the slopes the same? (yes\no).

11. Compute the slope of each line.

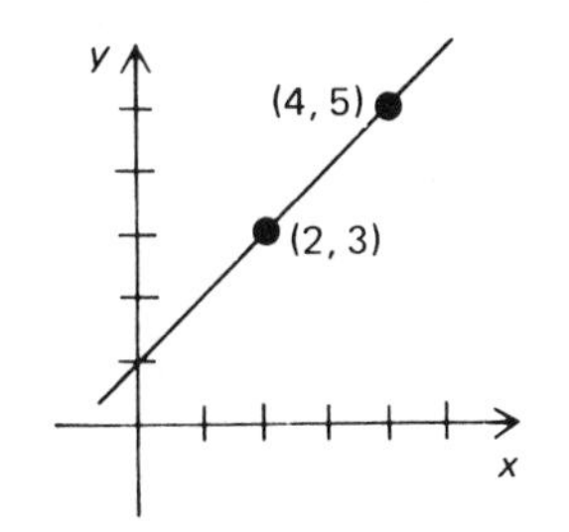

4 − 2; $^2/_2 = 1$

$$\text{slope} = \frac{5 - 3}{\_\_\_\_\_\_} = \_\_\_\_\_\_$$

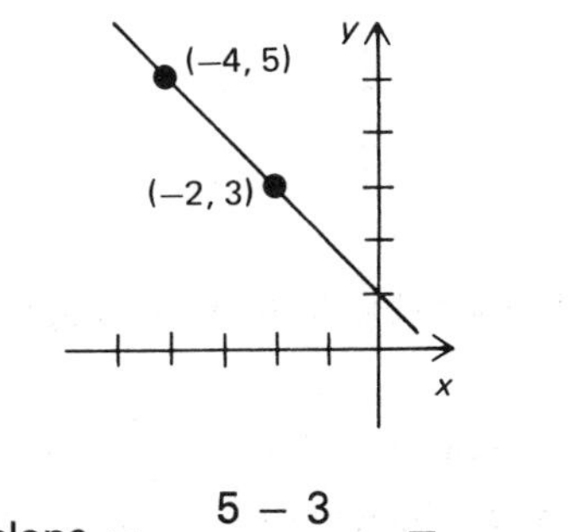

$\dfrac{2}{-2} = -1$

$$\text{slope} = \frac{5 - 3}{-4 - (-2)} = \_\_\_\_\_\_$$

Use the previous two graphs and slopes to answer the following: If the line goes *up* as we go to the right, the slope is (positive\negative). If the line goes *down* as we go to the right the slope is (positive\negative).

positive
negative

12. Is the slope of the line shown here positive or negative?

negative

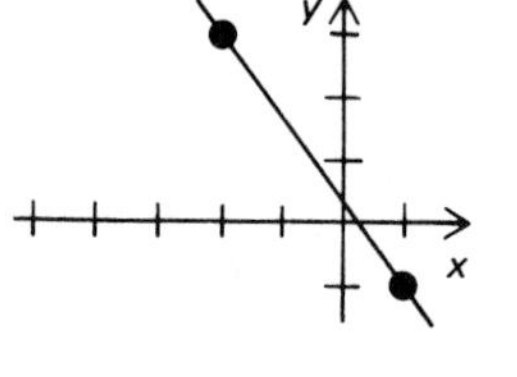

13. Graph the line $y = 3x + 2$.

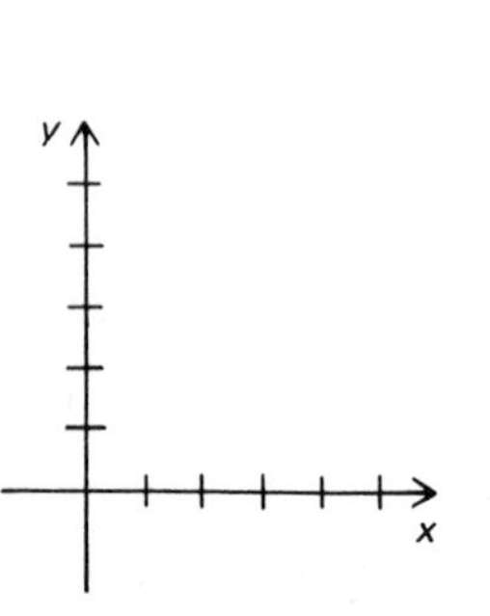

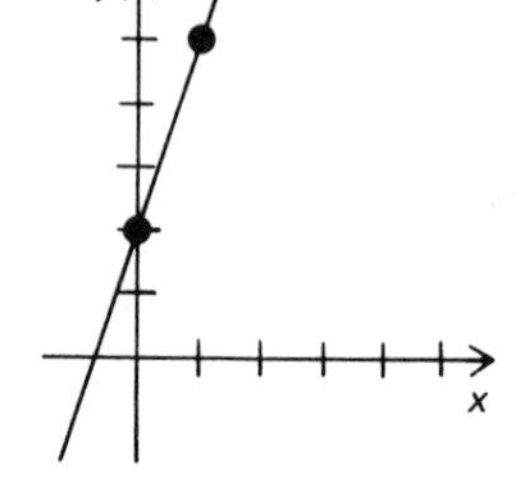

x	y
0	2
1	5

Use the points in the table to compute the slope. Let (0, 2) be the first point.

slope = ________ = ________

$\frac{2-5}{0-1}$; $\frac{-3}{-1} = 3$

The slope is _____. The coefficient of x in $y = 3x + 2$ is _____. Are these two values the same? (yes\no).

3; 3
yes

14. Graph the line $y = -2x + 1$:

x	y
0	—
3	−5

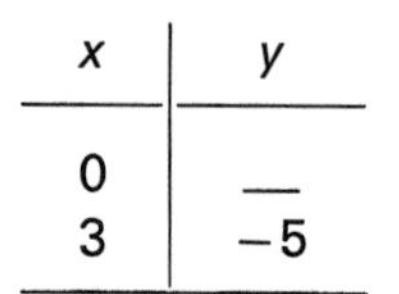

x	y
0	1
3	−5

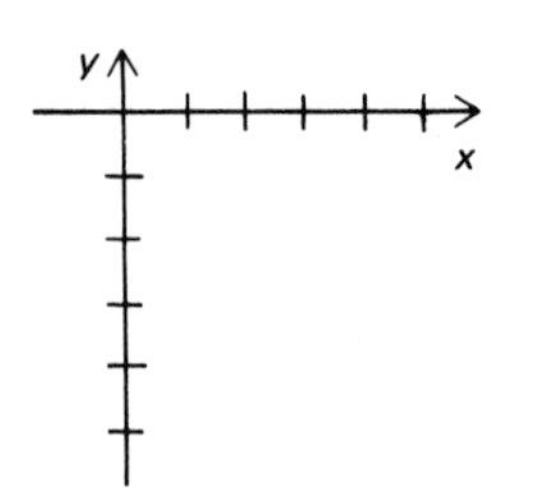

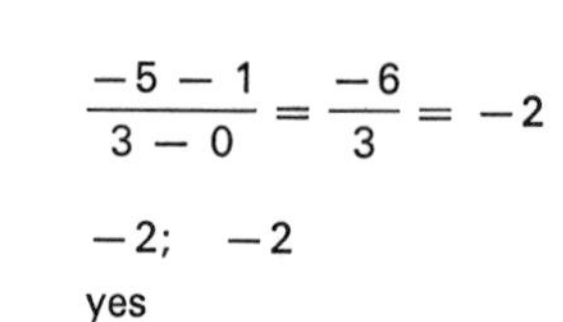

Use the two points in the table to find the slope. Let (3, −5) be the first point.

slope = ________

$\frac{-5-1}{3-0} = \frac{-6}{3} = -2$

The slope is _____. The coefficient of x in $y = -2x + 1$ is _____. Are these values equal? (yes\no).

−2; −2
yes

15. In addition to the slope, there is another important aspect of a line. It is called the y intercept.

The *y intercept* of a line is that point where the line crosses the y axis.

In each of the accompanying diagrams label the y intercept and find the x and y coordinates of the y intercept.

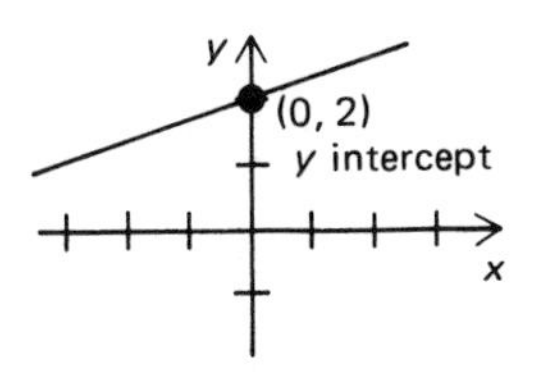

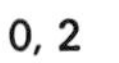

0, 2

The y intercept is (__, __).

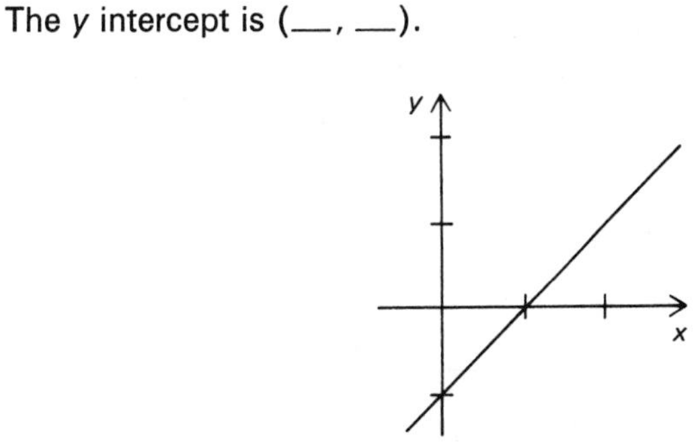

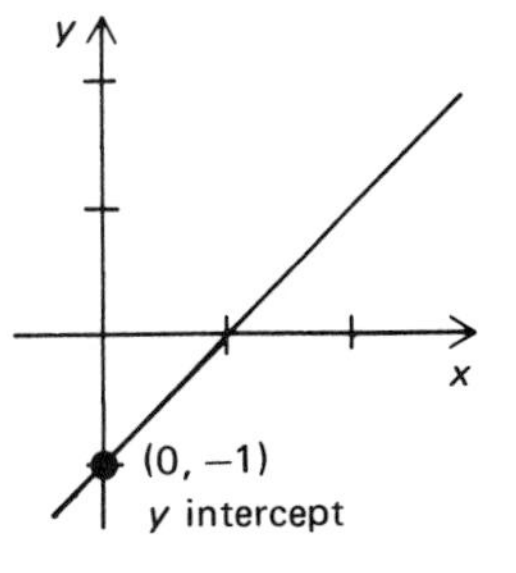

(0, −1)

The y intercept is ________.

16. If a line is not vertical (straight up and down), then the general equation corresponding to that line is $y = mx + b$. If we graph the equation $y = mx + b$, the resulting line will have slope m and the y value of the y intercept will be b. Consider this line, for example.

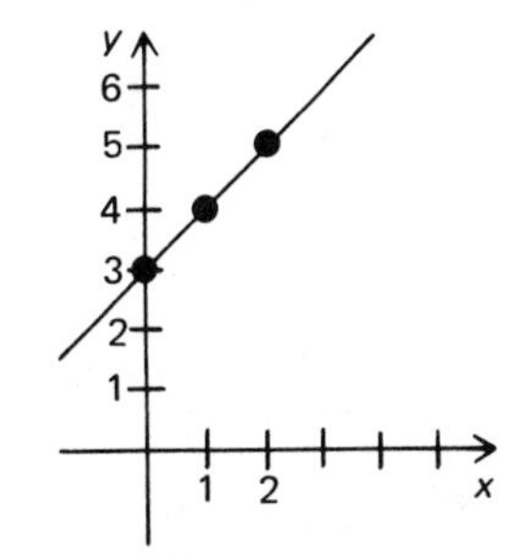

0, 3; 3

The y intercept is (__, __), so the y value of the y intercept is ____.

The points (1, 4) and (2, 5) are also on this line. If we call (1, 4) the first point and (2, 5) the second, then

4 − 5 or −1

$$\text{rise} = \_\_\_\_\_\_$$

1 − 2 or −1

$$\text{run} = \_\_\_\_\_\_$$

So the slope is given by

$\frac{-1}{-1}$ or 1

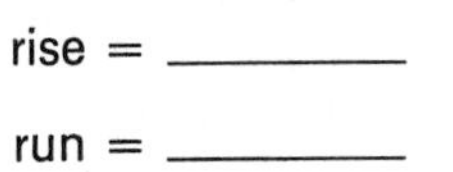

$$\text{slope} = \frac{\text{rise}}{\text{run}} = \_\_\_\_$$

slope

y intercept

1; 3

The general equation for a line is $y = mx + b$, where m is the ________ and b is the y value of the ____________. In our example the slope is $m =$ ____ and the y value of the y intercept is $b =$ ____. Therefore the equation corresponding to the graphed line is

1; 3

$$y = mx + b \qquad \text{or} \qquad y = \_\_x + \_\_\_\_$$

17. In the last frame we were given the line and we found the equation for that line. Now we are given the equation $y = {}^3/_2x + 1$, and are asked to find the graph. Let us compare $y = {}^3/_2x + 1$ with $y = mx + b$.

$$y = mx + b$$
$$y = {}^3/_2x + 1$$

The specific equation fits the form $y = mx + b$ if we let $m =$ ____ and $b =$ ____. So the graph (will be\need not necessarily be) a straight line. It will have slope $m =$ ____, and the y intercept will be the point

$$(0, b) = (0, \_\_)$$

${}^3/_2$

1; will be

${}^3/_2$

1

Since we know that the graph of our equation will be a straight line, we really only need to plot two points. For two points are enough to determine a straight line.

x	$y = {}^3/_2x + 1$	*Ordered pairs*
-2	$y = {}^3/_2(-2) + 1 = -2$	$(-2, -2)$
-1	$y = {}^3/_2(-1) + 1 =$ __	$(-1,$ __$)$

$-{}^1/_2$; $-{}^1/_2$

Now locate the ordered pairs on the xy plane and connect the points with a line.

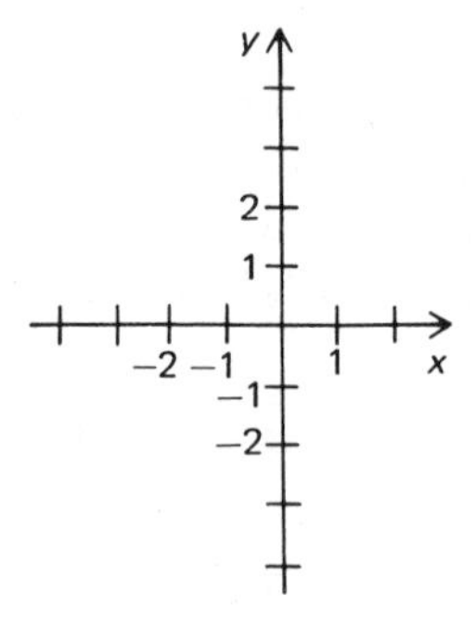

18. Graph $6y + 2x = 12$: Also find the slope and y intercept. First we solve this equation for y and see if it fits the form $y = mx + b$.

$$6y + 2x = 12$$
$$6y = -2x + 12$$
$$y = -{}^2/_6x + {}^{12}/_6$$
$$y = -{}^1/_3x + 2$$

The last equation fits the form $y = mx + b$ if we let $m =$ ____ and $b =$ ____. So the shape of the graph is a (circle\straight line\squiggle).

The slope of our line is $m =$ ____, and the y intercept is the point $(0, b)$ or $(0,$ ____$)$.

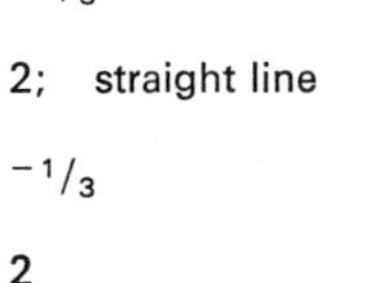

$-{}^1/_3$

2; straight line

$-{}^1/_3$

2

We need only plot two points to graph this line. Our work is organized in a table.

x	$y = -{}^1/_3x + 2$	*Ordered pairs*
0	2	(0, 2)
3	____	(__, __)

1; 3; 1

The point (0, 2) has been plotted here. Plot the other point and draw the graph of $6y + 2x = 12$.

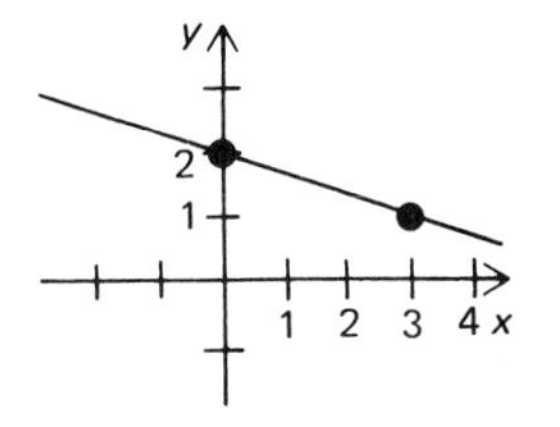

y
(0, 2)
x

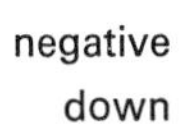
negative
down

Notice that the slope has been positive in our previous examples. In those examples the lines went up from left to right. But in our present example the slope is (positive\negative). In this example the line goes (up\down) from left to right.

19. Graph $2y + 3x = -4$. First we solve for y.

$-3/2$; -2

$$2y + 3x = -4$$

$$2y = -3x - 4$$

$$y = \_\_\_\_ x + \_\_\_\_$$

yes; $-3/2$

down

$(0, -2)$

Is this the equation of a line? (yes\no). What is the slope? ____. Does the line go up or down as we go from left to right on the xy plane? ________.

Give the y intercept. ________.

Since we have the y intercept, we already know one point on the line. It suffices to find one more ordered pair that satisfies the equation.

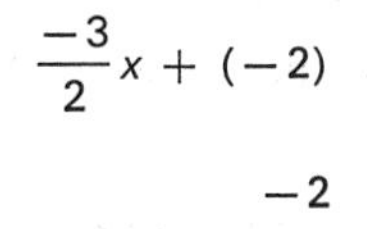
$\frac{-3}{2}x + (-2)$

-2
(any x value);
(corresponding y value)

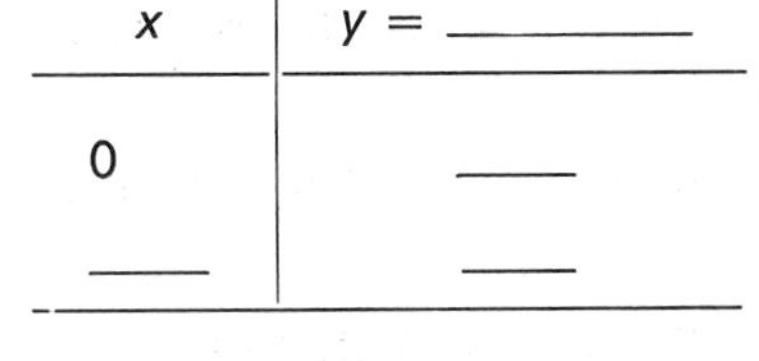

x	$y =$ ________
0	____
____	____

(Remember, you can pick any value you like for x.)

Now plot the two ordered pairs and draw the graph.

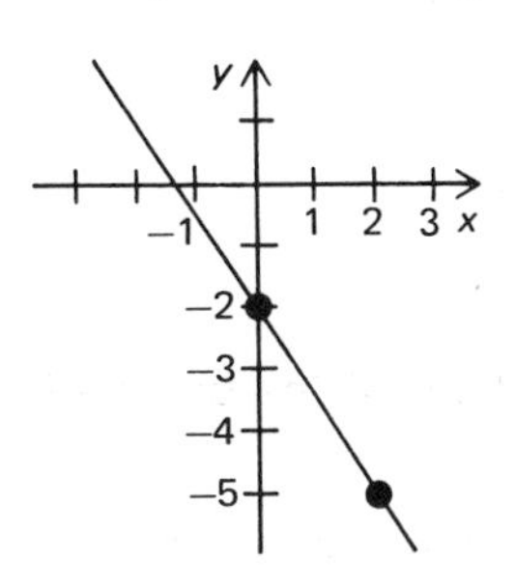

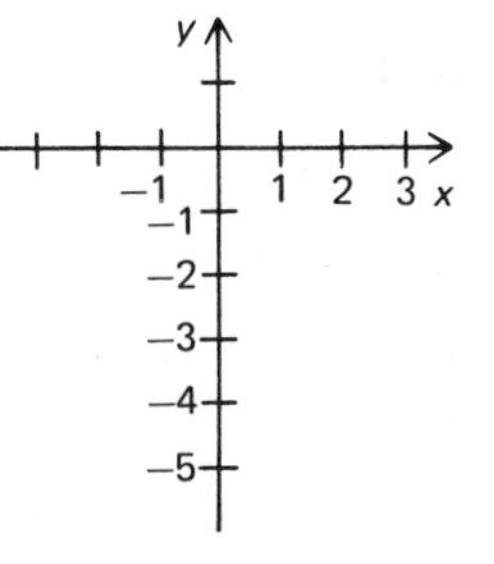

is not; no
no

20. If an equation in x and y does not fit the form $y = mx + b$, then the graph (is\is not) a straight line. Does $y = x^2$ fit the form $y = mx + b$? (yes\no). Is the graph a straight line. (yes\no).

Let's graph $y = x^2$. We will need several points to determine the shape. Fill in the table, then plot the points and connect them with a smooth curve.

x	y
0	0
1	1
−1	1
2	—
−2	—
3	—
−3	—

4

4

9

9

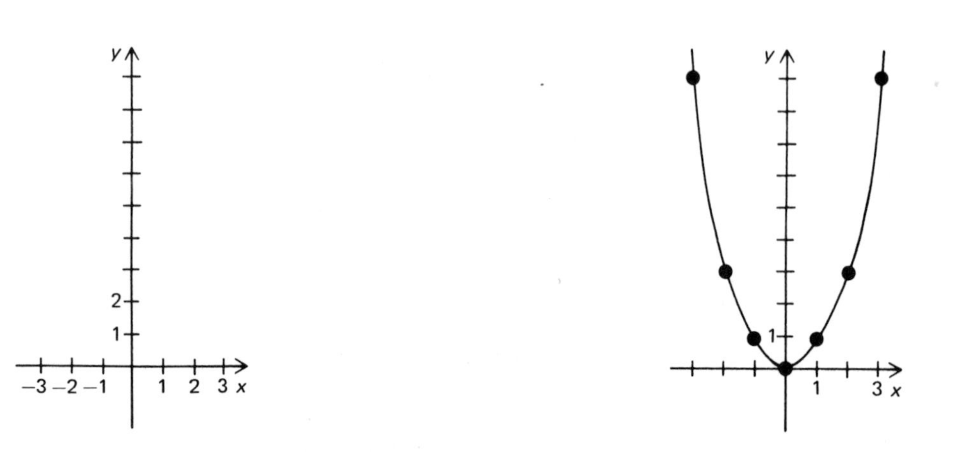

PRACTICE PROBLEMS

1. Plot the point (1, 3) and (5, 9). Indicate the rise and the run. Find the slope of the line connecting these points.
2. Plot the points (−2, 4) and (1, −1). Indicate the rise and run between the points. Find the slope.
3. Plot the points (−5, −1.2) and (−2, 3.2). What is the slope of the line connecting these points?
4. Plot the points (−2, 3) and (4, 0). What is the slope of the line connecting these points?

For Problems 5–10 use the slope formula to find the slope of the line through the given points.

5. (0, 1) and (2, 7)
6. (3, 6) and (4, 5)
7. (−1, −2) and (3, 4)
8. (−2, −4) and (6, 8)
9. (0, 1) and (1, 0)
10. (0, 3) and (1, 3)
11. Just look at the equation to determine the y intercept and the slope of the graph of $y = 2x + 3$.
12. Without plotting the graph indicate the slope and y intercept of the graph of

$$y = -2.5x + 18$$

13. Without looking at the graph, determine the slope and y intercept of the graph of

$$2y - 3x = 10$$

14. Two lines are parallel if and only if they have the same slope. Which of the following lines are parallel?

a. $y = 3x - 1$ *b.* $y = -1.5x + 4$ *c.* $y = 3x + 12$
d. $3x + 2y = 0$ *e.* $5y - 15x = 10$ *f.* $2y = 6x + 9$

15. Determine the equation of the line whose slope is 3 and whose y intercept is the point (0, 10). Graph this line.

16. Find the equation of a line with slope -4.5 and y intercept (0, 6). Graph this line.

17. A reservoir has 2,000 gallons of water in it. Water starts flowing into the reservoir at the rate of 100 gallons/hour. Let x be the number of hours passed since the water started flowing. Let y be the total number of gallons in the reservoir at time x.

a. Find a formula for y in terms of x.
b. What is the slope and y intercept of your formula?
c. Graph your formula.

SELFTEST

1. Plot the points (1, 2) and (5, 4). Indicate the rise and run between the points.

2. Find the slope of a line through the points (2, 4) and (1, 2).

3. Without graphing the equation $y = 2x + 1$, give
a. the slope. *b.* the y intercept.

4. *a.* Is the slope of $y = -3x + 2$ positive or negative?
b. As we move to the right on the x axis, will the line go up or down?

5. Give the equation of a line with slope 2 and y intercept (0, 3).

SELFTEST ANSWERS

1.

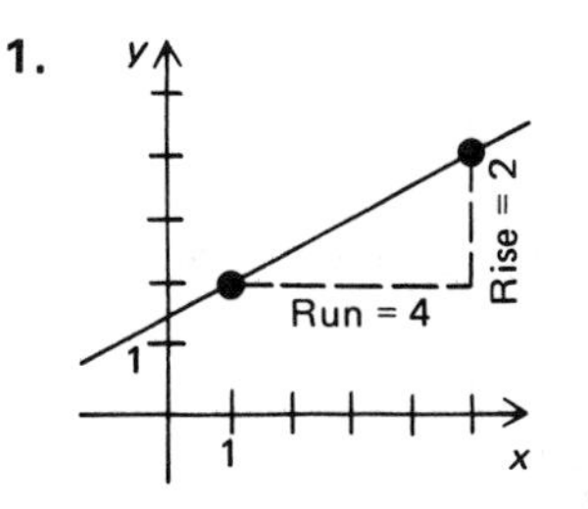

2. 2 3. *a.* 2 *b.* (0, 1) y intercept
4. *a.* negative *b.* down 5. $y = 2x + 3$

9.6 GRAPHING REGIONS OF A PLANE

In the last section we graphed functions corresponding to equations of the type $y = mx + b$. We found that the graph of such an equation was a line. In this section we will graph regions of the plane which are bounded by lines.

Example 1 Graph that region of the xy plane for which every point satisfies the inequalities $2 \leq x \leq 5$. The conditions $2 \leq x \leq 5$ is satisfied by all points (x, y) whose first coordinate lies in the x interval from 2 to 5. This region is the shaded column.

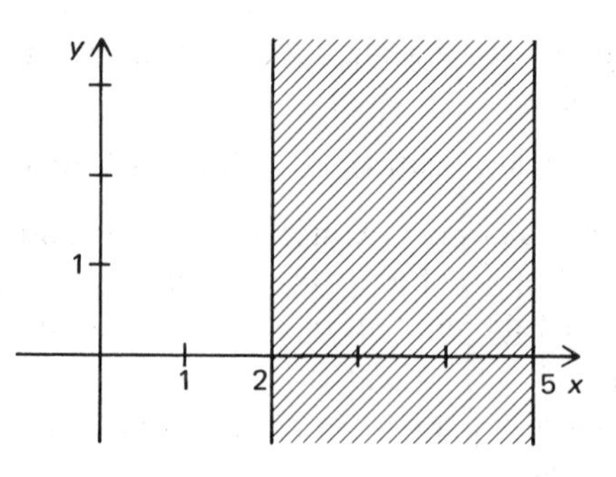

Notice that the vertical lines on the boundary of this region are included among the points which satisfy $2 \leq x \leq 5$.

Example 2 Graph that region of the xy plane for which every point satisfies both the inequalities $2 \leq x \leq 5$ and $1 \leq y \leq 2$. The condition $2 \leq x \leq 5$ was treated in the last example. The condition $1 \leq y \leq 2$ is satisfied by all points (x, y) whose second coordinate lies in the y interval from 1 to 2. This region is shaded in the figure. The horizontal lines on the boundary of this region are included among

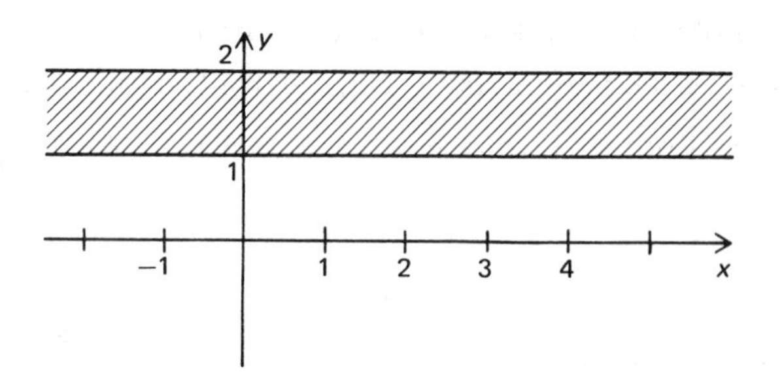

the points which satisfy $1 \leq y \leq 2$. The set of points which satisfy *both* conditions $2 \leq x \leq 5$ and $1 \leq y \leq 2$ is the set of points common to both the row and column. This is the shaded region. We see that the region satisfying both

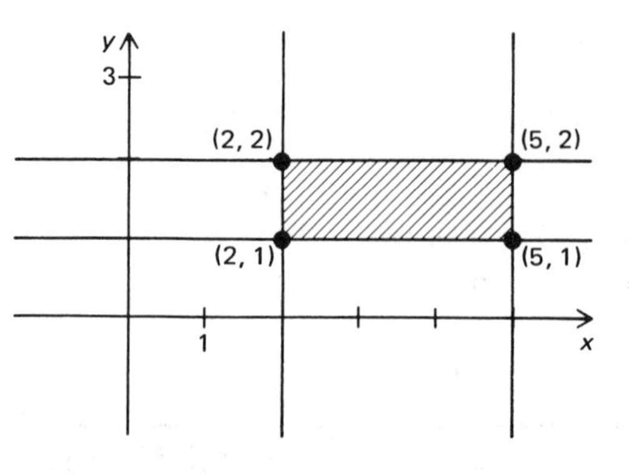

$2 \leq x \leq 5$ and $1 \leq y \leq 2$ is just the rectangle whose vertices are the points (2, 2), (5, 2), (5, 1), and (2, 1).

A line divides the plane into three parts. One part is the line itself. If a point (x, y) is not on

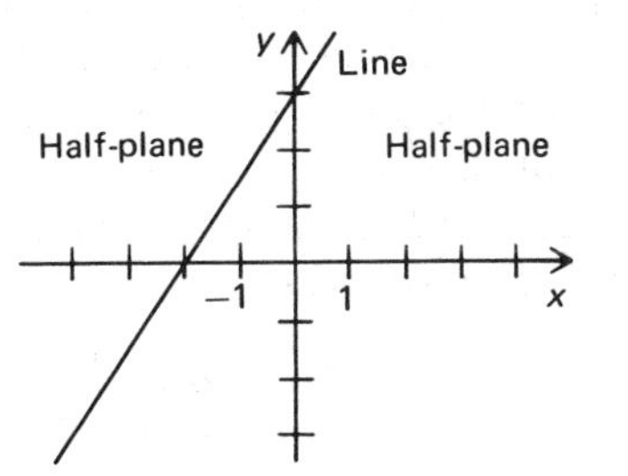

the line then it is in one of the half-planes on either side of the line. All the points in one of the half-planes satisfy the inequality $y < mx + b$, while all the points in the other half-plane satisfy the opposite inequality $y > mx + b$. Of course the points on the line satisfy the equality $y = mx + b$.

If we are to graph an inequality, how do we decide which half-plane to use? The following rule describes the situation.

Rule to graph inequalities: **First graph the line. To decide which half-plane to use, select a test point in one of the half-planes determined by the line. Substitute the coordinates of the test point in the original inequality. If the test point makes the inequality true, then use the half-plane that contains the test point. If the test point makes the inequality false, then use the *other* half-plane.**

Example 3 Graph the region of the plane which corresponds to the inequality $y \geq 2x + 1$. First we graph the line $y = 2x + 1$. To do this we need only two points on the line.

x	$y = 2x + 1$
0	1
1	3

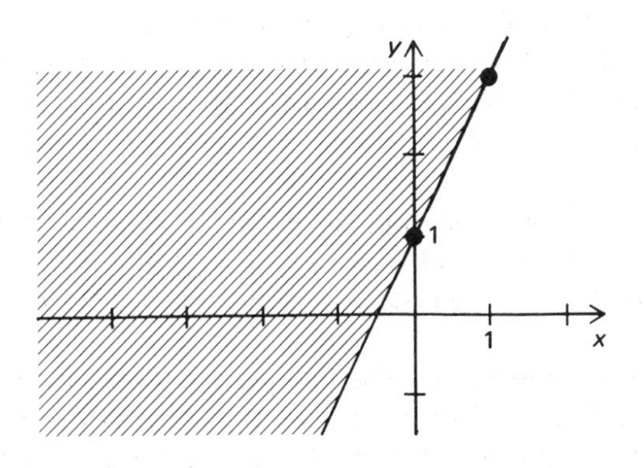

Now the line $y = 2x + 1$ divides the plane into two half-planes. One of these half-planes is included in the graph of $y \geq 2x + 1$. Which one? We use the rule to decide. If the inequality $y \geq 2x + 1$ is true for our selected test point, then the half-plane in which that test point is located is included in the graph of $y \geq 2x + 1$

Let us use the test point $(-1, 2)$. This point is located in the upper half-plane. Putting $x = -1$ and $y = 2$ in $y \geq 2x + 1$ we get $2 \geq 2(-1) + 1$ or $2 \geq -1$. The last statement is true. It follows that the upper half-plane is included in the graph of $y \geq 2x + 1$. The points on the line satisfy $y = 2x + 1$. So the graph of $y \geq 2x + 1$ is the upper half-plane with the line.

Example 4 Graph the region of the plane corresponding to the inequality $9x + 3y < 12$. First we must find the appropriate line with which to divide the plane. We use the line $9x + 3y = 12$. But no point on the line $9x + 3y = 12$ makes the inequality $9x + 3y < 12$ true. So we used a dashed line for the graph of $9x + 3y = 12$ rather than a solid line to indicate that the line itself is not part of the region that satisfies $9x + 3y < 12$.

To graph $9x + 3y = 12$, it is best to solve the equation for y. Then the equation will be in standard form for a line.

$$\begin{aligned} 9x + 3y &= 12 \\ 3y &= 12 - 9x \\ y &= 4 - 3x \\ y &= -3x + 4 \end{aligned}$$

x	$y = -3x + 4$
0	4
1	1

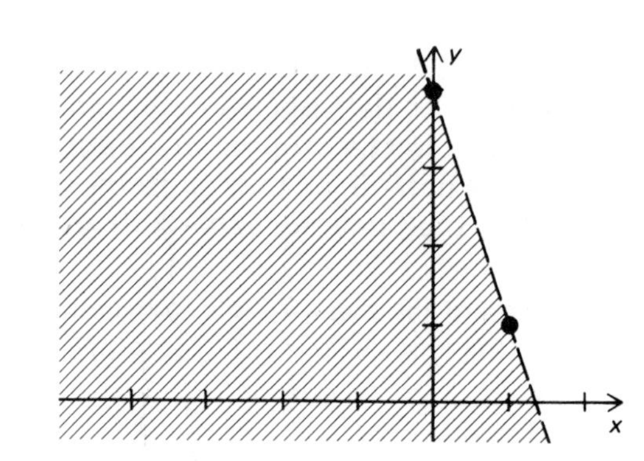

Which of the half-planes do we use for the graph of $9x + 3y < 12$? Let us use the test point (4, 5). This point is in the right half-plane. If we let $x = 4$ and $y = 5$ in $9x + 3y < 12$, we get $9(4) + 3(5) < 12$ or $51 < 12$. The last inequality is false. So the graph is *not* the right half-plane. The graph will include either the right or the left half-plane but not both. Our test point has shown that the right half-plane is not part of the graph. By simple deduction, the left half-plane is the graph.

In the examples we have seen inequalities which included the line and some which did not include the line. In general when the $>$ symbol or $<$ symbol is used the line is *not* included in the graph. But if either the $\leq$ symbol or $\geq$ symbol is used, the line is included in the graph.

Example 5 A club decides to make candy and sell it for a benefit drive. It costs \$3.00 to buy pots and pans. Then it costs \$0.25 to make one bar of candy. They decide to sell the candy at the rate of \$0.75 per bar. Find the break-even point and graph the region where the club will have a profit and the region where it will have a loss.

Let x be the number of bars of candy the club sells. Since they sell each bar for \$0.75, the amount received will be $R = 0.75x$. It costs \$0.25 to make one bar, so it will cost \$$0.25x$ to make x bars. But it also costs \$3.00 to buy the equipment, so the total cost function is $C = 0.25x + 3$. The break-even point is that place where the amount of money received equals the cost, that is, where $R = C$.

$$\begin{aligned} 0.75x = R = C &= 0.25x + 3 \\ 0.75x &= 0.25x + 3 \\ 0.50x &= 3 \\ x &= 6 \end{aligned}$$

So the club must sell 6 bars to break even. The club will lose money if the costs are greater than the receipts. That is if $C > R$. The loss will be $C - R$. The profit will occur when the receipts are greater than the costs or when $R > C$. In this case the profit will be $R - C$. Let's look at the graphs of the cost curve and the

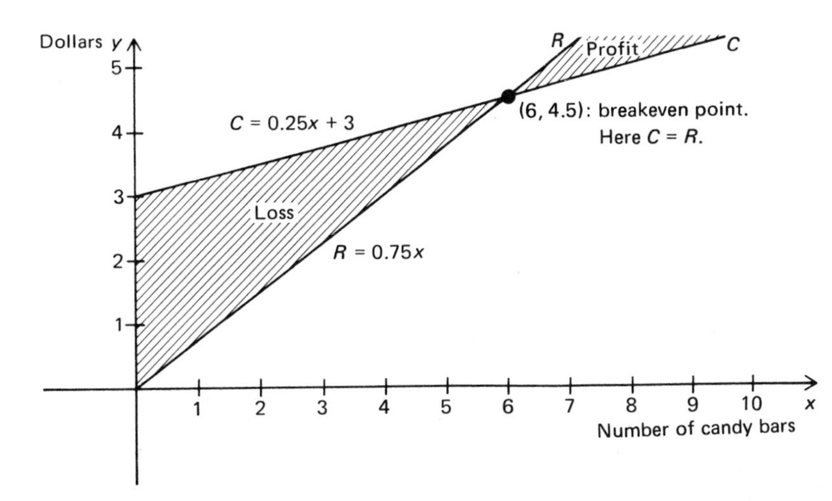

receipts curve. We see that the losses occur when the inequalities below are true.

Loss Region $\begin{cases} R \geq 0.75x & \text{above receipt curve} \\ C \leq 0.25x + 3 & \text{below cost curve} \end{cases}$

The profits occur when the opposite situation is true or when

Profit Region $\begin{cases} R \leq 0.75x & \text{below the receipts curve} \\ C \geq 0.25x + 3 & \text{above the cost curve} \end{cases}$

PROGRAM 9.6

1. In this section we will graph regions of the plane that are bounded by lines. A line divides the plane into three parts. One part is the line itself. The other parts are the half-planes on either side of the line.

   Rule to graph inequalities: First graph the line. To decide which half-plane to use, select a test point in one of the half-planes determined by the line. Substitute the coordinates of the test point in the original inequality. If the test point makes the inequality true, then use the half-plane that contains the test point. If the test point makes the inequality false, then use the *other* half-plane.

$-x + 3$

2. Graph the inequality $y < -x + 3$: First we graph the line $y =$ ________. Then we take any test point in one of the half-planes. The rule tells us if our test point lies in the correct half-plane.

   To graph $y = -x + 3$ we need only two solution pairs.

x	$y = -x + 3$	*Corresponding ordered pairs*
0	3	(0, 3)
1	—	(—, —)

2; 1, 2

Plot the two pairs and draw the graph of $y = -x + 3$. Now we pick a test point in one of the half-planes. Say we use the point (1, 10). Plot this point on your graph. The point (1, 10) is clearly in the (upper\lower) half-plane.

upper

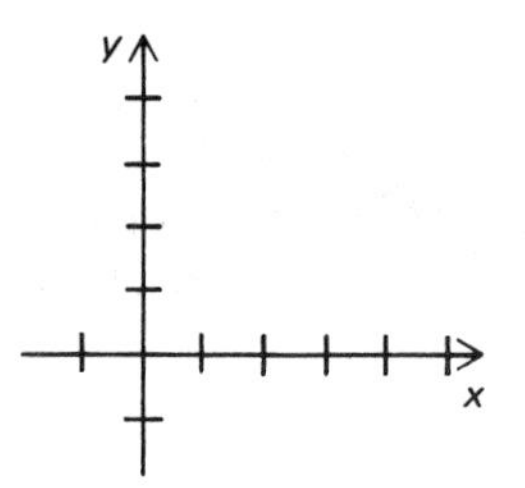

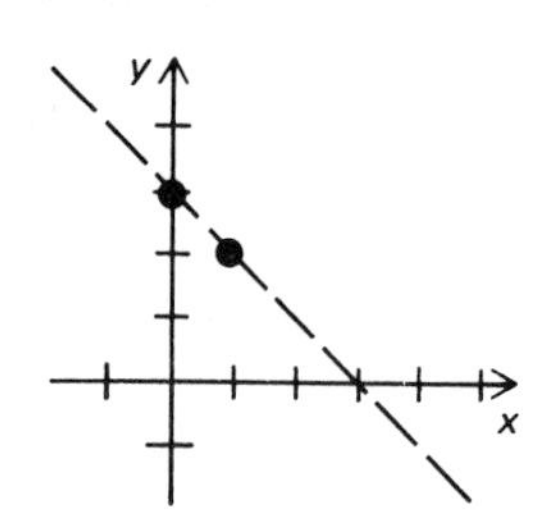

We use the rule to decide if the half-plane containing (1, 10) is the graph we want. We put the coordinates of (1, 10) into the inequality. If these coordinates make the inequality true then the half-plane that contains (1, 10) is our graph. Otherwise the other half-plane is our graph.

The x and y coordinates of (1, 10) are $x =$ ____ and $y =$ ____. Putting these into $y < -x + 3$ we get $10 < -1 + 3$, which is (true\false).

1; 10

false

Is the half-plane which contains (1, 10) the graph we are looking for? (yes\no). Which half-plane must be the graph of $y < -x + 3$? (upper half-plane\lower half-plane.) Now go back to the graph showing $y = -x + 3$ and shade the region that is the graph of $y < -x + 3$.

no

lower half-plane

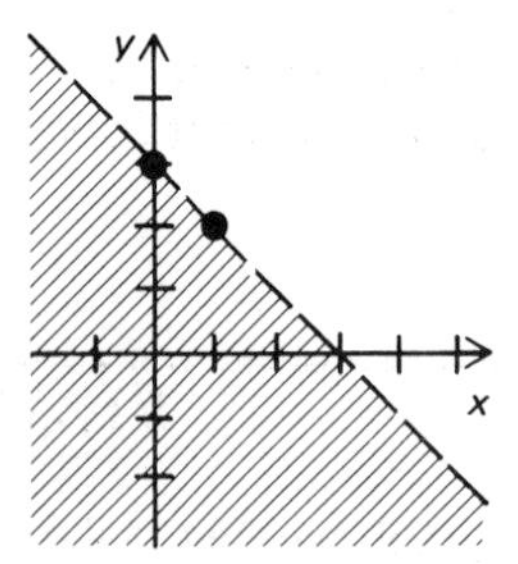

3. Graph $y > 4x - 1$. There are two main steps: First, graph the line $y = 4x - 1$. Second, pick any point on either side of the line, and substitute that point's coordinates into the inequality $y > 4x - 1$. If the inequality is true, shade the half-plane that contains the test point. If the inequality is false, shade the other half-plane.

So first we graph the line $y = 4x - 1$.

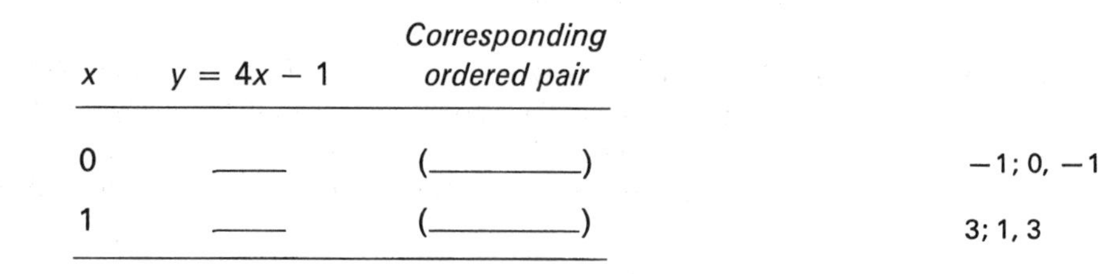

x	$y = 4x - 1$	*Corresponding ordered pair*
0	____	(________)
1	____	(________)

-1; 0, -1

3; 1, 3

Graph the ordered pairs and draw the graph of $y = 4x - 1$. Now pick any point on either side of the line. To be concrete let us take $(-5, 1)$ as our test point. Plot $(-5, 1)$ on the graph above. This point lies in the half-plane to the (left\right) of the line $y = 4x - 1$.

left

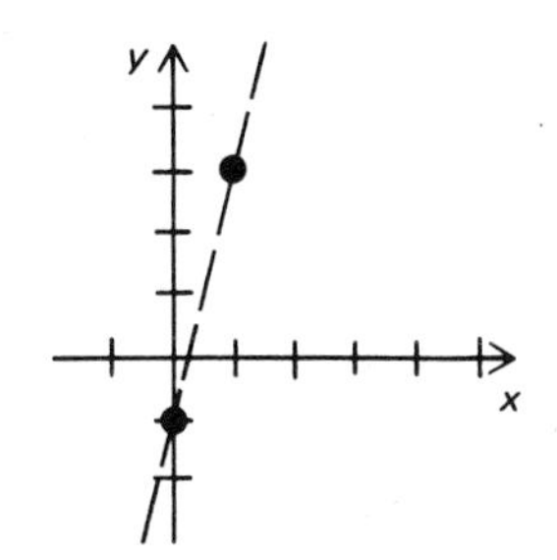

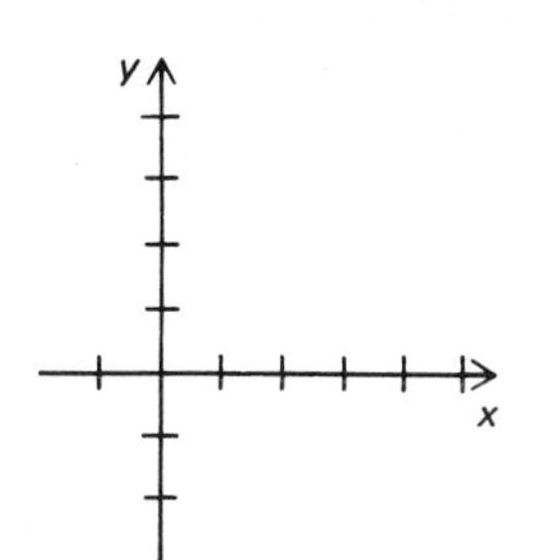

−5; 1
true

The x and y coordinates of $(-5, 1)$ are $x =$ ____ and $y =$ ____. If we put these coordinates in the inequality $y > 4x - 1$, we get the (true\false) statement $1 > 4(-5) - 1$.

true

The rule now assures us that $y > 4x - 1$ will remain (true\false) for any point in the half-plane containing $(-5, 1)$. So the graph of $y > 4x - 1$ (is\is not) the half-plane that contains our test point.

is

Now return to your graph of $y = 4x - 1$ and shade the region that corresponds to the graph of $y > 4x - 1$.

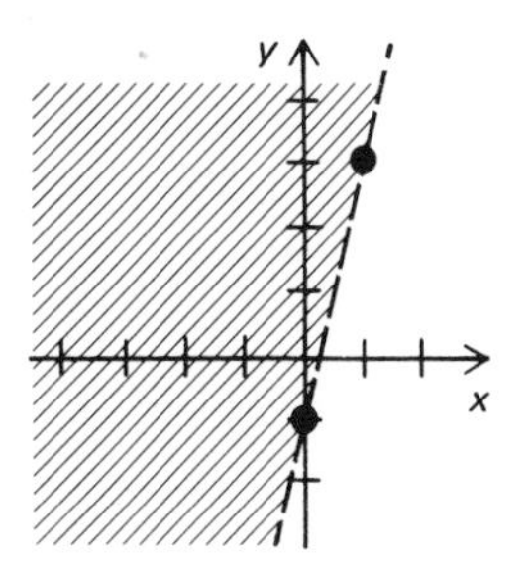

4. Graph $y \geq 4x - 1$: We recognize this as being almost the same as our last example. The "greater than" symbol has been replaced by the "greater than or equal" symbol. This means that all points that satisfy $y > 4x - 1$ or $y = 4x - 1$ are included in the graph. In this case the graph of $y \geq 4x - 1$ (includes\does not include) the line $y = 4x - 1$ as well as the half-plane to the left of the line. Graph $y \geq 4x - 1$.

includes

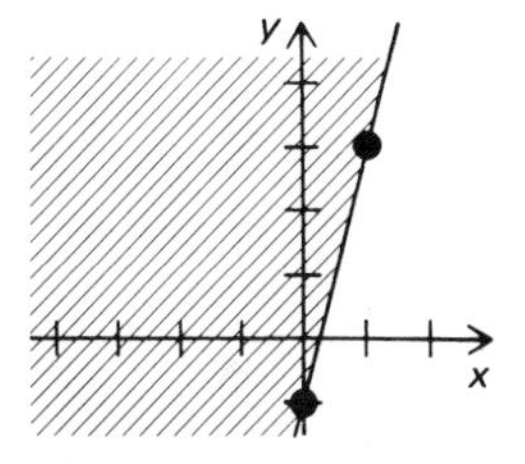

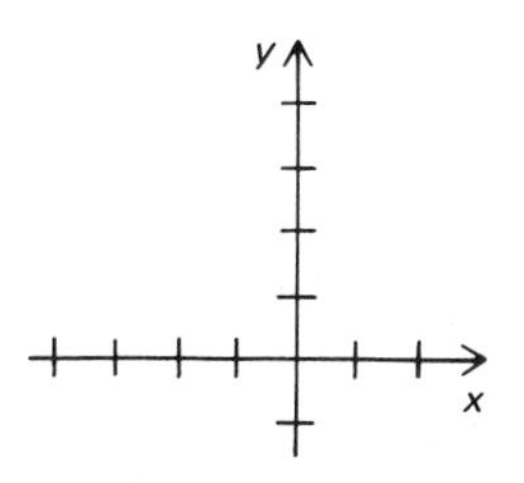

5. In general whenever the "greater than or equal" symbol occurs, the graph consists of the line as well as the half-plane. If only the "greater than" symbol occurs, the graph consists of the half-plane without the line.

$\geq$	line and half-plane
$>$	just half-plane
$\leq$	line and half-plane
$<$	just half-plane

does

The graph of $y \geq 5x + 1$ (does\does not) include the line $y = 5x + 1$, because the ________________ symbol is used.

greater than or equal
does not

The graph of $y > 5x + 1$ (does\does not) include the line $y = 5x + 1$, because only the ________________ symbol is used.

greater than

6. Fill in an appropriate symbol so that the graph includes the line $y = 2x - 5$ and one of the half-planes.

$$y \_\_\_\_ 2x - 5$$

$\geq$ or $\leq$

7. Graph $x + 2y > 8$. First we graph the line $x + 2y = 8$. If we solve $x + 2y = 8$ for y we obtain

$$x + 2y = 8$$

$$2y = \_\_\_\_\_\_$$

$8 - x$

$$y = \_\_\_\_\_\_$$

$4 - {}^{1}/_{2}x$

To graph the line we need two points. We use a dashed line to show that we (do\do not) include the line in the graph of $x + 2y > 8$.

do not

x	y	*Corresponding ordered pairs*
0	____	______
2	____	______

4; (0, 4)

3; (2, 3)

Graph $x + 2y = 8$.

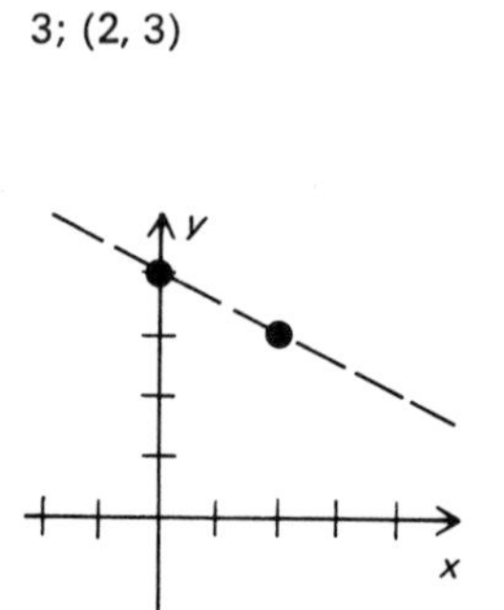

As a test point let's use (2, 5). This point lies in the (lower\upper) half-plane.

upper

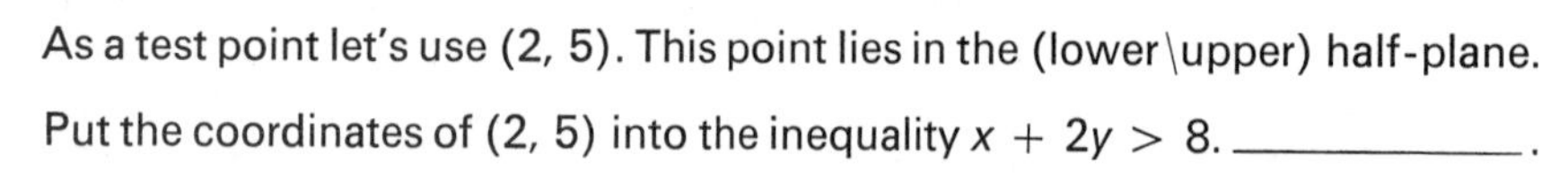

Put the coordinates of (2, 5) into the inequality $x + 2y > 8$. ______.

$2 + 2(5) > 8$

The resulting inequality is (true\false), since ____ > 8. Therefore the half-plane containing (2, 5) (is\is not) part of our graph.

true; 12

is

Is the line $x + 2y = 8$ also part of the graph? (yes\no). Shade the region corresponding to $x + 2y > 8$ on the graph.

no

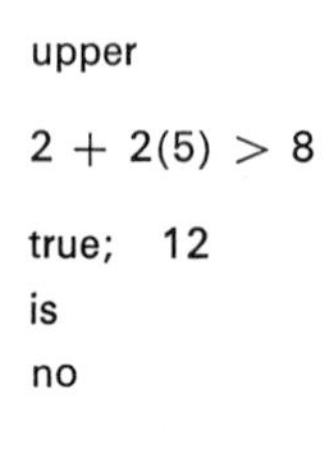

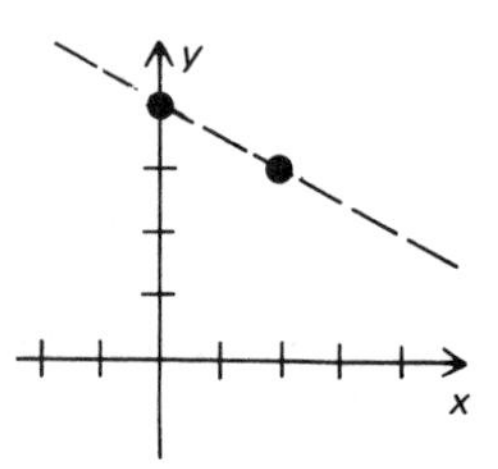

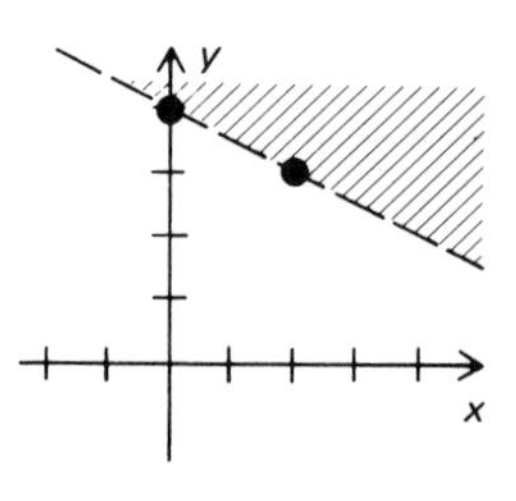

8. Graph $x < -2$. First we graph the line $x = -2$. This is a (vertical\horizontal) line through the point $(-2, 0)$. Plot this line. We use a (dashed\solid) line.

vertical

dashed

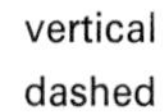

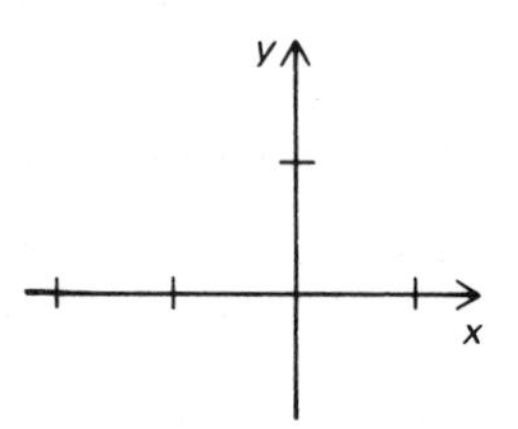

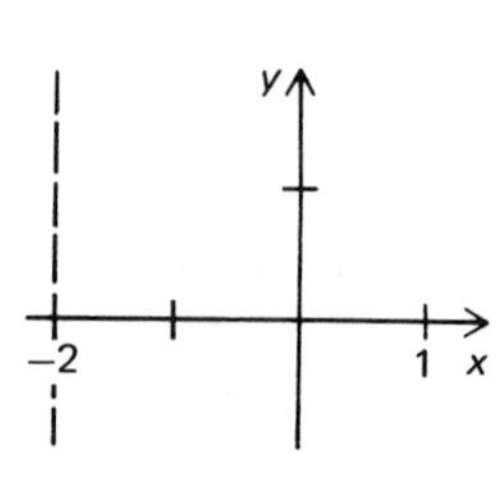

left

left

no

less than

The points (x, y) whose x coordinate is less than -2 lie to the (left\right) of the line $x = -2$. So the graph of $x < -2$ is the (left\right) half-plane.

Is the line $x = -2$ included in the graph of $x < -2$? (Yes\No), because the ________ symbol is used. Go back to your graph above and shade the graph of $x < -2$.

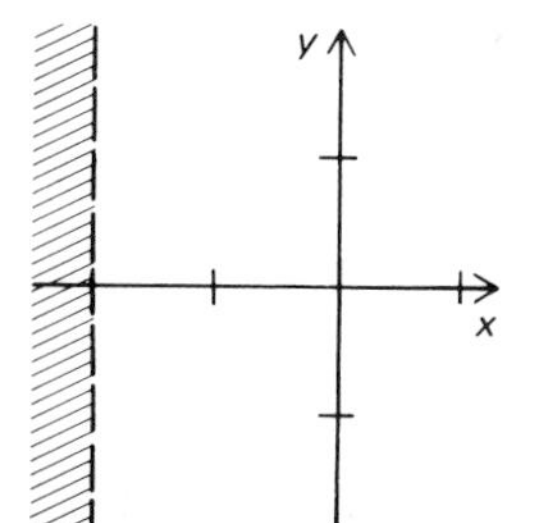

9. Graph that region which satisfies both inequalities $-3 < y \leq 5$. First we graph the lines $y = -3$ and $y = 5$. The line $y = -3$ has been plotted below. Plot the line $y = 5$. The condition $-3 < y \leq 5$ is satisfied by all points

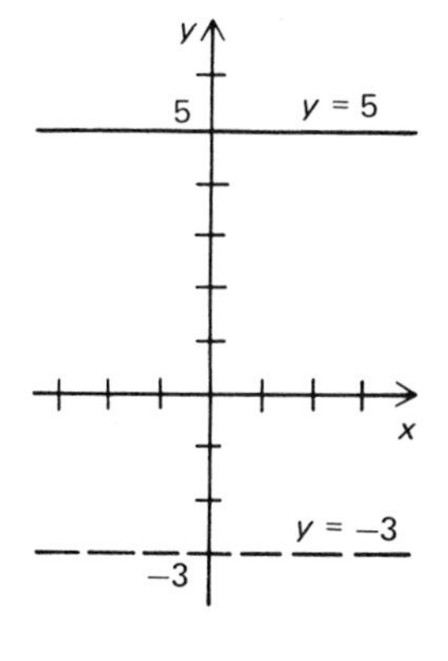

−3; 5

between

(x, y) whose y coordinate is between ____ and ____. Therefore the graph of $-3 < y \leq 5$ is the set of all points which lie (above\between\below) the lines $y = -3$ and $y = 5$.

$y = 5$;

One of the lines is included in the graph of the region $-3 < y \leq 5$. Which is it? The line ($y = -3$\$y = 5$).

On your graph above shade the region corresponding to $-3 < y \leq 5$.

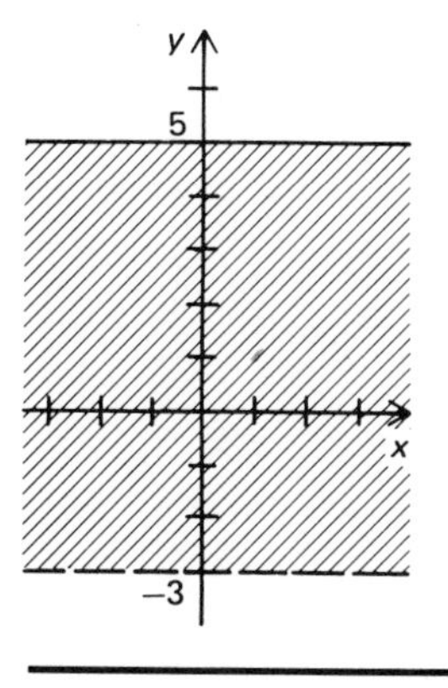

PRACTICE PROBLEMS

Graph the regions that satisfy the following inequalities and indicate if the lines are included in the region.

1. $1 \leq x$

2. $1 \leq x \leq 5$

3. $2 \leq y < 4$

4. $1 \leq x \leq 5$ and $2 \leq y \leq 4$

5. $2 \leq x \leq 3$ and $y < -1$

6. $y > 3x - 1$

7. $y \leq 9x + 2$

8. $y < x + 1$

9. $y > x + 2$

10. $y < 2x - 1$

11. $y \geq x + 4$

12. $y \leq 3x + 1$

13. $y > 2x$

14. $y < 4x$

15. $y > -4x - 7$

16. $y \geq {}^1/_2x - 5$

17. $10x + 5y < 20$

18. $16x - 2y \geq 8$

19. $7x > 21$

20. $9y + 1 \leq 19$

SELFTEST

1. Graph $y > 2x + 1$
2. Graph $y \leq 4 - 2x$
3. Graph $2x + y \geq 1$
4. Graph $x - y < 4$
5. Graph the region in which $x \geq 2$ and $y < 4$

SELFTEST ANSWERS

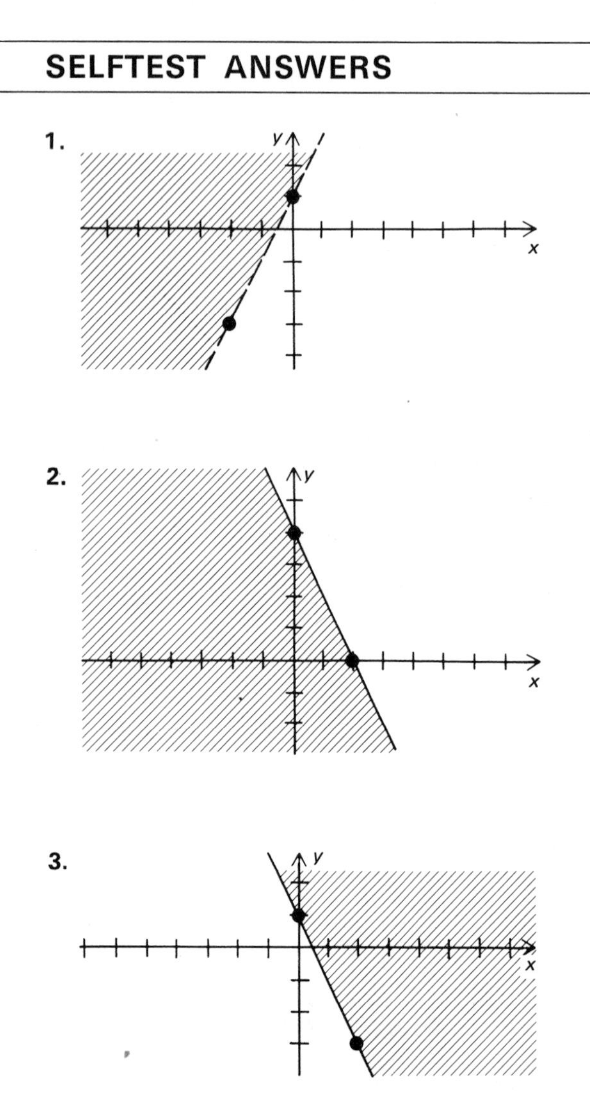

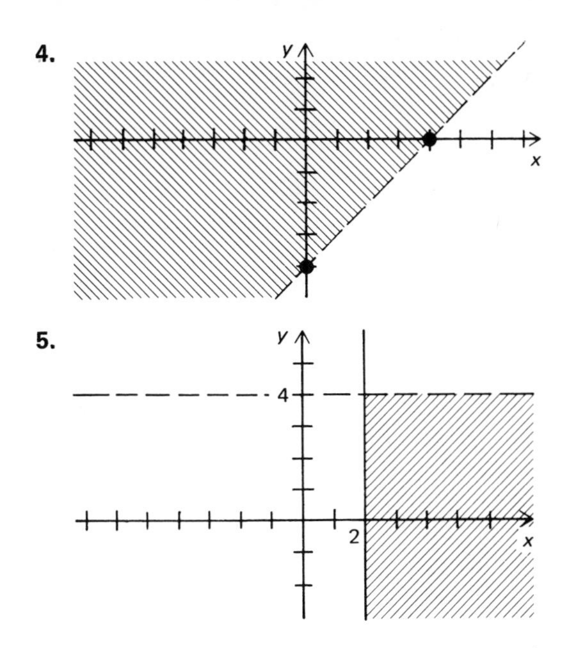
4.
y
x
5.
y
4
2
x

POSTTEST 9

1. If $(4, y)$ is to satisfy $2x - y = 7$, find y.
2. Shade the portion of the xy plane described by the conditions $x < 0$ and $y < 0$.
3. Consider the set $\{(0, 6), (3, 5), (2, 4), (8, 1)\}$.
   a. Is the set a relation?
   b. Is it a function? Why?
   c. What is the domain and range?
4. Graph $y = 4x - 2$.
5. Graph $x = 2$.
6. Find the slope of the line through $(0, 1)$ and $(2, 5)$.
7. Graph $2x + y = 4$.
   a. What is the slope?
   b. What is the y intercept?
8. Graph the region $y \leq 3x + 1$.
9. Graph the region in which $x \geq -1$ and $y > 3$.

Each problem above refers to a section in this chapter as shown in the table.

Problems	*Section*
1	9.1
2	9.2
3	9.3
4 and 5	9.4
6 and 7	9.5
8 and 9	9.6

$y = 4x - 2$

1. $(4, y)$
$2x - y = 7$
$8 - y = 7$
$-y = -1$
$y = 1$

POSTTEST 9 ANSWERS

1. 1

2. 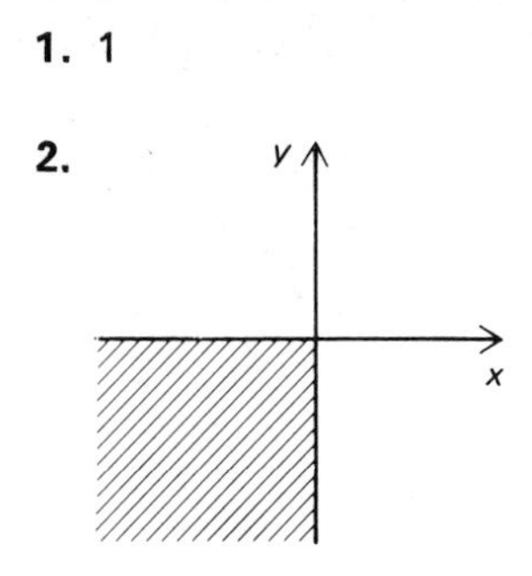

3. *a.* yes *b.* yes; each first element is paired with only one second element
   c. domain = {0, 3, 2, 8}; range = {5, 6, 4, 1}

4. 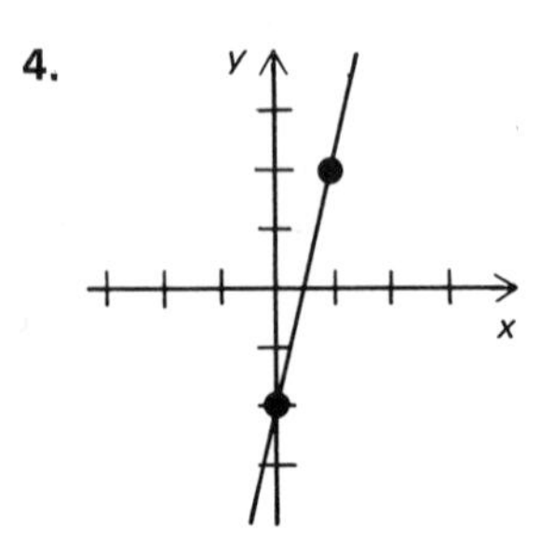

5. 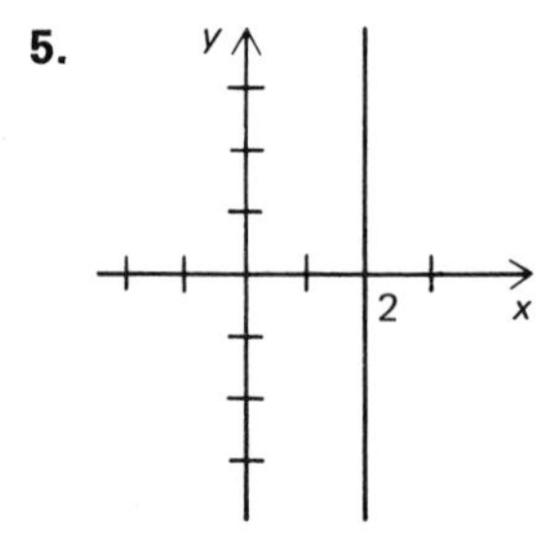

6. $^4/_2$ or 2

7. *a.* -2 *b.* (0, 4)

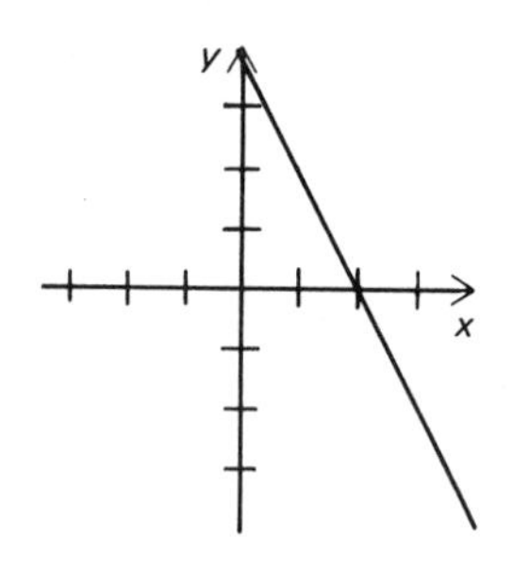

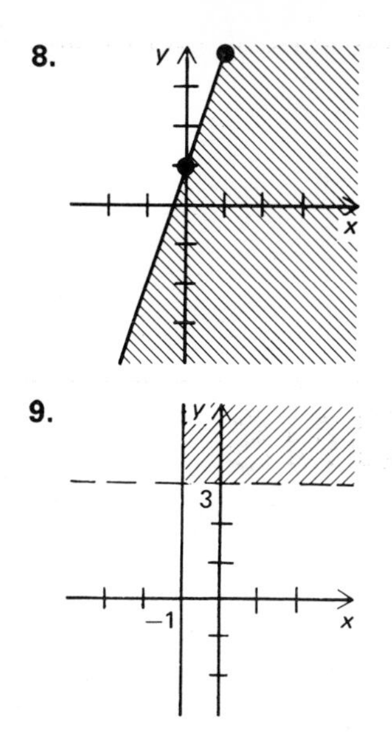
8.
y
x
9.
y
3
−1
x

10

SYSTEMS OF EQUATIONS

BASIC SKILLS

Upon completion of the indicated section the student should be able to:

10.1 *a.* Find the solution of a system of two linear equations in two unknowns by graphing the equations.
b. Identify an inconsistent system from the graph of the system.
c. Use the graph to identify a system of equations that has an infinite number of solutions.

10.2 *a.* Solve a system of two equations in two unknowns by the method of substitution.
b. Use the method of substitution to tell if a system is inconsistent or if it has infinitely many solutions.

10.3 *a.* Determine if two systems of two equations in two unknowns are equivalent.
b. Solve a system of two equations in two unknowns by using either of the following operations or a combination of them.
1. Multiply or divide an equation by a non-zero constant.
2. Add or subtract two of the equations of the system.

10.4 Apply the techniques of solving a system of two equations in two unknowns to solve stated problems.

PRETEST 10

1. Solve by graphing:

$$\begin{cases} x + y = 3 \\ 2x + y = 4 \end{cases}$$

2. Solve by graphing:

$$\begin{cases} x + y = 3 \\ 2x + 2y = 4 \end{cases}$$

Solve by algebraic techniques:

3. $\begin{cases} 2x + y = 5 \\ y = x - 1 \end{cases}$

4. $\begin{cases} x + y = 4 \\ -x + 2y = 5 \end{cases}$

5. $\begin{cases} 2x + y = 4 \\ x + y = 6 \end{cases}$

6. $\begin{cases} x + y = 3 \\ 2x + 2y = 6 \end{cases}$

7. What does it mean to say that two systems are not equivalent?

8. A hamburger and milk shake together cost \$1.30. If the hamburger is \$0.20 more than the milk shake, how much does each cost?

Each problem above refers to a section in this chapter as shown in the table.

Problems	*Section*
1 and 2	10.1
3	10.2
4–7	10.3
8	10.4

PRETEST 10 ANSWERS

1. (1, 2)

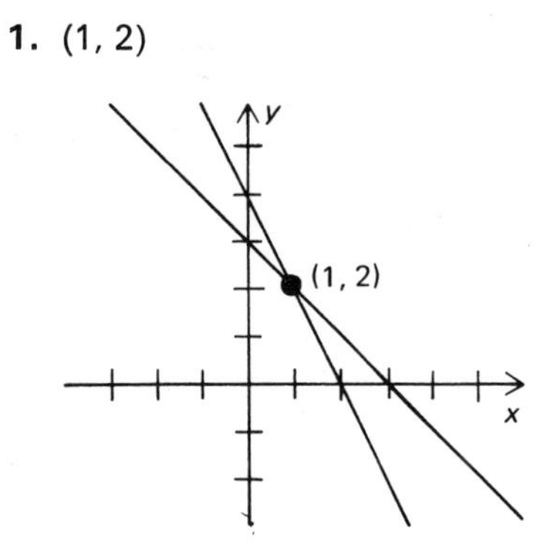

2. inconsistent

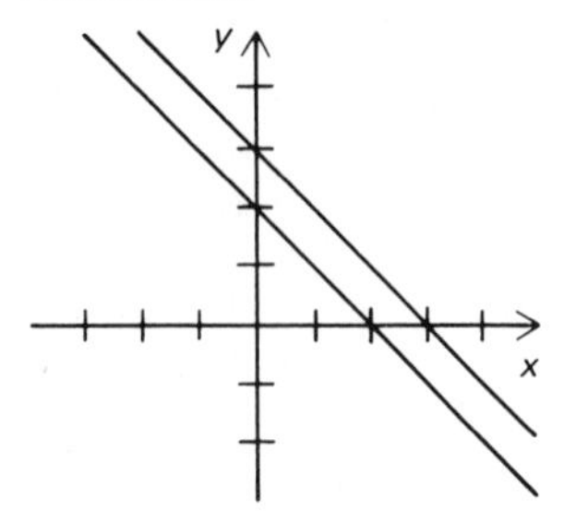

3. (2, 1) **4.** (1, 3) **5.** (−2, 8)

6. infinitely many **7.** they do not have the same solutions

8. milkshake cost \$0.55; hamburger cost \$0.75

10.1 GRAPHICAL SOLUTIONS OF LINEAR SYSTEMS

We say an equation is a *linear equation* if no term in the equation contains a product of two or more variables. For example, the equation $3x + 2y = 1$ is a linear equation. The term $3x$ has only one variable, x. The term $2y$ has y as its only variable. The 1 on the right hand side has no variable at all, but that is all right.

The equation $3x^2 = y$ is not linear, because the term $3x^2$ contains the product $x^2 = xx$ of the variable x. Likewise $xy = 1$ is not a linear equation. The term xy is a product of the two variables x and y.

In general, if we graph a linear equation in two variables we get a straight line. Hence the name linear equation. Linear equations are very important and occur in many applications of mathematics. Often it is necessary to find a solution to a pair on linear equations. The process is best introduced by using some examples.

Example 1 Suppose you are in the business of making staplers. Let x be the number of staplers you intend to produce in a year. It costs \$200 to buy equipment to set up your little factory. You find that it costs \$2 to make one stapler. You intend to sell your staplers at the rate of \$4 each. A good question to ask yourself is: How many staplers must I sell before I start to break even? In other words, what is the breakeven point for the business?

Let C represent your costs. It costs \$2 to make each stapler. If you make x staplers, that means it costs $2x$ dollars. But it also costs \$200 to get the business started. So your cost function is

$$C = 2x + 200$$

Let R represent your revenues, the money you take in from sales. You sell each stapler at \$4. So, if you sell x staplers, your revenue function is

$$R = 4x$$

From our work with graphing we know that both $C = 2x + 200$ and $R = 4x$ represent straight lines. The point where these lines intersect is the point where the cost equals the revenue. So the intersection of these lines is the breakeven point. The x coordinate of this point will tell us how many staplers we must make and sell to break even. The y coordinate or the \$ coordinate will tell us the amount in sales necessary to break even.

The pair of equations

$$C = 2x + 200$$
$$R = 4x$$

is called a *system of linear equations*. The point where the lines intersect is called a *solution of the system*.

Let us draw the graph of these two lines and find the solution to the system. Since both C and R represent money in dollars, we will let the vertical axis represent dollars. The horizontal axis represents the number x of staplers produced in a year.

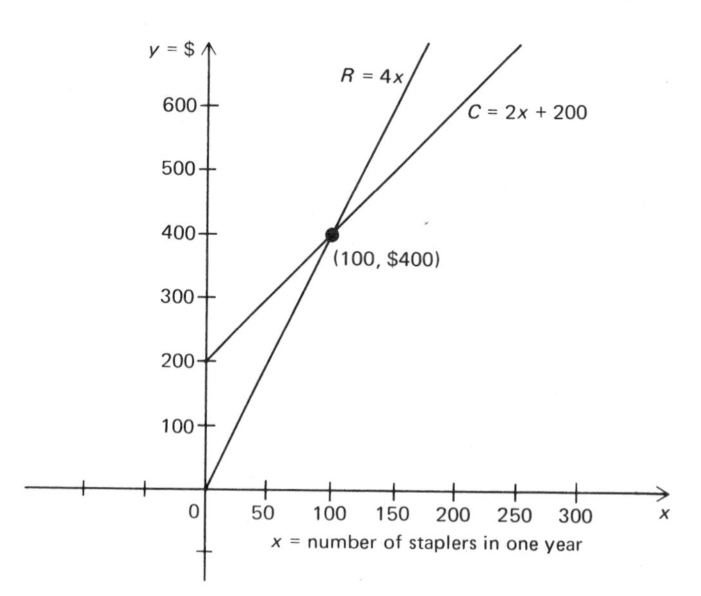

From the graph we see that the lines intersect at $x = 100$ and $R = C = \$400$. So you will break even if you sell 100 staplers. To make a profit you must sell more than 100 staplers.

Example 2 Use a graph to solve the system of linear equations.

$$\begin{cases} -x + y = 1 \\ 3x - y = -5 \end{cases}$$

We graph both equations on the same xy plane and find the point where the graphs intersect.

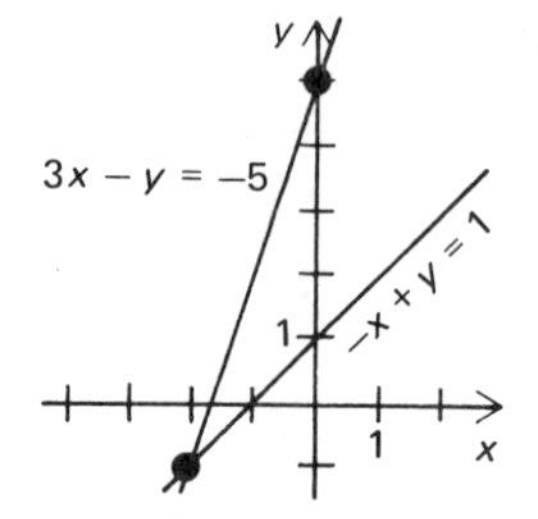

From the graph it looks as if the point $(-2, -1)$ is the solution. To check if $(-2, -1)$ is a solution, we see if $(-2, -1)$ satisfies *both* the equations $-x + y = 1$ and $3x - y = -5$.

Check: If $x = -2$ and $y = -1$, then

$-x + y = -(-2) + (-1) = 1$ and $3x - y = 3(-2) - (-1) = -5$

The point $(-2, -1)$ satisfies both equations and is in fact a solution of the system.

Example 3 Solve the system:

$$\begin{cases} x + y = 2 \\ 2x + 2y = 8 \end{cases}$$

The graphs of these lines are shown in the next figure. These lines do not intersect.

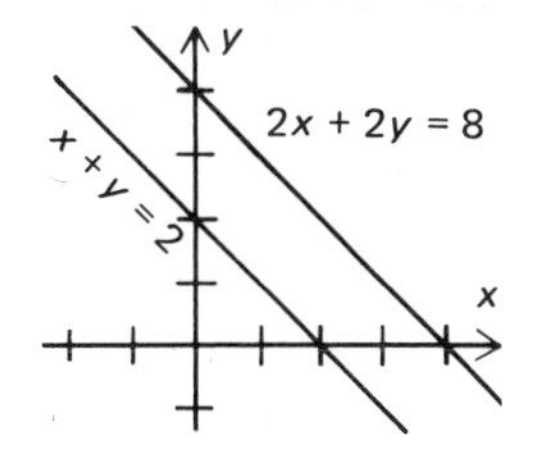

They are *parallel* lines. Therefore the given system of equations has no solution. How could we tell the lines were parallel? One way is to look at the slope of each line. Lines are parallel if and only if they have the *same* slope. The slope of the line $x + y = 2$ is -1. Likewise, the line $2x + 2y = 8$ has slope -1. So the lines are parallel and there is no solution to the system.

If a system of equations has no solution, we say the system is *inconsistent*. If the system does have a solution, we say it is *consistent*. In Examples 1 and 2 the systems were consistent. But in Example 3, the system was inconsistent. There was no solution to Example 3.

Example 4 Solve the system:

$$\begin{cases} y = 2x - 1 \\ 2y = 4x - 2 \end{cases}$$

If we graph these equations, we get a single-line graph. This is a reasonable

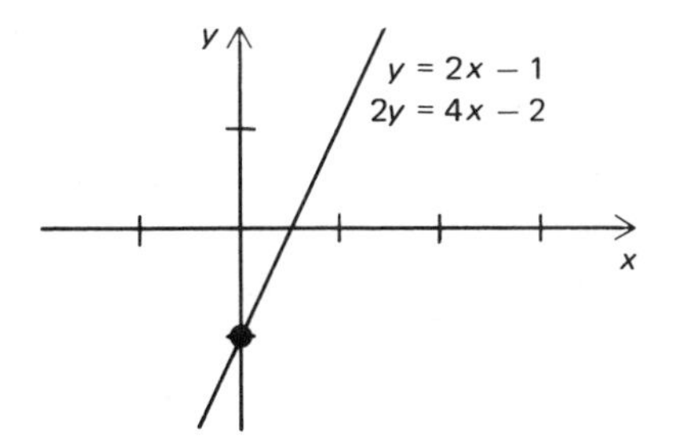

result; because if we divide both sides of the second equation by 2, we obtain the first equation. So the two equations are really equivalent and represent the same line. The solution to the system is the set of points where the lines intersect. So in this case the solution consists of the set of all points on the line graphed above. There are infinitely many solutions.

If a system of linear equations has more than one solution, it will in general have infinitely many solutions. Naturally, the system in Example 4 is consistent since it does have solutions.

Of course, we cannot draw a perfect line graph. Sometimes it will be hard to see exactly where lines intersect. In these cases, we can obtain only approximate solutions for the system by graphing. There are other ways to find a solution besides graphing. In the next section we present another way to solve a system of linear equations.

PROGRAM 10.1

1. An equation is called a *linear equation* if no term in the equation contains a product of two or more variables. Let's see if the equation $21x + 9y = 16$ is a linear equation. The term $21x$ has only one variable. It is x. The term $9y$ has ____ as its only variable. The 16 on the right side of the equation does not involve any variable. So $21x + 9y = 16$ (is\is not) a linear equation.

y

is

2. The term $-2y$ has only one variable. It is ____. How many variables does the term 4 have? (one\two\none). Therefore the equation $-2y = 4$ is a ________ equation.

y
none
linear

3. The term $3x^2$ contains a product of two variables, $3x^2 = 3 \cdot$ ____ $\cdot$ ____. Therefore the equation $3x^2 + 9x + 1 = 0$ (is\is not) a linear equation.

x; x
is not

4. In the expression $8xy + 3x - y$, the first term contains a product of (one\two\three) variables. Therefore, the equation $8xy + 3x - y = 6$ (is\is not) a linear equation.

two; is not

5. In general, the graph of a linear equation in two variables is a straight line. In fact, graphs that are straight lines come from linear equations. What is the shape of the graph of $3x + 2y = 5$? ____________. Is the graph of $2xy = 3$ a straight line? (Yes\No), because the equation is not________.

A straight line
no; linear

6. In this chapter, we will study systems of linear equations. A *system of linear equations* is a collection of two or more linear equations.

A system of linear equations may involve two, three, or more variables. A *solution* of the system is found if we can assign values to each of the variables in such a way that these values satisfy *each* equation of the system. The set of all solutions of a system is called the *solution set* for that system. For example,

$$-x + 2y = 7 \qquad \text{and} \qquad 4x + 3y = -6$$

is a system of two equations in ____ variables. The variables or unknowns are ____ and ____. We claim that the set of values $x = -3$ and $y = 2$ is a solution for this system. For when we substitute $x = -3$ and $y = 2$ into the equations, both equations are satisfied. Let's check the solution by substituting it into the equations.

two
x; y

$$-x + 2y = 7$$
$$-(-3) + 2(2) = 7$$

So the first equation is satisfied.

$$4x + 3y = -6$$
$$4(\_\_\_\_) + 3(\_\_\_\_) = -6$$

Therefore the second equation (is\is not) satisfied, for $-12 + 6 = -6$.

-3; 2
is

7. The important thing to notice is that our solution must work in (only one equation of the system\no equation of the system\all equations of the system).

all equations

8. Most of this chapter will deal with methods of solving a system of linear equations. The first method we will consider is the graphical method. As we said, a linear equation in two variables corresponds to the graph of a straight line. So a system of two such equations would give two lines. *The point of intersection is the solution of the system.* To solve the system

$$-2x + y = -5 \qquad \text{and} \qquad 4x + 2y = 6$$

we graph both lines and see where they meet. On the same xy plane below, graph both the lines of the system. Remember, in order to graph a line, we need only (one\two\three) points on that line.

two

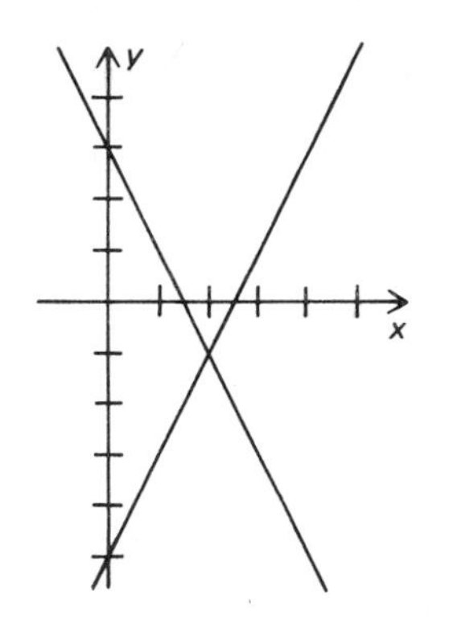

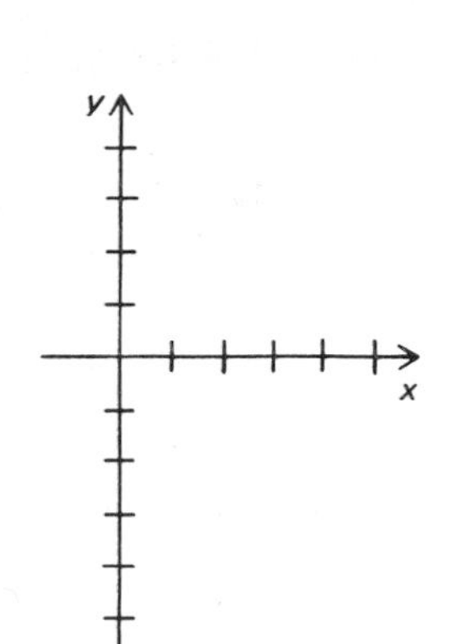

Look at the point of intersection of these lines. What are the x and y coordinates of this point? They are $x =$ ____ and $y =$ ____. So the solution of the system is $x =$ ____ and $y =$ ____.

2; -1

2; -1

9. It is always a good idea to check your solutions in the original system. Let's do that for the system we just solved. We will check the solution $x = 2$ and $y = -1$ in the system

$$-2x + y = -5 \qquad \text{and} \qquad 4x + 2y = 6$$

The first equation gives

$$-2(\_\_\_) + (\_\_\_) = -5$$

2; -1

So the first equation (is\is not) satisfied. The second equation gives

is

$$4(\_\_\_) + 2(\_\_\_) = 6$$

2; -1

Is it satisfied? (yes\no). Does $x = 2$ and $y = -1$ check as a solution of the system? (yes\no).

yes

yes

10. Look at the graph below. This graph represents the system

________ and ________

What is the solution of this system? ________.

$x + y = 3$; $x - y = 1$

$x = 2$; $y = 1$

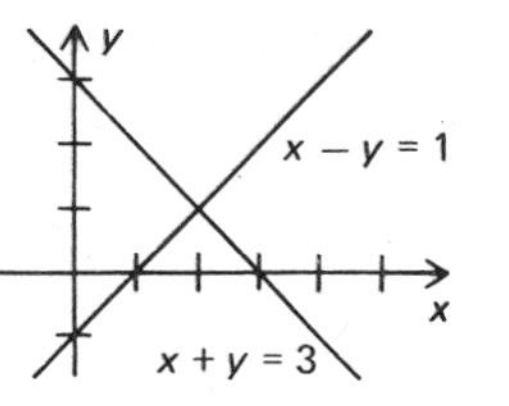

11. To find a solution of a system of two linear equations in two unknowns, we graph the equations and use the coordinates of the point where the lines ________.

intersect or meet

12. But what happens if the lines of a system don't meet? Then the system has *no* solution. Parallel lines never meet. What can we say about the solution of a system containing parallel lines? ________.

It has no solution

13. Two lines are parallel if and only if they have the same slope. We recall from our work with lines that if $y = mx + b$, then the slope of this line (is m\ is b).

m

14. To find the slope of $2x + 4y = 7$, we solve the equation for y.

$$2x + 4y = 7$$
$$4y = 7 - 2x$$
$$y = {}^{7}/_{4} - {}^{2}/_{4}x$$
$$y = \_\_\_\_x + {}^{7}/_{4}$$

So the slope is ____ .

$-{}^{2}/_{4}$ or $-{}^{1}/_{2}$

$-{}^{1}/_{2}$

15. To answer the question, "Are the lines $6x + 3y = 3$ and $2x + y = 9$ parallel?" we must find the slope of each line. If we solve the first equation for y, we get

$$y = \_\_\_\_x + 1$$

If we solve the second equation for y we get $y = \_\_\_\_x + 9$. Are the slopes of the lines the same? (yes\no). Therefore the lines (are\are not) parallel. Does this system have a solution? (yes\no).

-2

-2

yes; are

no

16. Draw the graph of the system

$$x + y = 4 \qquad \text{and} \qquad x + y = 1$$

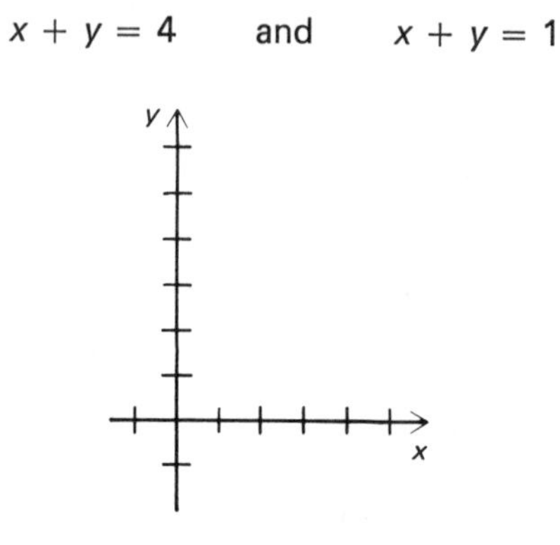

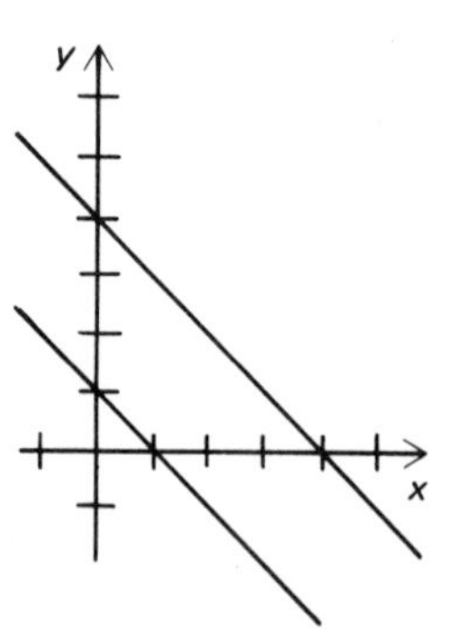

Do these lines appear to be parallel? (yes\no). Does this system appear to have a solution? (yes\no).

yes

no

17. Generally a graph that is carefully drawn will tell you if the lines in a system are parallel or not. But if your graph leaves you in doubt, check the slope of each line. The lines are parallel if the slopes are (equal\unequal).

equal

18. Compare the slopes of the lines in the system

$$y = 4x + 3 \qquad \text{and} \qquad y = 4x - 1$$

Are the slopes the same? (yes\no). Graph the system. What is the solution of the system? ______________________________.

yes

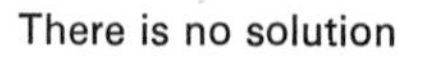

There is no solution

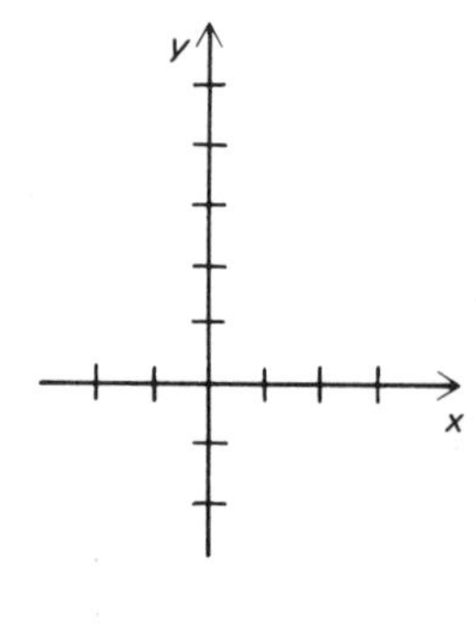

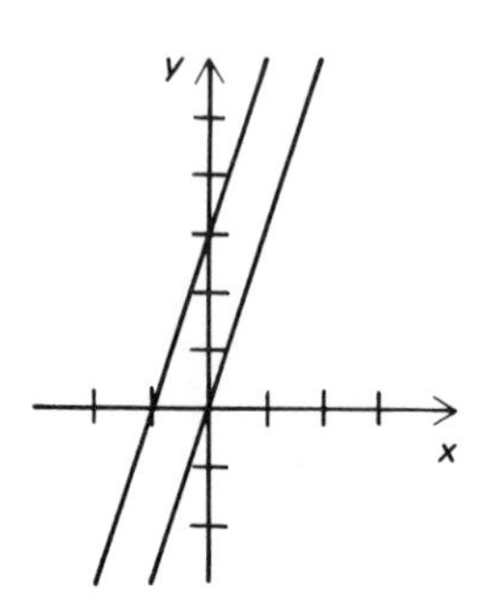

inconsistent
parallel

19. If a system has no solution, we say it is *inconsistent*. If the system does have a solution, we say it is *consistent*. In general, parallel lines come from a system that is (inconsistent\consistent) because there is no solution for a system with ________ lines.

20. Solve by graphing:

$$x - y = 3 \qquad \text{and} \qquad x + 2y = 3$$

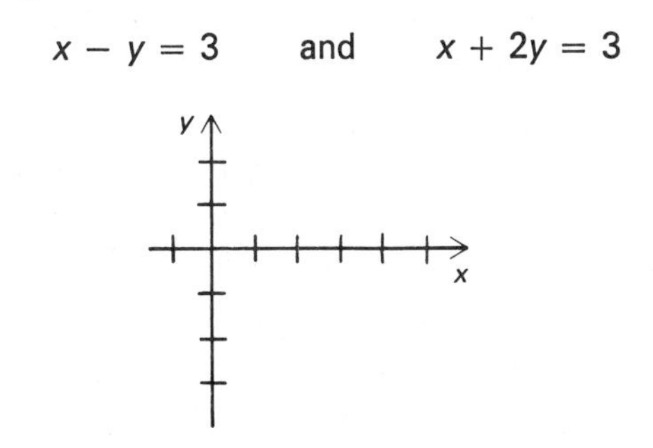

no (they cross)
yes; consistent
3; 0

Do the lines appear to be parallel? (yes\no). Does this system have a solution? (yes\no). The system is (consistent\inconsistent). The solution is $x =$ _____, $y =$ _____.

21. By graphing both lines on the xy plane, solve the system

$$5x - 2y = 8 \qquad \text{and} \qquad 10x - 4y = 16$$

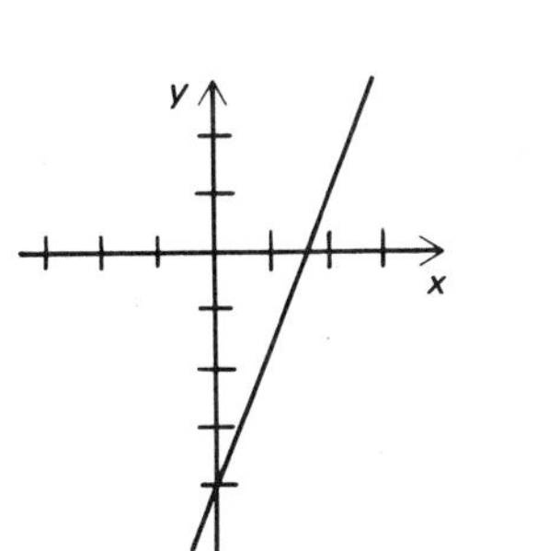

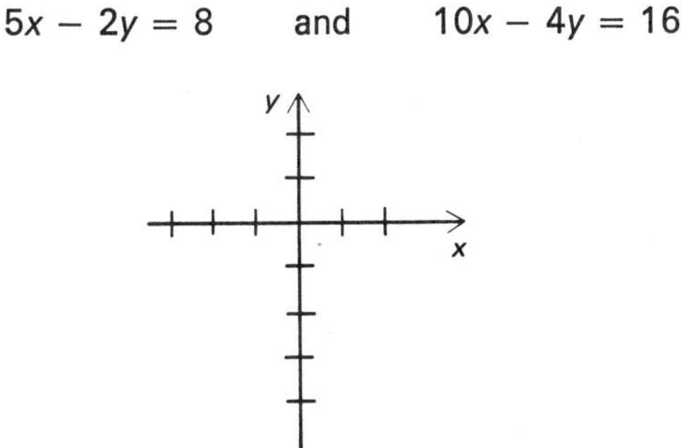

all points are in common

Where do these lines meet? (nowhere\at one point\all their points are in common).

one and the same line
$10x - 4y$;
16
yes
are; do

22. From the graph, we see that the lines $5x - 2y = 8$ and $10x - 4y = 16$ are really (two distinct lines\one and the same line). This makes sense because if we multiply the first equation by 2 (on both sides) we get ________ = _____. Is this new equation the same as the second equation of the original system? (yes\no). So the equations

$$5x - 2y = 8 \qquad \text{and} \qquad 10x - 4y = 16$$

(are\are not) equivalent and they (do\do not) represent the same line.

infinitely many
points

23. When two lines of a system are actually the same, how many solutions are there for the system? (none\one\two\infinitely many). There are infinitely many ________ on a line.

consistent; has

24. In general, it is true that if a system has more than one solution, it has infinitely many solutions. A system with infinitely many solutions is (consistent\inconsistent) because it (has\does not have) a solution.

PRACTICE PROBLEMS

Discuss each of the topics.

1. *a.* What is a linear equation?
   b. Give examples of equations that are linear equations.
   c. Give examples of equations that are not linear equations.
   d. If we graph a linear equation in two variables, what kind of graph do we get?
2. *a.* What is a system of linear equations?
   b. What do we mean by the term solution of a system of linear equations?
   c. What do the terms inconsistent and consistent systems mean?
   d. Does every system of linear equations have a solution?
3. *a.* Explain how to use a graph to solve a system of linear equations.
   b. If you are given a solution, how can you check to see if it is correct?
4. List some advantages and disadvantages of the graphical method of solving a system of linear equations.

Use a graph to solve each of the following systems. If the system is consistent, check your solution. If there are infinitely many solutions, just check that both equations represent the same line.

5. $\begin{cases} 1x + 3y = 3 \\ 5x + y = 1 \end{cases}$
6. $\begin{cases} 3x - 2y = 0 \\ x + y = 0 \end{cases}$
7. $\begin{cases} 2x + y = 3 \\ y = 3 \end{cases}$
8. $\begin{cases} x + 7y = -5 \\ x = 2 \end{cases}$
9. $\begin{cases} 4x - 3y = 6 \\ 8x - 6y = 0 \end{cases}$
10. $\begin{cases} x + y = 3 \\ 2x + 2y = 6 \end{cases}$
11. $\begin{cases} 6x + 15y = 12 \\ 2x + 5y = 4 \end{cases}$
12. $\begin{cases} x - y = 0 \\ 2x + y = -6 \end{cases}$
13. $\begin{cases} x + 3y = -8 \\ -4x + y = -7 \end{cases}$
14. $\begin{cases} 8x + 2y = 10 \\ 5x + y = 6 \end{cases}$
15. $\begin{cases} x - 3y = 0 \\ 2x - 6y = 10 \end{cases}$

SELFTEST

Solve each of the following systems of equations by graphing the two equations. Indicate if the system is inconsistent or if it has infinitely many solutions.

1. $\begin{cases} y = 3 - x \\ y = 4x - 7 \end{cases}$
2. $\begin{cases} 2x - y = 4 \\ 4x = 8 + 2y \end{cases}$
3. $\begin{cases} 2x + y = 0 \\ 2y + x = 6 \end{cases}$
4. $\begin{cases} y = 2x + 1 \\ y - 2x = 4 \end{cases}$

SELFTEST ANSWERS

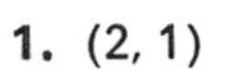

1. (2, 1)

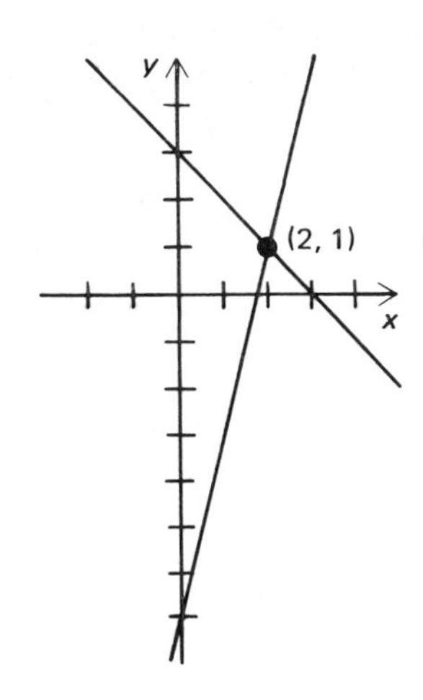

2. infinitely many solutions

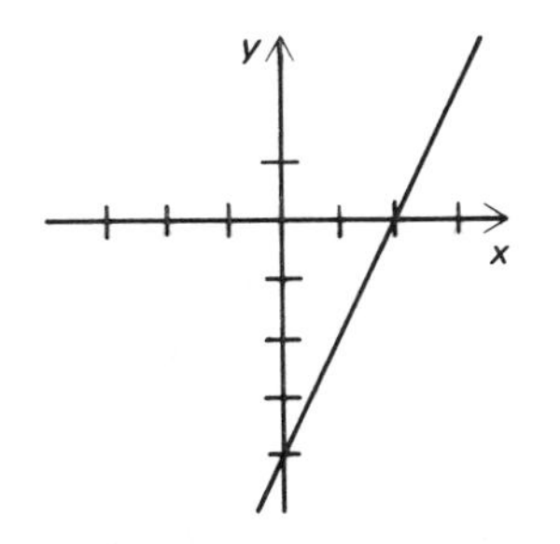

3. $(-2, 4)$

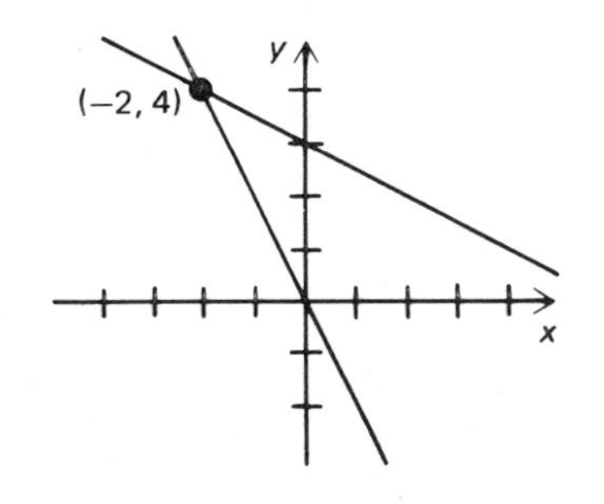

4. inconsistent

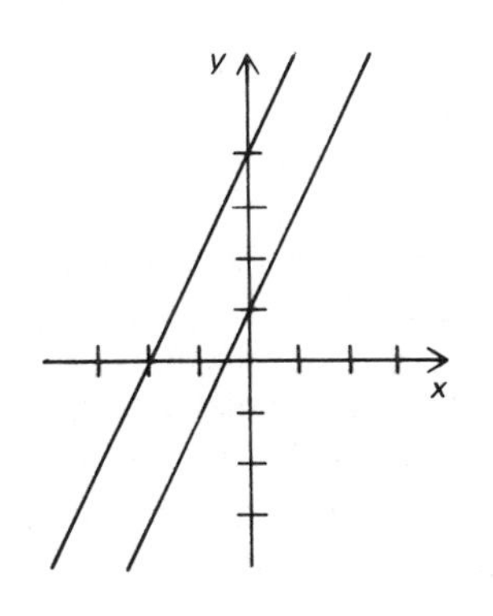

10.2 SOLVING SYSTEMS OF EQUATIONS BY SUBSTITUTION

In this section we will solve a system of linear equations by eliminating variables from some of the equations. The idea is that if we have enough equations, we might be able to eliminate variables until we wind up with a linear equation in just one unknown. We can solve this equation for the unknown and use the solution to find the other unknowns.

Example 1 Solve the system

$$3x - 4y = 7 \qquad \text{and} \qquad 2x + y = 1$$

Look at the second equation. It would be easy to solve this equation for y in terms of x. We get

$$y = 1 - 2x$$

A solution for the system must satisfy both equations. So, if $y = 1 - 2x$ is part of a solution, the first equation must be true if we replace y in that equation by $1 - 2x$. Let's do that.

$$\begin{aligned} 3x - 4y &= 7 \\ 3x - 4(1 - 2x) &= 7 \\ 11x - 4 &= 7 \\ 11x &= 11 \\ x &= 1 \end{aligned}$$

We conclude that $x = 1$; and since $y = 1 - 2x$, then

$$y = 1 - 2(1) = -1$$

So the point $(1, -1)$ is a solution for the system. To be sure let us check our solution in both of the given equations. If $x = 1$ and $y = -1$, then

$$3x - 4y = 3(1) - 4(-1) = 7 \qquad \text{and} \qquad 2x + y = 2(1) + (-1) = 1$$

The solution $(1, -1)$ checks.

Example 2 Solve the system

$$2y = x + 6 \qquad \text{and} \qquad 6y = 3x + 12$$

First we examine the equations. It would be easy to solve the first one for x. We get $x = 2y - 6$. This expression for x is substituted into the second equation.

$$\begin{aligned} 6y &= 3x + 12 \\ 6y &= 3(2y - 6) + 12 \\ 6y &= 6y - 18 + 12 \\ 6y &= 6y - 6 \\ 6y - 6y &= -6 \\ 0 &= -6 \end{aligned}$$

But the equation $0 = -6$ is ridiculous. So what happened? When we substituted $x = 2y - 6$ from the first equation we were going under the assumption that $x = 2y - 6$ was part of a solution for the system. But this system has no solution. It is inconsistent. The slope of each line is $1/2$, so the lines are parallel and do not intersect. Since the system is inconsistent we obtain the inconsistent equation $0 = -6$. In general, if the method of substitution results in an inconsistent equation, such as $0 = -6$, then the system itself is inconsistent. So there is no solution.

Inconsistent Equations: In the graph the lines are parallel. The *substitution method* yields an inconsistent or false equation, such as $0 = -6$.

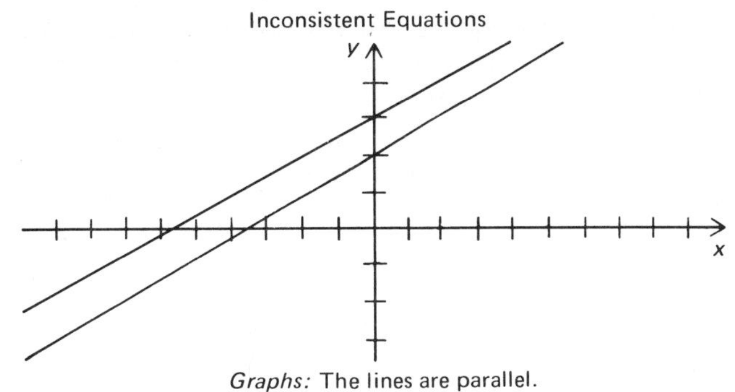

Graphs: The lines are parallel. *Substitution method* yields an inconsistent or false equation such as $0 = -6$.

Example 3 Solve the system

$$4x + 8y = 16 \qquad \text{and} \qquad 2x + 4y = 8$$

Look at the equations. If we solve the first equation for x we get

$$x = {}^1/_4(16 - 8y) \quad \text{or} \quad x = 4 - 2y$$

Now substitute this expression for x into the second equation.

$$\begin{aligned} 2x + 4y &= 8 \\ 2(4 - 2y) + 4y &= 8 \\ 8 - 4y + 4y &= 8 \\ 8 &= 8 \end{aligned}$$

The equation $8 = 8$ is certainly true, but it does not appear to tell us anything specific about the value of y or x. Look back at the given system of equations. If we divide the first equation by 2 we get the second equation. So the two equations are equivalent and their graph is one and the same line.

The solution of the system consists of all points that the lines have in common. Since the lines coincide, the solution is the set of all points on this line, or the set of all points that satisfy one of the equations.

In general, if the method of substitution gives an identity like $8 = 8$, then some equations in the system are equivalent equations and there are infinitely many solutions.

Equivalent Equations: In the graph the lines are exactly the same. The substitution method yields an identity or equation that is always true, such as $8 = 8$.

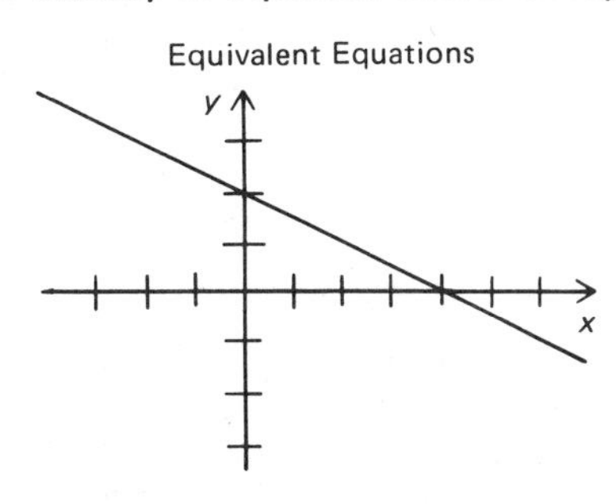

Graph: The lines are exactly the same. *Substitution method* yields an identity or equation which is always true such as $8 = 8$.

***Example 4** Solve the system

$$x + 3y + z = 0 \quad \text{and} \quad 2x - y + z = 1 \quad \text{and} \quad 4x + y + 2z = 5$$

Now we have three linear equations in three unknowns. The idea is to eliminate one of the unknowns, so we have only two equations in two unknowns. Then we solve the two equations in two unknowns as we did in the former examples.

Look at the first equation. It would be easy to solve this equation for z.

$$x + 3y + z = 0$$

$$\boxed{z = -x - 3y}$$

Let's put this expression for z into the remaining two equations.

$$\begin{aligned} 2x - y + z &= 1 \\ 2x - y + (-x - 3y) &= 1 \\ x - 4y &= 1 \end{aligned} \qquad \begin{aligned} 4x + y + 2z &= 5 \\ 4x + y + 2(-x - 3y) &= 5 \\ 2x - 5y &= 5 \end{aligned}$$

Now we have a system of two equations in two unknowns.

$$x - 4y = 1 \quad \text{and} \quad 2x - 5y = 5$$

We solve this system by the method described in Example 1. Please do this yourself on scratch paper. Your answers should be $x = 5$ and $y = 1$.

Next we want to find z. We put the formula for z in a box so that it would not get lost. Looking back we see $z = -x - 3y$. Since $x = 5$ and $y = 1$, then $z = -5 - 3(1) = -8$. Our answer is $x = 5$, $y = 1$, and $z = -8$. Let's check the answer in the original system of equations.

First equation: $x + 3y + z = 0$

$5 + 3(1) - 8 = 0$ checks

Second equation: $2x - y + z = 1$

$2(5) - 1 - 8 = 1$ checks

Third equation: $4x + y + 2z = 5$

$4(5) + 1 + 2(-8) = 5$ checks

So our answer checks out because it satisfies all three equations.

We could extend the method of Example 4 to solve a system of four equations in four unknowns. But the calculations get longer and more drawn out. For a system of two equations in two unknowns the method of substitution is quick and easy. For larger systems it can still be done, but it is complicated. Nevertheless, it is valuable to be able to solve larger systems. Many applications in physical science, business, social science, economics, etc., do involve such equations. It is therefore worth our while to know another method, one that is easier to apply for larger systems. Such a method will be presented next.

* Optional.

PROGRAM 10.2

1. We already know how to solve a system of linear equations by using a graph. However, no graph can be drawn perfectly. So most solutions obtained from graphs are only approximations.

   Now we will study another technique for solving a system of linear equations. It is called the *method of substitution*. Solutions obtained from this method are exact. The solution does not depend on how well we can draw a graph.

The main idea is to use one of the equations to isolate a variable. Then we substitute the expression for this variable into the other equations. The result is that these other equations contain one less variable. The process can be repeated until we have an equation with just one variable. We solve this equation and use the solution to find the other variables.

2. Let's find a solution of the system

$$x - y = 3 \qquad \text{and} \qquad 2x + y = 0$$

Looking at this system we see that it is not hard to solve the first equation for x.

$$x - y = 3$$

$3 + y$

$$\boxed{x = \underline{\qquad\qquad}}$$

We have put this expression for x in a box for easy reference because we will use it again.

If we substitute the above expression for x into the second equation of the system we obtain

$$2x + y = 0$$

$3 + y$

$$2(\underline{\qquad\qquad}) + y = 0$$

one

Now we have an equation in (one\two\three) variable(s). We solve this equation for y.

3

$$\underline{\qquad}y + 6 = 0$$

-2

$$y = \underline{\qquad}$$

We now have a value for y. Let's use that value to find x. Refer to the equation

$3 + y$

in the box, where $x =$ ________. But now we know $y = -2$, so

-2

$$x = 3 + (\underline{\qquad})$$

1

$$x = \underline{\qquad}$$

$1; -2$

Our solution is $x =$ ____ and $y =$ ____.

3. Now let's check our solution $x = 1$ and $y = -2$ in the original system

$$x - y = 3 \qquad \text{and} \qquad 2x + y = 0$$

yes

Our solutions give $1 - (-2) = 3$ in the first equation. Is this statement true? (yes\no). When we use $x = 1$ and $y = -2$ in the second equation $2x + y = 0$, we get

$2(1); (-2)$

$$\underline{\qquad} + \underline{\qquad} = 0$$

both; does

The solution checks in (one\both) equation(s) and therefore it (does\does not) check in the system.

4. In order to solve the system

$$8x + 12y = 28 \qquad \text{and} \qquad 4x + y = 9$$

We could start by solving either equation for either x or y. But in this case, we inspect the system and see that it would be easy to solve the second equation for y.

$$4x + y = 9$$

$9 - 4x$

$$\boxed{y = \underline{\qquad\qquad}}$$

(Since we will need to refer to this expression later we put it in a box.)

Now we substitute the above expression for y into the (first\second) equation of the system. [We have already used the (first\second) equation to get y.] — first; second

$$8x + 12y = 28$$

$$8x + 12\_\_\_\_\_\_ = 28$$ — $(9 - 4x)$

We simplify the left side of the above equation to get

$$-40x + \_\_\_\_ = 28$$ — 108

$$-40x = \_\_\_\_$$ — -80

$$x = \_\_\_\_$$ — 2

When we solved for y, we got $y = 9 - 4x$ (see the box on p. 387). But now we have $x = 2$. Therefore

$$y = 9 - 4x$$

$$y = 9 - 4(\_\_\_\_)$$ — 2

$$y = \_\_\_\_$$ — 1

Our solution is $x = \_\_\_\_$ and $y = \_\_\_\_$. You may check this solution in the original system to see that it works. — 2; 1

5. Solve the system

$$y = 4 - x \qquad \text{and} \qquad x + y = 5$$

The first equation is already solved for ____. So we use this expression in place of y in the (first\second) equation of the system. — y; second

$$x + y = 5$$

$$x + \_\_\_\_\_\_ = 5$$ — $(4 - x)$

If we simplify the left side we get the equation ____ $= 5$. But this is an inconsistent equation. Therefore, the original system must be (consistent\inconsistent). We can check that the original system has no solution in the following way. — 4; inconsistent

The slope of the line $y = 4 - x$ is ____. The slope of the line $x + y = 5$ is ____. So the lines (have\do not have) the same slope. These lines are parallel and do not intersect. So it makes sense that our system has no solution. — -1; -1; have

6. In general, if the method of substitution yields an inconsistent equation (like $4 = 5$), then the original system is also inconsistent. Therefore, the system (will\will not) have a solution. — will not

7. Given the system

$$9x - 2y = 4 \qquad \text{and} \qquad -27x + 6y = -12$$

let us solve the first equation for y.

$$9x - 2y = 4$$

$$-2y = 4 - \_\_\_\_$$ — $9x$

$$y = \frac{4}{-2} - \_\_\_\_$$ — $\frac{9x}{-2}$

$$y = -2 + \_\_\_\_$$ — $\frac{9x}{2}$

second — Recall $-(9x/-2) = +(9x/2)$. Next we substitute this expression for y into the (first\second\third) equation

$$-27x + 6y = -12$$

$-2 + \frac{9x}{2}$

$$-27x + 6(\_\_\_\_\_\_) = -12$$

-12 — If we simplify and collect like terms on the left, we get ______ $= -12$. The equation $-12 = -12$ is an identity. It is always true. This means that two equations in our system are equivalent equations. In fact, if we divide both sides of the second equation by -3 we get

$$\frac{-27x}{-3} + \frac{6y}{-3} = \frac{-12}{-3}$$

or

$9x$; $2y$; 4

$$\_\_\_ - \_\_\_ = \_\_\_$$

yes — Is this the first equation of the system? (yes\no).

Since the two equations of the system are equivalent they have (the same graph\a different graph). Therefore, the two lines of the system meet (nowhere\at only one point\at all points), and the solution is

the same graph

at all points

all points; common

______ on the ______ line.

one and the same

does

8. In general, if the method of substitution yields an identity (like $-12 = -12$), then two equations of the system are equivalent. It follows that the graphs of the two lines are (one and the same\two different) graph(s). So the solution consists of the infinitely many points that are common to both lines. Of course, the system is still consistent because it (does\does not) have a solution.

PRACTICE PROBLEMS

1. *a.* What is meant by the term inconsistent equation?
   b. Give an example of an inconsistent equation.
   c. If the method of substitution yields an inconsistent equation, what can you say about the original system of equations?

2. *a.* When is an equation an identity?
   b. If the method of substitution yields an identity, what can you say about the original system of equations.

Use the method of substitution to solve the following systems.

3. $\begin{cases} -2x + y = -7 \\ 3x - y = 12 \end{cases}$ **4.** $\begin{cases} 9x + 6y = 6 \\ 12x + 8y = 8 \end{cases}$ **5.** $\begin{cases} -3x + 15y = 6 \\ x - 5y = 1 \end{cases}$

6. $\begin{cases} 2x + y = 4 \\ -4x + 6y = 0 \end{cases}$ **7.** $\begin{cases} 6x - 2y = 0 \\ 4x + 10y = 17 \end{cases}$ **8.** $\begin{cases} -3x - 2y = -3 \\ x = 3 \end{cases}$

9. $\begin{cases} 8x - 2y = 16 \\ -4x + y = -8 \end{cases}$ **10.** $\begin{cases} {}^1/_2x + y = 2 \\ -x + 4y = 2 \end{cases}$ **11.** $\begin{cases} {}^1/_3x + 4y = 3 \\ x - y = 9 \end{cases}$

12. $\begin{cases} 8x - 3y = 7 \\ 2x + 5y = -4 \end{cases}$ **13.** $\begin{cases} 2x + 3y = 9 \\ 12x + 18y = 24 \end{cases}$ **14.** $\begin{cases} -3x - 2y = -3 \\ 4x + 7y = 4 \end{cases}$

15. $\begin{cases} 8x - 2y = 6 \\ -4x + y = -3 \end{cases}$ **16.** $\begin{cases} x + y = -16 \\ x - y = 14 \end{cases}$ **17.** $\begin{cases} 2x - 3y = 0 \\ 5x - 7y = 0 \end{cases}$

*18. $\begin{cases} 3x + y + z = 12 \\ \quad 2y + 4z = 8 \\ \quad\quad 5z = 5 \end{cases}$

*19. $\begin{cases} x + 3y - z = -2 \\ 2x + y + 3z = 11 \\ x - 2y + z = 4 \end{cases}$

*20. $\begin{cases} 2x + 0y + z = 4 \\ x + 0y + 3z = 2 \\ x + y + z = 3 \end{cases}$

* Optional.

SELFTEST

Use the method of substitution to solve the following systems. Indicate if a system is inconsistent or if it has infinitely many solutions.

1. $\begin{cases} 2x + y = 4 \\ y = x - 2 \end{cases}$

2. $\begin{cases} x + 4y = 10 \\ y - x = 0 \end{cases}$

3. $\begin{cases} 4x + 2y = 3 \\ y = -2x \end{cases}$

4. $\begin{cases} 2x = 2y + 8 \\ y + 4 = x \end{cases}$

5. $\begin{cases} 3x + 2y = 11 \\ 2x - 4y = 2 \end{cases}$

SELFTEST ANSWERS

1. (2, 0) **2.** (2, 2) **3.** inconsistent **4.** infinitely many solutions **5.** (3, 1)

10.3 SOLVING SYSTEMS OF EQUATIONS BY ADDITION AND SUBTRACTION

In this section we continue our study of systems of linear equations. We will learn a method of eliminating variables by adding or subtracting equations. The two rules below summarize the basic idea. The use of the rules will be explained in the examples that follow.

Rule 1: **If one of the equations in a system is multiplied or divided (on both sides) by a nonzero constant, it does not change the solution (or solutions) of the system.**

Rule 2: **If two of the equations in a system are added together or subtracted from each other (on both sides), it does not change the solution (or solutions) of the system.**

Of course, if we have a system of equations we are interested mainly in finding a solution for that system. Two systems of equations that have the same solution are *equivalent systems of equations*. We use the two rules to construct systems that are equivalent to a given system. The new system will be easier to solve.

Example 1 Do the two systems have the same solution? That is, are they equivalent systems?

$$\begin{cases} x - y = 1 \\ -2x + 8y = 7 \end{cases} \xrightarrow{\text{multiply by 2}} \begin{cases} 2x - 2y = 2 \\ -2x + 8y = 7 \end{cases}$$

The systems are equivalent. For the first equation in the right system was obtained by multiplying the first equation in the left system by 2.

Example 2 Are the following systems equivalent? Find the solution for each system.

$$\begin{cases} 8x + 2y = -3 \\ -8x + 5y = -4 \end{cases} \qquad \begin{cases} 8x + 2y = -3 \\ 7y = -7 \end{cases}$$

By rule **2** the two systems have the same solution. For if we add the two equations of the left system, we obtain the new second equation of the right system.

$$\begin{array}{rl} 8x + 2y = -3 & \\ -8x + 5y = -4 & \text{add} \\ \hline 7y = -7 & \text{new second equation} \end{array}$$

How do we find the solution? Which system should we use? Since they are equivalent, we could use either system. The system on the right,

$$8x + 2y = -3 \qquad 7y = -7$$

is easy to solve. Since $7y = -7$, then $y = -1$. Now put $y = -1$ in the first equation and solve for x.

$$\begin{aligned} 8x + 2y &= -3 \\ 8x + 2(-1) &= -3 \\ 8x &= -1 \\ x &= \frac{-1}{8} \end{aligned}$$

The solution for both systems is thus $x = {}^{-1}/_{8}$ and $y = -1$. Let's check the solution in both systems.

$$\begin{aligned} 8x + 2y &= -3 \\ 8({}^{-1}/_{8}) + 2(-1) &= -3 \\ -1 + (-2) &= -3 \quad \text{checks} \\ -8x + 5y &= -4 \\ -8({}^{-1}/_{8}) + 5(-1) &= -4 \\ 1 + (-5) &= -4 \quad \text{checks} \end{aligned}$$

$$\begin{aligned} 8x + 2y &= -3 \\ 8({}^{-1}/_{8}) + 2(-1) &= -3 \\ -1 + (-2) &= -3 \quad \text{checks} \\ 7y &= -7 \\ 7(-1) &= -7 \\ -7 &= -7 \quad \text{checks} \end{aligned}$$

Example 3 Solve the system

$$x - 3y = 1 \qquad \text{and} \qquad 2x + y = 4$$

We can eliminate either the x or y. Let's eliminate x. If we multiply the first equation by -2 we get the equivalent system on the right.

$$\text{System I} \left\{\begin{aligned} x - 3y &= 1 \\ 2x + y &= 4 \end{aligned}\right. \xrightarrow{\text{multiply by } -2} \text{System II} \left\{\begin{aligned} -2x + 6y &= -2 \\ 2x + y &= 4 \end{aligned}\right.$$

Now add the equations in system II to get an equivalent third system.

$$\begin{array}{rl} \text{System II} & \\ -2x + 6y = -2 & \\ 2x + y = 4 & \\ \hline 7y = 2 & \text{new second equation} \end{array} \qquad \text{System III} \left\{\begin{aligned} -2x + 6y &= -2 \\ 7y &= 2 \end{aligned}\right.$$

System III is easy to solve. The second equation, $7y = 2$, gives $y = {}^{2}/_{7}$. If we now put $y = {}^{2}/_{7}$ in the first equation we get

$$\begin{aligned} -2x + 6y &= -2 \\ -2x + 6({}^{2}/_{7}) &= -2 \\ -2x + {}^{12}/_{7} &= {}^{-14}/_{7} \\ -2x &= {}^{-26}/_{7} \\ x &= {}^{13}/_{7} \end{aligned}$$

Our solution is $x = {}^{13}/_{7}$ and $y = {}^{2}/_{7}$, which checks in all three systems.

The previous examples show us some basic techniques. To shorten our work a little more there is a third rule. It is in effect a combination of rules **1** and **2**.

Rule 3: **If one of the equations in a system is multiplied or divided by a nonzero constant and added to or subtracted from another equation, then the solution of the system is not changed.**

***Example 4** Solve the system:

$$\text{System I} \quad \begin{cases} 2x + 3y + z = 9 & \text{equation 1} \\ x - 2y - 3z = 1 & \text{equation 2} \\ 5x + 4y + 6z = 5 & \text{equation 3} \end{cases}$$

We can eliminate any of the variables. Let us eliminate z from the last two equations.

Step 1 Multiply equation 1 by 3 and add equation 2.

$$2x + 3y + z = 9 \xrightarrow{\text{multiply by 3}} \begin{array}{r} 6x + 9y + 3z = 27 \\ x - 2y - 3z = 1 \\ \hline \end{array}$$

$$\text{new second equation} \rightarrow 7x + 7y = 28$$

Step 2 Multiply equation 1 by -6 and add to equation 3.

$$2x + 3y + z = 9 \xrightarrow{\text{multiply by } -6} \begin{array}{r} 5x + 4y + 6z = 5 \\ -12x - 18y - 6z = -54 \\ \hline \end{array}$$

$$\text{new third equation} \rightarrow -7x - 14y = -49$$

To form our new system we will not change equation 1.

$$\text{System II} \quad \begin{cases} 2x + 3y + z = 9 & \text{original equation 1} \\ 7x + 7y = 28 & \text{new second equation} \\ -7x - 14y = -49 & \text{new third equation} \end{cases}$$

As you can see we have eliminated z from the last two equations. Now we can eliminate either x or y from the two new equations. Let's eliminate x. Again there are several ways to do this. We are just suggesting one of the ways.

Step 3 Add equations 2 and 3 of System II to obtain a new third equation.

$$\begin{array}{rl} 7x + 7y = 28 & \text{step 1 second equation} \\ -7x - 14y = -49 & \text{step 2 third equation} \\ \hline \text{new third equation} \rightarrow -7y = -21 & \end{array}$$

$$\text{System III} \quad \begin{cases} 2x + 3y + z = 9 & \text{original first equation} \\ 7x + 7y = 28 & \text{new second equation} \\ -7y = -21 & \text{newest third equation} \end{cases}$$

This last system of equations is easy to solve. For the third equation has only one variable, y. The second equation has only two variables, x and y. The first equation has all three variables.

We start with the bottom equation. Since $-7y = -21$, then $y = 3$. Now we use the second equation to find x. Since $y = 3$, the second equation, $7x + 7y = 28$, becomes $7x + 7(3) = 28$ or $7x = 7$ or $x = 1$.

Finally, we put $x = 1$ and $y = 3$ into the first equation to find z.

$$\begin{aligned} 2x + 3y + z &= 9 \\ 2(1) + 3(3) + z &= 9 \\ z &= -2 \end{aligned}$$

Our final solution is $x = 1$, $y = 3$, and $z = -2$. These values satisfy each equation in all three systems.

Rules **1**, **2**, and **3** can be used on a system of four equations in four unknowns or on even larger systems. Of course, the number of steps involved increases as the system gets larger.

* Optional.

Nevertheless, this method of elimination is one of the faster ways of solving a system of equations. It is used in many large electronic computers. These computers sometimes solve systems of a hundred equations in a hundred unknowns.

When solving a system of equations, it pays to be neat and organized. After applying the rules to eliminate variables, gather your equations together in a new (equivalent) system as we did in the examples. This step helps prevent disorganization and resulting sloppy errors.

Two special cases sometimes occur. Namely, the case where a system is inconsistent, or the case where there are infinitely many solutions.

Example 5 Solve the system

$$2x - y = 3 \qquad \text{and} \qquad 4x - 2y = 8$$

We might start in a number of ways. Let us begin by trying to eliminate y. If we multiply the first equation by -2 and add the result to the other equation, we will get a new second equation.

System I $\begin{cases} 2x - y = 3 \\ 4x - 2y = 8 \end{cases}$ $\xrightarrow{\text{multiply by } -2}$ $\begin{array}{r} -4x + 2y = -6 \\ 4x - 2y = 8 \\ \hline \end{array}$

new second equation $\rightarrow 0 = 2$

System II $\begin{cases} 2x - y = 3 & \text{original first equation} \\ 0 = 2 & \text{new second equation} \end{cases}$

The last equation, $0 = 2$, is certainly not true. What happened? When we get an inconsistent equation, such as $0 = 2$, it means that the whole system is inconsistent. That is, there is *no* solution for that system.

In general, if a system of equations is inconsistent, then using rules **1**, **2**, or **3** will eventually give an inconsistent equation. This tells us that there is no solution to the system.

Example 6 Solve the following system by eliminating y from the second equation.

System I $\begin{cases} 4x - y = -1 \\ 12x - 3y = -3 \end{cases}$ $\xrightarrow{\text{multiply by } -3}$ $\begin{array}{r} -12x + 3y = 3 \\ 12x - 3y = -3 \\ \hline \end{array}$ add

new second equation $\rightarrow 0 + 0 = 0$

System II $\begin{cases} 4x - y = -1 & \text{original first equation} \\ 0 = 0 & \text{new second equation} \end{cases}$

Now what do we do? The identity $0 = 0$ is always true, but it doesn't seem to tell us anything about x or y. In effect, we are left with just one equation, $4x - y = 1$. This means that *all* the infinitely many points on the line $4x - y = 1$ are solutions of the original system. The second equation in the original system gave us no new information. In fact, the second equation of the original system is just the first equation multiplied by 3.

In general, if the use of rules **1**, **2**, or **3** gives an identity, such as $0 = 0$, there are infinitely many solutions of the system.

PROGRAM 10.3

1. When studying a system of linear equations, we are mostly interested in finding a solution for that system. So, if two systems have the same solution, we will say they are *equivalent systems.*

In this section we will learn another method for solving a system of linear equations. The central ideas are summarized in three basic rules.

2. ***Rule 1:*** If one of the equations in a system is multiplied or divided (on both sides) by a nonzero constant, it does not change the solution (or solutions) of the system.

By rule **1**, the following systems are equivalent.

$$\begin{array}{l} 6x - 2y = 9 \\ 4x + y = 6 \end{array} \xrightarrow{\text{multiply by 2}} \begin{array}{l} 12x - 4y = 18 \\ 4x + y = 6 \end{array}$$

We multiplied the first equation in the left system by 2. Do the two systems have the same solution? (yes\no).

yes

3. Let us see if the following systems are equivalent.

$$\begin{array}{l} -x + 3y = 7 \\ 16x + 4y = 8 \end{array} \xrightarrow{\text{divide by 4}} \begin{array}{l} -x + 3y = 7 \\ \_\_x + \_\_y = \_\_\_ \end{array}$$

4; 1; 2

The second equation in the left system was (multiplied\divided) by 4 to get the new second equation in the right system.

divided

By rule **1** it follows that the two systems (are\are not) equivalent. So they (do\do not) have the same solution.

are

do

4. Consider the following two systems:

$$\begin{cases} 12x - 15y = 21 \\ 4x + 3y = -1 \end{cases} \quad \text{and} \quad \begin{cases} 4x - 5y = 7 \\ 20x + 15y = -5 \end{cases}$$

We changed the first equation by dividing both sides by _____. We changed

3

the second equation by (multiplying\dividing) both sides by _____.

multiplying; 5

In effect, we have made (one\two\no) applications of rule **1**. Therefore the systems (are\are not) equivalent.

two

are

5. Now let us look at the second basic rule.

Rule 2: If two of the equations in a system are added together or subtracted from each other (on both sides), it does not change the solution of the system.

Consider these two systems:

$$\begin{cases} 6x + 3y = -1 \\ 2x - 3y = 4 \end{cases} \quad \text{and} \quad \begin{cases} 6x + 3y = -1 \\ 8x + 0y = 3 \end{cases}$$

The systems (are\are not) equivalent. For to get the second equation in the right system we (added\subtracted\multiplied) the two equations in the left system.

are

added

6. Let's see if the following systems are equivalent.

$$\begin{cases} 10x + 4y = 8 \\ 6x + 4y = 1 \end{cases} \quad \text{and} \quad \begin{cases} 4x = 7 \\ 6x + 4y = 1 \end{cases}$$

To get the first equation in the right system we (subtracted\added\multiplied) the second equation of the left system from the first equation of the left system. Thus:

subtracted

$$\begin{array}{r} 10x + 4y = 8 \\ 6x + 4y = 1 \\ \hline \end{array} \quad \text{subtract}$$

new first equation → __________

$4x + 0y = 7$

yes; zero
yes
yes

Is it true that $0y = 0$? (yes\no). For zero times anything is ____. So is $4x + 0y$ the same as $4x$? (yes\no). Are the two systems in this frame equivalent? (yes\no).

7. The third basic rule is in effect a combination of rules **1** and **2**.

Rule 3: If one of the equations in a system is multiplied or divided by a nonzero constant and added to or subtracted from another equation, then the solution (or solutions) of the system are not changed.

To find if these systems,

$$\begin{cases} 3x - y = -1 \\ 2x + 2y = 10 \end{cases} \quad \text{and} \quad \begin{cases} 3x - y = -1 \\ 8x = 1 \end{cases}$$

are equivalent, we get a new second equation on the right by multiplying the first equation of the left system by 2 and adding the result to the second equation of the left system.

$$3x - y = -1 \xrightarrow{\text{multiply by 2}} \begin{array}{r} 6x - 2y = -2 \\ 2x + 2y = 10 \\ \hline \end{array} \quad \text{add}$$

$$\text{new second equation} \rightarrow 8x = 8$$

3; do

So the systems are equivalent by rule ____, and therefore (do\do not) have the same solution.

8. We just saw that the systems

$$\begin{cases} 3x - y = -1 \\ 2x + 2y = 10 \end{cases} \quad \text{and} \quad \begin{cases} 3x - y = -1 \\ 8x = 8 \end{cases}$$

one

are equivalent. The second equation of the right system is in (one\two) variable(s). It is easy to solve this equation for x.

1

$$8x = 8 \quad \text{so} \quad x = \_\_\_\_$$

Now we know the x value of the solution. To find the y value, we simply use this x value in the last equation.

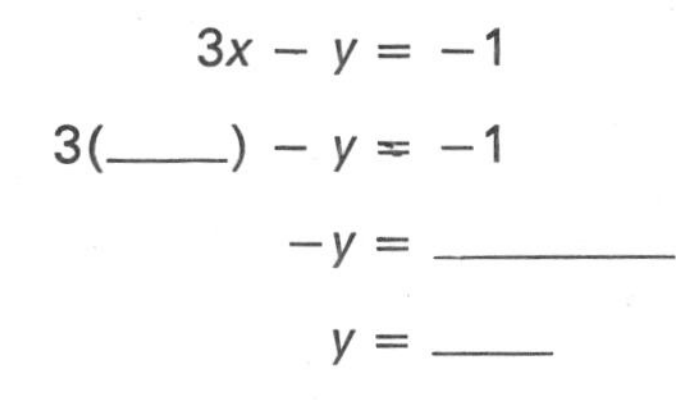

$$3x - y = -1$$

1

$$3(\_\_\_\_) - y = -1$$

$(-1 - 3)$ or -4

$$-y = \_\_\_\_\_\_$$

4

$$y = \_\_\_\_$$

1; 4

The solution for the original system is $x =$ ____ and $y =$ ____.

9. We use rules **1**, **2**, and **3** to create systems equivalent to the original system. But we want our new systems to be easier to solve. The following systems are equivalent. Which is the easier one to solve?

b.

a. $\begin{cases} 2x - 3y = 4 \\ 7x + 5y = 14 \end{cases}$ or *b.* $\begin{cases} 2x - 3y = 4 \\ \quad\;\; 31y = 62 \end{cases}$

10. Using the system

$$8x - y = 10 \qquad \text{and} \qquad x + 4y = 26$$

$8x + 32y = 208$

multiply the second equation by 8. What do you get? ____________.

Now subtract this result from the first equation in the system. What do you get?

$$\begin{array}{r} 8x - y = 10 \\ 8x + 32y = 208 \\ \hline \end{array} \quad \text{subtract}$$

______ 	$0x - 33y = -198$

Are the systems

$$\begin{cases} 8x - y = 10 \\ x + 4y = 26 \end{cases} \quad \text{and} \quad \begin{cases} 8x - y = 10 \\ -33y = -198 \end{cases}$$

equivalent? (yes\no). This is in accordance with rule ____. Which system will be easier to solve? (the left one\the right one). 	yes; 3 	the right one

11. The system represented by

$$4x - 3y = 2 \qquad \text{and} \qquad 2y = 6$$

is a type of system that is easy to solve. For the second equation contains only (one\two) unknown(s). We can solve for that unknown. 	one

$$2y = 6 \quad \text{so} \quad y = \_\_\_\_$$ 	3

Now we know $y =$ ____. We put this value of y into the first equation and solve for x. 	3

$$4x - 3y = 2$$

$$4x - 3(\_\_) = 2$$ 	3

$$4x - \_\_\_\_ = 2$$ 	9

$$4x = \_\_\_\_\_\_$$ 	$2 + 9$ or 11

$$x = \_\_\_\_$$ 	$^{11}/_4$

The solution is $x =$ ____ and $y =$ ____. 	$^{11}/_4$; 3

12. If we could somehow get this system,

$$\text{System I} \quad \begin{cases} 6x + y = 4 \\ 2x + y = 2 \end{cases}$$

to look like the system in the last frame, it would be fairly easy to solve. What we want is a system with one equation in one unknown. Let's subtract equation 2 from equation 1.

$$\begin{array}{r} 6x + y = 4 \\ 2x + y = 2 \\ \hline \end{array} \quad \text{subtract}$$

new first equation → ______ 	$4x + 0y = 2$

What happened to the y variable when we subtracted? ______ since $0y = 0$. We now have an equation in one unknown. We replace the first (or second—it does not matter) equation in System 1 by the new first equation $4x = 2$. 	It dropped out

$$\text{System II} \quad \begin{cases} 4x = 2 \\ 2x + y = 2 \end{cases}$$

Is System II equivalent to System I? (yes\no). Do they both have the same solution? (yes\no). 	yes 	yes

13. Now we see that System II is fairly easy to solve. Since $4x = 2$, then

$^{1}/_{2}$ — $x =$ ____. Next we put our solution for x into the second equation and solve for y.

$$2x + y = 2 \qquad \text{from System II}$$

$^{1}/_{2}$ — $2(\_\_\_\_) + y = 2$

1 — $\_\_\_\_ + y = 2$

1 — $y = \_\_\_\_$

$^{1}/_{2}$; 1 — Therefore, our solution is $x =$ ____ and $y =$ ____. Let's check these values in the original system.

$$6x + y = 4 \qquad \text{from System I}$$

$^{1}/_{2}$; 1 — $6(\_\_) + (\_\_) \stackrel{?}{=} 4$

3; 1; true — $(\_\_) + (\_\_) = 4$ (true\false)

$$2x + y = 2 \qquad \text{from System I}$$

$^{1}/_{2}$; 1 — $2(\_\_) + (\_\_) \stackrel{?}{=} 2$

1; 1; true — $(\_\_) + (\_\_) = 2$ (true\false)

yes — Do our solutions check in System I? (yes\no).

14. In solving the system

$$2x + 5y = 2 \qquad \text{and} \qquad x + 3y = {}^{3}/_{2}$$

we wish to eliminate one of the variables. Let's eliminate x. The coefficient of x in the first equation is 2 and in the second equation it is understood to

1 — be ____. If we multiply the second equation by 2 we get

$2x + 6y$; 3 — $\_\_\_\_\_\_ = \_\_\_\_$

We now have a new system.

2; 6; 3 — System II $\begin{cases} 2x + 5y = 2 \\ \_\_x + \_\_y = \_\_\_\_ \end{cases}$ new second equation

yes — What do you notice about the coefficients of x in the two equations of System II? Are they the same? (yes\no). If we subtract equation 2 from

They drop out — equation 1 in System II, what happens to the x terms? ________

$0x = 0$ — because ______.

$$\begin{array}{r} 2x + 5y = 2 \\ 2x + 6y = 3 \\ \hline \end{array} \qquad \text{subtract}$$

$0x - 1y = -1$ — ______

1 — From the equation $-y = -1$ we obtain $y =$ ____. Now we return to System II. We choose either equation and replace y with 1. Then we solve for x. Let us use the first equation in System II.

$$2x + 5y = 2$$

1 — $2x + 5(\_\_) = 2$

5 — $2x + \_\_\_\_ = 2$

$2 - 5$ or -3 — $2x = \_\_\_\_$

$^{-3}/_{2}$ or -1.5 — $x = \_\_\_\_$

Our solution is $x =$ ____ and $y =$ ____. Again, you may check this solution in the original system. It works.

−1.5; 1

15. To solve the system

$$2x + 5y = -2 \quad \text{and} \quad 3x - y = 14$$

let's first eliminate y. The coefficient of y in the first equation is ____. In the second equation it is −1. If we multiply the second equation by (3\5\7\−4) the coefficients of y will be opposites.

5

5

$$\begin{cases} 2x + 5y = -2 \\ 3x - y = 14 \end{cases} \xrightarrow{\text{multiply by 5}} \begin{cases} 2x + 5y = -2 \\ \underline{\qquad\qquad} \end{cases} \quad \text{add}$$

$\underline{\qquad\qquad}$

$15x - 5y = 70$

$17x + 0y = 68$

When we added the two equations of the right system, we got an equation in only one unknown. Now let's solve that equation for x.

$$17x = 68 \quad \text{so} \quad x = \underline{\quad}$$

4

Use this value of x in one of the equations of the original system to solve for y. The first equation will do.

$$2x + 5y = -2$$

$$2(\underline{\quad}) + 5y = -2$$

$$5y = \underline{\qquad}$$

$$y = \underline{\quad}$$

4

$-2 - 8$ or -10

-2

So the solution to our system is $x =$ ____ and $y =$ ____. These values will check in the original system.

4; −2

16. In order to solve

$$\text{System I} \quad \begin{cases} 8x + 3y = 16 \\ 5x + 2y = 10 \end{cases}$$

we may eliminate either x or y. Since the coefficients on y are smaller, we will eliminate y. That way we can deal with somewhat smaller numbers. If we multiply equation 1 by 2, we get

$$2 \cdot (8x + 3y = 16) = \underline{\qquad\qquad}$$

$16x + 6y = 32$

Next multiply equation 2 by 3. You get

$$3 \cdot (5x + 2y = 10) = \underline{\qquad\qquad}$$

$15x + 6y = 30$

Our new system of equations is

$$\text{System II} \quad \begin{cases} 16x + 6y = 32 \\ 15x + 6y = 30 \end{cases}$$

We constructed System II so that the coefficients on y would be the same. Now we eliminate y by subtracting the equations of System II.

$$\begin{array}{r} 16x + 6y = 32 \\ \underline{15x + 6y = 30} \end{array} \quad \text{subtract}$$

$\underline{\qquad\qquad}$

$x + 0y = 2$

2 — Part of our solution is $x =$ ____. We put this value of x into either equation of System II. Let's put $x = 2$ into the first equation.

$$16x + 6y = 32$$

2 — $$16(\_\_) + 6y = 32$$

32 — $$\_\_ + 6y = 32$$

32 − 32 or 0 — $$6y = \_\_\_\_$$

0 (that is, $^0/_6 = 0$) — $$y = \_\_\_\_$$

2; 0 — So our solution is $x =$ ____ and $y =$ ____. You may check this solution in the original system. It is correct.

17. The methods of this section can be used to solve much larger systems. In applications it is not uncommon to solve systems of three equations in three unknowns or even larger systems. The basic ideas do not change, but the work gets longer.

PRACTICE PROBLEMS

1. What does it mean to say that two systems of equations are equivalent?
2. *a.* In your own words state rules **1**, **2**, and **3** of this section.
   b. When we apply these rules, do we always obtain equivalent systems of equations?
3. Without going into much detail, explain how rules **1**, **2**, and **3** can help solve a system of linear equations.

Solve, using the methods of this section.

4. $x + 2y = 1$; $-x + 5y = -8$

5. $3x + y = 6$; $7x + y = 14$

6. $4x + 2y = -2$; $3x - 2y = -12$

7. $6x + 3y = 9$; $2x + y = 3$

8. $4x - 2y = 0$; $2x + y = 1$

9. $24x - 6y = 10$; $-12x + 3y = -5$

10. $3x - 5y = 7$; $6x - 10y = 9$

11. $12x + 8y = 8$; $6x - 10y = -10$

12. $14x + 6y = 20$; $21x - 4y = 17$

13. $4x - 7y = 2$; $-2x + 14y = -1$

14. $-^1/_2x + y = 1$; $x + y = 4$

15. $^1/_3x + 2y = 5$; $-x + ^1/_2y = -2$

16. $5x - 7y = -2$; $-3x + 2y = -1$

***17.** $3x + y + z = 4$; $-2x - y + 4z = 2$; $x - 5y + 3z = 4$

***18.** $6x + 9y + z = 21$; $4x + y + 3z = 9$; $x + 2y + 5z = 4$

***19.** $3x + 2y - 7z = 12$; $9y + z = 8$; $2z = -2$

***20.** $3x + 9y + z = 0$; $-x + y + 5z = 0$; $4x - 7y + 2z = 0$

* Optional.

SELFTEST

Decide if the given systems are equivalent and give a reason for your answer.

1. *a.* $\begin{cases} x + y = 4 \\ 3x - 2y = 8 \end{cases}$ *b.* $\begin{cases} 2x + 2y = 8 \\ 3x - 2y = 8 \end{cases}$

2. *a.* $\begin{cases} 2x + 4y = 7 \\ 3x - 2y = 8 \end{cases}$ *b.* $\begin{cases} 6x + 12y = 21 \\ -6x + 4y = -16 \end{cases}$

3. *a.* $\begin{cases} x + 2y = 5 \\ -x + y = 8 \end{cases}$ *b.* $\begin{cases} 3y = 13 \\ -x + y = 8 \end{cases}$

Solve the following systems by either multiplying or dividing an equation by a nonzero constant or by adding or subtracting two of the equations of the system.

4. $\begin{cases} x + y = 1 \\ 2x - y = 5 \end{cases}$

5. $\begin{cases} 2x + 3y = 4 \\ x - y = 2 \end{cases}$

6. $\begin{cases} 2x + y = 5 \\ 4x + 2y = 10 \end{cases}$

7. $\begin{cases} 2y + 4x = 8 \\ y + 2x = 10 \end{cases}$

8. $\begin{cases} 4x - 2y = 4 \\ 2x - 3y = -6 \end{cases}$

SELFTEST ANSWERS

1. yes; multiply equation 1 by 2.

2. yes; multiply equation 1 by 3 and equation 2 by -2.

3. yes; add equations 1 and 2 to get the new first equation.

4. $(2, -1)$ **5.** $(2, 0)$ **6.** infinitely many solutions **7.** inconsistent **8.** $(3, 4)$

10.4 APPLICATIONS

Systems of linear equations occur in many applications of mathematics. Although it is not within the scope of this book to give an in-depth treatment of all the different applications, it is interesting to look at a few problems involving systems of linear equations. If you are studying business, economics, biology, physical science, etc., you will no doubt encounter more applications in later courses.

Most applications are initially stated as word problems. When solving word problems we must remember to

1. write down clearly what is given;
2. clearly understand what is being asked;
3. assign variable names to the unknowns and set up a system of equations in these unknowns;
4. solve the system of equations;
5. check our answer in the original problem to see that it makes sense.

Example 1 A campaign manager must schedule his candidate's time between two districts—district A and district R. Each day in district A will yield 600 votes, while each day in district R will give 850 votes. The candidate must get at least 7,500 votes from these two districts. There are only 10 days to campaign. The candidate wants to spend as much time as possible in district A so that he can build his reputation there. How many days can he spare for district A and still get the necessary 7,500 votes?

Let A represent the number of days in district A and R represent the number of days in district R. The total number of days is 10, so

$$A + R = 10$$

The number of votes from district A is $600A$ since each day there gives 600 votes. Likewise, the number of votes from district R is $850R$. The sum of the votes must be 7,500. Therefore

$$600A + 850R = 7{,}500$$

The system we need to solve is

$$A + R = 10 \qquad \text{and} \qquad 600A + 850R = 7{,}500$$

Let's use the substitution method. From equation 1, we get

$$A = 10 - R$$

If we use this value for A in the second equation $600A + 850R = 7{,}500$, we get

$$\begin{aligned} 600(10 - R) + 850R &= 7{,}500 \\ 6{,}000 - 600R + 850R &= 7{,}500 \\ 250R &= 1{,}500 \\ R &= 6 \end{aligned}$$

Since $A = 10 - R$, $A = 10 - 6$ or 4. The candidate can spare 4 days for district A. He will then get 7,500 votes. For he will get 2,400 votes from district A and 5,100 votes from district R.

Example 2 An agricultural experiment station is experimenting with a new plant poison. The poision kills ragweed, but to a lesser extent it also kills alfalfa. At the experiment station, different concentrations of the poison are used on different plots. Each plot has both alfalfa and ragweed. As the concentration gets stronger the poison kills more ragweeds and more alfalfa plants. Let x be the concentration of the poison in ounces per ten gallons of water. Let R represent the number of ragweeds killed, and let A represent the number of alfalfa plants killed.

From the experimental plots, it is found that $R = 17x + 8$. For alfalfa, the result was $A = 5x + 35$.

Under any circumstance, what is the maximum concentration of the poison we should not exceed? That is, at what concentration are we killing as many alfalfa plants as ragweeds? We want to find that value of x for which $R = A$. Letting $R = A$, our system of equations becomes

$$R = 17x + 8 \qquad \text{and} \qquad R = 5x + 35$$

As we know, there are several ways to solve this system. The graphical method would not be good. Due to the steep slope of the first line, it would be hard to see exactly where the two lines meet on a graph. For two equations in two unknowns, the method of substitution is quick and easy. The second equation is already solved for R. Let's put this into the first equation.

$$\begin{aligned} 17x + 8 = R &= 5x + 35 \\ 17x + 8 &= 5x + 35 \\ 12x &= 27 \\ x &= 2.25 \end{aligned}$$

So 2.25 ounces per ten gallons is the maximum concentration to be used in any circumstances. You may check in the original equations that this concentration kills as many ragweeds as alfalfa plants.

***Example 3** The dietitian of a hospital wants each patient to have at least the following amounts of vitamins every day.

Vitamin A	25 milligrams
Vitamin B	20 milligrams
Vitamin C	35 milligrams

Each meal is made up of three basic foods which have the following vitamin content in milligrams of vitamins per ounce of food.

	Food I	*Food II*	*Food III*
Vitamin A	4	1	1
Vitamin B	1	1	2
Vitamin C	3	0	5

* Optional.

How much of each food should a patient eat in order to get his vitamins? That is, what is the minimal amount of these foods necessary to obtain the vitamins specified by the dietitian? Let x, y, and z represent the number of ounces of foods I, II, and III, respectively. From the table, we see that the amount of vitamins received is

$$\begin{aligned} 4x + 1y + 1z &= (\text{amount of vitamin A}) = 25 \\ 1x + 1y + 2z &= (\text{amount of vitamin B}) = 20 \\ 3x + 0y + 5z &= (\text{amount of vitamin C}) = 35 \end{aligned}$$

So we have

$$\text{System I} \quad \begin{cases} 4x + y + z = 25 \\ x + y + 2z = 20 \\ 3x + 0y + 5z = 35 \end{cases}$$

Looking at the system, it seems that it would be easiest to eliminate y from the second equation. Then the last two equations will involve only x and z.

Step 1: Subtract equation 2 from equation 1. This gives us a new second equation. Leave equations 1 and 3 as they are.

$$\begin{array}{r} 4x + y + z = 25 \\ -x - y - 2z = -20 \\ \hline 3x \qquad - z = 5 \end{array} \longrightarrow \quad \text{System II} \begin{cases} 4x + y + z = 25 \\ 3x - z = 5 \\ 3x + 5z = 35 \end{cases}$$

Step 2: Now we use equations 2 and 3 of system II to eliminate x. We subtract equation 3 from equation 2. This gives us a new equation in z only. We will use this equation in place of the original third equation.

$$\begin{array}{r} 3x - z = 5 \\ -3x - 5z = -35 \\ \hline -6z = -30 \end{array} \longrightarrow \quad \text{System III} \begin{cases} 4x + y + z = 25 \\ 3x - z = 5 \\ -6z = -30 \end{cases}$$

The equation $-6z = -30$ gives $z = 5$. If we put $z = 5$ into the second equation $3x - z = 5$, we get $3x - 5 = 5$ or $x = {}^{10}/_3$. Finally we put $x = {}^{10}/_3$ and $z = 5$ into the first equation, $4x + y + z = 25$. This yields $4({}^{10}/_3) + y + 5 = 25$ or $y = {}^{20}/_3$. Our solution is $x = {}^{10}/_3 = 3.33$, $y = {}^{20}/_3 = 6.67$ and $z = 5$. These values check in the original system. So a patient should get at least 3.33 ounces of food I, 6.67 ounces of food II, and 5 ounces of food III.

PROGRAM 10.4

1. In this section we will present a few applications of systems of linear equations. As with most applications, the initial problem is usually formulated as a word problem. When solving the problem it is our job to first set up a system of equations that represent the problem mathematically. Then we use the methods we have already learned to solve the system. As a final step it is always a good idea to check the solution in the context of the original problem to see if it makes sense.

These are some good general hints on how to solve word problems. But the only way to really learn applications is to work with some examples.

2. A ski club makes a deal for buying train tickets. They can get first-class coaches at the same price as second-class coaches provided they pay for at least 8 coaches each week. The first-class coaches hold 30 people and the second-class coaches hold 46 people. Since they must pay for 8 coaches anyway, they like to order as many first-class coaches as possible for the group size. One weekend they had 256 reservations. How many coaches of each type should they use?

The number of first-class and second-class coaches to be used.

The solution to this problem involves two unknowns. What are they? Let us give the unknowns a variable name: Let F stand for the number of first-class coaches used and let S stand for the number of second-class coaches used.

The problem imposes certain conditions. First of all, the club must use

8

at least ____ coaches each week. If we write this condition mathematically we get

(number of first-class coaches) + (number of second-class coaches)

8

= ____

Using our variables F and S we get

8

$$\boxed{F + S = \_\_\_\_}$$

We put this equation in a box for easy reference.

The other conditions imposed by the problem are: Each first-class coach

30; 46

holds ____ people? Each second-class coach holds ____ people?

If we have F first-class coaches, how many people will there be in the

30F

first-class coaches? ($30F \backslash 46F \backslash 100F$). Since each coach holds 30 people

30F

and there are F coaches, there will be ____ people in first-class coaches.

If we have S second-class coaches, how many people can we put into

46S

these coaches? ($30S \backslash 46S \backslash 100S$). That is, there are S coaches each with 46 people, so the total number of people in second-class coaches will be

46S

____ .

So we can put $30F$ people in first class and $46S$ people in second class.

256

How many people made reservations to go? ____. Therefore

256

(number in first class) + (number in second class) = ____

If we use our variables F and S we get

256

$$30F + 46S = \_\_\_\_$$

This equation together with the previous equation (in the box) make a system of two equations in two unknowns

$$\text{System I} \quad \begin{cases} F + S = 8 \\ 30F + 46S = 256 \end{cases}$$

30

Let's eliminate F. First we multiply equation 1 by ____ so that the coefficients of F in both equations are the same.

30; 240

$$\text{System II} \quad \begin{cases} 30F + \_\_S = \_\_\_\_ \\ 30F + 46S = 256 \end{cases}$$

Now subtract the first equation in System II from the second equation.

$$\begin{array}{r} 30F + 46S = 256 \\ 30F + 30S = 240 \\ \hline \end{array} \quad \text{subtract}$$

$16S = 16$

____

1

From the equation $16S = 16$ we get $S =$ ____.

We now put the value $S = 1$ into either equation in System II and solve for F. Let's put $S = 1$ into the second equation in System II.

$$30F + 46S = 256$$

$$30F + 46(\_\_) = 256 \qquad 1$$

$$30F = \_\_\_\_ \qquad 210$$

$$F = \_\_\_\_ \qquad 7$$

So our solution is $S =$ ____ and $F =$ ____. You may check this solution in System I. It works. 1; 7

We look to see if the solution makes sense in the original problem. There are 7 first-class coaches each of which holds 30 people. How many people are in first class? ____. There is one second-class coach that holds 46 people. So how many people are in second class? ____. Do these numbers total up so we can seat all of our 256 people? (yes\no). Explain.

7 · 30 or 210

46

yes

________________ 210 + 46 = 256

3. A developer wants to sell a house and lot for $35,000. He intends to have the house cost four times as much as the lot. There is a $10 tax for each thousand dollars paid for the lot. This tax is used to equip a park for the neighborhood. A prospective buyer wants to know the cost of the house, the cost of the lot, and the tax.

As we approach solution of the problem, we notice first that the tax is not included in the $35,000. It is extra. The tax cannot be computed until we have first found the cost of the lot.

Let L represent the cost of the lot, and let H represent the cost of the house. The cost of the lot and house together is ________. Therefore $35,000

$$L + H = \_\_\_\_\_\_\_\_ \qquad \$35,000$$

The house costs four times what the lot costs. Therefore

$$\_\_L = H \qquad 4$$

Now we have a system of two equations in two unknowns

$$L + H = 35,000 \quad \text{and} \quad 4L = H$$

Since the second equation is already solved for H, it will be easy to use the method of substitution. The second equation says

$$H = \_\_\_\_ \qquad 4L$$

Now we put $H = 4L$ into the first equation.

$$L + H = 35,000$$

$$L + (\_\_\_\_) = 35,000 \qquad 4L$$

$$\_\_\_\_L = 35,000 \qquad 5$$

$$L = \_\_\_\_\_\_\_\_ \qquad 7,000$$

Now that we know what L is we can put this value into either equation in the original system and solve for H. Let's put $L = 7,000$ into the second equation in the original system.

$$4L = H$$

$$4(\_\_\_\_\_\_\_\_) = H \qquad 7,000$$

$$\_\_\_\_\_\_\_\_ = H \qquad 28,000$$

28,000; 7,000

So our solution is H = ________ and L = ________. You may check this solution to see that it works.

10; lot
$70;
$7,000; each thousand dollars
10; 7

The tax is $_____ for every thousand dollars paid on the (lot\house\both). Therefore the total tax is ($280\$50\$70). That is, since the lot cost ________ and the tax is $10 on ____________, we have

$$\_\_\_\_ \cdot \_\_\_\_ = \$70$$

yes
yes

Let's see if our solution makes sense in the original problem. The cost of the house is $28,000, and the cost of the lot is $7,000. Do these total up to $35,000? (yes\no). Does the house cost four times what the lot costs? (yes\no).

PRACTICE PROBLEMS

1. A city manager needs to charter busses for 494 people. The standard bus and the luxury bus cost the same, but the luxury bus has seats for only 25 people while the standard bus seats 36. There is money for only 18 busses. The manager would like to use as many luxury busses as possible for his group. How many luxury busses can be chartered?

2. A brewer makes a solution of sugar water and yeast in a vat. The original solution has 10 parts of sugar per 100 parts water. During the fermentation process, on an average day the sugar concentration is decreased by 0.3 parts per 100 parts water; the alcohol concentration increases at the rate of 0.7 parts per 100 parts water. This strain of yeast is such that the fermentation stops when the alcohol concentration begins to exceed the sugar concentration. How many days will it take for the fermentation to be completed?

3. A program council is sponsoring a movie. The movie will be shown in an auditorium with 400 seats. The council must take in $450 to cover costs. Student tickets cost $1.00 and nonstudent tickets cost $1.50. What is the maximum number of student tickets that they should not exceed? That is, what is the maximum number of student tickets they can sell and still break even?

4. In a recent election the winning candidate received 1,521 votes more than his opponent. If there were 7,923 votes cast in all, how many votes did each candidate receive?

5. Two truck drivers are taking turns driving on a long trip. One driver found that he drove 3 hours more than the other. After the trip was over they had spent a total of 47 hours on the road. How many hours did each person drive?

6. The sum of the measures of the inside angles of any triangle is 180°. In a certain triangle one angle is 90°. The other two angles are such that one is four times as large as the other How large is each angle?

7. An isosceles triangle is a triangle where two sides have the same length. The perimeter of an isosceles triangle is 33 inches. The two equal sides are each 6 inches longer than the base. What is the length of each side?

   [Perimeter = (length of side 1) + (length of side 2) + (length of side 3)]

8. A rectangle has a perimeter of 46 feet. The length is 9 feet longer than the width. What are the dimensions of the rectangle? [Perimeter = 2(length) + 2(width).]

9. A rectangle is 15 feet longer than it is wide. Its perimeter is 322 feet. What are the dimensions of this rectangle?

10. Herman has volunteered to cook for lunch on the hiking club work day. He wants to cook meat loaf. He needs ground beef, veal, and pork. There should be 6 times as much beef as pork and twice as much veal as pork. He needs 18 pounds of meat. How many pounds of beef, veal, and pork should he buy?

11. A campaign manager must schedule his candidate's time between two districts. Each day spent in district A will give 350 votes, while each day spent in district B will give 500 votes. There are 20 days to campaign. The candidate wants to spend as much time as possible in district A so he can build his reputation there. He needs 8,200 votes to win. What is the maximum amount of time he can spare for district A and still win?

12. Suppose you are in the business of making back packs. It costs you $20 to make one pack. You plan to sell each pack for $45. It also costs $1,200 to buy equipment for your factory. How many packs must you sell before you start to make a profit?

13. A man's estate of $72,000 is to be divided among his wife, three sons, and three daughters. The wife is to receive four times as much as each son, and each son three times as much as each daughter. How much does each receive?

14. The sum of three numbers is 103. The second number is six times the first and the third is seven more than the first. What are the three numbers?

15. A bomber is sent toward a target 2,650 miles away. It is flying 500 miles/hour. Three hours later a fighter plane is sent out from the target area to intercept the bomber. It is flying 650 miles/hour. How far from the target will they meet? When will they meet?

SELFTEST

1. A rectangle has a perimeter of 300 feet. The length is 10 feet more than the width. Find the length and width. [Perimeter of a rectangle = 2(length) + 2(width).]

2. Two typists contract to type a manuscript. They are paid by the page. One types 51 more pages than the other. Together they typed the entire manuscript of 275 pages. How many pages did each type?

SELFTEST ANSWERS

1. length = 80 feet; width = 70 feet
2. one typed 112 pages; the other typed 163 pages

POSTTEST 10

1. Solve this system by graphing.

$$y - x = 2 \qquad \text{and} \qquad 2y + x = 7$$

2. Solve this system by graphing.

$$x - 3 = y \qquad \text{and} \qquad 2x - 2y = 6$$

Solve by algebraic techniques:

3. $\begin{cases} y - 3x = 7 \\ y = 4x + 2 \end{cases}$

4. $\begin{cases} x - y = 4 \\ x + y = 6 \end{cases}$

5. $\begin{cases} x + 2y = 4 \\ 2x - y = -2 \end{cases}$

6. $\begin{cases} x + y = 3 \\ 2x + 2y = 8 \end{cases}$

7. What does it mean to say that two systems are equivalent?

8. Two generators together produce 4,200 kilowatts of electricity in a certain time period. One generator has twice the capacity of the other. If they are both working at full capacity, how much does each produce in this period.

Each problem above refers to a section in this chapter as shown in the table.

Problems	*Section*
1 and 2	10.1
3	10.2
4–7	10.3
8	10.4

POSTTEST 10 ANSWERS

1. (1, 3)

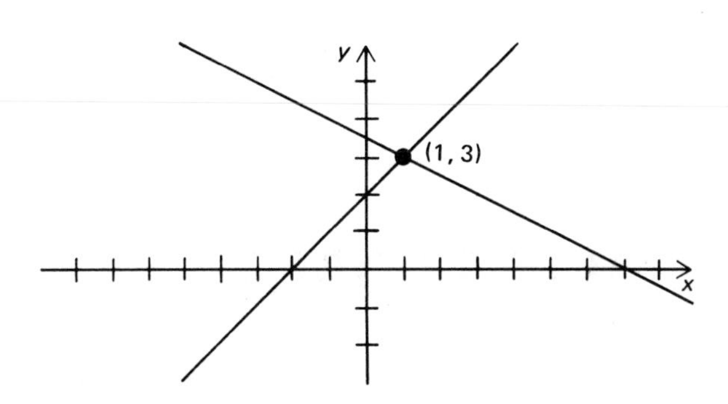

2. infinitely many solutions

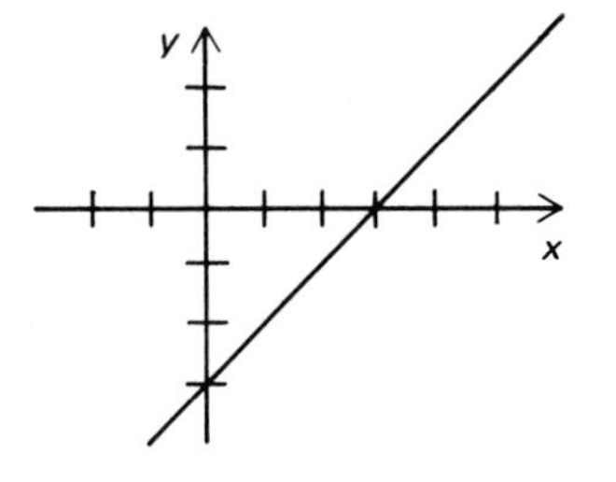

3. (5, 22) **4.** (5, 1) **5.** (0, 2) **6.** inconsistent **7.** they have the same solution set

8. small generator has capacity of 1,400 kilowatts; large generator has capacity of 2,800 kilowatts

QUADRATIC EQUATIONS

BASIC SKILLS

Upon completion of the indicated section the student should be able to:

11.1 *a.* Identify a quadratic equation in one variable and put it in standard form.
b. Use the techniques of factoring to find the solution set of a suitable quadratic equation in one unknown.

11.2 *a.* Identify a pure quadratic equation and put it in standard form.
b. Take the positive and negative square roots of both sides of an equation.
c. Solve pure quadratic equations by
1. isolating the square of the variable on one side and then taking the square roots on both sides, or
2. by using a general formula.

d. Write imaginary solutions in the form Mi where M is a real number and $i = \sqrt{-1}$.

11.3 *a.* State the quadratic formula and the equation to which it applies.
b. Use the discriminant to determine the nature of the solutions of a quadratic equation in one unknown.
c. Use the quadratic formula to solve quadratic equations in one variable.
d. Simplify the solutions obtained from the quadratic formula.
e. Judge which of the following solution techniques is easiest to apply to a given quadratic equation:
1. factoring,
2. technique for pure quadratics,
3. quadratic formula.

f. (Optional) Apply the technique of completing the square to solve quadratic equations in one unknown.

11.4 Solve stated problems that involve a quadratic equation in one unknown.

PRETEST 11

1. Which of the following equations are quadratic? Put those that are quadratic in standard form.

   a. $3x^2 + 2x - 4 = 0$
   b. $(x + 5)(x - 2) = 3x$
   c. $x^3 + 2x - 8 = 0$
   d. $4x^2 = 7$
   e. $5x^2 + 2x = 3x^2 - 4x$
   f. $2x + 3x - 4 = 0$

2. Find the solution set of $(x + 2)(x - 3) = 0$.

3. Find the solution set of $x^2 + 4x + 3 = 0$.

4. Find the solution set of $x^2 - 15 = 0$.

5. Find the solution set of $x^2 + 9 = 0$.

6. State the quadratic formula and the equation to which it applies.

7. Find the solution set of $x^2 - x + 1 = 0$.

8. Find the solution set of $3x^2 + 7x + 2 = 0$.

9. Compute the discriminant and discuss the nature of the solutions without actually finding the solutions of $2x^2 + 3x = 4$.

10. A rectangular conference table is 4 feet longer than it is wide. The top has an area of 32 square feet. How long is it? How wide is it? (Area of a rectangle = length × width).

Each problem above refers to a section in this chapter as shown in the table.

Problems	*Section*
1–3	11.1
4 and 5	11.2
6–9	11.3
10	11.4

PRETEST 11 ANSWERS

1. *a.* in standard form *b.* $x^2 + 0x - 10 = 0$
 d. $4x^2 + 0x - 7 = 0$ *e.* $2x^2 + 6x + 0 = 0$

2. $x = -2, x = 3$ 3. $x = -1, x = -3$ 4. $x = \pm\sqrt{15}$ 5. $x = \pm 3i$

6. $x = \dfrac{-b \pm \sqrt{b^2 - 4ac}}{2a}$; $ax^2 + bx + c = 0$ 7. $x = \dfrac{1 \pm \sqrt{3}\,i}{2}$ 8. $x = -2, \dfrac{-1}{3}$

9. discriminant = 41; the solutions are real and unequal 10. width = 4; length = 8

11.1 INTRODUCTION TO QUADRATIC EQUATIONS: SOLUTION BY FACTORING

Let a, b, and c be arbitrary real numbers. An expression of the type $ax^2 + bx + c$ where $a \neq 0$ is called a *quadratic expression in one variable* x. For example, each of $5x^2 + 2x + 1$ and $3x^2 - 5$ and $10x^2$ is a quadratic expression. But $5x - 1$ and $-8x$ are not quadratic expressions because they do not contain a term involving x^2. Neither is $x^3 + x^2 - 1$ a quadratic expression, for it contains x^3 as a term. In a quadratic expression the highest power of x that is permitted is the second power.

Example 1 Is $x(x + 2)$ a quadratic expression? Let us simplify $x(x + 2)$.

$$x(x + 2) = x^2 + 2x$$

If we let $a = 1$, $b = 2$, and $c = 0$ then

$$ax^2 + bx + c$$

becomes $1x^2 + 2x + 0$. So we see that when simplified, the expression $x(x + 2)$ is a quadratic expression.

Example 2 Is $(x + 1)(x + 3)$ a quadratic expression? We simplify the product.

$$(x + 1)(x + 3) = x^2 + x + 3x + 3 = x^2 + 4x + 3$$

If we let $a = 1$, $b = 4$, and $c = 3$ then

$$ax^2 + bx + c \qquad \text{becomes} \qquad 1x^2 + 4x + 3$$

The last expression is a quadratic expression. Since it is a simplified form of the original expression $(x + 1)(x + 3)$ we see that the original expression is quadratic.

Again let a, b, c be arbitrary real numbers with $a \neq 0$. Then an equation of the form

$$ax^2 + bx + c = 0$$

is called the *standard form for a quadratic equation in one variable x.*

Example 3 Write the equation $3x^2 + 2x = 0$ in the standard form for a quadratic equation. If we let $a = 3$, $b = 2$, $c = 0$ then

$$ax^2 + bx + c = 0 \quad \text{becomes} \quad 3x^2 + 2x + 0 = 0$$

The last equation is the standard form.

Example 4 Write $x^2 + 1 + 3x = -2 + x$ in the standard form: First we move everything to one side of the equation and simplify.

$$x^2 + 1 + 3x = -2 + x$$

$$x^2 + 1 + 3x + 2 - x = 0$$

$$x^2 + 3 + 2x = 0$$

Next we arrange the terms in descending powers of x from left to right.

$$x^2 + 2x + 3 = 0$$

We see that this will fit the standard form if we let $a = 1$, $b = 2$, and $c = 3$.

Example 5 Write $(x + 2)(3x - 1) = 0$ in the standard form:

$$(x + 2)(3x - 1) = 3x^2 + 5x - 2$$

Therefore the equation $(x + 2)(3x - 1) = 0$ becomes $3x^2 + 5x - 2 = 0$ in standard form. Referring back to the standard form we see that $a = 3$, $b = 5$, and $c = -2$ in this case.

Later in this chapter we will do more with the standard form of a quadratic equation.

Throughout this chapter our main interest will be to solve a quadratic equation in one variable. The *solution set* of a quadratic $ax^2 + bx + c = 0$ is the set of all numbers which "work in the equation." That is, if we substitute any member of the solution set for x in $ax^2 + bx + c = 0$ the equation is made true.

In general, a quadratic equation will have either one solution or two distinct solutions. But there will not be more than two solutions.

Example 6 Verify that both $x = 1$ and $x = 3$ are solutions of the equation $x^2 - 4x + 3 = 0$. If we let $x = 1$ in the equation, we get

$$\begin{aligned} x^2 - 4x + 3 &= 0 \\ 1^2 - 4(1) + 3 &= 0 \\ 0 &= 0 \qquad \text{checks} \end{aligned}$$

If we let $x = 3$, we get

$$\begin{aligned} x^2 - 4x + 3 &= 0 \\ 3^2 - 4(3) + 3 &= 0 \\ 0 &= 0 \qquad \text{checks} \end{aligned}$$

So the solution set of $x^2 - 4x + 3 = 0$ is $\{1, 3\}$.

There is a simple method of solving some quadratic equations. This method is based on the following observation: If the product of two algebraic expressions is zero then at least one of these expressions must be zero. If A and B are algebraic expressions and $AB = 0$, then either $A = 0$ or $B = 0$. If $x(x - 1) = 0$, then either $x = 0$ or $x - 1 = 0$. Likewise if

$$(2x + 3)(x + 1) = 0$$

then $2x + 3 = 0$ or $x + 1 = 0$.

We can use this observation together with our knowledge of factoring to solve many quadratic equations.

Example 7 Find the solution set for $x^2 - 5x - 6 = 0$: This is a quadratic equation in standard form. The idea is to factor the quadratic expression into two factors involving x. From our previous experience with this kind of factoring problem, we know to try factors of the form

$$(x + u)(x + v) = x^2 - 5x - 6$$

From inspection, and a little trial and error, we see that we should choose $u = -6$ and $v = 1$. If you have trouble with this step, go back to Chapter 4 and review factoring before you continue. So

$$x^2 - 5x - 6 = (x - 6)(x + 1)$$

and the equation $(x - 6)(x + 1) = 0$ is just the factored form of the equation $x^2 - 5x - 6 = 0$. Since $(x - 6)(x + 1) = 0$, then by our previous observation either $x - 6 = 0$ or $x + 1 = 0$. For if a product is zero, then one of the factors is zero.

$$\begin{aligned} &\text{If} \quad x - 6 = 0 \quad \text{then} \quad x = 6 \\ &\text{If} \quad x + 1 = 0 \quad \text{then} \quad x = -1 \end{aligned}$$

Let's check both solutions in the original equation.

$$\begin{aligned} x &= 6 \\ x^2 - 5x - 6 &= 0 \\ 6^2 - 5(6) - 6 &= 0 \\ 36 - 30 - 6 &= 0 \\ 0 &= 0 \qquad \text{checks} \end{aligned}$$

$$\begin{aligned} x &= -1 \\ x^2 - 5x - 6 &= 0 \\ (-1)^2 - 5(-1) - 6 &= 0 \\ 1 + 5 - 6 &= 0 \\ 0 &= 0 \qquad \text{checks} \end{aligned}$$

Example 8 Find the solution set of $2x^2 + 3x = x^2 - x + 5$: The first step is to write the equation in standard form. We do this by putting all the terms on one side of the equation and simplifying this side.

$$2x^2 + 3x = x^2 - x + 5$$
$$2x^2 - x^2 + 3x + x - 5 = 0$$
$$x^2 + 4x - 5 = 0$$

Next we factor the quadratic expression $x^2 + 4x - 5$. After a little thought we get

$$x^2 + 4x - 5 = (x + 5)(x - 1)$$

So the factored form of $x^2 + 4x - 5 = 0$ is $(x + 5)(x - 1) = 0$. Since the product is zero, one of the factors must be zero.

If $x + 5 = 0$ then $x = -5$
If $x - 1 = 0$ then $x = 1$

The solution set is $\{-5, 1\}$. You may check this solution in the original equation to see that it works.

Example 9 Solve $3x^2 = -x$: First we put everything on one side of the equation.

$$3x^2 = -x$$
$$3x^2 + x = 0$$

Since x is a factor of each term, $3x^2 + x$ factors as $x(3x + 1)$. So $3x^2 + x = 0$ is the same as $x(3x + 1) = 0$. Thus our two solutions come from the equations

$$x = 0 \qquad \text{and} \qquad 3x + 1 = 0$$

Our two solutions are $x = 0$ or $x = -1/3$. If we had divided both sides of the original equation $3x^2 = -x$ by x, we would have lost the solution $x = 0$ and the solution set would not be complete.

The method of solution by factoring is relatively easy provided you can see what the factors should be. In many problems it is not at all easy to determine the factors of a quadratic expression. For these problems we need a more general method than factoring by inspection. Such a method will be presented later in this chapter. The point is that many quadratic equations can be solved by the factoring method. This is less complicated than the general method. So we use the factorization method whenever we can.

PROGRAM 11.1

1. Let a, b, and c be any real numbers such that $a \neq 0$. An algebraic expression of the type $ax^2 + bx + c$ is called a *quadratic expression in one variable x*. Put in a different way, a quadratic expression is one involving x^2 but no higher power of x. For example, $x^2 + 1$, $3x^2 - x$, and $-5x^2 + 2x - 1$ are all quadratic expressions. They all involve x^2 but no higher power of x.

Is $x^2 + 3x$ a quadratic expression? (Yes\No), because it involves ____ and no higher power. Is $2x + 1$ a quadratic expression? (Yes\No), because it does not involve ____. Is $x^3 - x^2 + 5$ a quadratic expression? (yes\no).

Yes; x^2
No
x^2; no

2. In later work it will be important to identify the coefficients in a quadratic expression. In the general expression $ax^2 + bx + c$ the coefficient on x^2 is ____, the coefficient on x is ____ and the constant term is ____.

a; b; c

In the expression $12x^2 - 7x - 2$ the coefficient on x^2 is _____, the coefficient on x is _____, and the constant term is _____. So $12x^2 - 7x - 2$ fits the general form $ax^2 + bx + c$ if we let $a = 12$, $b =$ _____, and $c =$ _____.

12
-7; -2
-7
-2

3. The expression $-6x^2 + 10x - 2.5$ fits the general form $ax^2 + bx + c$ if we let $a =$ _____, $b =$ _____, and $c =$ _____.

-6; 10; -2.5

4. To see if $2x(3x - 1)$ is a quadratic, let us multiply the expression:

$$2x(3x - 1) = 6x^2 - 2x$$

The coefficient on x^2 is _____, the coefficient on x is _____, and the constant term is _____. Since, in the expression $6x^2 - 2x$, the constant term (is\is not) written, it is in effect 0.

6; -2
0
is not

So $6x^2 - 2x$ fits the general form $ax^2 + bx + c$ if we let $a =$ _____, $b =$ _____, and $c =$ _____. Therefore, when multiplied, the expression $2x(3x - 1)$ (is\is not) a quadratic expression.

6
-2; 0
is

5. Let us see if $(4x - 1)(x + 2)$ is a quadratic expression. Again, we must first multiply it out.

$$(4x - 1)(x + 2) = 4x^2 - x + 8x - 2$$
$$= \_\_\_\_\_\_\_\_$$

$4x^2 + 7x - 2$

This expression does fit the general form $ax^2 + bx + c$ if we let $a =$ _____, $b =$ _____, and $c =$ _____.

4
7; -2

6. An equation of the type $ax^2 + bx + c = 0$ (where $a \neq 0$) is called the *standard form for a quadratic equation in one variable x*. The equation $5x^2 + 17x - 3 = 0$ is a quadratic equation in standard form. The coefficients are $a =$ _____, $b =$ _____, and $c =$ _____.

5; 17; -3

7. The equation $5x^2 - 2 = 0$ is a quadratic equation, but it is not in standard form. In standard form the coefficients of x^2 and x and the constant term must all appear explicitly. Since x is not written in the expression $5x^2 - 2$, it must have a coefficient of _____, because ___$x = 0$. Therefore

0; 0

$$5x^2 - 2 = 5x^2 + 0x - 2$$

where the coefficient on x^2 is _____, the coefficient on x is _____, and -2 is the _________. And $5x^2 + 0x - 2 = 0$ fits the standard form $ax^2 + bx + c = 0$ if we let $a =$ _____, $b =$ _____, and $c =$ _____.

5; 0
constant term
5; 0; -2

8. To write $8x - 1 + 3x^2 = 0$ in standard form, we must rewrite the left side in descending powers of x. That is, the x^2 term comes first, the x term next, and the constant last. Now rewrite $8x - 1 + 3x^2 = 0$ in standard form.

_________ $= 0$. This fits the form $ax^2 + bx + c = 0$ if we let $a =$ _____, $b =$ _____, and $c =$ _____.

$3x^2 + 8x - 1$; 3
8; -1

9. Our main objective in this chapter is to learn ways of solving quadratic equations. A *solution* of a quadratic equation is any number that satisfies the equation. The *solution set* is the set of all numbers that satisfy the equation. In general, there will not be more than two solutions for any given quadratic equation.

We can check that $\{6, -1\}$ is the solution set of the equation

$$2x^2 - 10x - 12 = 0$$

if we let $x = 6$. We then get

6; 6 — $2(\_\_\_\_)^2 - 10(\_\_\_\_) - 12 = 0$

$2 \cdot 36$ or 72; $10 \cdot 6$ or 60 — $\_\_\_\_ - \_\_\_\_ - 12 = 0$

$0 = 0$

does — So $x = 6$ (does\does not) satisfy the equation.

Likewise, if we let $x = -1$ we get

$$2x^2 - 10x - 12 = 0$$

-1; -1 — $2(\_\_\_\_)^2 - 10(\_\_\_\_) - 12 = 0$

2; $10(-1)$ or -10 — $\_\_\_\_ - \_\_\_\_ - 12 = 0$

12 — $\_\_\_\_ - 12 = 0$

do — So both $x = 6$ and $x = -1$ (do\do not) check and are therefore solutions. And, since a quadratic equation has at most two solutions, the solution set

$\{6, -1\}$ — checks out to be ________.

10. Now we can use our previous knowledge of factoring to solve some quadratic equations. First we make the observation that *if the product of two algebraic expressions is zero, then at least one of the factors must be zero.* For example, if

$$(x - 3)(x + 2) = 0$$

then either

$$(x - 3) = 0 \quad \text{or} \quad (x + 2) = 0$$

Now let us solve

$$x^2 - x - 6 = 0$$

$(x - 3)$ and $(x + 2)$ — We recognize the factors of $x^2 - x - 6$ to be ____________. So,

$(x - 3)(x + 2) = 0$ — in factored form, $x^2 - x - 6 = 0$ becomes ____________.

As we have said, if the product is zero, then one of the factors must be

zero — ________. Therefore from $(x - 3)(x + 2) = 0$ we know that either

0; 0 — $x - 3 = \_\_\_\_$ or $x + 2 = \_\_\_\_$.

3 — If $x - 3 = 0$ then $x = \_\_\_\_$

-2 — If $x + 2 = 0$ then $x = \_\_\_\_$

3, -2 — So the solution set is $\{\_\_\_\_, \_\_\_\_\}$. You may check these in the original equation.

11. To find the solution set for

$$2x^2 + 5x + 3 = 0$$

we first factor $2x^2 + 5x + 3$. This involves a little trial and error with some educated guessing. One factor is $2x + 3$. Find the other factor.

$x + 1$ — $2x^2 + 5x + 3 = (2x + 3)(\_\_\_\_\_\_\_\_)$

If you have very much trouble with this step, review the Chapter 4 section on "Factoring $ax^2 + bx + c$" before you proceed.

In factored form the equation

$2x^2 + 5x + 3 = 0$ becomes $(2x + 3)(\_\_\_\_\_\_\_\_) = 0$ — $x + 1$

Since the product is zero, then at least one of the factors is ____. — zero

If $2x + 3 = 0$ then $x =$ ____ — $\frac{-3}{2}$

If $x + 1 = 0$ then $x =$ ____ — -1

Therefore the solution set is ________. — $\left\{\frac{-3}{2}, -1\right\}$

12. The solution set for

$$3x^2 - x = 15 + 2x^2 + x$$

is found by writing this quadratic equation in standard form. Moving all terms to the left side of the equation we obtain

________________ $= 0$ — $3x^2 - 2x^2 - x - x - 15$

which simplifies to ________ $= 0$. Next we find the factors of the left side. — $x^2 - 2x - 15$

$x^2 - 2x - 15 =$ ________ ________ — $(x - 5)$; $(x + 3)$

Therefore the equation in factored form becomes ________________. — $(x - 5)(x + 3) = 0$

Since the product is zero, then at least one of the factors is ____. — zero

If $x - 5 = 0$ then $x =$ ____ — 5

If $x + 3 = 0$ then $x =$ ____ — -3

So the solution set is ________. Let us check these solutions in the original equation. If $x = 5$, then — $\{5, -3\}$

$$3x^2 - x = 15 + 2x^2 + x$$

$3(\_\_)^2 - (\_\_) = 15 + 2(\_\_)^2 + (\_\_)$ — 5; 5; 5; 5

$75 - 5 = 15 +$ __ $+$ ____ — 50; 5

$70 =$ ____ — 70

So $x = 5$ (does\does not) check. And if $x = -3$, then — does

$$3x^2 - x = 15 + 2x^2 + x$$

$3(\_\_)^2 - (\_\_) = 15 + 2(\_\_)^2 + (\_\_)$ — -3; -3; -3; -3

$3(9) -$ ____ $= 15 + 2(\_\_) +$ __ — (-3); 9; (-3)

$30 =$ ____ — 30

Therefore both solutions (do\do not) check. — do

PRACTICE PROBLEMS

Which of the equations are quadratic equations? Write each quadratic equation in standard form.

1. $2x^2 + 3 = 0$
2. $x + 9 = 0$
3. $x + 12 = x^2$
4. $x^3 + 5x^2 + 2x - 1 = 0$

5. $-2x + 3x^2 + 1.2 = 0$

6. $5x - 3 + x^2 = 9x + 7$

7. $x(x - 1) = 0$

8. $2x(x^2 + 5x) = 0$

9. $(x - 3)(x + 2) = 0$

10. $x(x - 1)(x + 1) = 0$

Find the solution set of the given equation and check your solutions.

11. $x^2 = 0$

12. $x(2x - 6) = 0$

13. $15x(12 - 2x) = 0$

14. $(x - 3)(x + 2) = 0$

15. $(x - 6)(x + 6) = 0$

16. $(x - 5)(x + 1) = 0$

17. $x^2 + 2x + 1 = 0$

18. $x^2 - 3x = 0$

19. $23x^2 = 0$

20. $7x^2 + 21x = 0$

21. $x^2 - x - 20 = 0$

22. $21x - 7x^2 = 0$

23. $x^2 - 1 = 0$

24. $x^2 - 100 = 0$

25. $9x^2 - 1 = 0$

26. $x^2 + 2x - 3 = 0$

27. $4x^2 + 4x + 1 = 0$

28. $x^2 = 4x$

29. $2x^2 = x$

30. $x^2 = 2x - 1$

31. $3x^2 + 2x = -2x + x^2 - 4$

32. $5x^2 + x - 1 = 4x^2 - x - 1$

33. $x^2 = 10x - 25$

34. $x^2 - 12 = 2x - x^2$

35. $2x^2 + 11x = 4 - x^2$

SELFTEST

1. Which of the following are quadratic equations? Put those that are quadratics into standard form.

a. $2x^2 + x - 8 = 0$

b. $(x + 1)(x + 2) = 3$

c. $4x - 2x = 5$

d. $x^3 + 3x^2 = x$

e. $x^2 = 10$

Solve the following equations by factoring.

2. $x^2 + 3x + 2 = 0$

3. $x^2 + x = 12$

4. $2x^2 + 7x + 3 = 0$

5. $x^2 = 4x$

6. $x^2 - 16 = 0$

7. $2x^2 - 4x = x^2 + 2x - 5$

SELFTEST ANSWERS

1. *a.* in standard form *b.* $x^2 + 3x - 1 = 0$ *e.* $x^2 + 0x - 10 = 0$

2. $x = -1; x = -2$ 3. $x = -4; x = 3$ 4. $x = -1/2; x = -3$

5. $x = 0; x = 4$ 6. $x = 4; x = -4$ 7. $x = 5; x = 1$

11.2 SOLVING THE EQUATION $ax^2 + c = 0$

If $b = 0$ in the general quadratic expression $ax^2 + bx + c$, we get the expression $ax^2 + c$. An expression of the type $ax^2 + c$, where $a \neq 0$, is called a *pure quadratic expression in one variable* x. So a pure quadratic expression is just a quadratic expression which does not involve an x term. For example, $x^2 + 3x + 1$ and $x^2 - x$ are not pure quadratics since each involves an x term. But $x^2 - 3$, x^2, $5x^2 + 7$, and $-x^2 + 10$ are all pure quadratics because they are quadratics which have no x terms.

An equation of the type $ax^2 + c = 0$, where $a \neq 0$, is called a *pure quadratic equation in standard form.* In this section we will solve such equations.

Let us recall a few facts from our previous study of complex numbers. If N is a positive number, then a pure imaginary number is one of the form $\sqrt{-N}$. For example $\sqrt{-3}$, $\sqrt{-10}$, and $\sqrt{-25}$ are all pure imaginary numbers. We defined the pure imaginary number i as follows:

$$i = \sqrt{-1}$$
$$i^2 = -1$$

It follows that every pure imaginary number can be written as Mi where M is a real number.

$$\sqrt{-3} = \sqrt{3(-1)} = \sqrt{3}\sqrt{-1} = \sqrt{3}\,i \approx 1.732i \qquad \text{(by the tables)}$$
$$\sqrt{-10} = \sqrt{10(-1)} = \sqrt{10}\sqrt{-1} = \sqrt{10}\,i \approx 3.162i \qquad \text{(by the tables)}$$
$$\sqrt{-25} = \sqrt{25(-1)} = \sqrt{25}\sqrt{-1} = \sqrt{25}\,i = 5i$$

The method of solving pure quadratic equations is best explained by some examples.

Example 1 Solve $x^2 - 9 = 0$: We can immediately tell that the two solutions are 3 and -3.

$$(3)^2 - 9 = 9 - 9 = 0 \qquad \text{and} \qquad (-3)^2 - 9 = 9 - 9 = 0$$

But what technique will give us results when the solutions are not so obvious? Our new technique involves isolating x^2 on one side of the equation. Then we take the square root of both sides. Let us apply the new technique to $x^2 - 9 = 0$.

$$x^2 - 9 = 0$$
$$x^2 = 9 \qquad \text{add 9 to both sides}$$
$$\sqrt{x^2} = \sqrt{9} \qquad \text{take the square root of both sides}$$
$$x = 3$$

So far, we have only one solution. To get both solutions we must consider both the positive and negative square root. Thus we take both square roots of the right side. (Remember the $\pm$ notation means we consider the positive case and then the negative case.)

$$x = \pm\sqrt{9}$$
$$x = \pm 3$$

Example 2 Solve $x^2 - 5 = 0$: First we isolate x^2 on one side of the equation by adding 5 to both sides. Then we take the square root on both sides. Be sure to take both the $+$ and the $-$ roots.

$$x^2 - 5 = 0$$
$$x^2 = 5$$
$$x = \pm\sqrt{5}$$
$$x \approx \pm 2.236 \qquad \text{from the tables}$$

So the solution set is $\{\sqrt{5}, -\sqrt{5}\}$ or $\{2.236, -2.236\}$ from the tables.

Let us check our results. Remember that values of square roots from the tables are only approximations so when we substitute these values for x in the original equations, we will find that the solutions do not exactly check. It is usually easier to leave the square root in the square root form ($\sqrt{5}$). Recall that $(\sqrt{5})^2 = 5$ and $(-\sqrt{5})^2 = 5$. These facts are useful when we check our results.

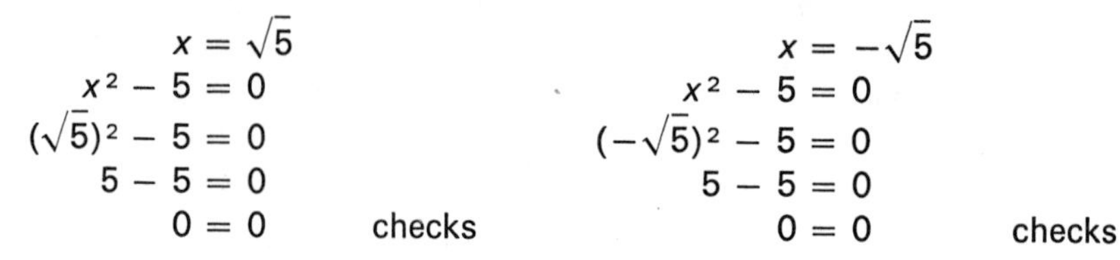

$$x = \sqrt{5} \qquad\qquad x = -\sqrt{5}$$
$$x^2 - 5 = 0 \qquad\qquad x^2 - 5 = 0$$
$$(\sqrt{5})^2 - 5 = 0 \qquad\qquad (-\sqrt{5})^2 - 5 = 0$$
$$5 - 5 = 0 \qquad\qquad 5 - 5 = 0$$
$$0 = 0 \quad \text{checks} \qquad\qquad 0 = 0 \quad \text{checks}$$

Example 3 Solve:

$$\begin{aligned} 3x^2 - 12 &= 0 && \text{given} \\ 3x^2 &= 12 && \text{add 12 to both sides} \\ x^2 &= 4 && \text{divide both sides by 3 and} \\ x &= \pm\sqrt{4} && \text{take the square root} \\ x &= \pm 2 && \text{on both sides} \end{aligned}$$

The solution set is $\{2, -2\}$. You may check these solutions in the original equation to see that they work.

Example 4 Solve:

$$\begin{aligned} x^2 + 2 &= 0 && \text{given} \\ x^2 &= -2 && \\ x &= \pm\sqrt{-2} && \text{take the square root} \\ x &= \pm\sqrt{2}\, i && \text{on both sides} \end{aligned}$$

So the solution set is $\{\sqrt{2}\, i, -\sqrt{2}\, i\}$. Let us check these solutions. Remember that

$$(-\sqrt{2}\, i)^2 = (-\sqrt{2}\, i)(-\sqrt{2}\, i) = 2i^2 = -2 \qquad \text{since } i^2 = -1$$

$$\begin{aligned} x &= \sqrt{2}\, i \\ x^2 + 2 &= 0 \\ (\sqrt{2}\, i)^2 + 2 &= 0 \\ -2 + 2 &= 0 \\ 0 &= 0 \quad \text{checks} \end{aligned} \qquad \begin{aligned} x &= -\sqrt{2}\, i \\ x^2 + 2 &= 0 \\ (-\sqrt{2}\, i)^2 + 2 &= 0 \\ -2 + 2 &= 0 \\ 0 &= 0 \quad \text{checks} \end{aligned}$$

Example 5 Solve:

$$\begin{aligned} 10x^2 + 3 &= 6x^2 - 9 && \text{given} \\ 10x^2 - 6x^2 + 3 &= -9 && \text{subtract } 6x^2 \text{ from both sides} \\ 4x^2 + 3 &= -9 && \text{simplify the left side} \\ 4x^2 &= -12 && \text{subtract 3 from both sides} \\ x^2 &= -3 && \text{divide both sides by 4} \\ x &= \pm\sqrt{-3} && \text{take the square root} \\ x &= \pm\sqrt{3}\, i && \text{on both sides} \end{aligned}$$

The solution set is $\{\sqrt{3}\, i, -\sqrt{3}\, i\}$. These solutions check.

Now let us look at the situation from a somewhat broader point of view. Suppose we are given an equation which we can put in pure quadratic form. That is, we can put the equation in the form $ax^2 + c = 0$ for some values of a and c and $a \neq 0$.

$$\begin{aligned} ax^2 + c &= 0 \\ ax^2 &= -c \\ x^2 &= -c/a \qquad a \neq 0 \\ x &= \pm\sqrt{-c/a} \end{aligned}$$

This is the general solution of the pure quadratic equation in standard form.

Example 6 Solve $5x^2 + 2 = 18 + x^2$ by using the general solution formula. The formula $x = \pm\sqrt{-c/a}$ works only for a pure quadratic equation in standard form $ax^2 + c = 0$. So our first job is to put the given equation into standard form.

$$\begin{aligned} 5x^2 + 2 &= 18 + x^2 && \\ 5x^2 - x^2 + 2 &= 18 && \text{subtract } x^2 \text{ from both sides} \\ 4x^2 + 2 &= 18 && \\ 4x^2 + 2 - 18 &= 0 && \text{subtract 18 from both sides} \\ 4x^2 - 16 &= 0 && \text{standard form for a pure quadratic} \end{aligned}$$

If we let $a = 4$ and $c = -16$, then we see that the equation $4x^2 - 16 = 0$ is in standard form. Now we use the formula.

$$x = \pm\sqrt{-c/a} = \pm\sqrt{-(-16)/4} = \pm\sqrt{4} = \pm 2$$

The solution set is $\{2, -2\}$, which you may check.

When solving a pure quadratic equation you may either use the method outlined in Examples 1–5 or you may use the general formula as explained in Example 6. Both ways are essentially equivalent and always give the same answers.

PROGRAM 11.2

1. An expression of the type

$$ax^2 + c$$

where a and c are real numbers and $a \neq 0$ is called a *pure quadratic expression in one variable x*. For instance $3x^2 + 2$ is a pure quadratic expression. So a pure quadratic is a quadratic with no x term. Which of the following are pure quadratic expressions?

a. $x^2 - 5$ *b.* $x^2 + 2x - 1$ *c.* $x^3 - 1$
d. $4x^2 + 6$ *e.* $x^3 - 4$ *f.* x^2

a., d., and *f.*

2. When we set a pure quadratic equal to zero, we have a *pure quadratic equation.* In *standard form* it looks like

$$ax^2 + c = 0 \qquad a \neq 0$$

We will solve such equations in this section. Identify the pure quadratic equations in the list below, and put them in standard form if necessary.

$x^2 + 4 = 0$ ________ pure quadratic

$2x^2 = 7$ ________ $2x^2 - 7 = 0$

$x^2 = 1 - x$ ________ not pure quadratic

$4x + 2 = 0$ ________ not a quadratic

$5 = x^2 + 1$ ________ $x^2 - 4 = 0$

3. We will see that some pure quadratic equations have solutions that are real valued, but others have imaginary-valued solutions.

Recall that the pure imaginary number i is defined as

$$i = \sqrt{-1} \qquad \text{and} \qquad i^2 = -1$$

Let's look at some possible solutions of $3x^2 + 3 = 0$. If we let $x = i$, we get

$3x^2 + 3 = 0$

$3(\_\_\_)^2 + 3 = 0$ i

$3(\_\_\_) + 3 = 0$ -1 since $i^2 = -1$

$\_\_\_ + 3 = 0$ -3 since $3(-1) = -3$

$\_\_\_ = 0$ 0

So $x = i$ (is\is not) a solution.   is

Let's investigate $-i$ as a possible solution. We need to know the value of $(-i)^2$.

$$(-i)^2 = (-i)(-i) = +i^2$$

-1 We already know that $i^2 =$ ____. Therefore $(-i)^2 = +i^2 = -1$, so it follows that

-1 $$(-i)^2 = \_\_\_\_$$

Let $x = -i$ in the equation

$$3x^2 + 3 = 0$$

$-i$ $$3(\_\_\_\_)^2 + 3 = 0$$

$-1; -1$ $$3(\_\_\_\_) + 3 = 0 \qquad (\text{since } (-i)^2 = \_\_\_\_)$$

0 $$\_\_\_\_ = 0$$

yes Is $x = -i$ a solution of $3x^2 + 3 = 0$? (yes\no). Since a quadratic equation has no more than two solutions, the solution set of $3x^2 + 3 = 0$ is

$\{i, -i\}$ ________.

4. We get pure imaginary numbers by taking the square root of a negative number. For instance $\sqrt{-5}$ will be a (real\pure imaginary) number. It turns out that we can write all pure imaginary numbers in the form Mi for some real number M. In the case of $\sqrt{-5}$, we have

pure imaginary

$$\sqrt{-5} = \sqrt{5(-1)} = \sqrt{5}\sqrt{-1} = \sqrt{5}\,i$$

Since $\sqrt{5}$ is a real number, if we let $M = \sqrt{5}$, then $\sqrt{5}\,i$ fits the form Mi. Likewise

$-1; 2$ $$\sqrt{-2} = \sqrt{2}\sqrt{\_\_\_\_} = \sqrt{\_\_\_\_}\,i$$

$16; 4$ $$\sqrt{-16} = \sqrt{\_\_\_\_}\sqrt{-1} = \_\_\_\_ i$$

5. If we square any pure imaginary number we get a negative number.

$9; -9$ $$(3i)^2 = (3i)(3i) = 3 \cdot 3ii = \_\_\_\_ i^2 = \_\_\_\_$$

Because $i^2 = -1$ Why is $9i^2 = -9$? ________________.

$16; -16$ $$(-4i)^2 = (-4i)(-4i) = \_\_\_\_ i^2 = \_\_\_\_$$

$7; -7$ $$(\sqrt{7}\,i)^2 = \sqrt{7}\sqrt{7}\,i^2 = \_\_\_\_ i^2 = \_\_\_\_$$

6. Let's see if $\sqrt{2}\,i$ is a solution of $x^2 + 2 = 0$. If we let $x = \sqrt{2}\,i$, we get

$$x^2 + 2 = 0$$

$\sqrt{2}$ $$(\_\_\_\_)^2 + 2 = 0$$

-2 $$\_\_\_\_ + 2 = 0$$

0 $$\_\_\_\_ = 0$$

is So $\sqrt{2}\,i$ (is\is not) a solution. Do you think $-\sqrt{2}\,i$ is also a solution?

yes; -2; also satisfies (yes\no). Because $(-\sqrt{2}\,i)^2 =$ ____, and therefore $-\sqrt{2}\,i$ (also satisfies\does not satisfy) the equation.

7. We just saw that $x^2 + 2 = 0$ has two solutions, $\sqrt{2}\,i$ and $-\sqrt{2}\,i$. In general the solutions of a pure quadratic equation will differ only in their sign. That is, one solution will be the negative of the other.

Since $\sqrt{6}$ is one solution of $x^2 - 6 = 0$ what is the other solution? _____. — $-\sqrt{6}$

8. Let's look at how we can solve $x^2 - 4 = 0$. If we isolate x^2 on the left, we get $x^2 =$ _____. We want to know x, not x^2. So if we take the square root of both sides we get $x = \sqrt{4}$ or $x =$ _____. But we know that two solutions differ only in their sign so another solution is _____. Therefore the solution set is __________.

4

2

-2

$\{2, -2\}$

9. In order to solve $x^2 + 16 = 0$, we first isolate the x^2 on one side of the equation. Then we take both the positive and negative square roots on both sides.

$$x^2 + 16 = 0$$

$$x^2 = \_\_\_\_\_ \qquad -16$$

$$x = \pm\sqrt{\_\_\_\_\_} \qquad -16$$

$$x = \pm 4\_\_\_\_\_ \qquad i$$

So the solution set is __________. Let's check these solutions. If $x = 4i$, then — $\{4i, -4i\}$

$$x^2 + 16 = 0$$

$$(\_\_\_\_\_)^2 + 16 = 0 \qquad 4i$$

$$\_\_\_\_\_ + 16 = 0 \quad \text{Explain.} \qquad -16;\ (4i)^2 = 16i^2 = -16$$

$$\_\_\_\_\_ = 0 \qquad 0$$

If $x = -4i$, then

$$x^2 + 16 = 0$$

$$(\_\_\_\_\_)^2 + 16 = 0 \qquad -4i$$

$$\_\_\_\_\_ + 16 = 0 \quad \text{Explain.} \qquad -16;\ (-4i)^2 = 16i^2 = -16$$

$$\_\_\_\_\_ = 0 \qquad 0$$

It follows that both solutions (do\do not) check. — do

10. To solve a pure quadratic equation in x we go through the following procedure: First, we isolate x^2 on one side of the equation. Second, we take square roots on both sides, being sure to take both the positive and negative roots. To solve $x^2 - 15 = 0$:

Step 1. We isolate x^2 to get

$$x^2 - 15 = 0$$

$$x^2 = \_\_\_\_\_ \qquad 15$$

Step 2. We take the positive and negative square roots on both sides.

$$x = \pm\sqrt{15}$$

The solution set is __________. — $\{\sqrt{15}, -\sqrt{15}\}$

adding 7

11. Solve $4x^2 - 7 = 5$: First we isolate x^2 by ________ to both sides.

$$4x^2 - 7 = 5$$

12

$$4x^2 = \_\_\_\_$$

3

$$x^2 = \_\_\_\_$$

positive and negative

Then we take the ________________ square roots, which gives us

$\pm\sqrt{3}$; $\{\sqrt{3}, -\sqrt{3}\}$

$x =$ ____. The solution set is ________.

12. The above process can be used on the standard form of a pure quadratic equation to obtain a formula for the general solution.

$$ax^2 + c = 0$$

First isolate x^2.

$$ax^2 + c = 0$$

$-c$

$$ax^2 = \_\_\_\_$$

subtracting c from

We obtained this answer by ________________ both sides. Then we

divide both sides by a

________________ to obtain

$\frac{-c}{a}$

$$x^2 = \_\_\_\_$$

Now we take the positive and negative square roots to get

$\pm\sqrt{\frac{-c}{a}}$

$$x = \_\_\_\_\_\_\_\_$$

$\left\{\sqrt{\frac{-c}{a}}, -\sqrt{\frac{-c}{a}}\right\}$

So the solution set for $ax^2 + c = 0$ is ________.

13. Use the above formula to solve $5x^2 - 20 = 0$: The equation fits the form

5; -20

$ax^2 + c = 0$ if we let $a =$ ____ and $c =$ ____. Therefore

$\frac{-(-20)}{5}$ or 4; 2

$$\frac{-c}{a} = \pm\sqrt{\frac{\_\_\_\_}{\_\_\_\_}} = \_\_\_\_$$

$\{2, -2\}$

So the solution set is ________.

14. Use the general formula to solve $9x^2 + 4 = 0$: We see that the equation

9; 4

fits the form $ax^2 + c = 0$ if we let $a =$ ____ and $c =$ ____. With these values of a and c we get

$\frac{-4}{9}$; $\frac{2i}{3}$

$$\pm\sqrt{\frac{-c}{a}} = \pm\sqrt{\_\_\_\_} = \pm\_\_\_\_$$

$\left\{\frac{2i}{3}, \frac{-2i}{3}\right\}$

The solution set is ________.

PRACTICE PROBLEMS

Which of these expressions are pure quadratic expressions?

1. x^2 **2.** $x^2 + 5$ **3.** $x^2 + 2x$ **4.** $x^2 + x - 1$

5. $x^3 - 1$ **6.** $5x^2 + 1$ **7.** $-18x^2 + 3$ **8.** $x + 1$

Which of these equations are pure quadratic equations? Write the pure quadratic equations in standard form.

9. $x^2 = 0$

10. $2x^2 + 10 = 0$

11. $5x^2 + 1 = 6$

12. $x^2 + 3x - 1 = 0$

13. $3 = x^2$

14. $8x^2 = 16 + x$

15. $12x^2 + 9 = -3$

16. $x^3 = 2x^2 - 1$

Write each number as a pure imaginary number of the form Mi where M is a real number and $i = \sqrt{-1}$.

17. $\sqrt{-1}$

18. $\sqrt{-16}$

19. $\sqrt{-36}$

20. $\sqrt{-2}$

21. $\sqrt{-25}$

22. $\sqrt{-9}$

23. $\sqrt{-8}$

24. $\sqrt{-12}$

Find the solution set of each equation and check your solutions.

25. $x^2 - 49 = 0$

26. $x^2 + 49 = 0$

27. $2x^2 - 50 = 0$

28. $-12x^2 + 49 = 0$

29. $4x^2 = 16$

30. $5x^2 + 20 = 0$

31. $2x^2 - 8 = 10$

32. $x^2 + 3 = 2x^2 - 1$

33. $6x^2 - 1 = 23$

34. $5x^2 + 2 = 2x^2 + 29$

Use the general solution formula to find the solution set. Check your solutions.

35. $3x^2 + 3 = 0$

36. $5x^2 - 45 = 0$

37. $-2x^2 + 18 = 0$

38. $17x^2 + 1 = 35$

39. $8 + 3x^2 = 23$

40. $x^2 + 9 = 2x^2 - 1$

41. $x^2 + 5 = 3x^2 - 123$

42. $7x^2 + 3 = 3 + 51x^2$

SELFTEST

1. Which of the following are pure quadratic equations? Put those that are pure quadratics in standard form.

   a. $x^2 - 3x = 4$ *b.* $(x + 2)(x - 2) = 5$ *c.* $4x^2 = 8$ *d.* $2x + 4 = 0$

Solve each of the following equations. Put pure imaginary solutions in the form Mi where M is a real number and $i = \sqrt{-1}$.

2. $3x^2 = 6$

3. $x^2 + 7 = 0$

4. $2x^2 - 7 = 11$

5. $4x^2 + 8 = 2x^2 - 10$

SELFTEST ANSWERS

1. *b.* $x^2 - 9 = 0$ *c.* $4x^2 - 8 = 0$ **2.** $x = \pm\sqrt{2}$ **3.** $x = \pm\sqrt{7}\,i$
4. $x = \pm 3$ **5.** $x = \pm 3i$

11.3 THE QUADRATIC FORMULA

In this section we will learn a method for solving *any* quadratic equation. In fact, we will introduce a formula which, when properly applied to a quadratic equation in standard form, always yields the solution set. This formula is appropriately called the quadratic formula. Mostly, we will just show you what the quadratic formula is, and how to use it. A derivation of the formula will be found at the end of this section. The method of proof is called *completing the square*. This method is used frequently in later mathematics courses.

Recall that when written in standard form, a quadratic equation in one variable looks like

$$ax^2 + bx + c = 0 \qquad a \neq 0$$

The quadratic formula says that the solutions of this equation are

$$\frac{-b + \sqrt{b^2 - 4ac}}{2a} \quad \text{and} \quad \frac{-b - \sqrt{b^2 - 4ac}}{2a}$$

If we use the symbols $\pm\sqrt{b^2 - 4ac}$ to mean both the positive and negative square roots then, in a more compact notation, we may say the following:

Quadratic Formula: **If a, b, and c are real numbers and $a \neq 0$, then the solutions of the quadratic equation in standard form $ax^2 + bx + c = 0$ are given by the following quadratic formula.**

$$\frac{-b \pm \sqrt{b^2 - 4ac}}{2a}$$

Let us notice a few things about this formula. First, it is clear that the restriction $a \neq 0$ is necessary. If $a = 0$ the quadratic formula would involve a division by zero; an operation which is never permitted. Second, the formula says that in general a quadratic equation will have two solutions. But in the special case that $\pm\sqrt{b^2 - 4ac} = 0$ there will be only one solution.

The quantity $b^2 - 4ac$ is called the *discriminant*. The discriminant can be obtained without actually solving the equation. And the discriminant yields information regarding the nature of the solutions.

Case 1. If $b^2 - 4ac = 0$, the solutions are real and equal.
Case 2. If $b^2 - 4ac > 0$, the solutions are real and unequal.
Case 3. If $b^2 - 4ac < 0$, the solutions are complex and unequal.

Example 1 Consider the equation $3x^2 + x - 1 = 0$. First compute the discriminant and discuss the nature of the solutions. Then use the quadratic formula to find the solutions. We notice immediately that our equation is already in standard form. If we let $a = 3$, $b = 1$, and $c = -1$ then

$$ax^2 + bx + c = 0 \qquad \text{becomes} \qquad 3x^2 + x - 1 = 0$$

Now we compute the discriminant with $a = 3$, $b = 1$, and $c = -1$.

$$b^2 - 4ac \qquad \text{becomes} \qquad 1^2 - 4(3)(-1) = 13 > 0$$

Since the discriminant is 13, our equation fits into case 2. The solutions will be real and unequal.

Next we use the quadratic formula to actually find the solutions.

First Solution	*Second Solution*
$\dfrac{-b + \sqrt{b^2 - 4ac}}{2a}$	$\dfrac{-b - \sqrt{b^2 - 4ac}}{2a}$
$\dfrac{-1 + \sqrt{1^2 - 4(3)(-1)}}{2(3)}$	$\dfrac{-1 - \sqrt{1^2 - 4(3)(-1)}}{2(3)}$
$\dfrac{-1 + \sqrt{13}}{6}$	$\dfrac{-1 - \sqrt{13}}{6}$
$\dfrac{-1}{6} + \dfrac{\sqrt{13}}{6}$	$\dfrac{-1}{6} - \dfrac{\sqrt{13}}{6}$

So the solutions are $-1/6 \pm \sqrt{13}/6$.

We could of course write the solutions to Example 1 in decimal form (by using the table in the back of our book). But in many ways the form $-1/6 \pm \sqrt{13}/6$ is better. For one thing, the number $\sqrt{13}$ is understood to be exactly the square root of 13, where as any decimal representation of $\sqrt{13}$ would of necessity be only approximate.

Example 2 Compute the discriminant and discuss the nature of the solutions of

$$x^2 - 2x + 2 = 0$$

Then find the solutions. The equation is already in standard form $ax^2 + bx + c = 0$ provided we let $a = 1$, $b = -2$, and $c = 2$. The discriminant is

$$b^2 - 4ac \qquad \text{or} \qquad (-2)^2 - 4(1)(2) \qquad \text{or} \qquad -4$$

Since the discriminant is negative, we are in case *3* and the solutions are complex and unequal. Using the quadratic formula with $a = 1$, $b = -2$, and $c = 2$ we get

$$\frac{-b \pm \sqrt{b^2 - 4ac}}{2a} = \frac{-(-2) \pm \sqrt{(-2)^2 - 4(1)(2)}}{2(1)}$$
$$= \frac{2 \pm \sqrt{-4}}{2(1)}$$
$$= \frac{2 \pm 2i}{2}$$
$$= \frac{2}{2} \pm \frac{2i}{2}$$
$$= 1 \pm i$$

So the solution set is $\{1 + i, 1 - i\}$. Let us check these.

For $1 + i$:

$$x^2 - 2x + 2 = 0$$
$$(1 + i)^2 - 2(1 + i) + 2 = 0$$
$$1 + 2i + i^2 - 2 - 2i + 2 = 0$$
$$1 + i^2 = 0$$

For $1 - i$:

$$x^2 - 2x + 2 = 0$$
$$(1 - i)^2 - 2(1 - i) + 2 = 0$$
$$1 - 2i + i^2 - 2 + 2i + 2 = 0$$
$$1 + i^2 = 0$$

And, since $i^2 = -1$,

$$\begin{aligned} 1 + i^2 &= 0 \\ 1 + (-1) &= 0 \\ 0 &= 0 \end{aligned}$$

so both solutions check.

Example 3 Given the equation

$$-2x + 5x^2 + 5 = -9 + 6x + 3x^2$$

compute the discriminant and discuss the nature of the solutions. Then find the actual solutions.

First we must put the given equation in standard form. To do this we move everything to one side of the equation.

$$\begin{aligned} -2x + 5x^2 + 5 &= -9 + 6x + 3x^2 \\ -2x - 6x + 5x^2 - 3x^2 + 5 + 9 &= 0 \\ -8x + 2x^2 + 14 &= 0 \end{aligned}$$

Next we write the expression on the left in decreasing powers of x from left to right. This equation in standard form is

$$2x^2 - 8x + 14 = 0$$

We see that we should choose $a = 2$, $b = -8$, and $c = 14$. The discriminant

$$b^2 - 4ac = (-8)^2 - 4(2)(14) = -48$$

So the solutions are complex and unequal (see case 3). Using the quadratic formula with $a = 2$, $b = -8$, and $c = 14$ we obtain

$$\begin{aligned} \frac{-b \pm \sqrt{b^2 - 4ac}}{2a} &= \frac{-(-8) \pm \sqrt{(-8)^2 - 4(2)(14)}}{2(2)} \\ &= \frac{8 \pm \sqrt{-48}}{4} \\ &= \frac{8 \pm \sqrt{16}\sqrt{3}\sqrt{-1}}{4} \\ &= \frac{8 \pm 4\sqrt{3}\,i}{4} \\ &= \frac{8}{4} \pm \frac{4\sqrt{3}\,i}{4} \\ &= 2 \pm \sqrt{3}\,i \end{aligned}$$

The solution set is $\{(2 + \sqrt{3}\,i), (2 - \sqrt{3}\,i)\}$.

The remaining material in this section is optional. It may be omitted without loss of continuity.

We will solve the next example in a somewhat different way. This method is called *completing the square*. In your next course in algebra you will do more examples involving this method. It is a good idea for you to see some examples using the method now, but we will reserve more ambitious work for a later course. The method of completing the square is powerful because it is so general. In fact we will soon see that the proof of the quadratic formula is nothing more than completing the square on the general quadratic equation in standard form.

Example 4 Use the method of completing the square to solve

$$x^2 + 6x - 1 = 0$$

The idea is to work with both sides of the equation until the left side is a perfect square. Hence the name "completing the square." Consider $x^2 + 6x - 1 = 0$. Move the constant term over to the right side.

$$x^2 + 6x = 1$$

The coefficient on x^2 is 1. This is important. If it were not 1 we would do something about it (see the next example). The coefficient on x is 6. One-half the coefficient on x is 3 and if we square this we get 9. Now 9 is the quantity we need to complete the square (see Section 4.2), and we add it to both sides.

$$x^2 + 6x = 1$$

Add 9 to both sides.

$$x^2 + 6x + 9 = 1 + 9$$
$$(x + 3)^2 = 10$$

Next we take the square roots on both sides to get

$$x + 3 = \pm\sqrt{10}$$
$$x = -3 \pm \sqrt{10}$$

You may check these solutions in the original equation.

Let's review what happened in the last example.

1. We moved the constant term to the other side of the equation. So

$$x^2 + 6x - 1 = 0 \qquad \text{became} \qquad x^2 + 6x = 1$$

2. We observed that the coefficient on x^2 was 1. This is important.
3. The coefficient on x was 6. To complete the square we took one half of 6 and squared the result.

$$(6/2)^2 = (3)^2 = 9$$

Then we added 9 to both sides of $x^2 + 6x = 1$ to get $x^2 + 6x + 9 = 10$.
4. At this point the expression $x^2 + 6x + 9$ was a perfect square. We factored it and got $x^2 + 6x + 9 = (x + 3)^2$. So our equation was

$$(x + 3)^2 = 10$$

5. We took square roots on both sides and solved for x to get

$$x = -3 \pm \sqrt{10}$$

Another example shows that the same process works again.

Example 5 Solve $3x^2 + 6x + 12 = 0$ by completing the square.

Step 1. Move the constant term to the other side of the equation. So

$$3x^2 + 6x + 12 = 0 \qquad \text{becomes} \qquad 3x^2 + 6x = -12$$

Step 2. The coefficient on x^2 is 3. We want it to be 1. So we divide both sides of the equation by 3.

$$\frac{3x^2 + 6x}{3} = \frac{-12}{3}$$

$$\frac{3x^2}{3} + \frac{6x}{3} = -4$$

$$x^2 + 2x = -4$$

Step 3. Now the coefficient on x is 2. To complete the square we take one-half of 2 and square the result.

$$({}^{2}/_{2})^2 = 1^2 = 1$$

Then we add 1 to both sides.

$$x^2 + 2x = -4$$
$$x^2 - 2x + 1 = -3$$

Step 4. Again the left side is a perfect square. If we factor it, we get $x^2 + 2x + 1 = (x + 1)^2$. So our equation is

$$(x + 1)^2 = -3$$

Step 5. We now take the square root of both sides and solve for x.

$$(x + 1)^2 = -3$$
$$x + 1 = \pm\sqrt{-3} = \pm\sqrt{3}\,i$$
$$x = -1 \pm \sqrt{3}\,i$$

The solution set is $\{(-1 + \sqrt{3}\,i), (-1 - \sqrt{3}\,i)\}$. These answers check in the original equation.

To show that the method of completing the square is very general, in our next example we will use it to prove the quadratic formula.

Example 6 Let a, b, and c be any real numbers with $a \neq 0$. Solve $ax^2 + bx + c = 0$: We use our outline for completing the square.

Step 1. Move the constant term to the other side of the equation to get

$$ax^2 + bx = -c$$

Step 2. The coefficient on x^2 is a. We want it to be 1. So we divide both sides of the equation by a.

$$\frac{ax^2 + bx}{a} = \frac{-c}{a}$$

$$\frac{ax^2}{a} + \frac{bx}{a} = \frac{-c}{a}$$

$$x^2 + \frac{bx}{a} = \frac{-c}{a}$$

Step 3. The coefficient on x is now b/a. To complete the square we take one-half of b/a and square the result.

$$\left(\frac{b}{2a}\right)^2 = \frac{b^2}{4a^2}$$

We add this result to both sides.

$$x^2 + \frac{b}{a}x + \frac{b^2}{4a^2} = \frac{-c}{a} + \frac{b^2}{4a^2}$$

Step 4. The expression on the left is a perfect square.

$$x^2 + \frac{b}{a}x + \frac{b^2}{4a^2} = \left(x + \frac{b}{2a}\right)^2$$

So our equation becomes

$$\left(x + \frac{b}{2a}\right)^2 = \frac{b^2}{4a^2} - \frac{c}{a}$$

If we put the right side over a common denominator, we get

$$\left(x + \frac{b}{2a}\right)^2 = \frac{b^2 - 4ac}{4a^2}$$

Step 5. Finally, we take the square root of both sides and solve for x.

$$x + \frac{b}{2a} = \pm\sqrt{\frac{b^2 - 4ac}{4a^2}}$$

$$x = \frac{-b}{2a} \pm \frac{\sqrt{b^2 - 4ac}}{2a}$$

$$x = \frac{-b \pm \sqrt{b^2 - 4ac}}{2a}$$

PROGRAM 11.3

1. If a, b, and c are real numbers and $a \neq 0$, then the solutions of the quadratic equation in standard form $ax^2 + bx + c = 0$ are given by the *quadratic formula* which is

$$\frac{-b \pm \sqrt{b^2 - 4ac}}{2a}$$

This formula is so important that we suggest you memorize both the formula and the equation to which it applies. Notice that the formula uses both the positive and negative square roots,

$$\pm\sqrt{b^2 - 4ac}$$

as part of the solution. So the formula as it stands actually represents two solutions.

2. In order to use the quadratic formula to solve $x^2 + 3x + 5 = 0$, we first determine the a, b, and c to be used in the formula. These are determined from the standard form of the equation.

We see that $x^2 + 3x + 5 = 0$ fits the standard form $ax^2 + bx + c = 0$

if we let $a =$ ____, $b =$ ____, and $c =$ ____. 1; 3; 5

Next we put these values of a, b, and c into the formula. Let us compute the quantity $b^2 - 4ac$ first.

$b^2 - 4ac = (\_\_\_\_)^2 - 4(\_\_\_\_)(\_\_\_\_)$ 3; 1; 5

$= \_\_\_\_ - \_\_\_\_$ 9; 20

$= \_\_\_\_$ −11

Therefore $\sqrt{b^2 - 4ac} =$ ____. It follows that $\sqrt{-11}$ or $\sqrt{11}\,i$

$$\frac{-b \pm \sqrt{b^2 - 4ac}}{2a} = \frac{-3 \pm \sqrt{11}\,i}{2}$$

$$= \frac{-3}{2} \pm \frac{\sqrt{11}\,i}{2}$$

So the two solutions are

$$\frac{-3}{2} + \frac{\sqrt{11}\,i}{2} \quad \text{and} \quad \_\_\_\_\_\_$$

$\frac{-3}{2} - \frac{\sqrt{11}\,i}{2}$

+ That is, we get the first solution when we use the ____ sign in the formula

− and the second solution when we use the ____ sign.

In general, unless otherwise said, the above method of writing the solutions is preferred. That is, we do not recommend that you convert the solution to a decimal approximation (by the use of the square root tables), unless there is a special reason to do so.

3. When using the quadratic formula in the foregoing problem we computed the quantity $b^2 - 4ac$ first. It is a good idea to do this because the quantity $b^2 - 4ac$ will tell us what the solutions should look like. Later, if our actual solutions turn out to be something unexpected, we know we have made an error and should check our work.

The quantity $b^2 - 4ac$ in the quadratic formula is called the *discriminant*. The following three cases describe the nature of the solutions.

Case 1. If $b^2 - 4ac = 0$ the solutions are real and equal.
Case 2. If $b^2 - 4ac > 0$ the solutions are real and unequal.
Case 3. If $b^2 - 4ac < 0$ the solutions are complex and unequal.

In problem 2 we have just computed $b^2 - 4ac = -11$. Which case does

3; complex / unequal / yes

this put us into? ____. So we expect the solutions to be (real\complex) and (equal\unequal). Did the actual solutions in the last problem live up to this expectation? (yes\no).

4. Below, for the equation $-2x^2 + x + 1 = 0$, we will compute the discriminant, discuss the nature of the solutions, and then find the actual solutions. We must first fit $-2x^2 + x + 1 = 0$ into the standard form

b; *c*; −2; 1 / 1

$ax^2 +$ ____$x +$ ____ $= 0$. We see that it fits if we let $a =$ ____, $b =$ ____, and $c =$ ____.

With these values of *a*, *b*, and *c* we can compute the discriminant.

1; −2; 1 / 1; −8 / 9

$$b^2 - 4ac = (\_\_\_\_)^2 - 4(\_\_\_\_)(\_\_\_\_)$$
$$= \_\_\_\_ - (\_\_\_\_)$$
$$= \_\_\_\_$$

positive / 2; real; unequal

So the discriminant is (negative\zero\positive). This puts us into case (*1**2**3*). We expect the solutions to be (real\complex) and (equal\unequal).

Now, to find the actual solutions, we use the quadratic formula with the above values of *a*, *b*, and *c*.

$\frac{-1 \pm \sqrt{9}}{2(-2)}$ / $\frac{-1 \pm 3}{-4}$

$$\frac{-b \pm \sqrt{b^2 - 4ac}}{2a} = \frac{-\_\_\_\_ \pm \sqrt{\_\_\_\_}}{2(\_\_\_\_)}$$
$$= \frac{\_\_\_\_\_\_}{\_\_\_\_}$$

If we simplify our last result we get

$\frac{3}{\ }$; $\frac{1}{-2}$ $\frac{3}{\ }$; 1

$$\frac{-1 + \_\_\_\_}{-4} = \_\_\_\_ \quad \text{and} \quad \frac{-1 - \_\_\_\_}{-4} = \_\_\_\_$$

$\{-1/2, 1\}$ / yes

So the solution set is ________. Are these solutions of the nature predicted by the discriminant? (yes\no).

5. Given the equation $4x^2 + 1 = 2x + 3x^2 - 1$: We will compute the discriminant and discuss the nature of the solutions. Then find the actual solutions.

First we must put the given equation into standard form. We begin by putting all terms on the left side of the equation and simplifying.

$$4x^2 + 1 = 2x + 3x^2 - 1$$
$$4x^2 - 3x^2 - 2x + 1 + 1 = 0$$
$$\underline{\hspace{5cm}} = 0$$

$x^2 - 2x + 2$

It is now in the standard form $ax^2 + bx + c = 0$ provided we let $a =$ ____,

1

$b =$ ____, and $c =$ ____.

-2; 2

The discriminant is ($c^2 - 4ab \backslash a^2 - 4ac \backslash b^2 - 4ac$) and when we put in the above values for a, b, and c, we find that the discriminant becomes ________________.

$b^2 - 4ac$

$(-2)^2 - 4 \cdot 1 \cdot 2 = 4 - 8 = -4$

The discriminant is (positive\negative\zero). This puts us into case (*1**2**3*). So we expect the solutions to be (real\complex) and (equal\unequal).

negative

3; complex

unequal

Now we use the quadratic formula to find the actual solutions. The quadratic formula is $x =$

$$\left(\frac{-b \pm \sqrt{c^2 - 4ab}}{2a} \middle\backslash \frac{c \pm \sqrt{b^2 - 4ac}}{2a} \middle\backslash \frac{-b \pm \sqrt{b^2 - 4ac}}{2a}\right)$$

$\dfrac{-b \pm \sqrt{b^2 - 4ac}}{2a}$

When we put $a = 1$, $b = -2$, and $c = 2$ into this formula we get

$$\left(\frac{-2 \pm \sqrt{-4}}{2} \middle\backslash \frac{2 \pm \sqrt{-4}}{2} \middle\backslash \frac{-2 \pm \sqrt{-12}}{2}\right)$$

$\dfrac{2 \pm \sqrt{-4}}{2}$

And, since $\sqrt{-1} =$ ____, the simplified solutions become

i

$$\frac{2 \pm \sqrt{-4}}{2} = \frac{2 \pm \sqrt{4}\sqrt{-1}}{2} = \frac{2}{2} \pm \frac{\underline{\hspace{1cm}}}{2} = 1 \pm \underline{\hspace{1cm}}$$

$2i$; i

Are the solutions $1 \pm i$ consistent with what we expected? (yes\no).

yes

The remaining part of this program deals with the method of completing the square. It may be omitted without loss of continuity.

6. Use the method of completing the square to solve $x^2 - 8x + 2 = 0$. The idea behind this method is to work with both sides of the equation until the left side of the equation is a perfect square. This is why it is called completing the square.

The method is outlined in five steps and it is recommended that you follow these steps when solving problems by completing the square.

Step 1. First get all terms involving x or x^2 on the left side of the equation. Move all constants to the right side. In this way our equation

$$x^2 - 8x + 2 = 0$$

should become ________________.

$x^2 - 8x = -2$

Step 2. Check that the coefficient on x^2 is 1. In our example, the coefficient on x^2 is ____. If it were not 1 we would so something to make it 1 (see Frame 7).

1

Step 3. Next we observe that the coefficient on x is ____. Now we take one-half of -8 and square the result to get

-8

$$(^{-8}/_2)^2 = \underline{\hspace{1cm}}$$

16

Then we add this result, which is 16, to both sides of our last equation, $x^2 - 8x = -2$, to get

$$x^2 - 8x + 16 = \underline{\hspace{1cm}}$$

14

$x - 4$

Step 4. At this point the left side of the last equation should be a perfect square. And it is.

$$x^2 - 8x + 16 = (\_\_\_\_\_\_\_\_)^2$$

In factored form

$$x^2 - 8x + 16 = 14 \qquad \text{becomes} \qquad (x - 4)^2 = 14$$

Step 5. Taking the square root of both sides and solving for x gives

$\pm\sqrt{14}$

$$x - 4 = \_\_\_\_\_\_\_\_$$

$4 \pm \sqrt{14}$

$$x = \_\_\_\_\_\_\_\_$$

$4 - \sqrt{14}$

The solutions are $4 + \sqrt{14}$ and ________. Of course we would get the same result if we used the quadratic formula.

7. Use the method of completing the square to solve $3x^2 + 21 = 36x$:

Step 1. Put all terms involving x or x^2 on the left. Put the constant terms on the right. What do you get?

$3x^2 - 36x$; -21

$$\_\_\_\_\_\_\_\_ = \_\_\_\_$$

3

divide

Step 2. The coefficient on x^2 is ____. We want it to be 1. What shall we do? (multiply\divide\subtract) on both sides by 3. Do this to the equation we got in step 1.

$$3x^2 - 36x = -21$$

$x^2 - 12x$; -7

$$\_\_\_\_\_\_\_\_ = \_\_\_\_$$

-12

Step 3. The new coefficient on x in this last equation is ____. Now we divide this number by 2 and square the result.

36

$$(^{-12}/_2)^2 = \_\_\_\_$$

To complete the square we add 36 to both sides of our equation so that

29

$$x^2 - 12x = -7 \qquad \text{becomes} \qquad x^2 - 12x + 36 = \_\_\_\_$$

Step 4. At this point the expression on the left side of the last equation should be a perfect square. In fact

$x - 6$

$$x^2 - 12x + 36 = (\_\_\_\_\_\_\_\_)^2$$

In factored form the equation

$(x - 6)^2$

$$x^2 - 12x + 36 = 29 \qquad \text{becomes} \qquad \_\_\_\_\_\_\_\_ = 29$$

Step 5. We take the square root of both sides and solve for x.

$$(x - 6)^2 = 29$$

$\pm\sqrt{29}$

$$x - 6 = \_\_\_\_$$

$6 \pm \sqrt{29}$

$$x = \_\_\_\_\_\_\_\_$$

$\{(6 + \sqrt{29}), (6 - \sqrt{29})\}$

The solution set is ____________.

8. Use the method of completing the square to solve $x^2 = x - 2$:

Step 1. After moving appropriate terms we should have the equation

$x^2 - x = -2$

____________.

1

Step 2. The coefficient on x^2 is ____. So what should we do in this step? ________________________.

Nothing, go to step *3*

-1

one-half

Step 3. The coefficient on x in the equation in step 1 is ____. We take (one-half\one-fourth\one-third) of this number and square the result. We get

$-1/2$; $1/4$

$$(\_\_\_\_)^2 = \_\_\_\_$$

We add ____ to both sides to complete the square, so that

$x^2 - x = -2$ becomes $x^2 - x +$ ____ $=$ ________

$^1/_4$

$^1/_4$; $-2 + {}^1/_4$ or $^{-7}/_4$

Step 4. Is the left side of the last equation a perfect square? It should be.

$$(x^2 - x + {}^1/_4) = (x - {}^1/_2)^2$$

Check these results for yourself by squaring the right side and seeing that it equals the left side. So our equation

$x^2 - x + {}^1/_4 = {}^{-7}/_4$ becomes $(x - {}^1/_2)^2 = {}^{-7}/_4$

Step 5. Now take square roots and solve for x.

$$\left(x - \frac{1}{2}\right)^2 = \frac{-7}{4}$$

$$x - \frac{1}{2} = \text{\_\_\_\_\_\_\_\_}$$

$$x = \text{\_\_\_\_\_\_\_\_}$$

$\pm\sqrt{\frac{-7}{4}}$ or $\frac{\pm\sqrt{7}\,i}{2}$

$\frac{1}{2} \pm \frac{\sqrt{7}\,i}{2}$

PRACTICE PROBLEMS

1. In this chapter we have studied several different methods of solving quadratic equations. List each method and give an example of the type of equation to which it applies.

2. Depending on the quadratic equation involved, some methods of solution are quicker and easier than others. For each method you listed in Problem 1 give an example of an equation for which that method is suited. Then give an example of an equation for which the method is not so well suited.

Compute the discriminant and discuss the nature of the solutions. Do not actually solve the equations.

3. $5x^2 + 2x + 1 = 0$

4. $x^2 + 10x - 1 = 0$

5. $6x^2 + 12x - 3 = 0$

6. $x^2 - 4x + 1 = 0$

7. $x^2 + 9x + 6 = 0$

8. $x^2 + 5 = 0$

9. $x^2 - 15 = 0$

10. $x^2 + 3x = 2x^2 - 1$

11. $1 - 2x + x^2 = 0$

12. $x^2 + 3x - 2x = 0$

First compute the discriminant and discuss the nature of the solutions. Then find the solution set by use of the quadratic formula.

13. $x^2 + 8x + 16 = 0$

14. $x^2 + 3x + 1 = 0$

15. $8x^2 + 2x - 1 = 0$

16. $4x^2 + 3x = 0$

17. $x^2 + 15 = 0$

18. $-3x^2 + 5 = -1$

19. $x^2 + 7 = 3$

20. $8x^2 + 4x - 1 = 0$

21. $-x^2 + 6x - 4 = 0$

22. $x^2 + 3 + 4x = 0$

23. $6 + 2x - x^2 = 0$

24. $5x^2 + 2x = 4x^2 - 1$

25. $2 - x^2 = x^2 - 2$

26. $x^2 = 8x + 1$

27. $7x = x^2$

28. $x^2 - 3x = -2$

29. $x^2 = 0$

30. $x^2 = 2x - 5$

31. $-49x = 7x^2$

32. $x^2 + 4x - 120 = 0$

33. $x^2 = 20 - 2x$

34. $x^2 + x + 3 = -2$

35. $4x^2 + x + 1 = -2x$

The remaining problems are optional. They may be omitted without loss of continuity. Find the solution set by completing the square.

36. $x^2 - 8x + 16 = 0$

37. $x^2 + 2x = 3$

38. $3x^2 + 12x = 9$

39. $2x^2 + 10x - 12 = 0$

40. $2x^2 - 6x + 10 = 0$

41. $x^2 - 6x + 1 = 0$

42. $5x^2 - 20x + 50 = 0$

SELFTEST

1. State the quadratic formula and the equation to which it applies.

Compute the discriminant of the quadratic equations and discuss the nature of the solutions. Then compute the solutions.

2. $x^2 + 3x - 4 = 0$

3. $2x^2 - x + 6 = 0$

4. $5x^2 + 8x = 7 + 2x^2$

5. $4x^2 = x$

6. $x^2 + 1 = -2x$

Optional: Find the solution set by completing the square.

7. $x^2 + 4x - 1 = 0$

8. $3x^2 + 6x = 9$

SELFTEST ANSWERS

1. The solutions for $ax^2 + bx + c = 0$ are given by

$$x = \frac{-b \pm \sqrt{b^2 - 4ac}}{2a}$$

2. Discriminant is 25 so the solutions are real and unequal. The solutions are 1, and -4.

3. Discriminant is -47. Solutions are complex and unequal. The solutions are $1/4 \pm (\sqrt{47}\,i)/4$.

4. Discriminant is 148. Solutions are real and unequal. Solutions are $(-8 \pm \sqrt{148})/6$ or $-4/3 \pm \sqrt{37}/3$.

5. Discriminant is 1. Solutions are real and unequal. Solutions are 0 and $1/4$.

6. Discriminant is 0. Solutions are real and equal. Solution is -1.

7. $\{-2 \pm \sqrt{5}\}$ **8.** $\{1, -3\}$

11.4 APPLICATIONS

In this section we will present several applications of quadratic equations. When solving problems of this nature it is always a good idea to

1. read the problem very carefully,
2. write down the important facts in equation form,
3. then solve the equations and check your answers.

As we have pointed out before, a quadratic equation will, in general, have two solutions. Usually only one solution makes sense in any given word problem. The other solution is mathematically correct, but does not apply in the context of the problem. Solutions of this type are called *extraneous solutions.*

Example 1 A pizza stand makes two sizes of pizza pie. The large size has twice the area of the small size. The pies are in the shape of a circle. If the large size has radius 12 inches, what is the radius of the small size?

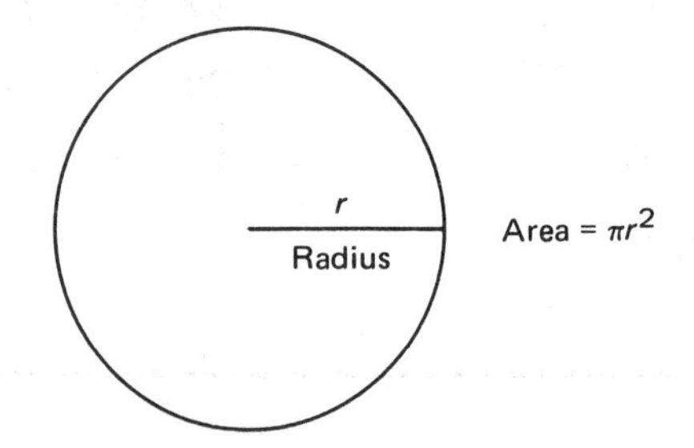

First we recall that a circle of radius r has area $A = \pi r^2$. What is the area of the large pie? Since the radius is 12 inches, the area must be $\pi(12)^2 = 144\pi$ square inches. And since the small pie has one-half the area of the large pie, its area is $144\pi/2 = 72\pi$ square inches. What is the radius of the small pie? Let r be the radius, then the area is πr^2. But we know that the area must be 72π. Therefore

$$\pi r^2 = 72\pi$$
$$r^2 = 72 \qquad \text{divide both sides by } \pi$$
$$r = \pm\sqrt{72} \qquad \text{take the square root on both sides}$$
$$r \approx 8.485 \text{ inches}$$

Of course, only the solution $r \approx 8.485$ makes sense. The other solution $r \approx -8.485$ is extraneous. You may check that $r \approx 8.485$ works.

It is interesting to note that if we just take half the radius of the large pie (that is $^{12}/_2$ or 6 inches), we do not get the correct answer. This is because the area of a circle grows as the *square* of its radius.

Example 2 The manager of a factory is looking at his production records. He is interested in three main quantities. The cost of raw material, the cost of labor, and the number of items produced. He organizes the entries in his books and scales the units so they will fit conveniently on a graph. On the x axis he puts the number of units produced and on the y axis he puts the cost in dollars. Above the x axis he plots two curves. One is L for labor costs, the other is R for the cost of raw material. He notices from his curves that as he buys a larger volume of raw material, the cost relative to the number of units produced decreases linearly. But the cost of labor goes up by the square of the number of units produced. After a little curve fitting he finds

$$R = 20 - 2x \qquad L = x^2$$

where x is the number of units produced.

At what volume of production will the raw material and labor costs be the same?

Let us draw a graph of the cost of labor and raw materials. Of course, in the

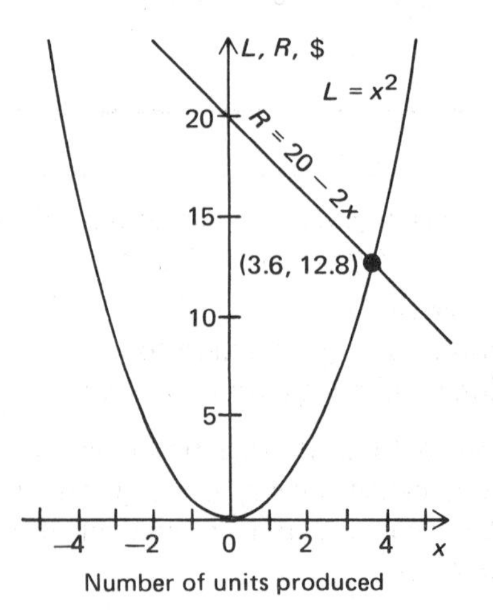

world of business it only makes sense to have $x > 0$. But for mathematical completeness we have drawn the graphs for the case $x < 0$ also.

The points at which the L graph and R graph meet are where $L = R$. At this point the cost of labor and raw material is the same, and, since $L = x^2$ and $R = 20 - 2x$, we have

$$x^2 = 20 - 2x$$
$$x^2 + 2x - 20 = 0$$

If we use the quadratic formula on this equation, we get

$$x = \frac{-2 \pm \sqrt{84}}{2}$$

From tables, $\sqrt{84} \approx 9.165$. Therefore the solution set of $x^2 + 2x - 20 = 0$ is $\{3.583, -5.583\}$ as decimal approximations. Only the solution $x \approx 3.583$ makes sense in the original problem. For the business world it does not make sense to talk about a negative production such as $x \approx -5.583$. So the solution $x \approx -5.583$ is extraneous. Of course both solutions are mathematically correct. Geometrically there should be two solutions since the complete L and R curves intersect at two points in our graph.

You may check that when the production volume is 3.583 units, then the raw material costs and labor costs are both about 12.84 monetary units.

Example 3 A submarine and a ship are 16 miles apart on routes which intersect at right angles. The submarine goes 4 miles/hour faster than the ship. How fast is each moving if they arrive at the intersection in 2 hours?

Let us draw a diagram. We let the point I represent the intersection, A represent the ship, and B represent the submarine. The vessels travel on straight-line routes which meet at right angles, and they are 16 miles apart.

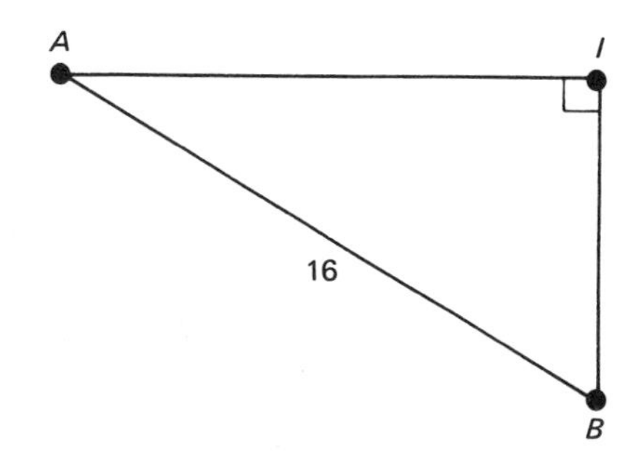

Let S represent the speed of the ship. The speed of the submarine is 4 miles/hour faster. So the submarine speed is $S + 4$.

They both arrive in 2 hours. Since distance = time × rate, in 2 hours the ship went $2S$ miles and the submarine went $2(S + 4)$ miles.

The distance from A to I is $2S$.
The distance from B to I is $2(S + 4)$.
The distance from A to B is 16.

From the Pythagorean theorem applied to the triangle in the figure we obtain

$$(2S)^2 + [2(S + 4)]^2 = 16^2$$
$$4S^2 + 4(S^2 + 8S + 16) = 256$$
$$8S^2 + 32S + 64 = 256$$

We can divide both sides by 8, so

$$S^2 + 4S + 8 = 128$$
$$S^2 + 4S - 120 = 0$$

From the quadratic formula we obtain

$$S = \frac{-4 \pm \sqrt{496}}{2}$$

$$S = -2 \pm 2\sqrt{31}$$

Again only the solution $-2 + 2\sqrt{31} \approx 9.136$ makes sense. The other, negative, solution is extraneous. The speed of the ship is $S \approx 9.136$. The speed of the submarine is $S + 4 \approx 13.136$.

PROGRAM 11.4

1. In this program we will examine some word problems. As always, when solving word problems you must first read the problem carefully. Then write down the important information in the form of mathematical equations.

Finally solve the equation and ________ the results in the original problem. — check

A quadratic equation usually has ________ solutions. In most applications only one of the solutions makes sense in the setting of the problem. — two

The other solution is mathematically correct but does not apply in the context of the problem. Such a solution is called an *extraneous solution*.

2. A given triangle is such that the base and height are the same. It has an area of 50 square inches. How do we find the base and the height? First we need some basic information about triangles.

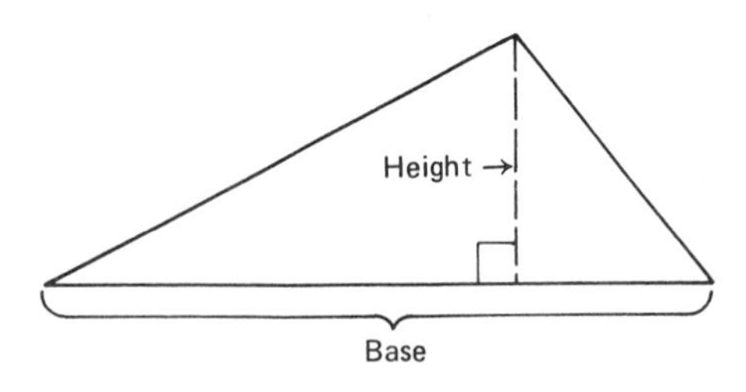

In our special case we know the (area\base\height) of a (triangle\square\circle). We also know that the base (equals\is four inches greater than\is half of) the height. When we put the known elements into the basic equation

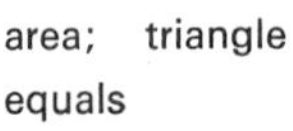
area; triangle
equals

$$\text{Area of triangle} = \frac{\text{base} \cdot \text{height}}{2}$$

we get

____ square inches = base · ________ ÷ 2 — 50; base or height

Since base = height, if we let b represent the base, we get the equation

____ = (____)^2 ÷ 2 — 50; b

This is a quadratic equation in the letter ____. We can solve this equation for b. — b

$50 = b^2/2$

________ $= b^2$ multiply both sides by 2 — 100

________ $= b$ take the square root on both sides — $\pm\sqrt{100}$ or ± 10

The two solutions are $b =$ ____ and $b =$ ____. — 10; -10

no

Can we have a negative length for the base? (yes\no). Is there an

yes; 10

equal

extraneous solution? (yes\no). Thus the base is ____ inches. Since the problem states that the height and base are (equal\both zero) the height

10 inches

must be ________. Do your answers make sense in the original problem? Check that our answers satisfy the conditions of the original problem. Are

yes; yes

the base and height equal? (yes\no). Is the area 50 square inches ?(yes\no).

3. A rectangular garage is 4 feet longer than it is wide. The owner decides to build storage cabinets along one length and one width. The cabinets are all 2 feet deep. The area of the remaining floor space is 192 square feet. What are the dimensions of the new floor space? Will a 14-foot boat still fit in the garage?

Let W represent the width of the garage, and let w represent the width of the floor space remaining after the cabinets are built. Let L represent the length of the garage, and let l represent the length of the remaining floor space.

The garage is 4 feet longer than it is wide so

4

$$L = W + \_\_\_\_$$

After the cabinets are installed the width is 2 feet less, so

2

$$w = W - \_\_\_\_$$

The length is also 2 feet less, so

2

$$l = L - \_\_\_\_$$

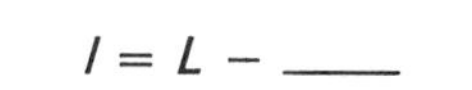

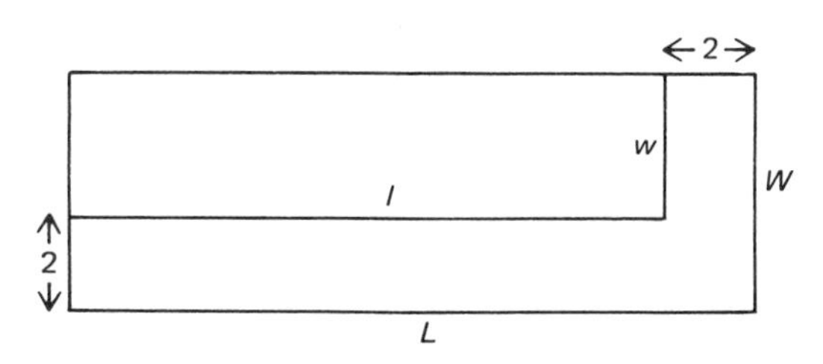

The main idea now is to express both w and l as a function of just one variable. We could express them as a function of either L or W, but it seems from the stated conditions that it is easier to use W. We already have $w = W - 2$. Since $L = W + 4$, then

$W + 4$; $W + 2$

$$l = L - 2 = (\_\_\_\_\_\_) - 2 = \_\_\_\_\_\_$$

Now we have both l and w expressed as functions of W. They are

$W - 2$

$$w = \_\_\_\_\_\_$$

$W + 2$

$$l = \_\_\_\_\_\_$$

Now we need to use another condition stated in the problem. The area

192

$l \cdot w$

of the remaining floor space is ____ square feet. But the area of the remaining floor space is equal to ($l \cdot w$ \ $L \cdot W$). That is, the remaining floor space

l; w; 192

is ____ feet long and ____ feet wide. Therefore $l \cdot w$ = ____ square feet.

$W + 2$; $W - 2$

So $(W + 2)(W - 2) = 192$, since l = ________ and w = ________.

$$(W + 2)(W - 2) = 192$$

$$W^2 - 4 = 192$$

196

$$W^2 = \_\_\_\_\_\_$$

$\pm\sqrt{196}$ or ± 14

$$W = \_\_\_\_\_\_$$

The number $W = -14$ does not make sense in this problem, because distance is always ______. Therefore $W = -14$ is an (extraneous\applicable) solution.

positive; extraneous

Now we know $W = 14$, but we are not done. The problem asks us to find ____ and ____. We have already found

l; w

$$l = W + 2 = \_\_\_\_ + 2 = \_\_\_\_ \text{ feet}$$

14; 16

$$w = W - 2 = \_\_\_\_ - 2 = \_\_\_\_ \text{ feet}$$

14; 12

Will a 14-foot boat still fit in the garage? (yes\no). The available length is ______.

yes

16 feet

4. An airplane is flying at an altitude of 2,000 feet above the ground. The speed of the plane relative to the ground is 200 feet/second. The pilot wants to eject a tracer smoke flare onto the sight of a crashed plane. He sights the crash and mistakenly ejects the flare straight down while he is directly over the crash. The flare is ejected with a downward speed of 40 feet/second. We want to know how far from the crash sight the flare will land. (Since the flare is small and heavy, we will ignore the effect of air resistance.)

We have to find a (distance\speed\size of object). But first we need to understand some of the physics of the situation and draw from two standard equations of physics.

distance

The flare continues to move horizontally at 200 feet/second until it hits the ground. It does not go straight down, but curves in a parabolic fashion. Let's look at a picture of the situation.

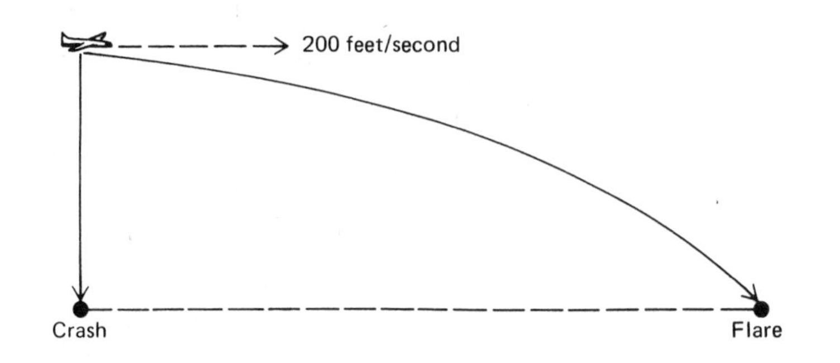

First we need to compute the time it takes the flare to hit the ground. An equation from physics says that for this falling object

$$s = 16t^2 + ut$$

where s is the distance the object falls, t is the time it takes to fall, and u is the initial speed in a downward direction.

In our particular problem we know s since that is the height of the plane and u since that is the speed at which the flare is ejected.

$$s = \_\_\_\_ \text{ feet}$$

2,000

$$u = \_\_\_\_ \text{ feet/second}$$

40

Substitute these values into the equation

$$\_\_\_\_ = 16t^2 + \_\_\_\_t$$

2,000; 40

This is now an equation in the letter ____.

t

Let's use the quadratic formula to solve for t. First put the equation in standard form.

$16t^2 + 40t - 2{,}000$

$\_\_\_\_\_\_\_\_\_\_ = 0$

16; 40; $-2{,}000$

$a = \_\_\_, \quad b = \_\_\_, \quad c = \_\_\_$

$$t = \frac{-b \pm \sqrt{b^2 - 4ac}}{2a}$$

-40; 40; 16; $-2{,}000$

$$t = \frac{\_\_\_ \pm \sqrt{(\_\_\_)^2 - 4(\_\_\_)(\_\_\_)}}{2(16)}$$

$$t = \frac{-40 \pm \sqrt{129{,}600}}{32}$$

$$t = \frac{-40 \pm \sqrt{100}\sqrt{1{,}296}}{32}$$

10; 36

$$t = \frac{-40 \pm (\_\_\_)(\_\_\_)}{32} \quad \text{use the root tables}$$

$$t = \frac{-40 \pm 360}{32}$$

10

$t = \_\_\_$ seconds

10

Thus the flare takes ___ seconds to reach the ground. During this time it is still traveling horizontally at the airplane's speed of 200 feet/second. How far does it travel horizontally? We use another equation from physics

Distance = speed · time

200; 10

= (___ feet/second)(___ seconds)

2,000 feet

= ______

2,000

Thus the flare lands ___ feet from the crash.

PRACTICE PROBLEMS

1. A baker needs to order pans to make a round layer cake that is 6 inches high but twice the volume of a cake 6 inches high and of diameter 8 inches. What diameter should the new pans be? (*Hint:* The diameter is 2 times the radius.)

2. Suppose you want to make two squares, one with twice the area of the other. If the smaller is 7 inches on a side, how many inches (to the nearest tenth) should the side of the larger square be?

3. A rectangle is such that the length is 4 feet more than the width. The area of the rectangle is 21 square feet. What are its dimensions?

4. The telephone company needs to run a line from a pole to a house. The line is attached to the pole at a point 40 feet above the ground. The inlet of the house is 10 feet above the ground. The line is stretched tight so it has no slack. If the house and pole are 60 feet apart how much line is necessary?

5. On a dart board the radius of the bull's eye is 2 inches. The area of the bull's eye and the next annular ring is 28.26 inches. What is the radius of the outer circle of this annular ring? How much greater is the area of the annular ring than that of the bull's eye? In this problem take π to be approximately 3.14.

6. A rectangular garage is 10 feet longer than it is wide. The owner decides to build storage cabinets along both lengths and one width. The cabinets are 2 feet deep. The area of the remaining exposed floor space is 108 square feet. What are the dimensions of the new floor space?

7. A right triangle is such that its two legs have the same length. The area of the triangle is 200 square inches. What are the lengths of the sides of this triangle?

8. In a certain triangle the base is twice the height. If the area of this triangle is 900 square inches what are the base and height?

9. A man hurls a rock into a canyon 174 feet deep. The initial velocity of the rock is 10 feet/second. How long does it take the rock to hit bottom? Recall that if u is the initial velocity, t is the time in seconds, and s is the distance of free fall in feet then

$$s = 16t^2 + ut$$

10. An airplane is moving in a straight line with ground speed 500 feet/second. The plane will eject a flare straight down with a speed of 50 feet/second. The plane is flying at 164 feet altitude. How far in front of a target (ground distance) should the pilot eject the flare. How long will it take the flare to reach the target?

11. Find three positive numbers $(x - 1)$, x, and $(x + 1)$ such that the sum of their squares is

   a. 29 *b.* 77 *c.* 149

12. The sides of a right triangle are x inches, x inches, and $x + 5$ inches. Find the lengths of the sides of this triangle.

13. The sides of a right triangle are x inches, $x + 1$ inches, and $x + 2$ inches. Find the area of this triangle.

14. Two people are going to mow a rectangular plot of grass. The plot is 60 feet by 40 feet. One of them starts on the edge and mows around the perimeter. How wide a strip should he mow from each side in order to do half the work?

15. Two runners are on straight-line routes which intersect at right angles. One runner goes 2 miles/hour faster than the other. How fast is each moving if they start out 6 miles apart and arrive at the intersection in 1 hour.

*16. Describe how you could apply the quadratic formula to solve an equation of the type

$$ax^4 + bx^2 + c = 0$$

*17. Describe how you could apply the quadratic formula to solve an equation of the type

$$\frac{a}{x^2} + \frac{b}{x} + c = 0$$

SELFTEST

1. You want to make two cardboard circles. One is to have three times the area of the other. If the smaller circle is to have radius 2 inches, what should the radius of the larger circle be? (Area of a circle $= \pi r^2$.)

2. A rectangular house lot is 7,650 square feet. It is only 5 feet longer than it is wide. What are the dimensions? (Area of a rectangle = length · width.)

SELFTEST ANSWERS

1. radius $= \sqrt{12} \approx 3.464$ inches 2. length = 90 feet; width = 85 feet

* You may find these problems more challenging.

POSTTEST 11

1. Which of the following equations are quadratic? Put those that are quadratic in standard form.

a. $2x^2 - 2x + 3 = 0$
b. $x^3 - 4x^2 + 3 = 0$
c. $(x + 3)(x - 1) = 4$
d. $3x^2 = 8$
e. $4x^2 + 2x - 1 = 3x^2$
f. $5x + 2x + 7 = 0$

2. Find the solutions of $(x - 4)(x + 2) = 0$.

3. Find the solutions of $x^2 + 2x + 1 = 0$.

4. Find the solutions of $x^2 - 5 = 0$.

5. Find the solutions of $x^2 + 16 = 0$.

6. State the quadratic formula and the equation to which it applies.

7. Find the solution set of $x^2 + x + 1 = 0$.

8. Find the solution set of $3x^2 - 7x + 2 = 0$.

9. Compute the discriminant and discuss the nature of the solutions without actually finding the solutions of $3x^2 + 6x = 2$.

10. An art student plans to make an abstract circular mosaic. He has ordered 1,256 tiles that are 1 square inch each. So he has enough tiles to cover an area of 1,256 square inches. What is the radius of the largest circle he can cover with these tiles? (Area of circle $= \pi r^2$; $r =$ radius, use $\pi \approx 3.14$.)

Each problem above refers to a section in this chapter as shown in the table.

Problems	*Section*
1–3	11.1
4 and 5	11.2
6–9	11.3
10	11.4

POSTTEST 11 ANSWERS

1. *a.* in standard form *c.* $x^2 + 2x - 7 = 0$ *d.* $3x^2 + 0x - 8 = 0$ *e.* $x^2 + 2x - 1 = 0$

2. $x = 4, x = -2$ **3.** $x = -1$ **4.** $x = \pm\sqrt{5}$ **5.** $x = \pm 4i$

6. $x = \dfrac{-b \pm \sqrt{b^2 - 4ac}}{2a}$; $ax^2 + bx + c = 0$ **7.** $x = \dfrac{-1 \pm \sqrt{3}\,i}{2}$ **8.** $x = 2, \dfrac{1}{3}$

9. discriminant $= 60$; the solutions are real and unequal **10.** radius $= 20$ inches

Table of Squares and Square Roots

N	N^2	$\sqrt{N}$	N	N^2	$\sqrt{N}$	N	N^2	$\sqrt{N}$	N	N^2	$\sqrt{N}$
1	1	1.000	**26**	676	5.099	**51**	2,601	7.141	**76**	5,776	8.718
2	4	1.414	**27**	729	5.196	**52**	2,704	7.211	**77**	5,929	8.775
3	9	1.732	**28**	784	5.292	**53**	2,809	7.280	**78**	6,084	8.832
4	16	2.000	**29**	841	5.385	**54**	2,916	7.348	**79**	6,241	8.888
5	25	2.236	**30**	900	5.477	**55**	3,025	7.416	**80**	6,400	8.944
6	36	2.449	**31**	961	5.568	**56**	3,136	7.483	**81**	6,561	9.000
7	49	2.646	**32**	1,024	5.657	**57**	3,249	7.550	**82**	6,724	9.055
8	64	2.828	**33**	1,089	5.745	**58**	3,364	7.616	**83**	6,889	9.110
9	81	3.000	**34**	1,156	5.831	**59**	3,481	7.681	**84**	7,056	9.165
10	100	3.162	**35**	1,225	5.916	**60**	3,600	7.746	**85**	7,225	9.220
11	121	3.317	**36**	1,296	6.000	**61**	3,721	7.810	**86**	7,396	9.274
12	144	3.464	**37**	1,369	6.083	**62**	3,844	7.874	**87**	7,569	9.327
13	169	3.606	**38**	1,444	6.164	**63**	3,969	7.937	**88**	7,744	9.381
14	196	3.742	**39**	1,521	6.245	**64**	4,096	8.000	**89**	7,921	9.434
15	225	3.873	**40**	1,600	6.325	**65**	4,225	8.062	**90**	8,100	9.487
16	256	4.000	**41**	1,681	6.403	**66**	4,356	8.124	**91**	8,281	9.539
17	289	4.123	**42**	1,764	6.481	**67**	4,489	8.185	**92**	8,464	9.592
18	324	4.243	**43**	1,849	6.557	**68**	4,624	8.246	**93**	8,649	9.644
19	361	4.359	**44**	1,936	6.633	**69**	4,761	8.307	**94**	8,836	9.695
20	400	4.472	**45**	2,025	6.708	**70**	4,900	8.367	**95**	9,025	9.747
21	441	4.583	**46**	2,116	6.782	**71**	5,041	8.426	**96**	9,216	9.798
22	484	4.690	**47**	2,209	6.856	**72**	5,184	8.485	**97**	9,409	9.849
23	529	4.796	**48**	2,304	6.928	**73**	5,329	8.544	**98**	9,604	9.899
24	576	4.899	**49**	2,401	7.000	**74**	5,476	8.602	**99**	9,801	9.950
25	625	5.000	**50**	2,500	7.071	**75**	5,625	8.660	**100**	10,000	10.000

ANSWERS TO EVEN-NUMBERED PROBLEMS

CHAPTER 1

Section 1.1

2. *a.* 2, 5, 8, 9
b. 2, 8
c. 5, 9
d. no element x in A satisfies $x + 1 = 4$; x can be 3

4. *a.* $\{1, 2, 3, \dots\}$ *b.* $\{0, 1, 2, 3, \dots\}$ *c.* $\{0, 2, 4, 6, \dots\}$ *d.* $\{1, 3, 5, \dots\}$

6. *a.* 0 *b.* 35 *c.* not def. *d.* 0
e. 0 *f.* not def. *g.* 1 *h.* 8
i. 3 *j.* 9 *k.* x *l.* 0
m. not def. *n.* x

Section 1.2

2. $x = 3$

4. they will balance; the one on the left will be heavier; the one on the left will be lighter.

Section 1.3

2. 2, 3, 5, 11, 29, 31

4. *a.* $2 \cdot 7$ *b.* $3 \cdot 5 \cdot 7$ *c.* $2 \cdot 11 \cdot 13$ *d.* $2 \cdot 2 \cdot 3 \cdot 5 \cdot 5$

6. *a.* 30 *b.* 44 *c.* 75
d. 42 *e.* 36 *f.* 3528

8. yes

Section 1.4

2. *a.* commutative + *b.* associative + *c.* commutative ·
d. distributive *e.* associative · *f.* distributive
g. commutative; associative; simplify

4. *a.* $5x + 15$ *b.* $4x + xy$ *c.* $4(x + y)$
d. $x(9 + 20)$ or $29x$ *e.* $2x + 8$ *f.* $15(x + 2)$
g. $2x + 6 + 2y$ *h.* $6a + 12 + 6b$ *i.* $3(x + 4 + y)$
j. $x(4 + 2 + 3)$ or $9x$ *k.* $230 + 23x$ *l.* $35 + 7x$

6. $(18 \div 6) \div 3 \neq 18 \div (6 \div 3)$ since $1 \neq 9$
$(16 - 4) - 3 \neq 16 - (4 - 3)$ since $9 \neq 15$
neither division nor subtraction is associative

CHAPTER 2

Section 2.1

2. *a.* $\dots, -3, -2, -1$ *b.* $\dots, -3, -2, -1, 0, 1, 2, 3, \dots$

4. *a.*

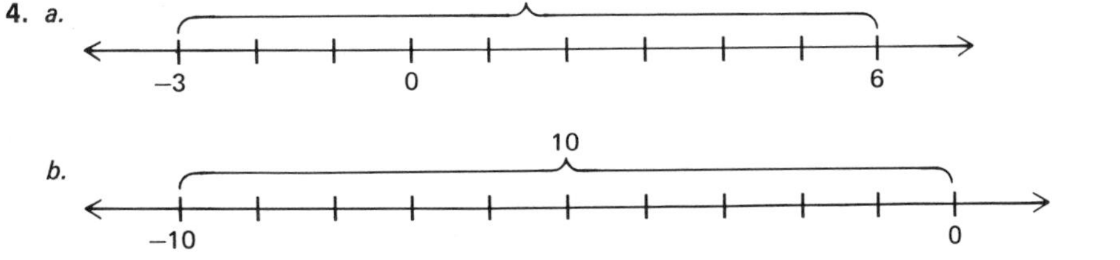

c.

[Number line from 0 to 25 marked 0, 5, 10, 15, 20, 25, with brace labeled 25 spanning 0 to 25]

d.

[Number line marked −7, 0, 6, with brace labeled 13 spanning −7 to 6]

6. *a.* 16 *b.* 25 *c.* 0 *d.* 3
e. 12 *f.* 3 *g.* 20 *h.* 6
i. 80 *j.* 30 *k.* 9 *l.* 5

8. *a.* −7 *b.* 9 *c.* −25 *d.* −14

Section 2.2

2. 59 **4.** −22 **6.** −10 **8.** −72 **10.** 10
12. 11 **14.** 15 **16.** −4 **18.** −14 **20.** 0
22. 32 **24.** −61 **26.** −211 **28.** −312 **30.** −93
32. 8 **34.** −7 **36.** −20 **38.** −101 **40.** 0
42. 10 **44.** 35 **46.** 0 **48.** 0 **50.** −50

Section 2.3

2. 50 **4.** −800 **6.** 94 **8.** 15 **10.** −7
12. 40 **14.** −400 **16.** 1346 **18.** 0 **20.** 78
22. −42 **24.** −4 **26.** 45 **28.** −100 **30.** 53
32. −70 **34.** 800 **36.** 0 **38.** −600 **40.** 3
42. 30 **44.** −110 **46.** 10 **48.** 9 **50.** 0

Section 2.4

2. −99 **4.** 0 **6.** −21 **8.** 24 **10.** 0
12. 32 **14.** −24 **16.** 16 **18.** −60 **20.** 170
22. 25 **24.** −42 **26.** 28 **28.** −7 **30.** 96
32. −43 **34.** −42 **36.** 24 **38.** 0 **40.** 60
42. −200 **44.** 84

Section 2.5

2. −1 **4.** 0 **6.** −22 **8.** −4 **10.** −1
12. 1 **14.** 3 **16.** −5 **18.** −1 **20.** 0
22. 3 **24.** −10 **26.** −41 **28.** −3 **30.** 0
32. −3 **34.** −69 **36.** 0 **38.** 10

CHAPTER 3

Section 3.1

2. *a.* $xxxxx$ *b.* yy *c.* x *d.* $(-x)(-x)(-x)(-x)$
e. $xxxyyyyyyyy$ *f.* $2 \cdot 2 \cdot 2 \cdot 2 \cdot 2 \cdot 2$ *g.* $(-3)(-3)(-3)$ *h.* $-8 \cdot 8$

4. y^6 **6.** $-4x^3$ **8.** $15x^6y^{12}$ **10.** $6z^4x^{14}y^2$ **12.** $147z^5x^9y^2$
14. $24x^9y^9$ **16.** y^3 **18.** $5x^6$ **20.** $5xy^3$ **22.** 7
24. $36x^7y^5z^2$ **26.** 27 **28.** 16 **30.** 81 **32.** −64
34. 2^3 or 8 **36.** 196 **38.** 23 **40.** 21 **42.** 3
44. −6 **46.** −9

48. *a.* $5y$ *b.* $25y^2$ *c.* $25y^2$
d. yes *e.* yes; for $(ax)^2 = (ax)(ax) = a^2x^2$

50. *a.* 16 *b.* 8 *c.* no

Section 3.2

2. *a.* 2 *b.* 3 *c.* 2 *d.* 8 *e.* 5
4. *a.* 4 *b.* −6 *c.* 3 *d.* 0 *e.* −1
6. *a.* binomial *c.* monomial
8. *a.* −5 *b.* 15 *c.* 15 *d.* 0
10. *a.* 1 *b.* 41

Section 3.3

2. $10x + 7$
4. $12x^2 + 2x - 3$
6. $x^4 - 2x^3 - 6x^2 + 15$
8. $105x^4 + 10x^3 + 3x - 6$
10. $-6x^4 + 83x^3 - 8x^2 - 116$
12. $9x^4 - x^3 + 7x - 8$
14. $2x$
16. $-6x^2 + 3x$
18. $-12y^3 + 4y^2 + 7$
20. $-28x^2 + 15x + 3$
22. $2x^4 + 2x^2 + x - 1$
24. $4x^4 - 8x^2 + 23x - 8$
26. $x - 43$
28. $14x - 5$
30. $-x^3 - 10x$
32. $4x^2 + 7x$
34. $-17x + 5$
36. $5x^2 + 10x + 5$

Section 3.4

2. $5x^2 - x$
4. $2x + 9$
6. $2x^5 - 4x^2 - 11$
8. $-12x^2 - 3x + 24$
10. $4x^2 - x + 8$
12. $2x^2 + 2x$
14. $36x - 54$
16. $-17x^2$
18. $-2x^2$
20. $-23x^3$
22. 0
24. $-27x^4 + 3x^3 - 25x + 10$
26. $-24x^3 + 3x^2 - 10x + 1$
28. $x^3 + 36x^2 + 55x + 9$
30. $x - 15$
32. $x + 1$
34. $x + 3$
36. $-2x^2 - 4x - 1$
38. $6x^2 - 18x + 6$
40. $x^2 - 2x - 6$
42. $95x^3 + 10x^2 + 82$
44. $15x^6 - 14x^5 - 3x^3 - 8x^2 + 66$
46. $-5y^2 + 6y - 25$
48. $6x^2 + 8x - 46$
50. $3x^2 - 4x + 80$
52. $13x^2 - 18x + 19$
54. $10x - 15$

Section 3.5

2. $6x^5y - 10x^2y$
4. $-10x^3 - 50x^2 + 20x$
6. $-8xy^3 + 20x^2y^2 - 5x^3y$
8. $30y^3 + 40y^2 + 35y + 6y^2 + 8y + 7 = 30y^3 + 46y^2 + 43y + 7$
10. $2x^3 + 8x^2 + 16x - 3x^2 - 12x - 24 = 2x^3 + 5x^2 + 4x - 24$
12. $-2x^3 + x^2 + x + 8x^2 - 4x - 4 = -2x^3 + 9x^2 - 3x - 4$
14. $3x^2 + 17x + 10$
16. $x^2 + 5x - 50$
18. $6x^2 - 21x + 18$
20. $4x^2 - 18xy - 10y^2$
22. $x^3 - 3x^2y + 2xy - 6y^2$
24. $x^2 + 4x + 4$
26. $25x^2 - 40x + 16$
28. $x^3 - 6x^2 + 12x - 8$

Section 3.6

2. $x - 4$
4. $2x + 1$
6. $x + 6$
8. $4x^2 - 1$, remainder $- 18$
10. $6x^3 + 5x^2 + 5x + 5$

CHAPTER 4

Section 4.1

2. $4(3x + 4)$
4. $9(2x^2 + 3x + 1)$
6. $4(10x^2 + 5x - 2)$
8. $11(2x^2 - 4x + 3)$
10. $5(5x^2 - 17x + 13)$
12. $15(3x^2 - x + 2)$
14. $4xy(4xy - 3)$
16. $2(9xy + 6x - 1)$
18. $33(x^2 + 2x - 3)$
20. $12(7x^2 - x - 1)$
22. $xy^2z(x - 1 + z)$
24. $16xy(4x - 8xy + y)$
26. $-1(8x^2 - x + 5)$
28. $-1(-17x + 8)$
30. $-1(2x + 5)$
32. $-1(-x - 1)$
34. $-1(-12x^2 - 10x - 6)$
36. $-1(25x^2 + 5x + 3)$
38. $-1(-16x^2 + 8x - 4)$

Section 4.2

2. $(x + 6)^2$
4. $(x + 2)^2$
6. $(x + 8)^2$
8. $(x + 10)^2$
10. not a perfect square
12. $(3x - 2)^2$
14. not a perfect square
16. not a perfect square
18. $(x^2 + 9)^2$
20. $(x + 1)(x - 1)$
22. $(x + 10)(x - 10)$
24. $(4x + 5)(4x - 5)$
26. $(8x + 1)(8x - 1)$
28. $(x^2 + 7)(x^2 - 7)$
30. not a difference of squares

Section 4.3

2. $(x + 2)(x + 2)$
4. $(x + 4)(x + 2)$
6. $(x + 3)(x - 2)$
8. $(x - 5)(x + 2)$
10. $(x + 6)(x - 2)$
12. $(x + 3)(x + 4)$
14. $(x - 1)(x - 18)$
16. $(x - 10)(x + 2)$
18. $(x - 5)(x - 1)$
20. $(x + 6)(x + 8)$
22. $(x + 9)(x - 4)$
24. $(x - 5)(x - 5)$
26. does not factor
28. $(x - 8)(x - 5)$
30. does not factor
32. $(x - 5)(x + 1)$
34. $x(x + 9)$
36. $(x + 12)(x + 3)$
38. $x(x + 1)$
40. $(x - 50)(x + 2)$
42. $(x - 9)(x - 9)$
44. $(x + 2)(x - 18)$
46. $(x + 64)(x + 1)$
48. $(x + 21)x$
50. $(x + 3)(x + 11)$

Section 4.4

2. $(2x + 1)(x + 3)$
4. $(3x - 1)(x + 5)$
6. $(2x - 1)(x - 3)$
8. $(2x + 1)(2x + 5)$
10. $3(2x + 1)(x - 1)$
12. $3(2x + 1)(x + 1)$
14. $(3x + 8)(x - 2)$
16. $(4x + 7)(2x - 1)$
18. $(3x + 4)(3x - 2)$
20. $(9x + 4)(x - 2)$
22. $3(2x + 5)(x - 1)$
24. $3(2x - 5)(x + 1)$
26. $(3x - 5)(x - 4)$
28. $-1(3x - 20)(x - 1)$
30. $-2(x - 3)(x + 1)$
32. $2(3x + 2)(x - 1)$

Section 4.5

2. $6xy$
4. $20x$
6. $12xy$
8. $70y^2x$
10. $8x(x - 4)$
12. $5x(2x - 1)$
14. $(x + 2)(x + 1)$
16. $5(x + 1)(x - 1)(x + 2)$
18. $(x + 2)(x - 2)^2$

CHAPTER 5

Section 5.1

2. $^{18}/_{7}$
4. $-^{1}/_{3}$
6. $^{8}/_{13}$
8. $6K$
10. $5/6x^2$
12. $3/2y^3$
14. $(x - 2)^2$
16. x
18. x
20. $2x + 4$
22. 4
24. $(2x^2 - 1)/2x$
26. $(4x - 1)/6$
28. $(10x + 2)/5x$
30. *a.* improper *b.* proper *c.* proper *d.* improper *e.* proper *f.* improper
32. *a.* and *b.* $x/2$ will be an integer if and only if x is an even integer.
34. *a.* and *b.* $12/x$ will be a proper fraction if and only if $12 < |x|$.

Section 5.2

2. *a.* $\frac{2}{6}, \frac{3}{6}$ *b.* $\frac{16}{28x}, \frac{35}{28x}$ *c.* $\frac{2}{10}, \frac{6}{10}$ *d.* $\frac{38}{14}, \frac{3}{14}$ *e.* $\frac{-8y}{18xy}, \frac{15x}{18xy}$ *f.* $\frac{12}{30}, \frac{-4}{30}$
4. *a.* $\frac{2 + 3x}{6x}$ *b.* $\frac{-7}{20}$ *c.* $\frac{-14x + 5y}{35xy}$ *d.* $\frac{4 + 2}{1}$ or 6 *e.* $\frac{-11}{15}$ *f.* $\frac{xw + yz}{yw}$ *g.* $\frac{2y + 3x}{xy}$ *h.* $\frac{4x + 5y}{xy}$ *i.* $\frac{28}{15x}$ *j.* $\frac{14x + 15y}{35xy}$ *k.* $\frac{3x + 3}{x(x + 3)}$ *l.* $\frac{11y - 12}{y(y - 2)}$
6. *a.* $\frac{2}{9}$ *b.* $\frac{24 - 5x}{4x}$ *c.* $\frac{1}{x}$ *d.* $\frac{8x - 35}{5x}$ *e.* $\frac{4y - 15x}{6xy}$ *f.* $\frac{12y - 5x}{10xy}$ *g.* $\frac{4x + 10}{x(x + 2)}$ *h.* $\frac{-2x - 12}{x(x + 3)}$
8. *a.* $\frac{9x^2 + 2y^2}{27x^4y^3}$ *b.* $\frac{5y^3 - 6x^2}{14x^3y^4}$ *c.* $\frac{3x - 6}{x^2 - 9}$ *d.* $\frac{3x + 7}{(x + 1)^2}$ *e.* $\frac{3x - 3}{(x + 1)(x + 2)(x - 2)}$

Section 5.3

2. *a.* $\frac{-7}{27}$ *b.* $\frac{2}{15}$ *c.* $\frac{14}{9}$ *d.* $\frac{-x}{2}$ *e.* 1 *f.* $\frac{-y^2}{3}$ *g.* $\frac{3}{2x}$ *h.* $\frac{2x^3}{25}$ *i.* $\frac{5}{3}$ *j.* $\frac{7x^2}{8y^2}$ *k.* $\frac{(x + 1)^2}{6}$ *l.* $\frac{(3x - 1)}{2x}$
4. *a.* $\frac{7}{30}$ *b.* -2 *c.* $\frac{13y}{48}$ *d.* 21

Section 5.4

2. $2x$
4. $\frac{35}{48}$
6. $\frac{8}{35x}$
8. $\frac{28}{9}$
10. 1
12. 49
14. $\frac{36 - 3x}{4x + 26}$
16. $\frac{x - 1}{3x + 3}$
18. $\frac{x + 1}{2x + 1}$

CHAPTER 6

Section 6.1

2. $\{0\}$
4. $\{5\}$
6. $\{10/3\}$
8. $\{1, -1\}$
10. $\{5, -6\}$
12. $\{1\}$
14. inconsistent
16. identity

Section 6.2

2. $\{55\}$
4. $\{-5\}$
6. $\{-4\}$
8. $\{-20\}$
10. $\{-1/5\}$
12. $\{19\}$
14. $\{-8\}$
16. $\{-8\}$
18. $\{1\}$
20. $\{6\}$
22. $\{3\}$
24. $\{-6\}$
26. $\{9\}$
28. $\{-130\}$
30. $\{-7\}$
32. $\{3\}$
34. $\{1\}$
36. $\{-7\}$
38. $\{-6\}$
40. $\{-18\}$
42. $\{-9\}$
44. $\{19\}$

Section 6.3

2. $\{-4\}$
4. $\{-5\}$
6. $\{1\}$
8. $\{9\}$
10. $\{8\}$
12. $\{-32\}$
14. $\{25\}$
16. $\{-81\}$
18. $\{15\}$
20. $\{16\}$
22. $\{8/7\}$
24. $\{-1\}$
26. $\{-3\}$
28. $\{-1\}$
30. $\{15\}$
32. $\{13/12\}$
34. $\{-1\}$
36. inconsistent
38. $\{-1\}$
40. $\{0, 4\}$
42. $\{-9, -4\}$
44. $\{0, 1\}$

Section 6.4

2. $z = x/y; y = x/z$
4. $r = q - s; q = r + s$
6. $x = \frac{C - 3B}{A}; B = \frac{C - Ax}{3}$
8. $T = \frac{QR}{S}; R = \frac{ST}{Q}$
10. $y = \frac{9}{5}(x - 32)$
12. $a = \frac{2y}{x}; y = \frac{xa}{2}$
14. $A = \frac{2}{B + C}; B = \frac{2 - AC}{A}; C = \frac{2 - AB}{A}$
16. $C = 3A - B; A = \frac{C + B}{3}; B = 3A - C$
18. $A = \frac{2V}{H} - B; H = \frac{2V}{A + B}; B = \frac{2V}{H} - A$

Section 6.5

2. $-15, -13, -11$
4. 165, 175 pounds
6. 1.5, 10.5 feet
8. 5 percent
10. 333.3 hours
12. 3,800 dollars
14. 10 pounds
16. total time 5 hours; average speed 1.4 miles/hour
18. 100 tickets
20. he ran 4.7 miles in $9/8$ hours; he rode 10.3 miles in $5/8$ hours

Section 6.6

2. false
4. true
6. true
8. true
10. true
12. $x < -3$
14. $x \geq 1/7$
16. $x > 1$
18. $x \geq 12$
20. $3 \geq x$
22. inconsistent

Section 6.7

2. $x < 2$
4. $x < -2$
6. $x < 1/6$
8. $11/5 < x$
10. $4/7 \leq x$
12. $x < 8$
14. $1 > x$
16. $7 \geq x$
18. $x \geq 3$
20. $x > -1/7$

CHAPTER 7

Section 7.1

2. $1/x$
4. x
6. $1/x$
8. $1/x^9$
10. $2x^4$
12. x^4/y^5
14. $1/x$
16. $-108xy^4$
18. $(x + y)^2$
20. x^2y^0 or x^2
22. $4(x + y)^4/x$
24. $1/(x + y)^3$

Section 7.2

2. 1×10^{10}
4. 5×10^6
6. 1×10^{-7}
8. 9.73×10^{-5}
10. 9.21×10^{-7}
12. 3.5×10^{-4}
14. 8.1×10^{-3}
16. 1.89×10^{-4}
18. 500
20. 7,300,000,000
22. 0.92
24. 0.000037
26. 13
28. 1.5
30. 0.00984
32. 0.00189
34. 8×10^9
36. 0.004
38. 6.70×10^8
40. 4×10^{-6}
42. 2.76×10^8
44. 1.318×10^{25}

Section 7.3

2. $10y^3$
4. x^{12}
6. $1/x^8$
8. $625y^{12}$
10. $49x^4$
12. $36x^6y^8$
14. x^8/y^{10}
16. x^5/y^6
18. $1/y^4$
20. 4
22. 25
24. x^{12}/y^{15}
26. $8x^6$
28. $1/x^{28}$
30. $81y^4/x$
32. $8x^6/y^3$
34. $x^6/4$
36. $1/27x^9$
38. 1
40. $y^8/4$

CHAPTER 8

Section 8.1

2. Not every point on the number line corresponds to a fraction; some points correspond to irrational numbers.
4. π, 2π, 3π, 4π, 5π, $\pi + 1$, $\pi + 2$, $\pi + 3$, $\pi + 4$, $\pi + 5$, $-\pi$, -2π, -3π, -4π, -5π; all of these are irrational
6. The set of all rational numbers together with the set of all irrational numbers make up the set of all real numbers.

Section 8.2

2. index is 2 radicand is 5
4. false
6. true
8. false
10. true
12. true
14. false
16. 2.646
18. 8.944
20. 12.449
22. 9.660
24. 3.391
26. *a.* 4 *b.* 8 *c.* 16 *d.* 32 *e.* 64
28. *a.* 16 *b.* 64 *c.* 256
30. 36, 49, 64, 81, 100, 121, 144, 169, 196, 225
32. 8
34. 3
36. 11
38. 3
40. 7
42. 14
44. 9

Section 8.3

2. 29
4. y^2
6. 26
8. x
10. $2y$
12. $x\sqrt{5}$
14. $3\sqrt{x}$
16. $2\sqrt{3}$
18. $5\sqrt{2}$
20. $40\sqrt{2}$
22. 30
24. $2\sqrt{2y}$
26. $4xy^2\sqrt{x}$
28. $y^2\sqrt{y}$
30. $5xy^2\sqrt{2}$

Section 8.4

2. $3\sqrt{7}$
4. $-\sqrt{y}$
6. $4\sqrt{2}$
8. $3\sqrt{10}$
10. $23\sqrt{3}$
12. $x\sqrt{5}$
14. $4x\sqrt{x}$
16. $7xy\sqrt{3}$
18. $2x^2y\sqrt{y}$
20. $4\sqrt{5}$
22. $-2\sqrt{x}$

Section 8.5

2. $\sqrt{15}$
4. $2a\sqrt{3}$
6. $15 + 8\sqrt{y} + y$
8. 4
10. $\sqrt{6}$
12. 5
14. $\frac{5}{2}\sqrt{3}$
16. 3
18. $\frac{2\sqrt{x}}{x}$
20. $\frac{\sqrt{2a}}{2}$
22. $\frac{6\sqrt{7}}{35}$
24. $\frac{20 + 5\sqrt{3}}{13}$

26. $\frac{18 - 3\sqrt{x}}{36 - x}$ 28. $\frac{3\sqrt{2x}}{2x}$ 30. $\frac{4\sqrt{6}}{9}$ 32. $\frac{3 + \sqrt{2}}{7}$

34. $\frac{(1 + \sqrt{5})}{4}$ 36. $\frac{\sqrt{x} - \sqrt{y}}{x - y}$

Section 8.6

2. $4(4 - 3\sqrt{2a})$ 4. $3 \cdot 3$ 6. $\sqrt{3}\,(\sqrt{5} + 5)$ 8. $\sqrt{2}\,(3 + \sqrt{5})$
10. $2\sqrt{5}\,(\sqrt{2} - 2)$ 12. $3ab(1 - 2\sqrt{b})$ 14. $^1/_4(2 + 3\sqrt{2})$ 16. $3 - \sqrt{2}$
18. $1 - \sqrt{b}$ 20. $^3/_2(1 + \sqrt{3})$ 22. $^1/_2(1 - \sqrt{2})$

Section 8.7

2. $6i$ 4. $\sqrt{7}\,i$ 6. $2\sqrt{5}\,i$ 8. $a\sqrt{b}\,i$
10. $3 + 0i$ 12. $-1 + 0i$ 14. $2 + 2i$ 16. $6 + 8i$
18. $11 + i$ 26. $-10 + 11i$ 28. $16 + 2i$ 30. $7 + 4i$
32. $-14 + 23i$
34. yes, every real number is also a complex number.
36. no, not every complex number is a real number.
38. $1 - i$ 40. $-1 + 2i$

CHAPTER 9

Section 9.1

2. *a.* true *b.* false *c.* false *d.* true *e.* true *f.* false
4. $x = 4$ 6. $y = 16$

Section 9.2

2. Drop a perpendicular from the point to the x axis. Let the point where this line meets the x axis be called x. Then drop a line from the point perpendicular to the y axis. Let the point where this line meets the y axis be called y. The point P corresponds to the ordered pair (x, y).

4.

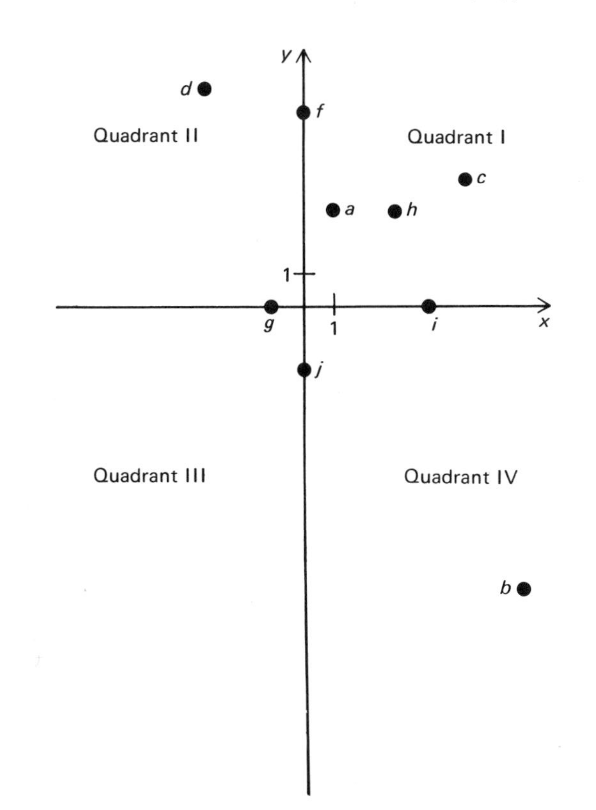

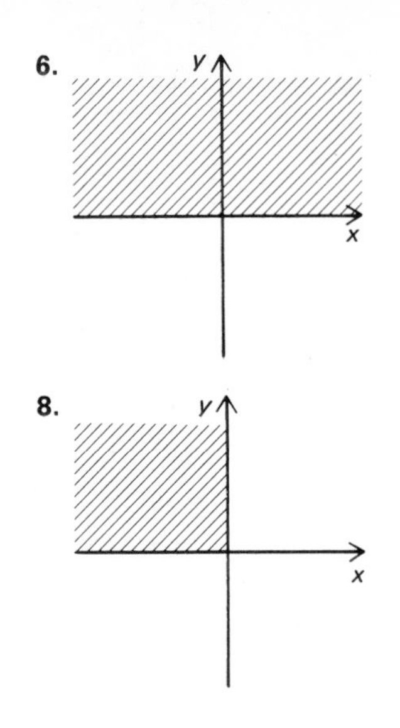

Section 9.3

2. *a.* domain is $\{1, 2, 3, 4, 5\}$ *b.* range is $\{2, 3, 4, 5, 6\}$

c. yes, every first element of an ordered pair is paired with only one second element of an ordered pair.

4. *a.* range is $\{0\}$ *b.* domain is $\{1, 2, 3, 4, 5\}$

c. yes, for the same reason as given in 2*c*.

6. *a.* 0 *b.* 3 *c.* 6 *d.* 9 *e.* 12

e. and *f.* Each ordered pair does satisfy $y = 3x$

8. yes

10. *a.* $g(4) = 2(4) = 8$

$g(5) = 2(5) = 10$

b. Two ordered pairs are (4, 8) and (5, 10)

12. (3, 8), (4, 11), (5, 14), (6, 17)

Section 9.4

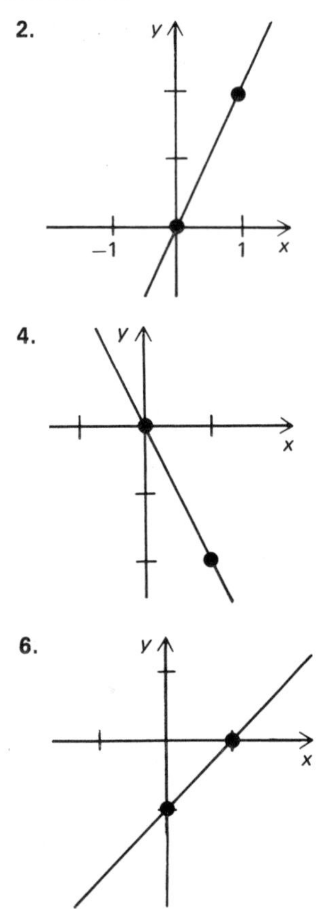

8.

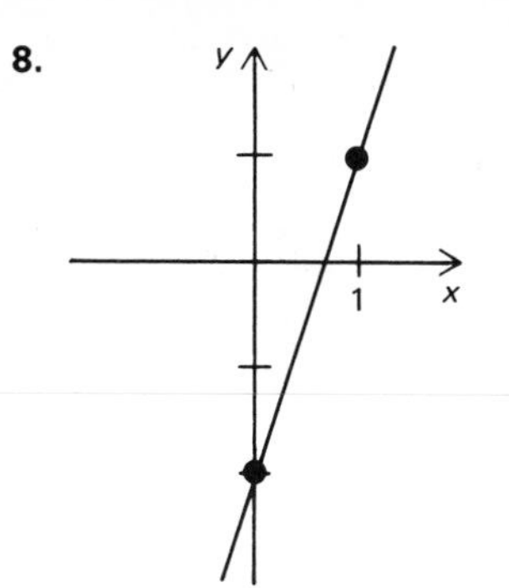

10.

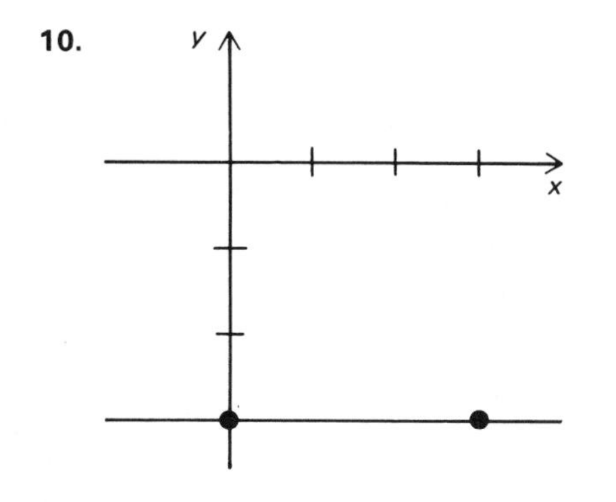

12.

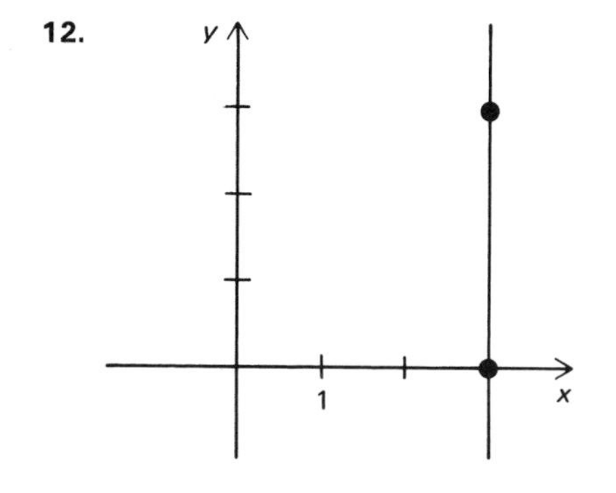

14.

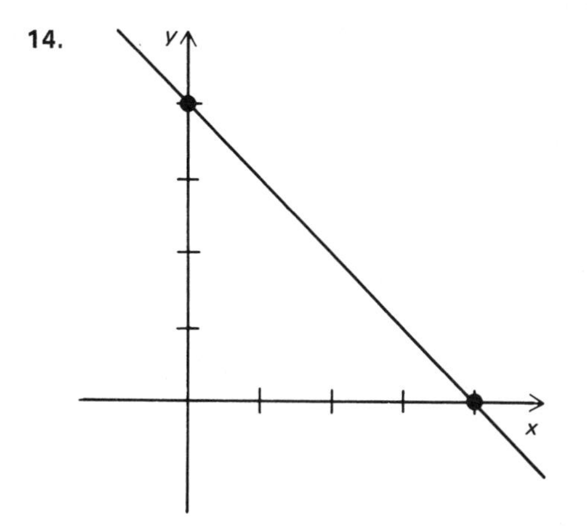

16.

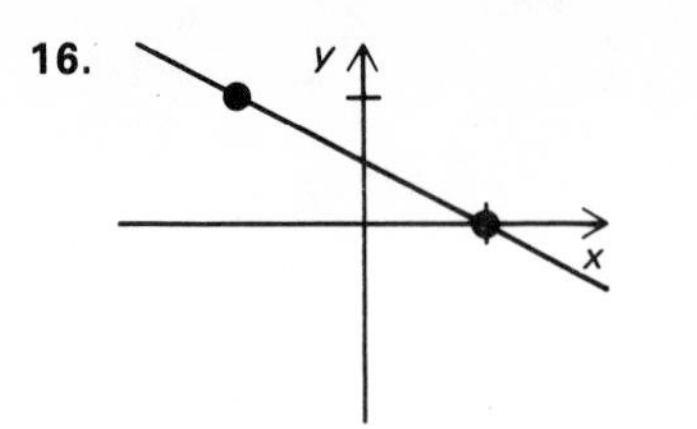

18.

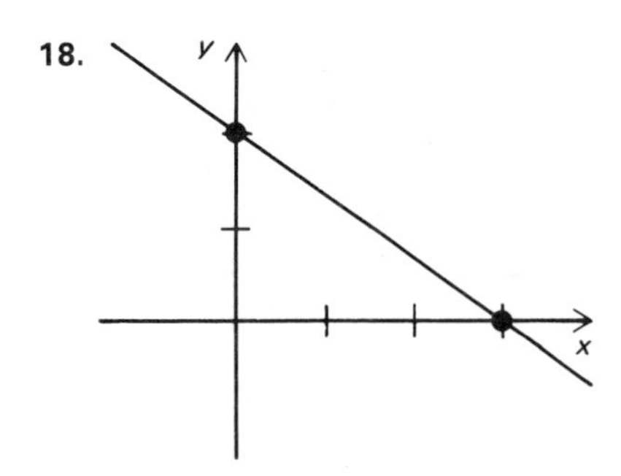

20.

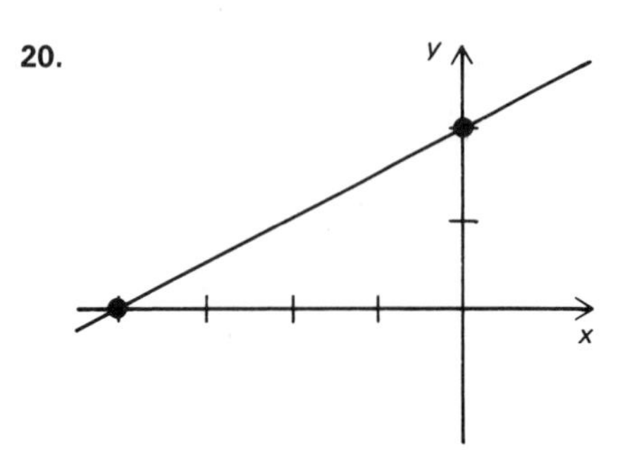

22.

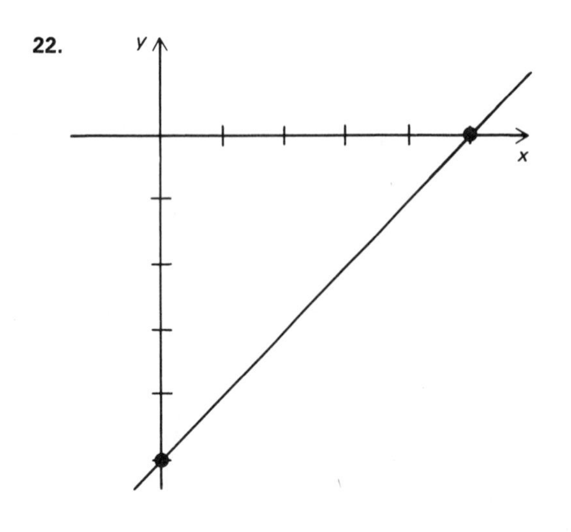

24.

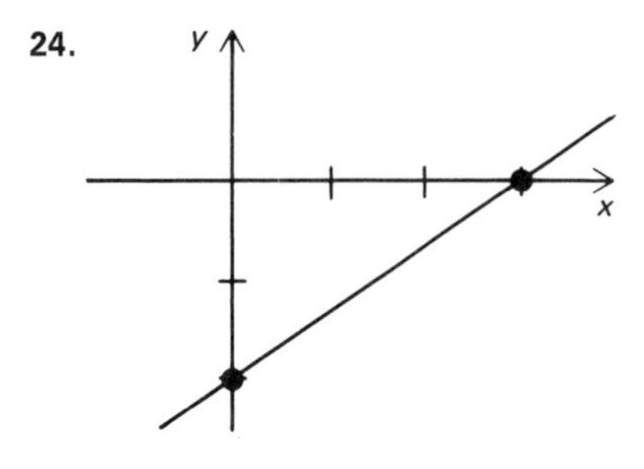

26. *a.* and *b.*

Section 9.5

2. Slope $= {}^{-5}/_{3}$

y

(−2, 4)

Rise = −1 − 4 = −5

x

(1, −1)

Run = 1 − (−2) = 3

4. Slope $= {}^{-3}/_{6} = {}^{-1}/_{2}$

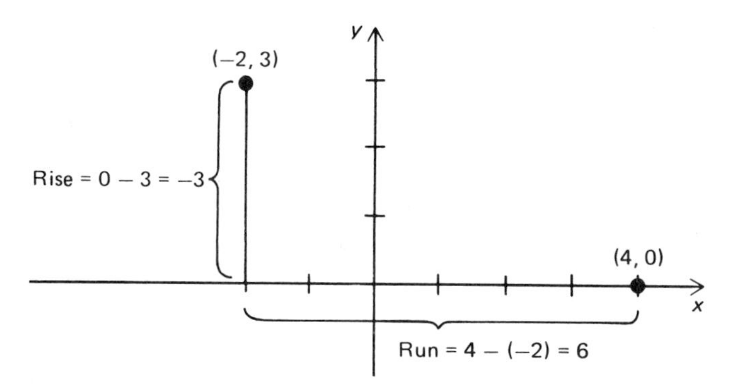

6. -1 **8.** ${}^{3}/_{2}$ **10.** 0

12. Slope is -2.5; y intercept is (0, 18)

14. The lines a, c, e, and f are parallel; the lines b and d are parallel

16.

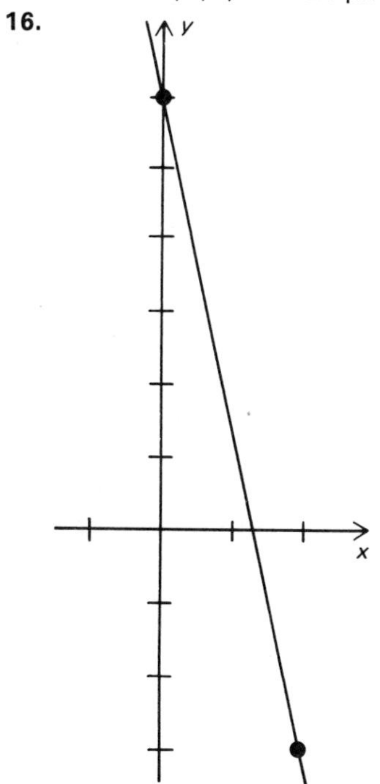

Section 9.6

2. Lines $x = 1$ and $x = 5$ are included.

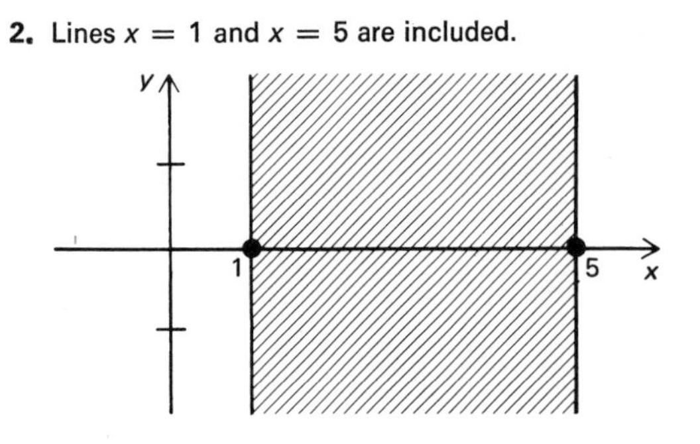

4. The edges are included.

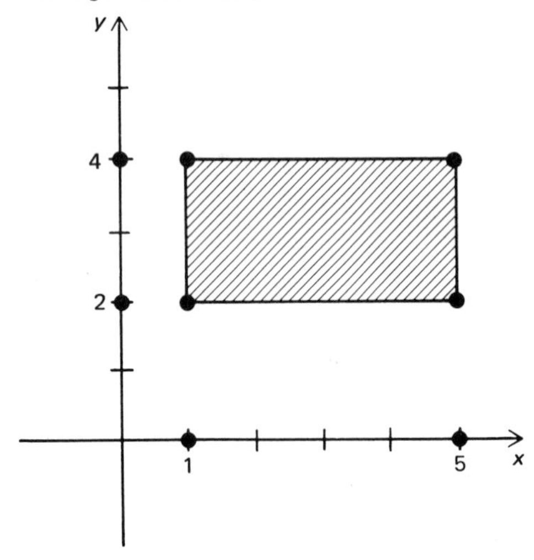

6. The line is not included.

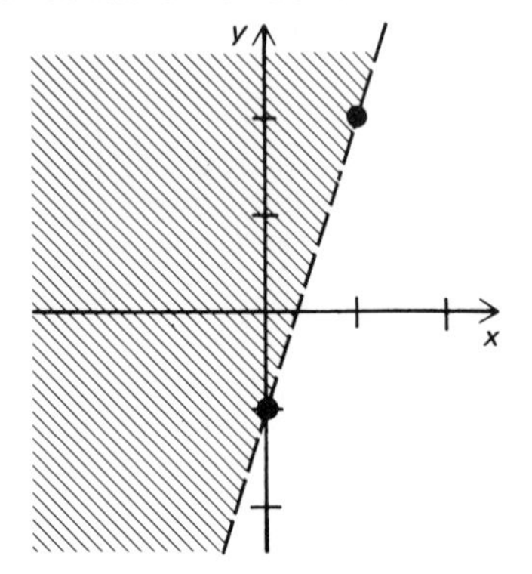

8. The line is not included.

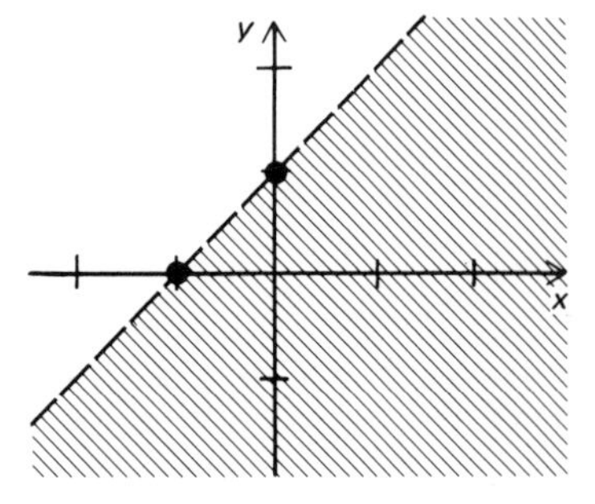

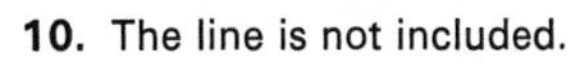

10. The line is not included.

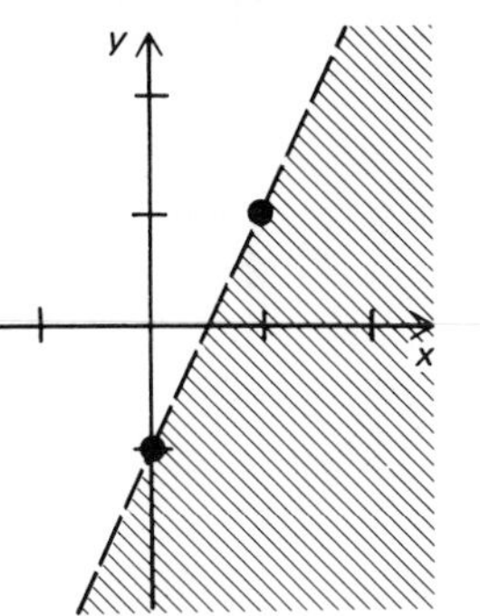

12. The line is included.

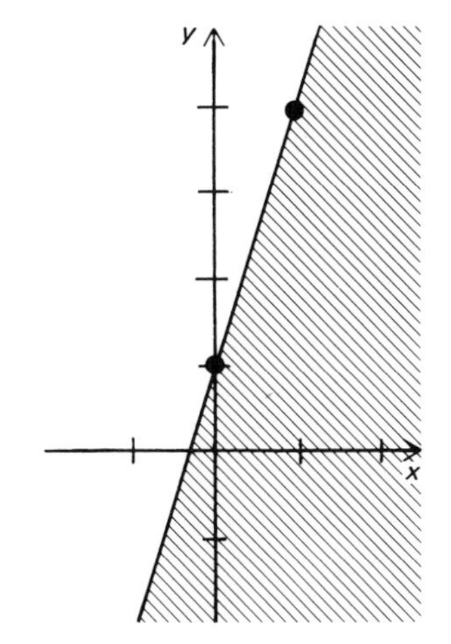

14. The line is not included.

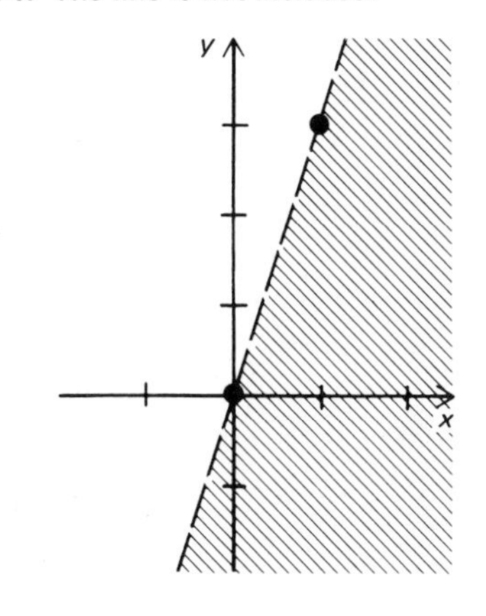

16. The line is included.

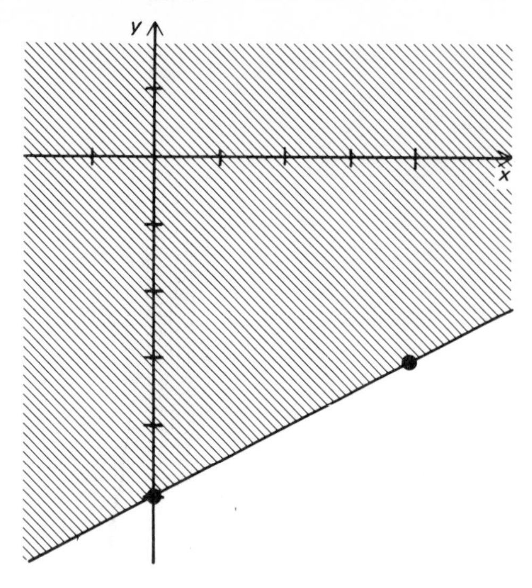

18. The line is included.

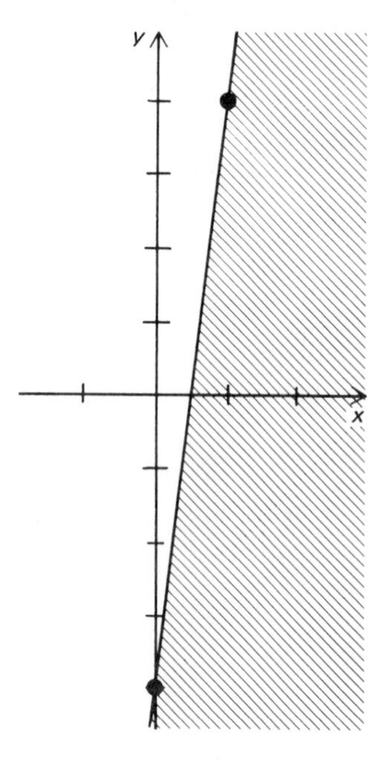

20. The line is included.

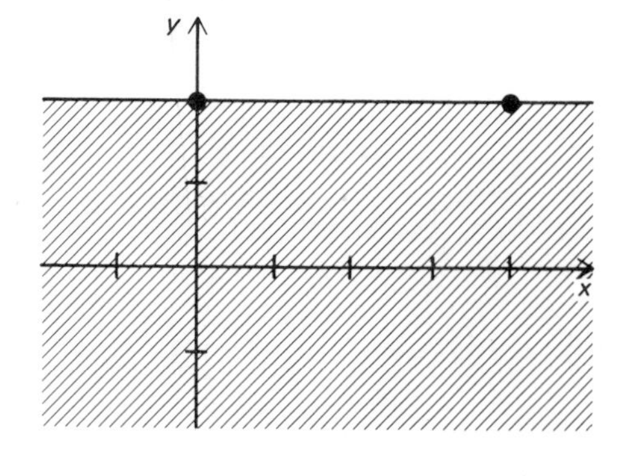

CHAPTER 10

Section 10.1

2. *a.* A system of linear equations is a collection of two or more linear equations.
   b. A solution is an assignment of values to the variables which satisfy *all* the equations in the system.
   c. If a system has no solution it is inconsistent. If it has a solution it is consistent.
   d. Not every system has a solution.
4. Advantage: you can see the solution graphically.
   Disadvantage: it may be hard to see exactly where two lines meet.

6. Solution is (0, 0) **8.** Solution (2, −1) **10.** The lines coincide
12. Solution is (−2, −2) **14.** Solution is (1, 1)

Section 10.2

2. *a.* An equation is an identity if it is always true (for example, $2x + x = 3x$).
   b. If the method yields an identity then there will be infinitely many solutions.

4. The lines coincide **6.** $(^3/_2, 1)$ **8.** (3, −3)
10. (2, 1) **12.** $(^1/_2, -1)$ **14.** (1, 0)
16. (−1, −15) **18.** $x = 3, y = 2, z = 1$ **20.** $x = 2, y = 1, z = 0$

Section 10.3

2. *a.* See rules **1**, **2**, and **3** in the text.
   b. These rules always give equivalent equations.

4. (3, −1) **6.** (−2, 3) **8.** $(^1/_4, ^1/_2)$
10. inconsistent **12.** (1, 1) **14.** (2, 2)
16. (1, 1) **18.** $x = 2, y = 1, z = 0$ **20.** $x = y = z = 0$

Section 10.4

2. 10 days **4.** 3,201 and 4,722 votes **6.** 18 and 72 degrees
8. 7 and 16 feet **10.** beef 12 pounds; veal 4 pounds; pork 2 pounds
12. 48 packs **14.** 12, 19, 72

CHAPTER 11

Section 11.1

2. not quadratic **4.** not quadratic **6.** $x^2 - 4x - 10 = 0$
8. not quadratic **10.** not quadratic **12.** {0, 3}
14. {3, −2} **16.** {5, −1} **18.** {0, 3}
20. {0, −3} **22.** {0, 3} **24.** {10, −10}
26. {1, −3} **28.** {0, 4} **30.** {1}
32. {0, −2} **34.** {3, −2}

Section 11.2

2. pure **4.** not pure **6.** pure
8. not quadratic **10.** pure quadratic in standard form **12.** not pure
14. not pure **16.** not quadratic **18.** $4i$
20. $\sqrt{2}\,i$ **22.** $3i$ **24.** $2\sqrt{3}\,i$
26. $\{7i, -7i\}$ **28.** {2, −2} **30.** $\{2i, -2i\}$
32. {2, −2} **34.** {3, −3} **36.** {3, −3}
38. $\{\sqrt{2}, -\sqrt{2}\}$ **40.** $\{\sqrt{10}, -\sqrt{10}\}$ **42.** {0}

Section 11.3

2. *a.* factoring method: good problem $x^2 + 2x + 1 = 0$; bad problem $x^2 + 3x + 10 = 0$
   b. pure quadratic method: good problem $x^2 + 1 = 0$; bad problem $x^2 + 3x + 10 = 0$
   c. quadratic formula: good problem $x^2 + 3x + 10 = 0$; bad problem (e.g., slower) $x^2 - 1 = 0$
   d. completing the square: same answer as part *c*

4. 104 distinct and real **6.** 12 distinct and real
8. −20 distinct and imaginary **10.** 13 distinct and real
12. 1 distinct and real **14.** distinct and real $(-3 \pm \sqrt{5}) \div 2$
16. distinct and real $0, -^3/_4$ **18.** distinct and real $\pm\sqrt{2}$

20. distinct and real $(-1 \pm \sqrt{3}) \div 4$

22. distinct and real $-1, -3$

24. real and equal -1

26. distinct and real $4 \pm \sqrt{17}$

28. distinct and real 2, 1

30. distinct and imaginary $1 \pm 2i$

32. distinct and real $-2 \pm 2\sqrt{31}$

34. distinct and imaginary $(-1 \pm \sqrt{19}i) \div 2$

36. 4

38. $-2 \pm \sqrt{7}$

40. $(3 \pm \sqrt{11}i) \div 2$

42. $2 \pm \sqrt{6}i$

Section 11.4

2. $7\sqrt{2} \approx 9.9$ inches

4. $30\sqrt{5} \approx 67.1$ feet

6. 6 feet by 18 feet

8. base is 60 inches; height is 30 inches

10. 1,000 feet before target; 2 seconds for the drop

12. 5 inches; 5 inches; 10 inches

14. $25 - 5\sqrt{13}$ or about 7 feet

16. Let $z = x^2$, then $z^2 = x^4$, so $ax^4 + bx^2 + c = 0$ becomes $az^2 + bz + c = 0$. Solve for z then take square roots to get $x = \pm\sqrt{z}$.

INDEX

Page numbers refer to the lecture portion of each section. Corresponding material can also be found in the programmed part accompanying the appropriate section.

BCDEFGHIJ–A–79876